Microbial Approaches for Sustainable Green Technologies

Microbial systems have a strong potential to develop green and sustainable technologies, including sources of renewable energy, alternative fuels, and biosynthetic materials for sustainable applications. Advances in these technologies are evolving to meet growing demand and industries are adapting to green technologies such as solar panels, bioethanol, hydroponics, and more. With the aid of sophisticated technology and integration strategies, these industries are moving toward being more environmentally friendly and sustainable. This book serves as a guide to the newest technologies that will enable the implementation of microbial technologies in fostering an eco-friendly industrial and environmental landscape, which will have widely positive impacts for generations to come.

- Provides recent insights on diverse technologies involved in green technologies.
- Explains the application of microbes via fungi to remediate pollutants and examines the latest treatment technologies in bioleaching and electronic waste treatment.
- Provides updated information on bioenergy and flexible fungal materials as alternatives to plastics.
- Discusses the application of IOT and communication electronics in the development of green technologies.

Jata Shankar, Ph.D., is a distinguished academician and researcher serving as a Professor in the Department of Biotechnology and Bioinformatics at Jaypee University of Information Technology, Waknaghat Solan, India. His academic journey includes earning a Ph.D. (2006) from the Institute of Genomics and Integrative Biology-Council of Scientific and Industrial Research in association with the Department of Biosciences, Jamia Millia Islamia, New Delhi, India. Notably, he pursued a postdoctoral fellowship at Stanford University, USA, and received additional postdoctoral support from an NIH-USA grant while working under mentorship with Prof. David A Stevens.

His mentorship extends to guiding five Ph.D. scholars, 40 B. Tech project theses, and 10 M.S. project theses. His research outputs include the submission of several hundred partial gene sequences to GenBank, substantial Next Generation Sequencing SRA-data, and a notable presence in protein databases. With over 50 peer-reviewed articles, including book chapters, and more than 10 as a lead author, Dr. Shankar is passionately involved in mentoring students to help them achieve their career goals. His multifaceted contributions underscore his commitment to advancing knowledge in biotechnology and bioinformatics.

Pradeep Verma completed his Ph.D. at Sardar Patel University in Gujarat, India in 2002. In the same year, he was selected as a UNESCO Fellow and joined the Czech Academy of Sciences in Prague, Czech Republic. He later moved to Charles University, Prague to work as a Post-Doctoral Fellow. In 2004, he joined the UFZ Centre for Environmental Research, Halle, Germany as a visiting scientist. He was awarded a DFG fellowship which provided him another opportunity to work as a Post-Doctoral Fellow at Gottingen University, Germany. He is also a recipient of various prestigious awards such as the Ron Cockcroft award by Swedish society and UNESCO Fellow ASCR, Prague. He has been awarded Fellow of the Mycological Society of India (MSI-2020); the Prof. P.C. Jain Memorial Award, Mycological Society of India 2020; and Fellow of the Biotech Research Society, India (2021). He is currently working as the Professor (former head and dean, the School of Life Sciences) at the Department of Microbiology, CURAJ. He is a member of various national and international societies and academies. He has completed two collaborated projects worth 150 million INR in microbial diversity and bioenergy.

Maulin P. Shah has been an active researcher and scientific writer in his field for over 20 years. He received a B.Sc. degree (1999) in Microbiology from Gujarat University, Godhra (Gujarat), India. He also earned his Ph.D. degree (2005) in Environmental Microbiology from Sardar Patel University, Vallabh Vidyanagar (Gujarat), India. His research interests include Biological Wastewater Treatment, Environmental Microbiology, Biodegradation, Bioremediation, and Phytoremediation of Environmental Pollutants from Industrial Wastewaters. He has published more than 240 research papers in national and international journals of repute on various aspects of microbial biodegradation and bioremediation of environmental pollutants. He is the editor of 65 books of international repute.

Microbial Approaches for Sustainable Green Technologies

Edited by
Jata Shankar, Pradeep Verma,
and Maulin P. Shah

CRC Press is an imprint of the
Taylor & Francis Group, an informa business

Designed cover image: Prof. Jata Shankar, Prof. Pradeep Verma, and Dr. Maulin P. Shah.

First edition published 2024
by CRC Press
2385 NW Executive Center Drive, Suite 320, Boca Raton FL 33431

and by CRC Press
4 Park Square, Milton Park, Abingdon, Oxon, OX14 4RN

CRC Press is an imprint of Taylor & Francis Group, LLC

ISBN: 978-1-032-52648-5 (hbk)
ISBN: 978-1-032-52649-2 (pbk)
ISBN: 978-1-003-40768-3 (ebk)

DOI: 10.1201/9781003407683

Typeset in Times LT Std
by KnowledgeWorks Global Ltd.

Contents

About the Editors

Professor Jata Shankar, Ph.D., is a distinguished academician and researcher serving as a Professor in the Department of Biotechnology and Bioinformatics at Jaypee University of Information Technology, Waknaghat Solan, India. His academic journey includes earning a Ph.D. (2006) from the Institute of Genomics and Integrative Biology-Council of Scientific and Industrial Research in association with the Department of Biosciences, Jamia Millia Islamia, New Delhi, India. Notably, he pursued a postdoctoral fellowship at Stanford University, USA and received additional postdoctoral support from an NIH-USA grant while working under mentorship with Prof. David A Stevens.

Throughout his career, Dr. Shankar has actively engaged in collaborative research, partnering with experts such as Thomas Wu from Genentech Industry and Jennifer Wortman at MIT-Harvard during his post-doc tenure. His academic contributions extend to his role as Assistant Professor at GGV-Central University, Bilaspur CG, and member DBT-builder grant. Dr. Shankar has made significant contributions to the field of microbiology, particularly in the area of fungal pathogens and the evaluation of phytochemicals as antifungal agents. His mentorship extends to guiding 5 Ph.D. scholars, 40 B. Tech project theses, and 10 M.S. project theses. His research outputs include the submission of several hundred partial gene sequences to GenBank, substantial Next Generation Sequencing SRA-data, and a notable presence in protein databases. With over 50 peer-reviewed articles, including book chapters, and more than 10 as a lead author, Dr. Shankar has established himself as a prolific researcher. He is an active reviewer and editorial member for peer-reviewed journals, has served as an invited speaker at prestigious institutions, and visited several countries for international for academic purposes. In addition to his research endeavors, Dr. Shankar is passionately involved in mentoring students to help them achieve their career goals. His multifaceted contributions underscore his commitment to advancing knowledge in biotechnology and bioinformatics.

Prof. Pradeep Verma completed his doctoral journey at Sardar Patel University, Gujarat, India, in 2002. In the same year, he started his international academic journey and was selected as a UNESCO fellow at the Czech Academy of Sciences in Prague, Czech Republic. His hard work and dedication toward research led him to Charles University, Prague, as a Post-Doctoral Fellow, following which he was a visiting scientist at UFZ Centre for Environmental Research, Halle, Germany, in 2004.

Prof. Verma's exceptional potential secured him a prestigious DFG fellowship, which paved the way for his future journey as a Post-Doctoral Fellow at Gottingen University, Germany. Following this he returned to his homeland India in 2007 and joined Reliance Life Sciences, Mumbai. During his tenure, the research work on biobutanol production resulted in several patents under his name. Following this he was awarded the prestigious JSPS Post-Doctoral Fellowship Programme which led him to the next landmark in his career at the Laboratory of Biomass Conversion at Kyoto University, Japan.

Prof. Verma's scientific career took flight in 2009 when he joined Assam University as a Reader and Founder Head. Following this crucial step was joining Guru Ghasidas Vishwavidyalaya, Bilaspur (A Central University) in 2011 as an Associate Professor until 2013. Currently, Prof. Verma holds the esteemed position of Professor at the School of Life Sciences at the Department of Microbiology, CURAJ, and was also the Former Dean and Head.

Prof. Verma's expertise covers specialization in Microbial Diversity, Bioremediation, Bioprocess Development, Lignocellulosic, and Algal Biomass-based Biorefinery. Recently for his contribution to the area of fungal microbiology, industrial biotechnology, and environmental bioremediations, he has been awarded the prestigious Fellow Award from the Mycological Society of India (2020), P.C. Jain Memorial Award (MSI), and Biotech Research Society of India (2021). In 2020 he was also awarded fellow of the Biotechnology Research Society of India (BRSI) and Fellow of the Academy of Sciences of AMI India 2021 (FAMSc). Furthermore, he has also been awarded the JSPS Bridge Fellow award in 2022 and a short-term visit to Kyoto University, Kyoto, Japan to strengthen ties between the two laboratories.

Prof Verma's research journey to date has 12 international patents as well. His scientific endeavors are visible through 85+ publications of which he has also contributed to the existing literature by contributing to several book chapters. His editorial potential is from his role as an editor for books with esteemed international publishers of high repute. He is a member of various national and international societies/academies and has also completed two collaborated projects worth 150 million INR in the area of microbial diversity and bioenergy. Further, Prof Verma also lends his expertise as a Guest Editor to prominent journals including Frontiers and MDPI. Additionally, he also is on the editorial board of "Current Nanomedicine" (Bentham Sciences).

Dr. Maulin P. Shah has emerged as a prominent researcher with his dedication to research spanning over two decades. He completed his BSc in Microbiology from Gujarat University, Godhra (Gujarat), India, in 1999. He further completed his PhD from Sardar Patel University, Vallabh Vidyanagar (Gujarat), India in 2005.

Dr. Shah currently holds the position of Chief Scientist and Head at the Industrial Wastewater Research Lab within the Division of Applied and Environmental Microbiology Lab at

Enviro Technology Ltd., Ankleshwar, Gujarat, India. His expertise and efforts are committed to investigating the complex relation between pollution and microbial diversity of wastewater via both cultivation-dependent and independent analyses.

His contributions to science and literature have advanced the scientific realms of environmental microbiology such as biological wastewater treatment, biodegradation, bioremediation, and phytoremediation of environmental pollutants. Dr. Shah's prolific scholarly journey is underscored by a staggering publication record, comprising more than 250 research papers across esteemed national and international journals. Beyond his research articles, he holds the distinguished role of editor for over 50 internationally renowned books, published by esteemed publishers such as Elsevier, Springer, RSC, and CRC Press.

His influence extends beyond publications, as he actively serves on the editorial boards of top-rated journals. Alongside his advisory role includes as the Advisory Board of "CLEAN—Soil, Air, Water" (Wiley), editorship of "Current Pollution Reports" (Springer Nature), "Environmental Technology and Innovation" (Elsevier), "Current Microbiology" (Springer Nature), "Journal of Biotechnology and Biotechnological Equipment" (Taylor & Francis), "Ecotoxicology (Microbial Ecotoxicology)" (Springer Nature), "Current Microbiology" (Springer Nature) along with associate editorships for "GeoMicrobiology" (Taylor & Francis) and "Applied Water Science" (Springer Nature). Dr. Shah's contributions thus continue to shape the landscape of environmental microbiology globally.

Contributors

Andi Solórzano Acosta
Dirección de Supervisión y Monitoreo en las Estaciones Experimentales Agrarias
Instituto Nacional de Innovación Agraria (INIA)
Lima, Perú

Fiorella Maité Arquíñego-Zárate
Faculty of Pharmacy and Biochemistry
Biotechnology and Omics in Life Sciences Research Group
Universidad Nacional Mayor de San Marcos
Limam, Perú

Sampan Attri
RT-PCR Laboratory
District Hospital
Fatehgarh Sahib, Punjab, India

Divya Bajaj
Department of Zoology
Hindu College
University of Delhi
Delhi, India

Fiorella Gomez Barrientos
Faculty of Pharmacy and Biochemistry
Biotechnology and Omics in Life Sciences Research Group
Universidad Nacional Mayor de San Marcos
Limam, Perú

Khushboo Bhange
Department of Biochemistry
Pt. J.N.M. Medical College Raipur
Raipur, Chhattisgarh, India

Pradeep Bhatnagar
Department of Microbiology and Biotechnology
IIS (Deemed to be University)
Jaipur, Rajasthan, India

Milagros Estefani Alfaro Cancino
Faculty of Pharmacy and Biochemistry
Biotechnology and Omics in Life Sciences Research Group
Universidad Nacional Mayor de San Marcos
Limam, Perú

Heidy Mishey Aguirre Catalan
Faculty of Pharmacy and Biochemistry
Biotechnology and Omics in Life Sciences Research Group
Universidad Nacional Mayor de San Marcos
Lima, Perú

Andrea León Chacón
Faculty of Pharmacy and Biochemistry
Biotechnology and Omics in Life Sciences Research Group
Universidad Nacional Mayor de San Marcos
Limam, Perú

Nagendra Kumar Chandrawanshi
School of Studies in Biotechnology
Pt. Ravishankar Shukla University
Raipur, Chhattisgarh, India

Harsh Charak
Department of Microbiology
Ram Lal Anand College
University of Delhi
Delhi, India

Payal Chaturvedi
Department of Biotechnology
IIS (Deemed to be University)
Jaipur, Rajasthan, India

Abhishek Chaudhary
Department of Biotechnology and Bioinformatics
Jaypee University of Information Technology Waknagaht
Solan, Himachal Pradesh, India

Sandeep Choudhary
Department of Biomedical Engineering
Central University of Rajasthan
Ajmer, Rajasthan, India

Sandhya Chouhan
Department of Biomedical Engineering
Central University of Rajasthan
Ajmer, Rajasthan, India

Saurabh Sudha Dhiman
Civil and Environmental Engineering
South Dakota Mines
Rapid City, South Dakota

Composite and Nanocomposite Advanced Manufacturing – Biomaterials (CNAM-Bio) Center
South Dakota Mines
Rapid City, South Dakota

Chemistry, Biology and Health Sciences
South Dakota Mines
Rapid City, South Dakota

Anamika Dhyani
Hematology & Hemotherapy Foundation (HEMOAM)
Amazon, Brazil

Kaushiki Dutta
Department of Microbiology
Ram Lal Anand College
University of Delhi
Delhi, India

Garima
Department of Medical Laboratory Science
Lovely Professional University
Jalandhar, Punjab, India

Utsha Ghosh
Department of Biotechnology
IIS (Deemed to be University)
Jaipur, Rajasthan, India

Aishani Gupta
Department of Microbiology
Ram Lal Anand College
University of Delhi
Delhi, India

Department of Psychology
Indraprastha Women's College
University of Delhi
Delhi, India

Priya Gupta
Amity Institute of Biotechnology
Amity University Chhattisgarh
Raipur, Chhattisgarh, India

Saurabh Gupta
Amity Institute of Biotechnology
Amity University Chhattisgarh
Raipur, Chhattisgarh, India

Vandana Gupta
Department of Microbiology
Ram Lal Anand College
University of Delhi
Delhi, India

Edwin Hualpa-Cutipa
School of Environmental Engineering
Universidad Continental
Lima, Peru

Jorge Johnny Huayllacayan Mallqui
Faculty of Pharmacy and Biochemistry
Biotechnology and Omics in Life Sciences Research Group
Universidad Nacional Mayor de San Marcos
Limam, Perú

Kanika Jangir
Department of Biotechnology
IIS (Deemed to be University)
Jaipur, Rajasthan, India

Shreya Kapoor
Department of Microbiology
Ram Lal Anand College
University of Delhi
Delhi, India

Department of Biotechnology
Delhi Technological University
New Delhi, India

Sumanpreet Kaur
Department of Medical Laboratory Science
Lovely Professional University
Jalandhar, Punjab, India

Khemraj
School of Studies in Biotechnology
Pt. Ravishankar Shukla University
Raipur, Chhattisgarh, India

M. Nithya Kruthi
Amity Institute of Biotechnology
Amity University
Noida, Uttar Pradesh, India

Abhishek Kumar
School of Bioengineering and Food Technology
Shoolini University
Solan, Himachal Pradesh, India

Yadira Karolay Ravelo Machari
Faculty of Pharmacy and Biochemistry
Biotechnology and Omics in Life Sciences Research Group
Universidad Nacional Mayor de San Marcos
Limam, Perú

Pramod Kumar Mahish
Department of Biotechnology
Govt. Digvijay Autonomous PG College
Rajnandgaon, Chhattisgarh, India

Parikshana Mathur
Department of Biotechnology
Central University of Rajasthan
Ajmer, Rajasthan, India

María José Mayhua
Faculty of Pharmacy and Biochemistry
Biotechnology and Omics in Life Sciences Research Group
Universidad Nacional Mayor de San Marcos
Lima, Perú

Varsha Meshram
School of Studies in Biotechnology
Pt. Ravishankar Shukla University
Raipur, Chhattisgarh, India

Rashi Nagar
Department of Biotechnology
Central University of Rajasthan
Ajmer, Rajasthan, India

Priyanka Narad
Amity Institute of Biotechnology
Amity University
Noida, Uttar Pradesh, India

Lalit Mohan Pandey
Indian Institute of Technology Guwahati
Guwahati, Assam, India

Sebika Panja
School of Bioengineering and Food Technology
Shoolini University
Solan, Himachal Pradesh, India

Sombir Pannu
Indian Institute of Technology Guwahati
Guwahati, Assam, India

Arun Parashar
Faculty of Pharmaceutical Science
Shoolini University of Biotechnology and Management Sciences
Solan, Himachal Pradesh, India

Palak Patel
Department of Biomedical Engineering
Central University of Rajasthan
Ajmer, Rajasthan, India

Tarun Kumar Patel
Department of Biotechnology
Sant Guru Ghasidas Government Post Graduate College
Dhamtari, Chhattisgarh, India

Sandeep Patra
Department of Microbiology
Ram Lal Anand College
University of Delhi
Delhi, India

Sneh Priya
Department of Microbiology
Ram Lal Anand College
University of Delhi
Delhi, India

Nikol Gianella Julca Santur
Faculty of Pharmacy and Biochemistry
Biotechnology and Omics in Life Sciences Research Group
Universidad Nacional Mayor de San Marcos
Lima, Perú

Abhishek Sengupta
Amity Institute of Biotechnology
Amity University
Noida, Uttar Pradesh, India

Charu Sharma
Department of Biotechnology
IIS (Deemed to be University)
Jaipur, Rajasthan, India

Deepak Sharma
Department of Biotechnology
Chandigarh College of Technology
Chandigarh Group of Colleges Landran
Mohali, Punjab, India

Department of Biotechnology and Bioinformatics
Jaypee University of Information Technology Waknagaht
Solan, Himachal Pradesh, India

Deepika Sharma
School of Health Sciences and Technology
UPES
Dehradun, Uttarakhand, India

Sangeeta Devendrakumar Singh
Amity Institute of Biotechnology
Amity University Chhattisgarh
Raipur, Chhattisgarh, India

Nalini Soni
Amity Institute of Biotechnology
Amity University Chhattisgarh
Raipur, Chhattisgarh, India

Swati Srivastava
Civil and Environmental Engineering
South Dakota Mines
Rapid City, South Dakota

Lucero Katherine Castro Tena
Universidad Nacional José Faustino Sánchez Carrión
Huacho, Peru

Raman Thakur
Department of Medical Laboratory Science
Lovely Professional University
Jalandhar, Punjab, India

Pankaj Tiwari
Indian Institute of Technology Guwahati
Guwahati, Assam, India

Balasubramanian Velramar
Amity Institute of Biotechnology
Amity University Chhattisgarh
Raipur, Chhattisgarh, India

Nidhi Verma
Department of Microbiology
Ram Lal Anand College
University of Delhi
Delhi, India

Neetu Kukreja Wadhwa
Department of Zoology
Hindu College
University of Delhi
Delhi, India

Varsha Yadav
Gochar Mahavidyalaya
Maa Shakumbhari University
Saharanpur, Uttar Pradesh, India

Preface

Microbial approaches play a pivotal role in advancing sustainable green technologies, forging an interdisciplinary realm at the crossroads of microbiology and evolving technologies. This convergence aims to address the sustainable development goals outlined by the United Nations. By harnessing the power of microorganisms, this innovative approach offers multifaceted solutions for environmental challenges.

In the pursuit of sustainable development, microbial technologies contribute significantly to diverse sectors such as agriculture, waste management, energy production, and environmental remediation. Microbes, including bacteria, fungi, and algae, exhibit unique capabilities that can be harnessed to mitigate ecological issues and promote green practices.

In agriculture, for instance, microbial biofertilizers enhance soil fertility and nutrient uptake, reducing the reliance on chemical inputs. Moreover, the use of microorganisms in waste management processes facilitates efficient decomposition and recycling, minimizing the environmental impact of organic waste.

Microbial fuel cells exemplify the intersection of microbiology and energy production. These cells harness the metabolic activity of microorganisms to generate electricity from organic matter. This not only provides a sustainable energy source but also aids in waste treatment.

Additionally, the application of microbiology in environmental remediation involves the use of microbial communities to break down pollutants and contaminants. This eco-friendly approach aligns with the goals of sustainable development, contributing to the restoration and preservation of ecosystems.

As a researcher, delving into the intricacies of microbial driven green technologies offers a fertile ground for exploration and discovery. The synergy between microbiology and technology propels advancements that hold promise for a more sustainable future.

In conclusion, the amalgamation of microbiology and developing technologies for sustainable green solutions represents a compelling frontier in research and innovation. By understanding and harnessing the potential of microorganisms, we pave the way for transformative developments that align with the global agenda for sustainable progress.

Preface

1 Microbial Sensors
A Tool for Accelerating Sustainable Green Technologies

Deepak Sharma, Arun Parashar, Deepika Sharma, Sampan Attri, and Abhishek Chaudhary

1.1 INTRODUCTION

In the pursuit of sustainable green technology, innovative solutions that minimize environmental impact while fostering economic growth are gaining popularity worldwide. Among these solutions, the production and application of microbial sensors have emerged as a particularly intriguing and promising topic. These minuscule yet powerful devices capitalize on the unique abilities of microbes to recognize and respond to various stimuli, providing valuable insights and facilitating efficient monitoring and control in industries such as environmental monitoring, agriculture, healthcare, and bioprocessing (Cui et al., 2019; Jaiswal and Shukla, 2020; Liu and Choi, 2021). Microbial sensors are an essential tool in the quest for sustainability since they have a number of benefits over conventional sensors. They make use of microbes, amazing flexibility and specificity, which allows them to recognize and react to even the smallest alterations in their environment (Jiang et al., 2019; Simoska et al., 2021). At the core of microbial biosensors lies a living organism or a component derived from a microorganism that can identify and react to specific analytes. These organisms can include bacteria, yeast, algae, or even genetically engineered cells designed for sensing purposes. When the target analyte is recognized, it triggers a physiological or biochemical response within the microorganism, resulting in a measurable output signal that can be detected and quantified. A basic block diagram of a biosensor shown in Figure 1.1. The detection mechanisms employed by microbial biosensors vary depending on the application and target analyte, ranging from changes in cell viability, enzyme activity, and gene expression to the production of specific biomolecules like proteins or metabolites (Clemons et al., 2016). The measurable output signal can be taken in the form of optical, electrical, electrochemical, and even changes in the growth or behaviour of the organism (Gao et al., 2020; Lei et al., 2006a; Ronald and Beutler, 2010).

One of the key advantages of microbial biosensors is their exceptional selectivity and sensitivity. Microorganisms possess the ability to identify specific analytes with remarkable precision, even at very low concentrations. This selectivity

DOI: 10.1201/9781003407683-1

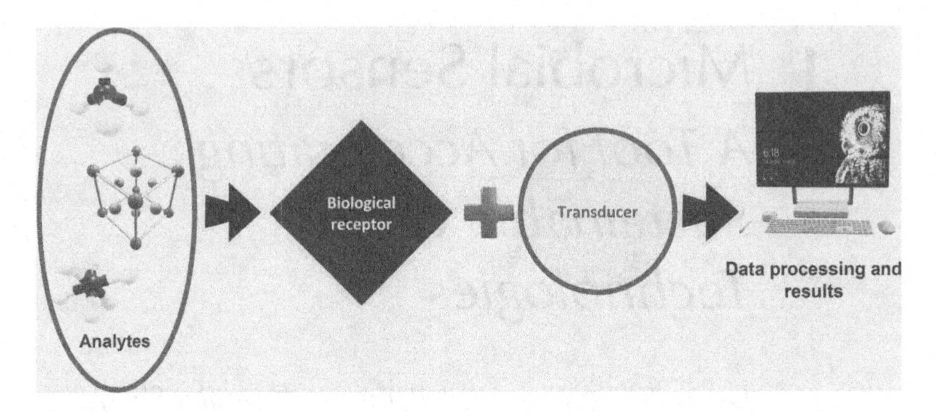

FIGURE 1.1 A basic block diagram of a biosensor.

is particularly valuable in applications such as medical diagnostics, environmental monitoring of pollutants, and food safety testing. Additionally, microbial biosensors offer other benefits, including cost-effectiveness and relative simplicity in development and operation. By utilizing microorganisms as the sensing component, the need for complex synthetic receptors or antibodies is eliminated, reducing overall costs and complexity. Moreover, microbial biosensors can be seamlessly integrated into existing systems and technologies, enabling real-time and continuous monitoring of target analytes (Abrevaya et al., 2015; Ivars-Barceló et al., 2018; Zhang and Keasling, 2011). Microbial sensors leverage these inherent capabilities to provide real-time, high-resolution data, enabling prompt decision-making and focused actions. They can be deployed in remote or challenging environments, offering versatility and scalability at an affordable cost. Furthermore, they expand the range of applications compared to traditional sensors (Bilal and Iqbal, 2019; Neethirajan et al., 2018).

Microbial biosensors are used in the healthcare industry for drug discovery, illness biomarker monitoring, and medical diagnostics. They help with early illness identification, personalized treatment, and therapy monitoring by quickly and sensitively detecting infections, poisons, and other biomarkers. Microbial biosensors are used in environmental monitoring to find contaminants such as heavy metals, pesticides, and organic compounds in soil, water, and air. Their real-time monitoring capabilities allow for quick source location and risk evaluation for environmental pollution (Hill et al., 2020; Tremblay and Hallenbeck, 2009). Microbial biosensors aid in process improvement, quality assurance, and waste management in industrial operations. Critical factors, including nutrition levels, pH, temperature, and product concentrations, may be monitored by them, enabling effective manufacturing operations and lowering waste output (Grote et al., 2014; Park et al., 2013).

1.2 TYPES OF MICROBIAL SENSORS

Microbial sensors, commonly referred to as biosensors, include a wide range of methodologies and technology. They make use of the distinct traits and capacities

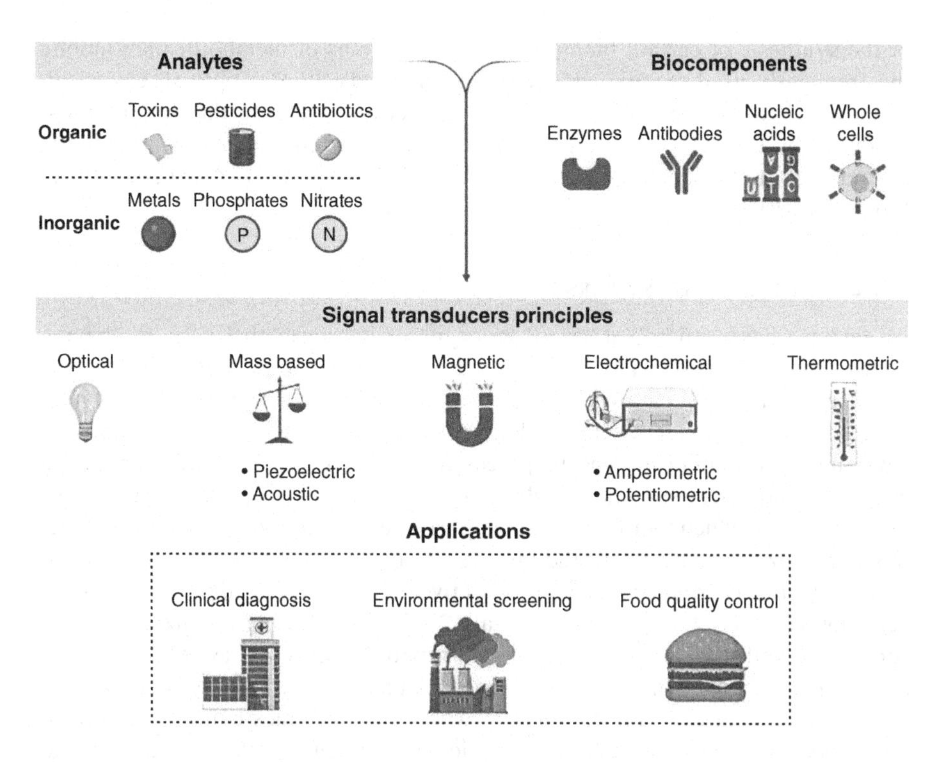

FIGURE 1.2 Classification of microbial sensors with transducer principles and application.

of microorganisms to identify and quantify a wide range of analytes, from environmental contaminants to biomolecules. A schematic presentation of the classification of microbial biosensors is given in Figure 1.2. In the subsequent sections, we will go into depth about some of the most popular kinds of microbial sensors.

1.2.1 Whole-Cell Sensors

Whole-cell microbial sensors are a form of biosensor that use whole microorganisms as the sensing element to find and react to certain analytes. These sensors are adaptable instruments for a variety of applications because they take advantage of microbes' innate capacity to recognize and respond to environmental changes (Chu et al., 2021). Living cells, often bacteria, yeast, or algae, are at the heart of whole-cell microbial sensors because they have been carefully chosen or manipulated to demonstrate a particular reaction to a target analyte. The presence of the desired analyte can be detected by these cells thanks to genetic or receptor components. Target analytes interact with sensor cells, causing the cells to undergo a physiological or biochemical reaction that results in the generation of a quantifiable output signal (Abbasian et al., 2018). Whole-cell microbial sensors' modes of operation might change based on the application and kind of the target analyte. Changes in cell viability, enzyme activity, gene expression, or the synthesis of certain biomolecules like proteins or metabolites are among the frequently used techniques. Numerous methods, like as optical, electrical, or electrochemical ones, can be used to detect and measure the output signal produced by the cells. Whole-cell sensors offer advantages such as simplicity, cost-effectiveness, and robustness (Belkin, 2003; Cai et al., 2018; Hu et al., 2018; Woo et al., 2020).

1.2.2 Enzyme-Based Sensors

An enzyme-based microbial sensor is a type of biosensor that uses an enzyme as the sensing component to identify and measure certain analytes. These sensors use the enzymes' catalytic abilities to selectively and sensitively detect target molecules, making them useful tools in a variety of fields including healthcare, environmental monitoring, and bioprocessing (Kim et al., 2018). When creating enzyme-based microbial sensors, the enzyme is immobilized on the surface of the transducer, which can be a solid support like an electrode or a nanoparticle. Since the enzyme is kept near the transducer during the immobilization process, the enzymatic reaction products are effectively transferred to the transducer for detection (Kim et al., 2018; Pundir et al., 2019). By selecting an enzyme that is compatible with the target, enzyme-based microbial sensors may be made to be more selective. Due to their capacity to identify and catalyse certain reactions with the substrates that they are designed to work with, enzymes exhibit remarkable selectivity. The sensor can achieve exceptional specificity by choosing an enzyme that has a high affinity for the target analyte, reducing interference from other substances in the sample (Bachosz et al., 2022; Bollella and Katz, 2020; Kurbanoglu et al., 2020). Table 1.1 represents the classification of enzyme-based biosensors along with their working principle.

The enzymatic reaction between the target analyte and the enzyme provides the basis for the detection method of enzyme-based microbial sensors. Usually, this reaction alters an electrical, electrochemical, or optical characteristic that can be measured and quantified (Gavrilaş et al., 2022). For instance, an enzyme-catalysed reaction in an electrochemical-based sensor may result in an electrical current or potential that may be measured with electrodes. An optical-based sensor may experience a change in absorbance, fluorescence, or luminescence as a result of the reaction, which may be detected optically and quantified (Leonard et al., 2003; Nigam and Shukla, 2015; Yunus, 2018). Microbial sensors built using enzymes have various benefits. First of all, because enzymes are so highly specific, they can detect just certain target analytes, even in tricky sample matrices. This specificity supports the sensor's accuracy and dependability. Second, high catalytic efficiency in enzymes enables the sensor to reach sensitive detection limits. Enzymatic processes amplifying the signal increase the sensor's sensitivity, enabling it to detect analytes at low concentrations (Bose et al., 2022; Liu et al., 2012b). Enzyme-based microbial sensors may also be readily customized for diverse analytes by choosing and immobilizing the proper enzyme, making them adaptable for a variety of applications.

TABLE 1.1
Classification of Various Enzyme-Based Biosensors

Enzyme-Based Biosensors			
Enzyme Bioreceptor		**Enzyme Transducer**	
Enzyme-based antibody biosensors	ELISA and ELFA are examples of enzyme-based antibody biosensors	Optical biosensors	Absorption, reflection, refraction, Raman, infrared, chemiluminescence, dispersion, fluorescence, and phosphorescence phenomena are considered for sensing
		Fibre-optic biosensors	Uses phenomena of total internal reflection
		Bioluminescence biosensors	Utilizes enzymes that have the ability to radiate photons
		Electrochemical biosensors	Uses bio-electrochemical components
		Amperometric biosensors	It works on enzyme structure that catalytically transforms electrochemical nonactive analytes into products that can be oxidized at a working electrode
		Potentiometric biosensor	Uses an ion-selective electrode
		Conductometric biosensors	These are based on the correlation between conductance and a biological element
		Calorimetric biosensor	It works on the basics of the calorimetric principle and changes in temperature

1.2.3 Microbial Fuel Cells (MFCs)

The MFC sensor represents a groundbreaking advancement in the realm of electrochemical microbial biosensors, offering distinct advantages over biosensors employing conductometric, amperometric, potentiometric, or voltammetric transducers. At the heart of its functionality lies the utilization of electroactive microorganisms as the sensing component. These microorganisms, through their electron transfer processes, generate an electric signal that responds to changes in the presence or concentration of target analytes. Unlike many other applications of MFCs, the primary objective of the MFC sensor is to detect variations in the cell's output under different environmental conditions (Su et al., 2011). MFCs operate by harnessing the metabolic activity of microorganisms to convert organic matter into electricity. Within the anode chamber, microorganisms oxidize organic compounds, liberating electrons in the process. These electrons then travel through an external circuit to the cathode chamber, thereby generating an electrical current. The microorganisms involved in MFCs can be naturally occurring or genetically engineered to enhance their electron transfer capabilities (Jiang et al., 2017, 2018; Olias and Di Lorenzo, 2021). A basic representation of MFC is displayed in Figure 1.3. The versatility of

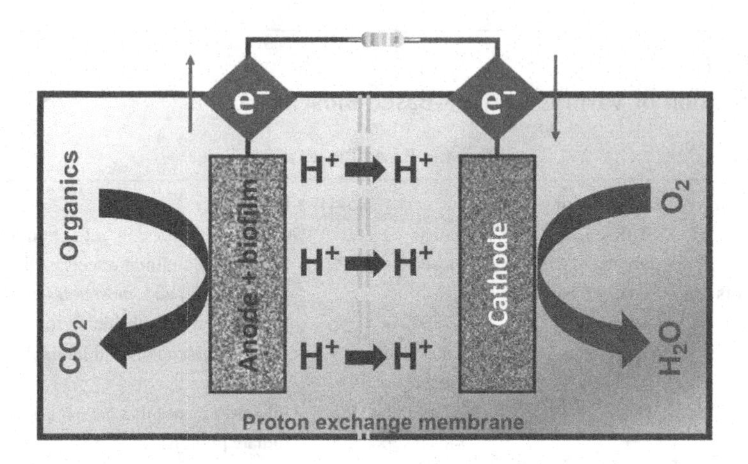

FIGURE 1.3 A basic microbial fuel cell (MFC) showing organic compound degradation and simultaneous electron transfer.

MFCs extends to diverse applications such as wastewater treatment, energy production, and biosensing. With its unique approach, the MFC sensor offers tremendous potential for numerous fields. By capitalizing on the inherent abilities of electroactive microorganisms, this sensor provides a novel pathway for biosensing, enabling valuable insights and practical contributions to various research domains (Olias and Di Lorenzo, 2021; Su et al., 2011).

One of the most exciting prospective applications of MFC technologies is the identification of harmful compounds in wastewater using biosensors based on MFC technology. As a result, these sensors have evolved through time from offline to online monitoring systems (Sonawane et al., 2020). Many different metrics, including volatile fatty acids (VFA), chemical oxygen demand (COD), dissolved oxygen (DO), biological oxygen demand (BOD), and numerous hazardous components found in wastewater, have been analysed using the MFC (Jin et al., 2017; Xu et al., 2015). Kumar and Sundramoorthy have developed an electrochemical biosensor using acetylcholinesterase for detecting the organophosphate pesticide methyl parathion (Kumar and Sundramoorthy, 2019). The biosensor was constructed through a three-step process, viz. cross-linking acetylcholinesterase using glutaraldehyde, immobilizing the cross-linked enzyme onto a glassy carbon electrode that was modified with semiconducting single-walled carbon nanotubes, and treating it with bovine serum albumin to prevent non-specific binding. This biosensor exhibited a linear detection range of 1×10^{-10} to 1×10^{-6} M for methyl parathion, with a limit of detection of 3.75×10^{-11} M. This innovative biosensor provides a sensitive and reliable method for detecting methyl parathion, which is an organophosphate pesticide. A number of additional investigations have shown that MFC-based biosensors are capable of detecting organic chemicals including *p*-nitrophenol (PNP), formaldehyde, and levofloxacin as well as heavy metals like copper, chromium, or zinc (Jin et al., 2017; Kumar and Sundramoorthy, 2019; Olias and Di Lorenzo, 2021; Su et al., 2011). A few other MFC sensors are listed in Table 1.2.

1.2.4 Genetically Engineered Sensors

Genetically engineered microbial whole-cell biosensors have emerged as a powerful tool for environmental assessment. This technology harnesses the capabilities of living microbial cells, whether prokaryotic or eukaryotic, to provide valuable insights into the composition, safety, and quality of chemicals present in the environment and their impact on living organisms (Tolosa et al., 1999). These biosensors, particularly the "lights on" biosensors, are intricately designed to generate signals such as electric current, potential, heat, conductivity, chromagen, luminescence, or fluorescence. The activation of these signals is controlled by promoters that are specifically responsive to certain chemicals or environmental stresses. Transcriptional regulators play a crucial role in interacting with these chemicals or stresses to activate the promoters, thus triggering the desired output signals. By detecting and accurately calibrating these output signals, it becomes possible to interpret the presence and relative quantity of the specific activator or chemical being monitored (Camsund et al., 2011; Wan et al., 2021). In essence, genetically engineered microbial whole-cell biosensors enable us to gain valuable information about the environmental conditions and potential hazards by utilizing living cells as biosensors, thereby facilitating effective environmental monitoring and assessment. Genetically engineered sensors offer the advantage of tailoring the microorganism's properties to improve sensitivity, selectivity, and response dynamics for specific applications (He et al., 2016; Riding et al., 2013). A couple of examples highlighting the significance of engineered sensors are described below.

Fluorescent biosensors that are genetically encoded allow for real-time visualization of glucose changes in astrocytes and neurons. However, it is difficult to precisely translate fluorescence observations to brain tissue glucose concentrations. Imaging using fluorescence lifetimes overcomes this problem (Díaz-García et al., 2019). *Thermus thermophilus* glucose-binding protein and a circularly permuted T-Sapphire fluorescent protein variant are combined in the iGlucoSnFR-TS sensor to track neuronal glucose levels. For accurate glucose determination, calibration at the experimental temperature is necessary. In ex vivo brain tissue, intracellular glucose concentrations vary from 2 to 10 mM, or around 20% of the external concentration, but awake mouse cortical neurons' in vivo cytosolic glucose concentrations fall between 0.7 and 2.5 mM (Díaz-García et al., 2019). A couple of recombinant microbial sensors are shown in Table 1.2.

1.2.5 Optical Sensors

Optics is a commonly utilized technique in microbial biosensors, where optical detection relies on measuring luminescent, fluorescent, colorimetric, or other optical signals resulting from microorganism-analyte interactions. These optical signals are then correlated with the concentration of target compounds. Genetically engineered microorganisms play a significant role in optical whole-cell biosensors, where a reporter gene is combined with an inducible gene. Upon encountering the target analyte, the inducible gene is activated, subsequently regulating the expression of the reporter gene. This regulation either triggers the production of a measurable optical

TABLE 1.2
List of Microbial Sensors Utilized for Various Applications

Technique	Microorganism (Promoter/ Reporter Construct)	Target Analyte	Sample	Detection Range/Limit	Reference
Whole-cell microbial sensor	*E. coli* DH5α (P*ars*/*ars*R-*phi*YFP)	Arsenite and arsenate	–	Up to 8 μmol/L arsenite Up to 25 μmol/L arsenate	Sato and Kobayashi, 1998
	E. coli JM109 (P*ars*/*ars*R-*lac*Z)	Arsentate	Drinking water	<10 μg/L	De Mora et al., 2011
	E. coli species (*ars*R/*lux*CDABE)	Arsenic	Contaminated water	0.74–60 μg/L	Sharma et al., 2013
	D. radiodurans (*cad*R-*crt*I for red pigment production and *cad*R-*lac*Z pH change)	Cadmium	Lab experimental conditions	50 nM to 1 mM and 1–10 mM	Joe et al., 2012
	E. coli	Mercuric chloride, mercurous chloride, and mercuric ammonium chloride	Contaminated milk and chicken extract	50 nM to 10 μM	Guo et al., 2020
	E. coli bioreporter cells	Ciprofloxacin	Milk	7.2 ng/mL	Lu et al., 2019
	Probiotic *E. coli*	Nitrate	Mice with native gut microbiota	–	Woo et al., 2020
	LuxAB biosensor system from *Photorhabdus luminescens* was implemented in *E. coli*	Aldehydes	Lab experimental conditions	–	Bayer et al., 2021
	Recombinant *Hansenula polymorpha*	L-Lactate	Lab experimental conditions	2.5-fold higher sensitivity than control	Karkovska et al., 2015
	Pichia angusta VKM Y-2518	Ethanol	Lab experimental conditions	0.012 mM	Voronova et al., 2008

(*Continued*)

TABLE 1.2 (*Continued*)
List of Microbial Sensors Utilized for Various Applications

Technique	Microorganism (Promoter/ Reporter Construct)	Target Analyte	Sample	Detection Range/Limit	Reference
MFC-based sensors	Natural microorganisms	BOD	Artificial wastewater	250 mg/L	Liu et al., 2012a
	Natural microorganisms	BOD	Artificial wastewater	200 mg/L	Kim et al., 2003
	Activated sludge	BOD	Municipal wastewater	129 mg/L	Wang et al., 2010
	Activated sludge	BOD	Municipal wastewater	0.088 mg O_2/L	Liu et al., 2013
	Aeromonas hydrophila	BOD	Wastewater from the textile plant	692 mg/L	Chen et al., 2018
	Natural sediments	Oil spills	Lake and urban stream water	30.78 and 27.29 µA/mL of sensitivity, respectively	Dai et al., 2021
	Wild-type *Shewanella oneidensis* MR1	Formaldehyde	Water samples	For 0.001%, 0.01%, and 0.02% of formaldehyde, inhibition ratios of 7.88%, 16.08%, and 23.14% were obtained, respectively	Cho et al., 2020
	Hyphomicrobiaceae and *Cloacibacillus*	Lead	Lab experimental conditions	–	Xu et al., 2021
	–	Mercury, avermectin, and chlortetracycline hydrochloride	Lab experimental conditions	Twice higher inhibition ratios (36%, 15%, and 9%)	Zhao et al., 2019

signal ("turns on") or suppresses it ("turns off"). The appeal of optical sensing techniques lies in their ability to simultaneously monitor multiple analytes, making them highly suitable for high-throughput screening applications (Su et al., 2011). The fluorescence characteristics of microbes are also utilized by microbial biosensors in a variety of sensing applications. The fluorescence intensity, which is directly inversely proportional to the concentration of the target analyte, is measured using the fluorescent sensing approach. With this method, the fluorescent material is excited at a lower wavelength before being detected at a longer wavelength. Analytical chemistry has seen a considerable increase in the use of fluorescent biosensors because of how easily they are built using common molecular biology methods (Ibraheem and Campbell, 2010).

Microbial biosensors use a bioluminescent sensing approach to measure changes in luminescence released by live microorganisms, which react to the target analyte in a dose-dependent way. The most often utilized reporter gene in bioluminescent microbial biosensors is the *lux* gene, which in bacteria encodes luciferase (Galluzzi and Karp, 2006). Flavin mononucleotide and long-chain fatty aldehyde are oxidized by oxygen with the help of luciferase, which causes the release of blue-green light. In contrast to fluorescence biosensors that measure protein, bioluminescence intensity represents enzyme activity. As a result, compared to fluorescence, bioluminescence offers a quicker and more accurate detection method. When the whole lux operon (*lux* CDABE) is used as the reporter system, bioluminescent microbial biosensors also have the benefit of not requiring external substrates (Galluzzi and Karp, 2006; Lei et al., 2006).

In colorimetric sensing methods used in microbial biosensors, a chromogen substrate is metabolically changed into a coloured molecule by the microbial sensing element (Azevedo et al., 2005). A spectrophotometer can be used to measure the coloured product or be used to visually identify it (Lei et al., 2006). The extensive use of colorimetric methods in the creation of microbial biosensors is due to the measuring set-up's affordability and simplicity.

1.2.6 Microbial Biosensors with Transducers

Microbial biosensors with transducers are cutting-edge instruments that bring together the special traits of microbes with transduction technology to enable sensitive and selective analyte detection. These biosensors have a number of benefits, including high specificity, quick reaction times, and the capacity to identify a broad spectrum of target compounds. Transducers can include electrodes, microfluidic devices, or nanomaterial-based systems. For example, microorganisms can be immobilized on electrodes, and the electron transfer generated by their metabolic activities is converted into an electrical signal. These biosensors with transducers enable the conversion of microbial reactions into electrical, optical, or acoustic signals for detection and quantification (Lim et al., 2015; Plekhanova and Reshetilov, 2019). In order to identify and interact with certain analytes, microbial biosensors with transducers make use of the biological components of microorganisms. A signal is produced by the microbe as a result of the encounter, which causes a metabolic reaction. A measured output is then produced from this signal once it has been transduced by the integrated transducer. The transducer functions as an interface between the biological recognition component and the detection system, making it possible to transform the biochemical signal into a format that can be quickly measured and quantified (Hassan et al., 2016; Lim et al., 2015; Plekhanova and Reshetilov, 2019; Ponomareva et al., 2011). The transducer in biosensors can be electrochemical transducers, optical transducers, mass-sensitive transducers, or thermal transducers (Hassan et al., 2016; Ponomareva et al., 2011; Xu and Ying, 2011).

1.3 APPLICATIONS OF MICROBIAL SENSORS IN SUSTAINABLE GREEN TECHNOLOGIES

1.3.1 Environmental Monitoring and Pollution Control

In environmental research, microbial biosensors have become increasingly popular, notably in the monitoring of general toxicity. Numerous microorganisms have been thoroughly investigated and used for biosensing, including yeast, bacteria, microbial consortia, genetically modified bacteria, and altered cells. *Trichosporon cutaneum*, *Arxula adeninivorans* LS3, *Pseudomonas fluorescens*, *Pseudomonas putida*, and *Saccharomyces cerevisiae* are a few examples of microbes frequently used in microbial biosensors (Xu and Ying, 2011). Table 1.2 lists some of the microbial biosensors that are employed in environmental applications. In order to analyse and track environmental variables for a better knowledge of ecological health and the effects of contaminants, these biosensors are essential.

The dissolved oxygen needed for the oxidation of organic compounds during a 5-day period is measured using the standard biochemical oxygen demand (BOD) technique. For active engagement in environmental monitoring, this time span is too long, though. For estimating BOD, biosensors provide a quick substitute. BOD measurement is possible, for instance, in around 3 minutes using tiny oxygen electrodes covered with *Trichosporon cutaneum* yeast (Xu and Ying, 2011; Yang et al., 1997). With this method, prompt interventions in environmental monitoring and process management are made possible for both wastewater samples and standard BOD solutions. In addition, MFC biosensors can detect BOD and require maintenance during periods of inactivity. Hibernating electroactive bacteria were used to make a viable method for maintaining MFC biosensors (Guo et al., 2021). After hibernation, the biosensors quickly recovered voltage output and accurately detected BOD concentrations of 500 and 200 mg/L. The bacteria's ability to generate current remained intact, as indicated by consistent anode potentials. Cyclic voltammetry analysis revealed the presence of redox couples and cytochromes, which supported EAB metabolism and acted as temporary electron sinks during hibernation (Guo et al., 2021). This method simplifies and enhances the flexibility of BOD detection in MFC biosensors, with potential applications in other biosensing instruments. In another work, a strategy for assessing BOD_5 using bacterial indicators was developed, resulting in a bioassay that provided reliable and accurate BOD_5 measurements in just 3 hours compared to the reference method's 5 days (Jouanneau et al., 2019). The method expanded the measurement range, eliminating the need for dilution in most wastewater samples, unlike the reference method. The results demonstrated a strong correlation ($r^2 = 0.85$) with reference to BOD_5 values, but the method had limitations in assessing low BOD_5 levels below 25 mg/L according to the reference method. A list of a few other microbial sensors for BOD monitoring is given in Table 1.2.

Engineered microbial technology has been used by Karbelkar et al. to develop a novel method for remediating organophosphate (OP) pesticides (Karbelkar et al., 2021). With two modified strains, the system uses microbial electrochemistry. The first strain (*E. coli*) breaks down the pesticide, whereas the second strain (*S. oneidensis*), without the aid of labels or external electrochemical stimulation, produces current in reaction to the breakdown product. Notably, this method is distinct since *E. coli* serves as a scaffold for the OPs' enzyme breakdown, negating the requirement to maintain the survival of two microbial strains concurrently. The device exhibits great sensitivity, outperforming current colorimetric and fluorescence sensors and detecting OP breakdown products at submicromolar levels. Previously, an amperometric microbial biosensor that used a carbon paste electrode implanted with genetically modified cells that expressed organophosphorus hydrolase (OPH) on their cell surfaces to directly detect organophosphate nerve poisons has also been reported (Mulchandani et al., 2001). However, their use in environmental monitoring applications is constrained by the poor sensitivity of potentiometric and optical biosensors as well as the insufficient selectivity of amperometric biosensors towards phenolic chemicals.

1.3.2 Bioremediation and Waste Treatment

Microbial sensors are essential for tracking and enhancing the efficiency of bioremediation and waste treatment processes. In order to enhance the sensing performance of cell-based biosensors for detecting toxic and pathogenic contaminants in water, Wan et al. focused on developing a modular, cascaded signal amplification approach (Wan et al., 2019). Their research focused on the detection of arsenic and mercury, and they produced bacterial sensors with extreme sensitivity. They were able to expand the detection limit by up to 5,000 times while also enhancing the output signal by a factor of 750. Likewise, Cho et al. constructed a paper-based MFC that may be used as a mobile, one-time, on-site water quality monitor (Cho et al., 2019). Two layers of paper were combined to create the sensor, which had a straightforward, affordable, and disposable design. The bacterial cells were air-dried and pre-inoculated onto the sensor to make it easier for the on-site applications. The voltage the microbe produced before and after the air-drying procedure was measured, and the inhibition ratio was computed to account for any differences. The inhibition ratios obtained with the addition of various formaldehyde concentrations (0%, 0.001%, and 0.02%) were 7.88%, 16.08%, and 23.14%, respectively (Table 1.2). In another study, Adekunle et al. built an environmental biosensor that tracks toxicity in real time using the great sensitivity of a MFC to changes in electron donors and acceptors (Adekunle et al., 2019). In tests conducted in the lab, the biosensor demonstrated an R^2 of 0.95–0.97, indicating a quick and reliable reaction to changes in total heavy metal concentration. This reaction is a result of heavy metals interfering with electroactive bacteria's ability to function.

1.3.3 Renewable Energy Production

MFCs are being studied scientifically because of their capacity to operate under a variety of pressures and weather conditions. *Geobacter sulphurreducens*, a kind of bacteria used in MFCs, forms biofilms on electrodes (Kurniawan et al., 2022). These biofilms allow electrons to go more easily to the electrodes and discharge c-type cytochromes that build up at the biofilm-electrode interface, facilitating electron transmission. Bacteria attaching to the MFC eliminates the need for expensive separated enzymes since the microorganisms function as low-cost operating substrates for the MFC. This procedure is carried out in a small bioreactor where the bacteria are preserved as biofilms, lowering operating and maintenance expenses. For MFC applications, this novel technique offers considerable benefits in terms of efficacy, sustainability, and affordability (De Vela, 2021). MFCs work as electrochemical energy producers that resemble batteries and generate power. They have the potential for excellent operating efficiency because of their capability of direct conversion of chemical energy into electricity. MFCs are an environmentally favourable choice since they use their electron transfer abilities to produce energy without having to burn the organic molecules found in wastewater. Due to their dual functioning, MFCs may treat wastewater and generate electricity at the same time, which lessens their dependency on conventional energy sources and lowers costs for end users. It's significant that MFCs display equivalent effectiveness in removing organic matter to that of traditional wastewater treatment facilities, guaranteeing efficient and dependable wastewater treatment operations (Hernández-Fernández et al., 2016; Pal et al., 2017).

1.4 ADVANCES IN MICROBIAL SENSOR TECHNOLOGIES

Environmental monitoring and biomedical applications have greatly benefited from developments in microbial sensor technology. The establishment of genetically modified microbial sensors, which are intended to detect certain target chemicals or environmental circumstances, is one noteworthy breakthrough. These sensors use microbial cells as live sensing components that provide a quantifiable signal in response to the presence of the target analyte. Enhancing the sensitivity, selectivity, and reaction time of microbial sensors has been the focus of recent advances. This involves modifying microbial cells with better receptor proteins, signal amplification systems, and signal transduction pathways using synthetic biology approaches. Additionally, the combination of microbial sensors with nanomaterials, such as carbon nanotubes and nanoparticles, has improved signal detection and boosted stability. As a result of the miniaturization of microbiological sensors, portable and wearable devices have also been created, allowing for real-time and on-site monitoring. The creation of extremely sensitive and multiplexed microbial sensor systems has also been aided by developments in microfluidics and lab-on-a-chip technologies. Additionally, real-time data analysis and remote monitoring have been made possible by the integration of microbiological sensors with wireless communication systems and data processing algorithms. These developments in microbial sensor technologies show significant potential for a variety of applications, such as environmental monitoring, food safety, medical diagnostics, and industrial process control, opening the door to more efficient and long-lasting solutions to tackle numerous problems.

1.5 CHALLENGES AND FUTURE DIRECTIONS

Microbial sensors, as they advance and find widespread applications, require addressing regulatory and ethical considerations (Figure 1.4). The use of genetically engineered microorganisms raises concerns about environmental release and unintended consequences, necessitating the establishment of regulatory frameworks to ensure safe and responsible deployment. Ethical considerations encompass issues like privacy, data ownership, and informed consent in healthcare. Scaling up microbial sensor production and implementation for industrial applications poses challenges in terms of cost-effectiveness, reproducibility, and stability. Strategies for mass production, quality control, and long-term performance monitoring are necessary to seamlessly integrate microbial sensors into industrial processes. Integration with existing technologies is crucial to maximize the impact of microbial sensors. Collaborations between scientists, engineers, and industry professionals can develop interdisciplinary approaches that combine microbial sensing with data analytics, automation, and control systems. Integration with IoT platforms, AI, and machine learning enhances capabilities for real-time decision-making, predictive analytics, and autonomous control. Standardization of protocols is vital for the widespread adoption and comparability of microbial sensing technologies. Development of

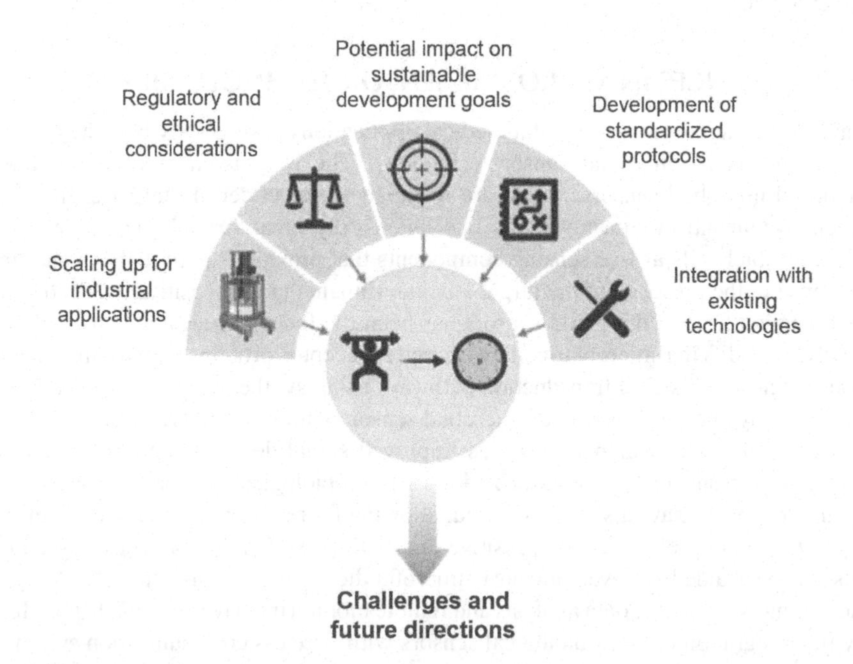

FIGURE 1.4 Schematic representation of various challenges to microbial sensors to accomplish sustainable green technologies.

standardized procedures, calibration methods, and performance metrics ensures consistency and reliability across platforms and applications. International standards and collaborative efforts between academia, industry, and regulatory bodies facilitate robust microbial sensing protocols. Microbial sensors have significant potential to contribute to the achievement of the United Nations' Sustainable Development Goals (SDGs). Enabling real-time monitoring and control, they support SDGs related to environmental protection, resource efficiency, and human health. Applications include sustainable agriculture, water and air quality monitoring, waste management, and disease control. Harnessing the full potential of microbial sensors accelerates progress towards a sustainable and resilient future.

1.6 CONCLUSION

Microbial sensors represent a powerful tool for accelerating sustainable green technologies. Overcoming the challenges related to regulatory and ethical considerations, scaling up for industrial applications, integration with existing technologies, development of standardized protocols, and alignment with the sustainable development goals will be the key to unlocking their full potential. With continued research, innovation, and collaboration, microbial sensors are poised to revolutionize various industries, promoting sustainability and driving us closer to a greener and more prosperous future.

REFERENCES

Abbasian, F., E. Ghafar-Zadeh, and S. Magierowski. 2018. Microbiological sensing technologies: A review. *Bioengineering* 5. doi: 10.3390/bioengineering5010020.

Abrevaya, X.C., N.J. Sacco, M.C. Bonetto, A. Hilding-Ohlsson, and E. Cortón. 2015. Analytical applications of microbial fuel cells. Part II: Toxicity, microbial activity and quantification, single analyte detection and other uses. *Biosens Bioelectron* 63: 591–601. doi: 10.1016/j.bios.2014.04.053.

Adekunle, A., V. Raghavan, and B. Tartakovsky. 2019. On-line monitoring of heavy metals-related toxicity with a microbial fuel cell biosensor. *Biosens Bioelectron* 132: 382–390. doi: 10.1016/j.bios.2019.03.011.

Azevedo, A.M., D.M.F. Prazeres, J.M.S. Cabral, and L.P. Fonseca. 2005. Ethanol biosensors based on alcohol oxidase. *Biosens Bioelectron* 21: 235–247. doi: 10.1016/j.bios.2004.09.030.

Bachosz, K., M.T. Vu, L.D. Nghiem, J. Zdarta, L.N. Nguyen, and T. Jesionowski. 2022. Enzyme-based control of membrane biofouling for water and wastewater purification: A comprehensive review. *Environ Technol Innov* 25: 102106. doi: 10.1016/J.ETI.2021.102106.

Bayer, T., A. Becker, H. Terholsen, I.J. Kim, I. Menyes, S. Buchwald, K. Balke, S. Santala, S.C. Almo, and U.T. Bornscheuer. 2021. Article luxab-based microbial cell factories for the sensing, manufacturing and transformation of industrial aldehydes. *Catalysts* 11. doi: 10.3390/catal11080953.

Belkin, S. 2003. Microbial whole-cell sensing systems of environmental pollutants. *Curr Opin Microbiol* 6: 206–212. doi: 10.1016/S1369-5274(03)00059-6.

Bilal, M., and H.M.N. Iqbal. 2019. Microbial-derived biosensors for monitoring environmental contaminants: Recent advances and future outlook. *Process Saf Environ Prot* 124: 8–17. doi: 10.1016/j.psep.2019.01.032.

Bollella, P., and E. Katz. 2020. Enzyme-based biosensors: Tackling electron transfer issues. *Sensors (Switzerland)* 20: 1–32. doi: 10.3390/s20123517.

Bose, S., S. Maity, and A. Sarkar. 2022. Review of microbial biosensor for the detection of mercury in water. *Environ Quality Manag* 31: 29–40. doi: 10.1002/tqem.21742.

Cai, S., Y. Shen, Y. Zou, P. Sun, W. Wei, J. Zhao, and C. Zhang. 2018. Engineering highly sensitive whole-cell mercury biosensors based on positive feedback loops from quorum-sensing systems. *Analyst* 143: 630–634. doi: 10.1039/c7an00587c.

Camsund, D., P. Lindblad, and A. Jaramillo. 2011. Genetically engineered light sensors for control of bacterial gene expression. *Biotechnol J* 6: 826–836. doi: 10.1002/biot.201100091.

Chen, C.Y., T.H. Tsai, P.S. Wu, S.E. Tsao, Y.S. Huang, and Y.C. Chung. 2018. Selection of electrogenic bacteria for microbial fuel cell in removing Victoria blue R from wastewater. *J Environ Sci Health A Tox Hazard Subst Environ Eng* 53: 108–115. doi: 10.1080/10934529.2017.1377580.

Cho, J.H., Y. Gao, and S. Choi. 2019. A portable, single-use, paper-based microbial fuel cell sensor for rapid, on-site water quality monitoring. *Sensors (Switzerland)* 19. doi: 10.3390/s19245452.

Cho, J.H., Y. Gao, J. Ryu, and S. Choi. 2020. Portable, disposable, paper-based microbial fuel cell sensor utilizing freeze-dried bacteria for in situ water quality monitoring. *ACS Omega* 5: 13940–13947. doi: 10.1021/acsomega.0c01333.

Chu, N., Q. Liang, W. Hao, Y. Jiang, P. Liang, and R.J. Zeng. 2021. Microbial electrochemical sensor for water biotoxicity monitoring. *Chem Eng J* 404. doi: 10.1016/j.cej.2020.127053.

Clemons, K.V., J. Shankar, and D.A Stevens. 2016. Mycologic endocrinology. In *Microbial endocrinology: Interkingdom signaling in infectious disease and health* (Vol. 874, pp. 337–363).

Cui, Y., B. Lai, and X. Tang. 2019. Microbial fuel cell-based biosensors. *Biosensors* 9. doi: 10.3390/bios9030092.

Dai, Z., R. Yu, X. Zha, Z. Xu, G. Zhu, and X. Lu. 2021. On-line monitoring of minor oil spills in natural waters using sediment microbial fuel cell sensors equipped with vertical floating cathodes. *Sci Total Environ* 782. doi: 10.1016/j.scitotenv.2021.146549.

De Mora, K., N. Joshi, B.L. Balint, F.B. Ward, A. Elfick, and C.E. French. 2011. A pH-based biosensor for detection of arsenic in drinking water. *Anal Bioanal Chem* 400: 1031–1039. doi: 10.1007/s00216-011-4815-8.

De Vela, R.J. 2021. A review of the factors affecting the performance of anaerobic membrane bioreactor and strategies to control membrane fouling. *Rev Environ Sci Biotechnol* 20: 607–644. doi: 10.1007/s11157-021-09580-2.

Díaz-García, C.M., C. Lahmann, J.R. Martínez-François, B. Li, D. Koveal, N. Nathwani, M. Rahman, J.P. Keller, J.S. Marvin, L.L. Looger, and G. Yellen. 2019. Quantitative in vivo imaging of neuronal glucose concentrations with a genetically encoded fluorescence lifetime sensor. *J Neurosci Res* 97: 946–960. doi: 10.1002/jnr.24433.

Galluzzi, L., and M. Karp. 2006. Whole cell strategies based on lux genes for high throughput applications toward new antimicrobials. *Comb Chem High Throughput Screen* 9: 501–514.

Gao, Y., F. Yin, W. Ma, S. Wang, Y. Liu, and H. Liu. 2020. Rapid detection of biodegradable organic matter in polluted water with microbial fuel cell sensor: Method of partial coulombic yield. *Bioelectrochemistry* 133. doi: 10.1016/j.bioelechem.2020.107488.

Gavrilaş, S., C. Ştefan Ursachi, S. Perţa-Crişan, and F.D. Munteanu. 2022. Recent trends in biosensors for environmental quality monitoring. *Sensors* 22. doi: 10.3390/s22041513.

Grote, M., M. Engelhard, and P. Hegemann. 2014. Of ion pumps, sensors and channels – Perspectives on microbial rhodopsins between science and history. *Biochim Biophys Acta Bioenerg* 1837: 533–545. doi: 10.1016/j.bbabio.2013.08.006.

Guo, F., Y. Liu, and H Liu. 2021. Hibernations of electroactive bacteria provide insights into the flexible and robust BOD detection using microbial fuel cell-based biosensors. *Sci Total Environ* 753. doi: 10.1016/j.scitotenv.2020.142244.

Guo, M., J. Wang, R. Du, Y. Liu, J. Chi, X. He, K. Huang, Y. Luo, and W. Xu. 2020. A test strip platform based on a whole-cell microbial biosensor for simultaneous on-site detection of total inorganic mercury pollutants in cosmetics without the need for predigestion. *Biosens Bioelectron* 150. doi: 10.1016/j.bios.2019.111899.

Hassan, S.H.A., S.W. Van Ginkel, M.A.M. Hussein, R. Abskharon, and S.E. Oh. 2016. Toxicity assessment using different bioassays and microbial biosensors. *Environ Int* 92–93: 106–118. doi: 10.1016/j.envint.2016.03.003.

He, W., S. Yuan, W.H. Zhong, M.A. Siddikee, and C.C. Dai. 2016. Application of genetically engineered microbial whole-cell biosensors for combined chemosensing. *Appl Microbiol Biotechnol* 100: 1109–1119. doi: 10.1007/s00253-015-7160-6.

Hernández-Fernández, F.J., A.P. De Los Ríos, F. Mateo-Ramírez, M.D. Juarez, L.J. Lozano-Blanco, and C. Godínez. 2016. New application of polymer inclusion membrane based on ionic liquids as proton exchange membrane in microbial fuel cell. *Sep Purif Technol* 160: 51–58. doi: 10.1016/j.seppur.2015.12.047.

Hill, A., S. Tait, C. Baillie, B. Virdis, and B. McCabe. 2020. Microbial electrochemical sensors for volatile fatty acid measurement in high strength wastewaters: A review. *Biosens Bioelectron* 165. doi: 10.1016/j.bios.2020.112409.

Hu, J., K. Fu, and P.W. Bohn. 2018. Whole-cell *Pseudomonas aeruginosa* localized surface plasmon resonance aptasensor. *Anal Chem* 90: 2326–2332. doi: 10.1021/acs.analchem.7b04800.

Ibraheem, A., and R.E. Campbell. 2010. Designs and applications of fluorescent protein-based biosensors. *Curr Opin Chem Biol* 14: 30–36. doi: 10.1016/j.cbpa.2009.09.033.

Ivars-Barceló, F., A. Zuliani, M. Fallah, M. Mashkour, M. Rahimnejad, and R. Luque. 2018. Novel applications of microbial fuel cells in sensors and biosensors. *Appl Sci* 8. doi: 10.3390/app8071184.

Jaiswal, S., and P. Shukla. 2020. Alternative strategies for microbial remediation of pollutants via synthetic biology. *Front Microbiol* 11. doi:10.3389/fmicb.2020.00808.

Jiang, Y., N. Chu, and R.J. Zeng. 2019. Submersible probe type microbial electrochemical sensor for volatile fatty acids monitoring in the anaerobic digestion process. *J Clean Prod* 232: 1371–1378. doi: 10.1016/j.jclepro.2019.06.041.

Jiang, Y., P. Liang, P. Liu, D. Wang, B. Miao, and X. Huang. 2017. A novel microbial fuel cell sensor with biocathode sensing element. *Biosens Bioelectron* 94: 344–350. doi: 10.1016/j.bios.2017.02.052.

Jiang, Y., X. Yang, P. Liang, P. Liu, and X. Huang. 2018. Microbial fuel cell sensors for water quality early warning systems: Fundamentals, signal resolution, optimization and future challenges. *Renew Sust Energ Rev* 81: 292–305. doi: 10.1016/j.rser.2017.06.099.

Jin, X., X. Li, N. Zhao, I. Angelidaki, and Y. Zhang. 2017. Bio-electrolytic sensor for rapid monitoring of volatile fatty acids in anaerobic digestion process. *Water Res* 111: 74–80. doi: 10.1016/j.watres.2016.12.045.

Joe, M.H., K.H. Lee, S.Y. Lim, S.H. Im, H.P. Song, I.S. Lee, and D.H. Kim. 2012. Pigment-based whole-cell biosensor system for cadmium detection using genetically engineered *Deinococcus radiodurans*. *Bioprocess Biosyst Eng* 35: 265–272.

Jouanneau, S., E. Grangé, M.J. Durand, and G. Thouand. 2019. Rapid BOD assessment with a microbial array coupled to a neural machine learning system. *Water Res* 166. doi: 10.1016/j.watres.2019.115079.

Karbelkar, A.A., E.E. Reynolds, R. Ahlmark, and A.L. Furst. 2021. A microbial electrochemical technology to detect and degrade organophosphate pesticides. *ACS Cent Sci* 7: 1718–1727. doi: 10.1021/acscentsci.1c00931.

Karkovska, M., O. Smutok, N. Stasyuk, and M. Gonchar. 2015. L-Lactate-selective microbial sensor based on flavocytochrome b2-enriched yeast cells using recombinant and nanotechnology approaches. *Talanta* 144: 1195–1200. doi: 10.1016/j.talanta.2015.07.081.

Kim, H.J., H. Jeong, and S.J. Lee. 2018. Synthetic biology for microbial heavy metal biosensors. *Anal Bioanal Chem* 410: 1191–1203. doi: 10.1007/s00216-017-0751-6.

Kim, M., S.M. Youn, S.H. Shin, J.G. Jang, S.H. Han, M.S. Hyun, G.M. Gadd, and H.J Kim. 2003. Practical field application of a novel BOD monitoring system. *J Environ Monitor* 5: 640–643. doi: 10.1039/b304583h.

Kumar, T.H.V., and A.K. Sundramoorthy. 2019. Electrochemical biosensor for methyl parathion based on single-walled carbon nanotube/glutaraldehyde crosslinked acetylcholinesterase-wrapped bovine serum albumin nanocomposites. *Anal Chim Acta* 1074: 131–141. doi: 10.1016/j.aca.2019.05.011.

Kurbanoglu, S., C. Erkmen, and B. Uslu. 2020. Frontiers in electrochemical enzyme based biosensors for food and drug analysis. *TrAC Trends Anal Chem* 124: 115809. doi: 10.1016/J.TRAC.2020.115809.

Kurniawan, T.A., M.H.D. Othman, X. Liang, M. Ayub, H.H. Goh, T.D. Kusworo, A. Mohyuddin, and K.W. Chew. 2022. Microbial fuel cells (MFC): A potential game-changer in renewable energy development. *Sustainability* 14. doi: 10.3390/su142416847.

Lei, Y., W. Chen, and A. Mulchandani. 2006. Microbial biosensors. *Anal Chim Acta* 568: 200–210. doi: 10.1016/j.aca.2005.11.065.

Leonard, P., S. Hearty, J. Brennan, L. Dunne, J. Quinn, T. Chakraborty, and R. O'kennedy. 2003. Advances in biosensors for detection of pathogens in food and water. *Enzyme Microb Technol* 32: 3–13.

Lim, J.W., D. Ha, J. Lee, S.K. Lee, and T. Kim. 2015. Review of micro/nanotechnologies for microbial biosensors. *Front Bioeng Biotechnol* 3. doi: 10.3389/fbioe.2015.00061.

Liu, L., and S. Choi. 2021. Miniature microbial solar cells to power wireless sensor networks. *Biosens Bioelectron* 177. doi: 10.1016/j.bios.2021.112970.

Liu, Y., Z. Matharu, M.C. Howland, A. Revzin, and A.L. Simonian. 2012b. Affinity and enzyme-based biosensors: Recent advances and emerging applications in cell analysis and point-of-care testing. *Anal Bioanal Chem* 404: 1181–1196. doi: 10.1007/s00216-012-6149-6.

Liu, C., H. Zhao, S. Gao, J. Jia, L. Zhao, D. Yong, and S. Dong. 2013. A reagent-free tubular biofilm reactor for on-line determination of biochemical oxygen demand. *Biosens Bioelectron* 45: 213–218. doi: 10.1016/j.bios.2013.01.041.

Liu, C., H. Zhao, L. Zhong, C. Liu, J. Jia, X. Xu, L. Liu, and S. Dong. 2012a. A biofilm reactor-based approach for rapid on-line determination of biodegradable organic pollutants. *Biosens Bioelectron* 34: 77–82. doi: 10.1016/j.bios.2012.01.020.

Lu, M.Y., W.C. Kao, S. Belkin, and J.Y. Cheng. 2019. A smartphone-based whole-cell array sensor for detection of antibiotics in milk. *Sensors* 19. doi: 10.3390/s19183882.

Mulchandani, P., W. Chen, A. Mulchandani, J. Wang, and L Chen. 2001. Amperometric microbial biosensor for direct determination of organophosphate pesticides using recombinant microorganism with surface expressed organophosphorus hydrolase. *Biosens Bioelectron* 16: 433–437.

Neethirajan, S., V. Ragavan, X. Weng, and R. Chand. 2018. Biosensors for sustainable food engineering: Challenges and perspectives. *Biosensors* 8. doi: 10.3390/bios8010023.

Nigam, V.K., and P. Shukla. 2015. Enzyme based biosensors for detection of environmental pollutants – A review. *J Microbiol Biotechnol* 25: 1773–1781. doi: 10.4014/jmb.1504.04010.

Olias, L.G., and M. Di Lorenzo. 2021. Microbial fuel cells for in-field water quality monitoring. *RSC Adv* 11: 16307–16317. doi: 10.1039/d1ra01138c.

Pal, S., J. Handa, and U.K. Jain. 2017. Chemical hydrolysis optimization for release of sugars from wheat bran. *Int J Innov Sci Eng Technol* 4: 68–72.

Park, M., S.L. Tsai, and W. Chen. 2013. Microbial biosensors: Engineered microorganisms as the sensing machinery. *Sensors* 13: 5777–5795. doi: 10.3390/s130505777.
Plekhanova, Y. V., and A.N. Reshetilov. 2019. Microbial biosensors for the determination of pesticides. *J Anal Chem* 74: 1159–1173. doi: 10.1134/S1061934819120098.
Ponomareva, O.N., V.A. Arlyapov, V.A. Alferov, and A.N. Reshetilov. 2011. Microbial biosensors for detection of biological oxygen demand (a review). *Appl Biochem Microbiol* 47: 1–11. doi: 10.1134/S0003683811010108.
Pundir, C.S., and A. Malik. 2019. Bio-sensing of organophosphorus pesticides: A review. *Biosens Bioelectron* 140. doi: 10.1016/j.bios.2019.111348.
Riding, M.J., K.J. Doick, F.L. Martin, K.C. Jones, and K.T. Semple. 2013. Chemical measures of bioavailability/bioaccessibility of PAHs in soil: Fundamentals to application. *J Hazard Mater* 261: 687–700. doi: 10.1016/j.jhazmat.2013.03.033.
Ronald, P.C., and B. Beutler. 2010. Plant and animal sensors of conserved microbial signatures. *Science* 330: 1061–1064. doi: 10.1126/science.1189468.
Sato, T., and Y. Kobayashi. 1998. The ars operon in the skin element of *Bacillus subtilis* confers resistance to arsenate and arsenite. *J Bacteriol* 180: 1655–1661.
Sharma, P., S. Asad, and A. Ali. 2013. Bioluminescent bioreporter for assessment of arsenic contamination in water samples of India. *J Biosci* 38: 251–258. doi: 10.1007/s12038-013-9305-z.
Simoska, O., E.M. Gaffney, S.D. Minteer, A. Franzetti, P. Cristiani, M. Grattieri, and C. Santoro. 2021. Recent trends and advances in microbial electrochemical sensing technologies: An overview. *Curr Opin Electrochem* 30. doi: 10.1016/j.coelec.2021.100762.
Sonawane, J.M., C.I. Ezugwu, and P.C. Ghosh. 2020. Microbial fuel cell-based biological oxygen demand sensors for monitoring wastewater: State-of-the-art and practical applications. *ACS Sens* 5: 2297–2316. doi: 10.1021/acssensors.0c01299.
Su, L., W. Jia, C. Hou, and Y. Lei. 2011. Microbial biosensors: A review. *Biosens Bioelectron* 26: 1788–1799. doi: 10.1016/j.bios.2010.09.005.
Tolosa, L., I. Gryczynski, L.R. Eichhorn, J.D. Dattelbaum, F.N. Castellano, G. Rao, and J.R. Lakowicz. 1999. Glucose sensor for low-cost lifetime-based sensing using a genetically engineered protein. *Anal Biochem* 2671: 114–120.
Tremblay, P.L., and P.C. Hallenbeck. 2009. Of blood, brains and bacteria, the Amt/Rh transporter family: Emerging role of Amt as a unique microbial sensor. *Mol Microbiol* 71: 12–22. doi: 10.1111/j.1365-2958.2008.06514.x.
Voronova, E.A., P. V. Iliasov, and A.N. Reshetilov. 2008. Development, investigation of parameters and estimation of possibility of adaptation of Pichia angusta based microbial sensor for ethanol detection. *Anal Lett* 41: 377–391. doi: 10.1080/00032710701645729.
Wang, J., Y. Zhang, Y. Wang, R. Xu, Z. Sun, and Z. Jie. 2010. An innovative reactor-type biosensor for BOD rapid measurement. *Biosens Bioelectron* 25: 1705–1709. doi: 10.1016/j.bios.2009.12.018.
Wan, J., W. Peng, X. Li, T. Qian, K. Song, J. Zeng, F. Deng, S. Hao, J. Feng, P. Zhang, Y. Zhang, J. Zou, S. Pan, M. Shin, B.J. Venton, J.J. Zhu, M. Jing, M. Xu, and Y Li. 2021. A genetically encoded sensor for measuring serotonin dynamics. *Nat Neurosci* 24: 746–752. doi: 10.1038/s41593-021-00823-7.
Wan, X., F. Volpetti, E. Petrova, C. French, S.J. Maerkl, and B. Wang. 2019. Cascaded amplifying circuits enable ultrasensitive cellular sensors for toxic metals. *Nat Chem Biol* 15: 540–548. doi: 10.1038/s41589-019-0244-3.
Woo, S.G., S.J. Moon, S.K. Kim, T.H. Kim, H.S. Lim, G.H. Yeon, B.H. Sung, C.H. Lee, S.G. Lee, J.H. Hwang, and D.H Lee. 2020. A designed whole-cell biosensor for live diagnosis of gut inflammation through nitrate sensing. *Biosens Bioelectron* 168. doi: 10.1016/j.bios.2020.112523.
Xu, M., J. Li, B. Liu, C. Yang, H. Hou, J. Hu, J. Yang, K. Xiao, S. Liang, and D. Wang. 2021. The evaluation of long term performance of microbial fuel cell based Pb toxicity shock sensor. *Chemosphere* 270. doi: 10.1016/j.chemosphere.2020.129455.

Xu, X., and Y. Ying. 2011. Microbial biosensors for environmental monitoring and food analysis. *Food Rev Int* 27: 300–329. doi: 10.1080/87559129.2011.563393.

Xu, Z., B. Liu, Q. Dong, Y. Lei, Y. Li, J. Ren, J. McCutcheon, and B Li. 2015. Flat microliter membrane-based microbial fuel cell as "on-line sticker sensor" for self-supported in situ monitoring of wastewater shocks. *Bioresour Technol* 197: 244–251. doi: 10.1016/j.biortech.2015.08.081.

Yang, Z., H. Suzuki, S. Sasaki, S. Mcniven, and I. Karube. 1997. Comparison of the dynamic transient-and steady-state measuring methods in a batch type BOD sensing system. *Sens Actuators B Chem* 45: 217–222.

Yunus, G. 2018. Biosensors: An enzyme-based biophysical technique for the detection of foodborne pathogens. In *Enzymes in food biotechnology* (pp. 723–738). Academic Press.

Zhang, F., and J. Keasling. 2011. Biosensors and their applications in microbial metabolic engineering. *Trends Microbiol* 19: 323–329. doi: 10.1016/j.tim.2011.05.003.

Zhao, T., B. Xie, Y. Yi, and H. Liu. 2019. Sequential flowing membrane-less microbial fuel cell using bioanode and biocathode as sensing elements for toxicity monitoring. *Bioresour Technol* 276: 276–280. doi: 10.1016/j.biortech.2019.01.009.

2 Biosensors
Role and Application in Green Technologies

Sandhya Chouhan, Palak Patel, Parikshana Mathur, and Sandeep Choudhary

2.1 INTRODUCTION

When analysing the profound impact on the built environment, green technology is becoming a crucial component of sustainable building initiatives. The term "Green technology," which also goes by the names "clean technology" and/or "environmental technology," is used to describe the creation and use of cutting-edge methods that attempt to improve environmental conditions, encourage long-term sustainability, and lessen human effects on the earth. It includes a wide variety of practices, technologies, and goods that aim to preserve natural resources, lessen the effects of pollution and greenhouse gas emissions, and encourage a greener way of life (Qamar et al. 2021). Biosensors are utilized by green technology to revolutionize sustainable practices in several industries. It is a device that combines the best of biology, chemistry, and engineering. They are designed to react with enzymes, antibodies, or nucleic acids and then transform the biochemical reaction into an observable signal. Biosensors are effective instruments for environmental monitoring and permit immediate identification and measurement of pollutants in soil, water, and air (Rodríguez-Mozaz et al. 2004). When compared to more traditional analytical methods, biosensors have several benefits, such as the ability to be miniaturized and carried around easily, as well as their rapid reaction times with high specificity. Biosensors may be used in a variety of fields which include medications, ecology, dietary sciences, and biotechnology.

The bioreceptor, a biological component, is the central mechanism in biosensor technology, which functions by recognizing and binding a target analyte. Typically, this bioreceptor is very selective, meaning it can only detect its intended target. When a molecule binds to another, it triggers a transduction pathway that ultimately results in an electrical, optical, or electrochemical output signal from the biological contact. The strength of the signal is then determined and linked to the concentration of the target analyte (Figure 2.1) (Karunakaran, Rajkumar, and Bhargava 2015).

Over time, there have been significant advancements and milestones in the development of biosensors (Table 2.1). Green technology-based biosensors have paved the way in establishing eco-friendly and sustainable techniques. Owing to this, research and innovation have pushed the creation of biosensors that are more ecologically friendly and have a smaller footprint (Cova et al. 2022).

DOI: 10.1201/9781003407683-2

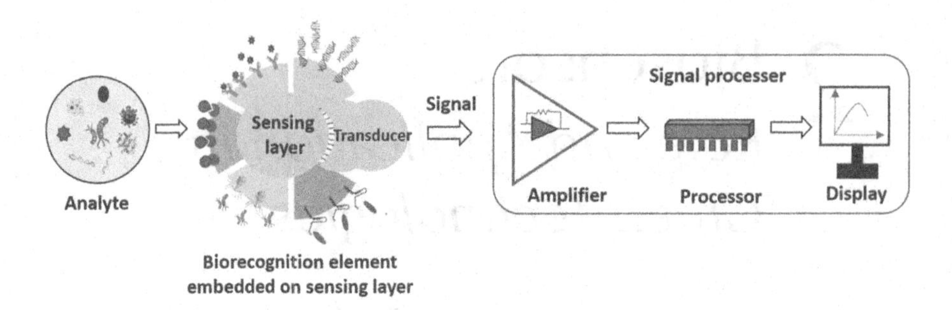

FIGURE 2.1 Schematic diagram of the biosensor.

TABLE 2.1
Development of Green Technology-Based Biosensors during 1980–1999

Year	Discovery	Reference
1984	Scheller and his fellow researchers were able to detect urea using an enzyme-based biosensor that they designed	Kirstein, Kirstein, and Scheller 1985
1987	The biosensor was developed by David M. Rawson, Allison J. Willmer, and Marco F. Cardosi for online testing of surface water pesticide contamination	Rawson, Willmer, and Cardosi 1987
1989	Anthony P. P. Turner developed a whole-cell biosensor for environmental monitoring	Rawson, Willmer, and Turner 1989
1990	Fibre-optic biosensor with time-resolved detection based on antibodies was created by T. Vo-Dinh, T. Nolan, Y. F. Cheng, J. P. Alarie, and M. J. Sepaniak	Vo-Dinh et al. 1990
1992	Research chemist KR. Rogers and research investigator J. N. Lin collaborated on constructing a biosensor for the use in surveillance of the environment	Rogers and Lin 1992
1993	Bioluminescent sensors for the ecological detection of bioavailable mercury(II) were created by O. Selifonova, R. Burlage, and T. Barkay	Selifonova, Burlage, and Barkay 1993
1996	Chlorophyll fluorescence was used as the basis for the biosensor that was created by Daniela Merz, Michael Geyer, D. A. Moss, and Hans-Joachim Ache to detect herbicides	Merz et al. 1996
1997	Acetylcholine mini sensors based on metal-supported lipid bilayers were created by M. Rehák, M. Šnejdárková, and T. Hianik. These sensors were designed to determine the presence of environmental contaminants	Rehák, Šnejdárková, and Hianik 1997
1998	Immobilised algae cells were used in the creation of a sensor that can detect environmental toxins, which was created by Dieter Frense, Adrian Muller, and Dieter Beckmann	Frense et al. 1998
1999	In order to determine volatile organic molecules, Martine Naessens and Canh Tran-Minh created a biosensor that utilised immobilised *Chlorella* microalgae as the active ingredient	Naessens and Tran-Min 1999

2.2 CHARACTERISTICS OF BIOSENSORS

2.2.1 Selectivity

Biosensors are distinguished by their selectivity, or their capacity to discriminate between a target molecule or analyte and background chemicals. This selectivity is achieved by employing biological recognition components like enzymes, antibodies, or nucleic acids that bind to or otherwise interact with the target analyte. Biosensor's superior selectivity allows for trustworthy detection even in very heterogeneous sample matrices (Ahmed et al. 2019).

2.2.2 Sensitivity

Biosensors are very sensitive and can detect analytes at extremely low concentrations. This quality is essential for environmental monitoring and medical diagnostics that depend on detecting minute concentrations of target molecules. The detection of signal and signal-to-noise ratio of biosensors are improved by a few methods. These include the employment of extremely sensitive biological identification components, signal amplification techniques, and cutting-edge transducer technologies (Ahmed et al. 2019).

2.2.3 Specificity

Biosensors' sensitivity and specificity come from the carefully selected biological recognition components, each of which has its own set of binding characteristics and chemical recognition capabilities. These recognition components engage preferentially with the analyte of interest, allowing for precise detection while reducing the impact of confounding molecules (Peveler, Yazdani, and Rotello 2016). In applications like medical diagnostics and food safety, specificity is essential for generating dependable and exact findings.

2.2.4 Rapid Response Time

Biosensors' quick turnaround time makes them ideal for continuous monitoring and detection in real-time. Applications where prompt action or intervention is needed, such as a toxin or pathogen detection, benefit greatly from the fast reaction time. To facilitate prompt action and decision-making, biosensors are engineered to provide a quantifiable signal or reaction upon interacting with the target analyte (Cahuantzi-Muñoz, González-Fuentes, and Ortiz-Frade 2019).

2.2.5 Accuracy

Providing exact quantitative data, biosensors provide reliable readings. This quality is crucial in situations when it is necessary to precisely monitor analyte concentrations or identify certain biomarkers. Biosensor accuracy is affected by a few parameters, including recognition element specificity, signal transduction pathways, and calibration and validation procedures (Cahuantzi-Muñoz et al. 2019).

2.2.6 Portability

Many biosensors are small and lightweight, making them ideal for use in the field. Being portable makes monitoring easier and timelier in regions with limited infrastructure or access. Point-of-care testing and real-time monitoring in various places are made possible by portable biosensors, which also find use in environmental monitoring and medical diagnostics, as well as in the food safety industry (Eyvazi et al. 2021).

2.2.7 Reusability

Unlike disposable sensors, biosensors may be used again and over again, saving money and preventing waste. Biosensors can be used more than once because their biological recognition parts can be repaired or replaced while the sensor platform remains untouched. In addition to saving money, the reduced amount of trash produced by reusable biosensors makes them an important tool for greener practices (Bocanegra-Rodríguez et al. 2020).

2.2.8 Stability

The performance and dependability of biosensors remain consistent over time and throughout a wide range of operating situations, demonstrating their stability. Over the course of its useful life, a biosensor's stability ensures dependable and precise results. The robustness of the sensor platform, the storage and handling conditions, and the stability of the recognition components all have an impact on the stability of the biosensor (Song et al. 2021).

2.2.9 Miniaturization

Biosensors may be made smaller so that they can be used in more compact systems. The miniaturization of biosensors paves the way for their expansion into fields like continuous monitoring and real-time tracking of analytes through their use in portable, wearable, or implantable forms. The miniaturization of biosensors has several potential advantages in healthcare settings, including enhanced patient convenience and safety (Derkus 2016).

2.3 TYPES OF BIOSENSORS

Biosensors can be differentiated based on whether they primarily rely on the physical or biological components for their sensing mechanism.

2.3.1 Biological Biosensor

In order to detect and respond to the target analyte, biological biosensors use biological agents such as enzymes, cells, and nucleic acids as sensing elements (Mehrotra 2016). These sensing elements exhibit high specificity and selectivity towards the target analyte. It can reach great sensitivity owing to biological systems' intrinsic amplification processes. Here are some examples of biological biosensors.

2.3.1.1 Enzyme-Based Biosensor

An enzymatic biosensor is a device consisting of an enzyme that recognizes the desired analyte and interacts with it to generate a chemical signal. A transducer then converts the chemical signal into an acoustic signal, and an electrical amplifier amplifies it further. Together, these three components work to produce the biosensor's output (Kaur et al. 2019) as shown in Figure 2.2a. Clark and Lyons developed the first glucose sensor in the early 1960s, which used the enzyme glucose oxidase (Gox) as a receptor (Vyas et al. 2023).

2.3.1.2 Immunosensor

Immunosensor detects the actual immunoreaction between the antibody and its target antigen and forms a stable immunocomplex. Immunosensor is divided into two categories: labelled and non-labelled. Non-labelled immunosensors are constructed in such a manner that the immunocomplex, also known as the antigen-antibody complex, may be identified instantly by detecting the physical changes induced by the complex's development. This allows for the immunocomplex to be detected without

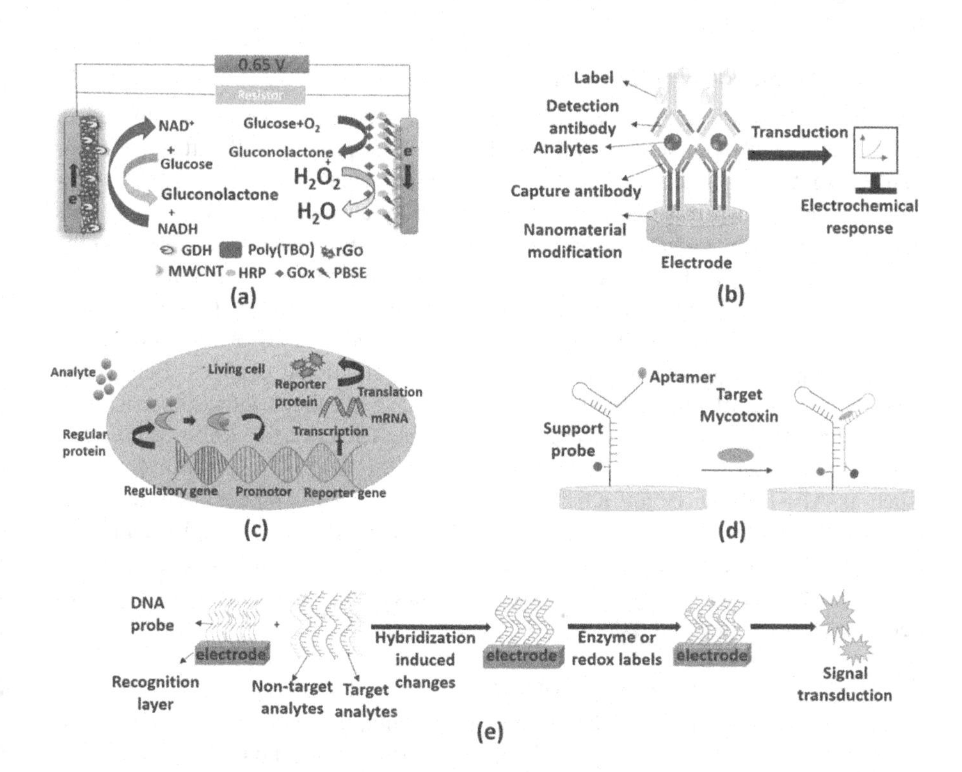

FIGURE 2.2 Type of biological biosensor: (a) enzyme-based biosensor, (b) immunosensor, (c) whole-cell-based sensor, (d) aptamer-based sensor, and (e) DNA sensor.

the need for labelling the sensor (Aizawa 1994). On the other hand, a labelled immunosensor is one that contains a label that can be detected with great sensitivity. The immunocomplex is therefore determined in a sensitive manner using label measurement (Figure 2.2b). An immunocomplex is formed when an antigen or antibody that must be identified is dissolved in a solution and then combined with an antibody or antigen that is coupled to a complementary matrix. Because of this, the surface's physical characteristics, which include optical quality, inherent piezo frequency, transmembrane potential, or potential of the electrode, may be changed and monitored. Other examples of these attributes include the transmembrane potential.

2.3.1.3 Whole-Cell-Based Biosensor

These biosensors utilize microbial cells as the receptor and transducer to create detectable output signals in response to chemical or physiological stimuli. These biosensors may be used in a wide variety of applications (Figure 2.2c). The cells are often genetically modified, and they include a sensing module and a reporting module that together make it possible to detect, record, and quantify the target analyte. Low cost, mobility, and environmental compatibility are some of the benefits that come with using a biosensor that is based on entire cells (Saltepe et al. 2017). It is said that the recently developed biosensor that is based on entire cells can identify the dangerous bacteria *Pseudomonas aeruginosa* and *Burkholderia pseudomallei* that contaminate water.

2.3.1.4 Aptamer-Based Biosensor

Aptamers are short single-stranded nucleic acid oligomers that are composed of either RNA or DNA molecules and have the ability to preferentially attach to a particular target (Kong and Byun 2013). As a recognition element, aptamers offer a number of benefits that are superior to those offered by classical antibodies. Aptamers are known for their great sensitivity and selectivity, which may be attributed to the amazing flexibility and ease with which their structures can be designed. After they have been selected, aptamers may be synthesized from commercial sources in a manner that is very repeatable and extremely pure. When they attach to their targets, aptamers frequently go through severe conformational changes. Because of this, the creation of innovative biosensors that have excellent detection sensitivity and selectivity is made significantly more flexible (Song et al. 2008). Biosensors based on aptamers have seen extensive usage in the field of mycotoxin detection (such as AFB1 (Bennett and Klich 2003)), one of the most toxic contaminants in food and agriculture (Figure 2.2d). Ultrasensitive techniques, on the other hand, are difficult to construct using simple apta sensor recognition. As a result, a new generation of apta sensor with signal amplification and enhancement for mycotoxins has been developed (Guo et al. 2020).

2.3.1.5 DNA Hybridization Biosensor

The complementary pairing of DNA bases in hybridization biosensors are the mechanism that underpins the biorecognition process. Immobilized on the electrode surface are single-stranded DNA segments that are between 20 and 40 base pairs in length and have a high degree of target selectivity. It is necessary to immobilize the

DNA fragments in a manner that ensures they retain their optimal orientation, availability to the desired analyte, stability, and responsiveness. During the hybridization process, target DNA binds to the sequence that is complementary to that of the capture or probe DNA, which results in the production of an electrical signal as shown in Figure 2.2e. An example of an electrochemical indicator is ferrocenyl naphthalene diimide. This indicator attaches to the DNA duplexes, which then produces a quantifiable electrochemical signal (Tichoniuk, Ligaj, and Filipiak 2008).

2.3.2 Physical Biosensor

Physical biosensors are devices that are self-contained and integrated, and they monitor the electricity that is generated because of an electroactive biological material being reduced or oxidized. Some examples of physical biosensors include the following.

2.3.2.1 pH Sensor

A pH sensor can be used to identify whether a solution is acidic or alkaline. The chemical industry, the water and wastewater industry, the food and beverage industry, the pharmaceutical industry, power plants, primary industries, and the oil and gas industry all use pH sensors. A pH sensor is a pH probe that has two electrodes, one of which serves as a reference electrode and the other of which is a sensor electrode. These electrodes measure hydrogen ion activity in a solution as mentioned in Figure 2.3a. Ion exchange results in the production of a voltage, which is then measured by the pH meter and converted into a pH value that is understandable (Saari and Seitz 1982). For example, a fluorophore-based pH sensor can detect an alteration in milk's pH, which will reveal the cause of milk deterioration (Choudhary et al. 2019).

2.3.2.2 Temperature Sensors

Temperature sensors are sensors that measure and monitor changes in temperature in biological systems or surroundings. It is intended to offer precise and real-time temperature data for a variety of applications. It uses temperature-sensitive material that undergoes changes in their physical and chemical characteristics. The change in temperature is converted into measurable signals including electrical voltage, resistance, or frequency as shown in Figure 2.3b. Common temperature-sensitive materials used in temperature sensors are thermocouples, thermistors, and resistance temperature detectors (RTD). One of the examples of temperature biosensors is DS18B20 [13], widely used in environmental monitoring, patient monitoring, and medical devices (Vasuki, Varsha, and Mithra 2019).

2.3.2.3 Oxygen Sensor

Oxygen sensors are used to measure the level of oxygen in liquid media. Dissolved oxygen sensors (Sosna et al. 2007) based on Clark electrodes as shown in Figure 2.3c have played a major role in detecting solution-based pollution. Monitoring oxygen in ambient settings, particularly dissolved oxygen, is required in the medical, food processing, and waste management sectors (Ramamoorthy, Dutta, and Akbar 2003).

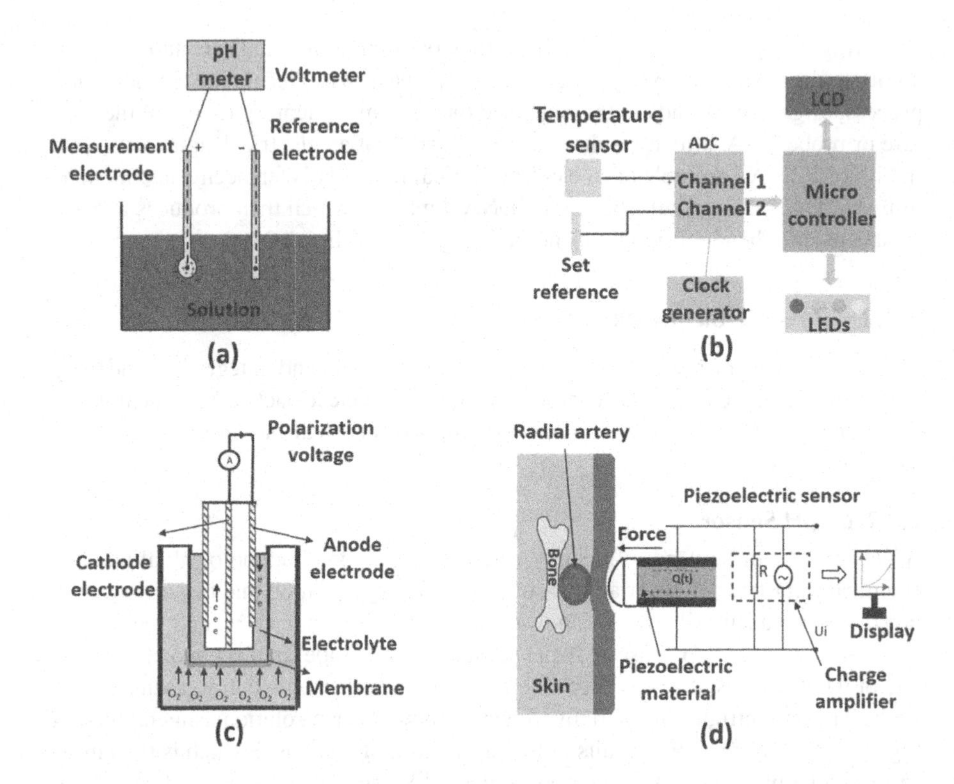

FIGURE 2.3 Type of physical biosensor: (a) pH sensor, (b) temperature sensor, (c) oxygen sensor, and (d) pressure sensor.

2.3.2.4 Pressure Sensor

Pressure sensors are used in bioprocessing equipment such as bioreactors and filtration systems to monitor and control pressure. Next-generation monitoring systems utilize pressure sensors to identify biological signals related to health statuses such as tendon healing, activity levels, blood flow, electronic or artificial skin, the development of intraocular glaucoma, catheters, and point-of-care immunoassays. These biological signals include blood flow, tendon healing, activity levels, and intraocular glaucoma. Utilizing piezoelectric material on the surface of the transducer enables the arterial pulse blood flow pressure sensor to be employed for the early identification of failed vascular anastomosis as shown in Figure 2.3d (Yu et al. 2021)

2.3.3 Commercially Available Biosensor

There are several commercially available biosensors that are widely used in various fields keeping sustainable development in check. A few such biosensors are as follows.

2.3.3.1 Soil Moisture Sensor

It is used in agriculture and landscaping to optimize water consumption and irrigation practices. These sensors assess the moisture level of the soil and give real-time data to assist farmers and gardeners in making intelligent water management decisions. Soil moisture sensors are available from commercial manufacturers such as Decagon Devices (formerly METRE Group) and Delta-T Devices for a variety of agricultural and environmental applications (Yu et al. 2021).

2.3.3.2 Air Quality Sensor

Pollutants in the atmosphere are monitored and measured using air quality monitors. In both indoor and outdoor contexts, these sensors can detect different gases, particulate matter, and volatile organic compounds. Commercial air quality sensors from Aeroqual, Sensirion, and Alphasense are frequently used in environmental monitoring and green building applications (Rawal 2019).

2.3.3.3 Water Quality Sensor

Water quality sensors are critical for monitoring and assuring water source safety. They can detect pH, dissolved oxygen, conductivity, turbidity, and a variety of pollutants. Water quality sensors and monitoring systems are available from companies such as YSI, Hach, and Thermo Fisher Scientific for environmental monitoring, wastewater treatment, and aquatic research (Pule, Yahya, and Chuma 2017).

2.3.3.4 Solar Radiation Sensor

Sensors for solar radiation measure the intensity and spectral distribution of solar energy. These sensors are critical in determining solar energy potential, optimizing solar panel placement, and analysing the performance of solar energy systems. Commercial solar radiation sensors for renewable energy applications are available from companies such as Kipp & Zonen, Hukseflux, and EKO Instruments (Nugraha and Adriansyah 2022).

2.3.3.5 Breathalyzer

Breathalyzers are biosensors that examine the breath of a person in order to identify the amount of alcohol in their blood. Commercial breathalyzer brands include BACtrack and AlcoMate, among others. BACtrack and others employ fuel cell sensor technology to determine the amount of alcohol in a person's blood. Breathalyzers that use fuel cell sensors offer astounding precision and responsiveness, utilize the identical cutting-edge fuel cell innovation, that the drug treatment centres, clinics, and corporations, as well as law enforcement, use for roadside alcohol testing. Breathalyzers with fuel cell sensors also have an extraordinarily long battery life (Borkenstein and Smith 1961).

2.4 CURRENT ADVANCEMENTS IN BIOSENSORS FOR GREEN TECHNOLOGY

Biosensing-based analytical methods are extremely essential due to indisputable benefits such as cheap cost, high sensitivity and selectivity, field use, and miniaturization. The demand for analysis of many compounds that may be significant for the environment, food, health, and other sectors makes biosensor technology more relevant (Sezgintürk 2020). In recent years, biosensor technology has permitted identification of a variety of biomolecules, including harmful microbes, in a quick, sensitive, simple, and inexpensive manner as mentioned in Table 2.2. Emerging

TABLE 2.2
New Age Technology-Based Biosensors

Biosensor	Analyte	Biorecognition Element	Transduction System	Research Year	Reference
Nano biosensor	Opioids (morphine, codeine, oxycodone)	Gold nanoparticle	Calorimetric	2021	Razlansari, Ulucan-Karnak, and Kahrizi 2022
Electrochemical sensor	Norepinephrine	Fe_2O_3@CeO_2 core–shell NPs	Voltametric	2019	Mazloum-Ardakani et al. 2020
Surface plasmon sensor (SPR)	Sample that flows through the chip	Gold chip	Optical	2021	Idil et al. 2021
Quantum dot-based nano biosensor	Metal ions	Quantum dots	Calorimetric and conductometric	2021	Singh et al. 2018
Graphene-based biosensor	DNA and protein	Graphene	Fluorescence	2020	Bai, Xu, and Zhang 2020

technologies such as nanotechnology, lab-on-a-chip microdevices, and so on drive future research to produce biosensors with much improved performance.

2.5 CHALLENGES IN DEVELOPING BIOSENSORS

While biosensors have significant promise for green technology applications, there are various hurdles that must be overcome throughout their development. Here are some of the most significant issues.

2.5.1 Selectivity and Sensitivity

In order to detect specific target analytes correctly and reliably, biosensors must have high selectivity and sensitivity. However, obtaining selectivity and sensitivity can be difficult, especially in complicated sample matrices containing a variety of interfering chemicals. Improving biosensor specificity and sensitivity while reducing false-positive and false-negative readings remains a serious issue (Bhalla et al. 2016).

2.5.2 Miniaturization and Integration

Biosensors must be miniaturized, portable, and integrated into numerous devices or systems for many green technology applications. Miniaturization without sacrificing performance or sensitivity might be difficult. Furthermore, connecting biosensors with other technologies, such as wireless communication or data analysis platforms, necessitates the resolution of compatibility and interface challenges (Liu et al. 2020).

2.5.3 Longevity and Stability

Biosensors must retain their performance and stability over long periods of time. Temperature, humidity, and pH are all environmental conditions that can affect the stability and durability of biosensors. Ensuring biosensors' long-term stability is critical to their practical application and widespread adoption (Panjan, Virtanen, and Sesay 2017).

2.5.4 Cost-Effectiveness

The cost of developing, manufacturing, and deploying biosensors is a considerable problem. Biosensors must be cost-effective, scalable, and accessible to a wide range of users, including individuals, corporations, and governmental organizations, in order to be widely used in green technology applications. It is critical for commercial viability to reduce total costs while preserving performance and dependability (Sadana 2005).

2.5.5 Standardization and Regulatory Compliance

It is critical to establish standardized techniques and guidelines for biosensor creation, testing, and validation. Consistent quality control and regulatory compliance guarantee that biosensors are accurate, reliable, and safe. Addressing the problems of standardization and regulatory compliance is critical for biosensor acceptability and incorporation into green technology practices (Kisaalita 1992).

2.6 APPLICATIONS OF BIOSENSORS IN GREEN TECHNOLOGY

Environmental pollution is one of the major concerns of the modern world which is caused by release and accumulation of various harmful chemical substances due to industrial development, population growth, and urbanization (Rogers and Lin 1992). Over the course of the past few years, several initiatives and legislative acts for the control of environmental pollution have been adopted. These developments have coincided with an increased level of scientific and societal interest in this sector (Gavrilaş et al. 2022). Biosensors are innovative analytical instruments that have been developed as a direct result of field monitoring. These sensors are able to provide readings that are quick, reliable, and sensitive at a reduced cost (Rodríguez-Mozaz et al. 2004). It can detect and quantify biological and chemical analytes in an efficient and eco-friendly manner. Some of the key fields that involve the utilization of biosensors are discussed below.

2.6.1 Environmental Monitoring

Biosensors are used in critical monitoring of soil, water, and air samples to detect pollutants such as pathogens, pesticides, heavy metals, and other harmful chemical compounds.

2.6.1.1 Heavy Metal Detection

Because heavy metals cannot be broken down by living organisms and because their pervasive existence in the biosphere indicates that they are highly hazardous, they endanger the environment as well as human health. Conventional methods such as inductively coupled plasma mass spectrometry and cold vapor atomic absorption spectrometry are both expensive and need the expertise of trained professionals. As a consequence, the development of biosensors to monitor heavy metal concentrations in environmental samples was necessary (Velusamy, Periyasamy, and Kumar 2022).

2.6.1.2 Biochemistry Oxygen Demand

The biological oxygen demand (BOD) of the water is a metric that shows the quantity of biodegradable organic minerals present. The traditional method for determining BOD takes a lot of time, but a biosensor-based method can provide an accurate reading in a short amount of time (Chee 2011).

2.6.1.3 Nitrogen Compound

Nitrides are used in food preservation, but the continuous usage of nitrogen compounds is causing serious health issues, the increment of nitrides in the groundwater and surface water is harmful to the aquatic environment. Ryu et al. (2020) created a biosensor for amperometric nitrite detection using cytochrome *c* nitrite reductase (ccNiR) immobilized and electrically coupled on a glassy carbon electrode by immobilization in redox active complex compound [ZnCrAQS] double-layered hydroxide comprising anthraquinone-2-sulphonate (AQS). A linear spectrum between analyte concentrations ranging from 0.015 to 2.35 M, an admissible limit of 4 nM, and a quick reaction to nitrite (5 s) were all displayed by the device.

2.6.2 Agriculture and Crop Management

The introduction of pesticides and fertilizers throughout the contemporary phase of industrial growth brought about significant improvements to agricultural productivity; nevertheless, the adverse effects of their residues in food and the deterioration of the environment cannot be understated and should not be ignored. The use of biosensors for the biological diagnostics of crops and soil paves the way for more reliable early-stage prevention of soil disease and purification of infected soil (Wang et al. 2022).

2.6.2.1 Biosensor in the Detection of Crop Disease

Crop disease refers to various diseases and illnesses in the plant and these diseases can be caused by various pathogens, including fungi, viruses, bacteria, nematodes, and other microorganisms. SPR (surface plasmon resonance) based biosensor which is typically based on an enzyme-linked immunosorbent assay (ELISA) technique employs a label (fluorescent marker, antibody) which is a highly sensitive biosensor that can identify the pathogen even at minimal concentration. The early stages of the illness that causes soybean rust can be diagnosed with its assistance (Khater, de la Escosura-Muñiz, and Merkoçi 2017).

2.6.2.2 Detection of Pathogens in Plants

Plant pathogen identification is essential for managing plant diseases and guaranteeing the health and yield of crops. QCM (quartz crystal microbalance) is an acoustic-based biosensor used to detect pathogens like *Ralstonia solanacearum*, *Pseudomonas syringae* pv. tomato, and *Xanthomatas campestris* pv. vesicatoria (Fang and Ramasamy 2015).

2.6.3 Renewable Energy Production

Biosensors have various applications in renewable energy production, especially in biofuel (e.g., ethanol and biogas) and bioenergy production. Here is the list of applications of biosensors in renewable energy production.

2.6.3.1 Fermentation Process Monitoring

Biofuels which are naturally produced by microbial cell factories are ethanol and *n*-butanol (in acetone-butanol-ethanol fermentation). Using biosensors to monitor the fermentation process for biofuel generation is an efficient way to assure optimal process control and product quality. Key parameters that can be measured by biosensors in the fermentation process for biofuel production are substrate concentration, biomass and cell viability, metabolite production, pH, and temperature (Choudhary and Joshi 2022).

2.6.3.2 Microorganism Detection and Monitoring

Microorganism fuel cells (MFCs) are cheaper and can be produced by bacteria populating sludge and soil. Microbial fuel cells generate electricity from renewable sources such as wastewater, biomass, or organic waste. MFCs may use these plentiful and frequently underutilized resources to generate clean, sustainable energy, decreasing dependency on fossil fuels and contributing to a more ecologically friendly energy mix (Lim et al. 2015).

2.6.4 Water Quality Assessment

With rising concerns about water shortage and pollution, the capacity to continually monitor and analyses the composition of water is becoming increasingly important. Instead of relying on traditional sensors, the quality of water could be monitored by enzymes and natural microbes (Hui et al. 2022). Let's consider some emerging water quality biosensors.

2.6.4.1 Graphene Nano Sensor

Under the US Environmental Protection Agency (EPA), researchers at Lowa State University have developed a biosensor that can detect contaminants as small as 0.6 nm in length and thus can detect organophosphates such as pesticides. It is a low-cost method of detecting pathogens and other pollutants in not just water but also soil and livestock (Sundramoorthy, Vignesh Kumar, and Gunasekaran 2018).

2.6.4.2 Bacterial Sensor

This sensor works on the principle of three-dimensional recognition, and the resulting picture is examined using algorithms that take into account 59 different quantified image properties. The sensor consists of the following components: (1) a water sample is held in an optical flow cell while an assessment is being performed, (2) a system for dark field imaging with a complementary metal oxide semiconductor (CMOS)-based camera set-up, a magnifying lens, and a LED, or light-emitting diode light source, and (3) a system for image analysis that can recognize and categories certain water sample constituents (Højris et al. 2016).

2.6.5 Food Quality Monitoring

Quality and safety, food product upkeep, and processing are all difficult issues in the food processing sector. To resolve this, development of biosensors for simple, real-time, selective, and low-cost food monitoring procedures seems advantageous (Choudhary, Vyas, and Joshi 2022). In the process of monitoring the quality of food, enzyme-based biosensors may be used to test a wide variety of chemical components including carboxylic acids, inorganic ions, gases, alcohols, amides, amines, carbohydrates, amino acids, phenols, cofactors, and heterocyclic compounds, among other things (Choudhary and Joshi 2022). Biosensors may also be used to evaluate and analyse products such as wine, beer, and yoghurt. Enzymes were able to be immobilized on electrodes by first being encased in a photo-cross-linkable polymer and then being engulfed in that polymer. An automated flow-based biosensor might be used to determine the concentrations of the three organophosphate pesticides found in milk (Mehrotra 2016).

2.7 CONCLUSION

Incorporation of technologically advanced resources into industrial processes helps to maintain a high level of living but at the cost of environmental deterioration. Green technology-based biosensors have been created to reduce the deterioration caused by the advancement of technology. These biosensors use eco-friendly components like cellulose to lessen their environmental footprint and encourage long-term sustainability. They also reduce the usage of synthetic and potentially dangerous chemicals by recognizing targets with naturally occurring substances like enzymes and biopolymers. Green biosensors use low-energy, cost-effective manufacturing technologies. Biomaterials may be deposited on bendable and biodegradable substrates using methods including inkjet printing, screen printing, and electrochemical deposition. This helps make these biosensors more commercially viable and scalable by decreasing the energy required during production. Green biosensors benefit from the incorporation of nanoparticles since they increase their efficiency and sensitivity. Nanomaterials, such as carbon nanotubes, graphene, and quantum dots, have a significant surface-area-to-volume ratio in addition to other desired features, such as their exceptional conductivity. Because of their enhanced signal transduction, rapid reaction times, and high sensitivity, green biosensors are great instruments for detecting a wide variety of analytes. This is because green biosensors are made from plant-based materials. Because of how well these biosensors pair with mobile technology, they may be used for real-time, on-the-spot analysis. Data transmission and analysis have now become easier using a wireless connection and smartphone technology, allowing remote monitoring and data exchange. Green biosensors are especially helpful in locations with few resources or that are geographically isolated because of their mobility. Environmental monitoring, healthcare, food safety, agriculture, and renewable energy are some of the areas where these biosensors are finding use. Ongoing research attempts to expand the range of analytes that green biosensors can identify, improve the accuracy and dependability of the biosensors, and offer applications tailored to certain industries.

ABBREVIATIONS

AQS	anthraquinone-2-sulphonate
BOD	biological oxygen demand
ccNiR	cytochrome *c* nitrite reductase
CMOS	complementary metal oxide semiconductor
DNA	deoxyribonucleic acid
EPA	environmental Protection Agency
GDH	glutamate dehydrogenase
Gox	glucose oxidase
H_2O	water
H_2O_2	hydrogen peroxide
HRP	horseradish peroxidase
LED	light emitting diode
MFCs	microorganism fuel cells
MWCNT	multi-walled carbon nanotube
NADH	nicotinamide adenine dinucleotide
PBSE	1-pyrenebutyric acid *N*-hydroxysuccinimide ester
Poly(TBO)	poly(toluidine-blue O)
rGO	reduced graphene oxide
RNA	ribonucleic acid
SPR	surface plasmon resonance

REFERENCES

S. Ahmed, N. Shaikh, N. Pathak, A. Sonawane, V. Pandey, and S. Maratkar. 2019. "Chapter 3 – An Overview of Sensitivity and Selectivity of Biosensors for Environmental Applications." In *Tools, Techniques and Protocols for Monitoring Environmental Contaminants*, 53–73.

M. Aizawa, 1994. "Immunosensors for Clinical Analysis." *Advances in Clinical Chemistry* 31: 247–75.

Y. Bai, T. Xu, and X. Zhang. 2020. "Graphene-Based Biosensors for Detection of Biomarkers." *Micromachines* 11 (1): 60.

J. W. Bennett, and M. Klich. 2003. "Mycotoxins." *Clinical Microbiology Reviews* 16 (3): 497–516.

N. Bhalla, P. Jolly, N. Formisano, and P. Estrela. 2016. "Introduction to Biosensors." *Essays in Biochemistry* 60 (1): 1–8.

S. Bocanegra-Rodríguez, N. Jornet-Martínez, C. Molins-Legua, and P. Campíns-Falcó. 2020. "New Reusable Solid Biosensor with Covalent Immobilization of the Horseradish Peroxidase Enzyme: In Situ Liberation Studies of Hydrogen Peroxide by Portable Chemiluminescent Determination." *ACS Omega* 5 (5): 2419–27.

R. F. Borkenstein, and H. W. Smith. 1961. "The Breathalyzer and Its Applications." *Medicine, Science and the Law* 2 (1): 13–22.

S. Cahuantzi-Muñoz, M. González-Fuentes, L. Ortiz-Frade, et al. 2019. "Electrochemical Biosensor for Sensitive Quantification of Glyphosate in Maize Kernels." *Electroanalysis* 31(5): 927–35.

G. J. Chee, 2011. "Biosensor for the Determination of Biochemical Oxygen Demand in Rivers." *Environmental Biosensors* 257–76.

S. Choudhary, and A. Joshi. 2022. "Development of an Embedded System for Real-Time Milk Spoilage Monitoring and Adulteration Detection." *International Dairy Journal* 127: 105207.

S. Choudhary, B. Joshi, G. Pandey, and A. Joshi. 2019. "Application of Single and Dual Fluorophore-Based pH Sensors for Determination of Milk Quality and Shelf Life Using a Fibre Optic Spectrophotometer." *Sensors and Actuators B: Chemical* 298: 126925.

S. Choudhary, T. Vyas, and A. Joshi. 2022. "Fluorescence-Based Sensing Assay for Point of Care Detection of Healthcare Parameters in Food Samples." *ECS Transactions* 107 (1): 8495.

C. M. Cova, E. Rincón, E. Espinosa, L. Serrano, and A. Zuliani. 2022. "Paving the Way for a Green Transition in the Design of Sensors and Biosensors for the Detection of Volatile Organic Compounds (VOCs)." *Biosensors* 12 (2): 51.

B. Derkus, 2016. "Applying the Miniaturization Technologies for Biosensor Design." *Biosensors and Bioelectronics* 79: 901–13.

S. Eyvazi, B. Baradaran, A. Mokhtarzadeh, and M. de la Guardia. 2021. "Recent Advances on Development of Portable Biosensors for Monitoring of Biological Contaminants in Foods." *Trends in Food Science & Technology* 114: 712–21.

Y. Fang, and R. P. Ramasamy. 2015. "Current and Prospective Methods for Plant Disease Detection." *Biosensors* 5 (3): 537–61.

D. Frense, A. Müller, and D. Beckmann. 1998. "Detection of Environmental Pollutants Using Optical Biosensor with Immobilized Algae Cells." *Sensors and Actuators B: Chemical* 51 (1): 256–60.

S. Gavrilaş, C. Ştefan Ursachi, S. Perţa-Crişan, and F.-D. Munteanu. 2022. "Recent Trends in Biosensors for Environmental Quality Monitoring." *Sensors (Basel, Switzerland)* 22 (4): 1513.

X. Guo, F. Wen, N. Zheng, M. Saive, M.-L. Fauconnier, and J. Wang. 2020. "Aptamer-Based Biosensor for Detection of Mycotoxins." *Frontiers in Chemistry* 8: 195.

B. Højris, S. C. B. Christensen, H.-J. Albrechtsen, C. Smith, and M. Dahlqvist. 2016. "A Novel, Optical, On-line Bacteria Sensor for Monitoring Drinking Water Quality." *Scientific Reports* 6 (1): 23935.

Y. Hui, Z. Huang, M. E. E. Alahi, A. Nag, S. Feng, and S. C. Mukhopadhyay. 2022. "Recent Advancements in Electrochemical Biosensors for Monitoring the Water Quality." *Biosensors* 12 (7): 551.

N. Idil, M. Bakhshpour, S. Aslıyüce, A. Denizli, and B. Mattiasson. 2021. "A Plasmonic Sensing Platform Based on Molecularly Imprinted Polymers for Medical Applications." In *Plasmonic Sensors and Their Applications*, 87–102.

C. Karunakaran, R. Rajkumar, and K. Bhargava. 2015. "Chapter 1 – Introduction to Biosensors." In *Biosensors and Bioelectronics*, 1–68.

J. Kaur, S. Choudhary, R. Chaudhari, R. D. Jayant, and A. Joshi. 2019. "Enzyme-Based Biosensors." *Bioelectronics and Medical Devices*, 211–40.

M. Khater, A. de la Escosura-Muñiz, and A. Merkoçi. 2017. "Biosensors for Plant Pathogen Detection." *Biosensors & Bioelectronics* 93: 72–86.

D. Kirstein, L. Kirstein, and F. Scheller. 1985. "Enzyme Electrode for Urea with Amperometric Indication: Part I—Basic Principle." *Biosensors* 1 (1): 117–30.

W. S. Kisaalita. 1992. "Biosensor Standards Requirements." *Biosensors and Bioelectronics* 7 (9): 613–20.

H. Y. Kong, and J. Byun. 2013. "Nucleic Acid Aptamers: New Methods for Selection, Stabilization, and Application in Biomedical Science." *Biomolecules & Therapeutics* 21 (6): 423–34.

J. W. Lim, D. Ha, J. Lee, S. K. Lee, and T. Kim. 2015. "Review of Micro/Nanotechnologies for Microbial Biosensors." *Frontiers in Bioengineering and Biotechnology* 3: 61.

D. Liu, J. Wang, L. Wu, Y. Huang, Y. Zhang, M. Zhu, Y. Wang, Z. Zhu, and C. Yang. 2020. "Trends in Miniaturized Biosensors for Point-of-Care Testing." *TrAC Trends in Analytical Chemistry* 122: 115701.

M. Mazloum-Ardakani, Z. Alizadeh, F. Sabaghian, B. Mirjalili, and N. Salehi. 2020. "Novel Fe_2O_3@CeO_2 Coreshell-Based Electrochemical Nanosensor for the Voltammetric Determination of Norepinephrine." *Electroanalysis* 32 (3): 455–61.

P. Mehrotra, 2016. "Biosensors and Their Applications – A Review." *Journal of Oral Biology and Craniofacial Research* 6 (2): 153–59.

D. Merz, M. Geyer, D. A. Moss, and H.-J. Ache. 1996. "Chlorophyll Fluorescence Biosensor for the Detection of Herbicides." *Fresenius' Journal of Analytical Chemistry* 354 (3): 299–305.

M. Naessens, and C. Tran-Minh. 1999. "Biosensor Using Immobilized Chlorella Microalgae for Determination of Volatile Organic Compounds." *Sensors and Actuators B: Chemical* 59 (2): 100–102.

M. Nugraha, and A. Adriansyah. 2022. "Development of a Solar Radiation Sensor System with Pyranometer." *International Journal of Electrical and Computer Engineering (IJECE)* 12: 1385.

P. Panjan, V. Virtanen, and A. M. Sesay. 2017. "Determination of Stability Characteristics for Electrochemical Biosensors via Thermally Accelerated Ageing." *Talanta* 170: 331–36.

W. J. Peveler, M. Yazdani, and V. M. Rotello. 2016. "Selectivity and Specificity: Pros and Cons in Sensing." *ACS Sensors* 1 (11): 1282–85.

M. Pule, A. Yahya, and J. Chuma. 2017. "Wireless Sensor Networks: A Survey on Monitoring Water Quality." *Journal of Applied Research and Technology* 15 (6): 562–70.

M. Z. Qamar, M. Noor, W. Ali, and M. Qamar. 2021. "Green Technology and Its Implications Worldwide." *The Inquisitive Meridian Multidisciplinary Journal* 3: 10.

R. Ramamoorthy, P. Dutta, and S. A. Akbar. 2003. "Oxygen Sensors: Materials, Methods, Designs and Applications." *Journal of Materials Science* 38: 4271–82.

R. Rawal. 2019. "Air Quality Monitoring System." *International Journal of Computational Science and Engineering* 9: 1–9.

D. M. Rawson, A. J. Willmer, and M. F. Cardosi. 1987. "The Development of Whole Cell Biosensors for On-line Screening of Herbicide Pollution of Surface Waters." *Toxicity Assessment* 2 (3): 325–40.

D. M. Rawson, A. J. Willmer, and A. P. P. Turner. 1989. "Whole-Cell Biosensors for Environmental Monitoring." *Biosensors* 4 (5): 299–311.

M. Razlansari, F. Ulucan-Karnak, M. Kahrizi, et al. 2022. "Nanobiosensors for Detection of Opioids: A Review of Latest Advancements." *European Journal of Pharmaceutics and Biopharmaceutics* 179: 79–94.

M. Rehák, M. Šnejdárková, and T. Hianik. 1997. "Acetylcholine Minisensor Based on Metal-Supported Lipid Bilayers for Determination of Environmental Pollutants." *Electroanalysis* 9 (14): 1072–77.

S. Rodríguez-Mozaz, M.-P. Marco, M. Lopez de Alda, and D. Barceló. 2004. "Biosensors for Environmental Applications: Future Development Trends." *Pure and Applied Chemistry* 76: 723–52.

K. R. Rogers, and J. N. Lin. 1992. "Biosensors for Environmental Monitoring." *Biosensors and Bioelectronics* 7 (5): 317–21.

M. K. Ryu, D. Thompson, Y. Huang, B. Li, and Y. Lei. 2020. "Electrochemical Sensors for Nitrogen Species: A Review." *Sensors and Actuators Reports* 2 (1): 100022.

L. A. Saari, and W. R. Seitz. 1982. "pH Sensor Based on Immobilized Fluoresceinamine." *Analytical Chemistry* 54 (4): 821–23.

M. K. Sadana, 2005. "Market Size and Economics for Biosensors." *Fractal Binding and Dissociation Kinetics for Different Biosensor Applications*, 265–99.

B. Saltepe, E. Sahin Kehribar, S. Yirmibesoglu, and U. Seker. 2017. "Cellular Biosensors with Engineered Genetic Circuits." *ACS Sensors* 3: 13–26.

O. Selifonova, R. Burlage, and T. Barkay. 1993. "Bioluminescent Sensors for Detection of Bioavailable Hg(II) in the Environment." *Applied and Environmental Microbiology* 59 (9): 3083–90.

M. K. Sezgintürk, 2020. "Chapter One – Introduction to Commercial Biosensors." In *Commercial Biosensors and Their Applications*, 1–28.

R. D Singh, R. Shandilya, A. Bhargava, R. Kumar, R. Tiwari, K. Chaudhury, R. K. Srivastava, I. Y. Goryacheva, and P. K. Mishra. 2018. "Quantum Dot Based Nano-Biosensors for Detection of Circulating Cell Free MiRNAs in Lung Carcinogenesis: From Biology to Clinical Translation." *Frontiers in Genetics* 9: 616.

M. Song, X. Lin, Z. Peng, S. Xu, L. Jin, X. Zheng, and H. Luo. 2021. "Materials and Methods of Biosensor Interfaces with Stability." *Frontiers in Materials* 7: 583739.

S. Song, L. Wang, J. Li, C. Fan, and J. Zhao. 2008. "Aptamer-Based Biosensors." *TrAC Trends in Analytical Chemistry* 27 (2): 108–17.

M. Sosna, G. Denuault, R. W. Pascal, R. D. Prien, and M. Mowlem. 2007. "Development of a Reliable Microelectrode Dissolved Oxygen Sensor." *Sensors and Actuators B: Chemical* 123 (1): 344–51.

A. K. Sundramoorthy, T. H. Vignesh Kumar, and S. Gunasekaran. 2018. "Chapter 12 - Graphene-Based Nanosensors and Smart Food Packaging Systems for Food Safety and Quality Monitoring." *Graphene Bioelectronics*, 267–306.

M. Tichoniuk, M. Ligaj, and M. Filipiak. 2008. "Application of DNA Hybridization Biosensor as a Screening Method for the Detection of Genetically Modified Food Components." *Sensors* 8 (4): 2118–35.

S. Vasuki, V. Varsha, R. Mithra, et al. 2019. "Thermal Biosensors and Their Applications." *American International Journal of Research in Science Technology, Engineering and Mathematics* 2328–3491.

K. Velusamy, S. Periyasamy, P. S. Kumar, et al. 2022. "Biosensor for Heavy Metals Detection in Wastewater: A Review." *Food and Chemical Toxicology* 168: 113307.

T. Vo-Dinh, T. Nolan, Y. F. Cheng, J. P. Alarie, and M. J. Sepaniak. 1990. "A Fiberoptic Antibody-Based Biosensor with Time-Resolved Detection." In *Chemical, Biochemical, and Environmental Fiber Sensors*, 1172: 266–72.

T. Vyas, S. Choudhary, N. Kumar, and A. Joshi. 2023. "Point-of-Care Biosensors for Glucose Sensing." In *Nanobiosensors for Point-of-Care Medical Diagnostics*, 107–36.

X. Wang, Y. Luo, K. Huang, and N. Cheng. 2022. "Biosensor for Agriculture and Food Safety: Recent Advances and Future Perspectives." *Advanced Agrochem* 1 (1): 3–6.

Z. Yu, G. Cai, X. Liu, and D. Tang. 2021. "Pressure-Based Biosensor Integrated with a Flexible Pressure Sensor and an Electrochromic Device for Visual Detection." *Analytical Chemistry* 93 (5): 2916–25.

L. Yu, W. Gao, R. R. Shamshiri, S. Tao, Y. Ren, Y. Zhang, and G. Su. 2021. "Review of Research Progress on Soil Moisture Sensor Technology." *International Journal of Agricultural and Biological Engineering* 14 (4): 32–42.

3 Microorganism-Mediated Microplastic Degradation Methods and Mechanism

Nalini Soni, Saurabh Gupta, Priya Gupta, Sangeeta Devendrakumar Singh, Khushboo Bhange, and Balasubramanian Velramar

3.1 INTRODUCTION

Anthropogenic activities are now observed as the primary causes of declining biodiversity and ecosystem services (Abel et al. 2017). The massive production and consumption of plastics is a typical indicator of human activity. Plastics are synthetic polymeric materials with a wide range of chemical properties and are used in a wide variety of modern applications (Gewert, Plassmann, and Macleod 2015). More than 80% of all manufactured plastics are believed to be thermoplastics, which are industrialised through polymerisation to produce high-molecular-weight polymers from low-molecular-weight monomers (Abel et al. 2017). Physical techniques like extrusion, melting, and palletising, as well as chemical techniques like blending with antioxidants, copolymers of polycarbonate, plasticisers, and colourants, can change and strengthen their physical and chemical properties (Bittner et al. 2014). Because of this, plastic materials exhibit intricate chemical characteristics in addition to a solid physical structure. The demand for and use of plastics has been steadily rising over the past several decades because of their affordability, effective malleability, and durability; now, 8300 plastics are produced worldwide (Geyer, Jambeck, and Law 2017). The European Union (EU) received 27.1 million metric tonnes (Mt) of plastic garbage in 2016, wherein 27.3% were waste materials dumped in landfills, 31.1% were recycled, and 41.6% were used for energy recovery (Gaylor et al. 2013). Plastic trash, particularly biodegradable plastics, is more prone to physical fragmentation (fracture) than to mineralisation, which reduces the size of the polymers. The term "microplastics" (MPs) refers to the created plastics with particle sizes less than 5 mm (Li, Liu, and Chen 2018). The widespread existence of MPs as new environmental contaminants is receiving more and more attention (Figure 3.1) (Cole et al. 2011). Primary MPs and secondary MPs are the two subgroups of MPs. Primary MPs are extremely small (5 mm in diameter) and directly infiltrate the environment. These MPs are created through extrusion or grinding, either for use directly or as a feed material for other products. For instance, microbeads in air-blasting media and cosmetics, as well as cleaning goods. The secondary MPs result from the breaking up of bigger plastic detritus (Napper and Thompson 2020). Additionally, MPs serve

DOI: 10.1201/9781003407683-3

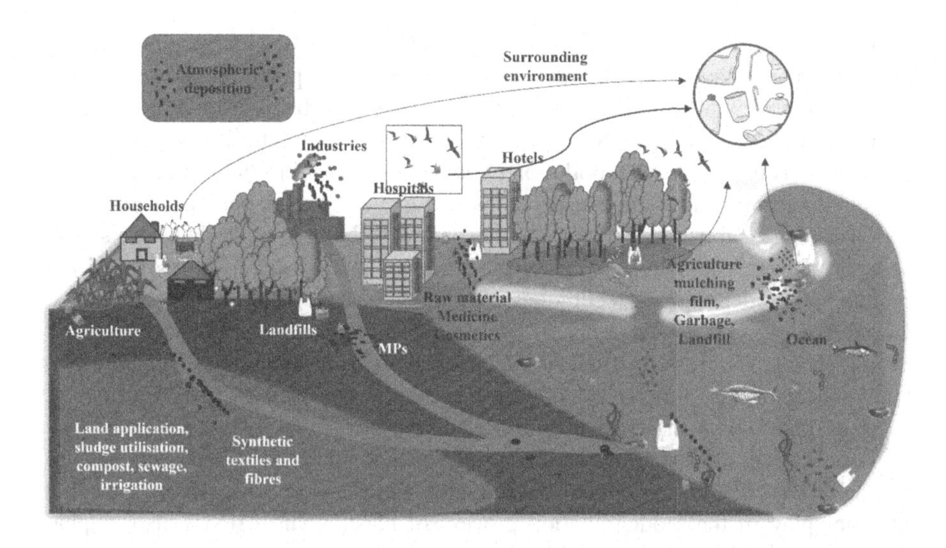

FIGURE 3.1 Different sources of MPs contamination finally reaching the aquatic ecosystem.

as a vehicle for the passage of hazardous substances, particularly metals and persistent organic pollutants, from the environment to the biota.

The main environmental source of MPs, counting the fragmentation of macroplastics, is the straight release of microfibers or microbeads from personal care items (Atugoda et al. 2021). They are ubiquitous pollutants that are causing increasing harm to both aquatic and terrestrial ecosystems (Zhang et al. 2021). Frequent release of MPs into the environment could be harmful to living creatures. Further, they are particularly harmful to the soil and have an effect on plant growth and development, and they may also be harmful to human health. Sludge composts may carry MPs into soils where they then interact with soil biota and spread to other parts of the environment (Zhu et al. 2020). With the explosion in plastic trash over the past few decades, plastic pollution is quickly evolving into a severe global Eco-environmental issue. The United Nations Environment Programme also lists MP (with a diameter of 5 mm) waste as one of the major environmental problems. MPs disperse quicker than conventional plastics because of their tiny size and bioretention ability (Zhang et al. 2021). The persistence of MPs and their possible negative effects on biota make MPs pollution a serious environmental issue. Most scientific studies have concentrated on the dispersion, ingestion, fate, behaviour, amount, and impacts of MPs. However, just a few research have discussed how techniques for MPs removal and remediation have been developed. In this chapter, we summarise the most current studies on the microorganisms-based degradation of MPs and investigate the underlying characteristics and mechanisms. Briefly, this chapter documented the microorganisms involved in MPs degradation and the environmental elements influencing MPs breakdown such as pH, temperature, and microbial strain action. Further, the chapter also emphasises the related degradation effects (such as degradation rate, weight loss, and molecular weight alteration) along with the mechanisms behind the breakdown of MPs by microbes. Finally, projections for MPs degradation utilising microorganisms have been elaborated, in addition to the potential future research areas. This review offers a major systematic overview of the microorganisms based breakdown of MPs and serves as a guide for upcoming investigations looking into efficient ways to reduce MPs pollution (Yuan et al. 2020). Plastic pollution is a significant environmental issue that becomes worse every year, especially with regard to non-biodegradable residual plastic films and MPs. Numerous studies have shown how plastic pieces degrade; however, little is known about the condition of the physicochemical biodegradation techniques utilised to treat plastics and how effectively they degrade polymers. As a result, this review investigates the impact of several physicochemical variables, such as mechanical micronised, intense temperature, ultraviolet radiation, and pH value, on the breakdown of plastics/MPs (Prata et al. 2020). Additionally, this chapter summarises the effectiveness of these elements' physicochemical degradation under various settings while also discussing numerous methods of physicochemical degradation. Furthermore, the crucial part that enzymes play in the biodegradation process of plastics and MPs is also emphasised. The subjects included in this chapter collectively offer a strong foundation for future studies on the processes used to degrade plastics and MPs and their outcomes. Because they pose a serious threat to biota, MPs pollution has gained international attention as an environmental problem. However, there have only been a few investigations on the removal of MPs contamination. Due to their smaller size than plastic goods, MPs were not suited for typical treatment methods. So many therapy options for MPs have been researched. This review enumerated the most recent reports on MPs-degrading processes, counting AOPs (photocatalytic oxidation, direct photodegradation, and electrochemical oxidation), biodegradation, matching degradation mechanisms, and current development status (Du, Xie, and Wang 2021). MPs (plastics smaller than 5 mm and nanoplastics that are smaller than 0.1 mm) originate from either the straight emission of plastic into the surrounding or the disintegration of larger plastic waste. Although effects on marine life have been extensively examined, little research has been done on potential implications on terrestrial ecosystems. Most plastics produced, used, and usually abandoned on land are what end up in the oceans. Therefore, the first significant interactions between MPs and biota in terms of ecology might take place in terrestrial systems. This chapter addresses the physical and chemical underpinnings of the corresponding known consequences and explores the pervasive toxicity of MNPs polymers as well as the persistent MP pollution as a potential agent of global change in terrestrial systems. Even in particle-rich habitats like soils, significant shifts in continental settings are feasible. An increasing body of research also shows that terrestrial species, including soil-dwelling invertebrates, terrestrial fungi, and plant-pollinators, that mediate crucial ecosystem services and functions interact with MPs. To understand the effects of MPs on the terrestrial environment, more research is required. We propose that MP pollution may represent an increasing global change danger to terrestrial ecosystems because of its pervasiveness, environmental persistence, and varied interactions with continental biota (Abel et al. 2017). Plastics are still being introduced into the ocean through a variety of paths due to increased industrialisation, population growth, and economic development; this could potentially lead to connected environmental, economic, and health problems.

3.2 POLLUTION CAUSED BY MPs WASTE: THE EXISTING SITUATION AND ITS FATE IN THE ENVIRONMENT

Global concern over new persistent pollutants such as MPs is rising. Even though MPs are the subject of substantial research in aquatic systems, little is known about their existence or ultimate fate in agricultural systems. The application of biosolids and compost, mulching film, wastewater irrigation, polymer-based fertilisers and pesticides, and atmospheric deposition are the main sources of MPs pollution in agricultural soils. The fate and distribution of MPs in the soil environment are principally determined by the characteristics of the soil, farming practices, and variety of the soil biota. Despite the rising MPs pollution of the soil environment, there are no standardised methods for detection and quantification. Studies now emphasise the most recent advancements in MP detection and quantification techniques while also reviewing the origins, fate, and dispersion of MPs in the soil environment. They also focus on how MPs interact with and affect soil biota. Future research areas could focus on cytotoxic effects on people and animals, biomagnification potency, nonlinear behaviour in the soil environment, standardised analytical methods, optimal management practices, and international agricultural industry regulations for sustainable development (Kumar et al. 2020). Due to their adaptability, robustness, and affordability, a variety of industries employ plastics for construction, packaging, electronics, agriculture, and automotive works. The manufacturing of plastic has drastically expanded in recent decades as a result of the extensive and widespread use of these synthetic materials, leading to an alarming level of plastic pollution in many environmental contexts (Elsamahy et al. 2023). The COVID-19 epidemic has resulted in a notable increase in the creation of plastic trash, including single-use masks, gloves, tissue papers, and PPE. Single-use throwaway masks made of various polymeric substances, including polyacrylonitrile, polyethylene, polystyrene, polyurethane, and polypropylene, could have a significant negative influence on the environment, human health, and veterinary health.

Practices of healthcare waste management are improperly managed and disposed. They possess major health risks and present additional difficulties for municipal governments responsible for managing solid trash. Presently, incineration is used to manage the bulk of COVID-19-related medical wastes, which results in the production of MP particles, dioxin, furans, and other dangerous metals, including cadmium and lead. It poses a serious threat to aquatic life and contaminates food, endangering the safety of the world's supply of food. Additionally, the existence of plastic is viewed as a problem due to the rise in carbon emissions and presents a serious threat to the world's food supply. The breakdown of MPs by axenic and mixed culture microorganisms, such as bacteria, fungus, microalgae, etc., is an environmentally viable way to lower the harm posed by MPs. This research focuses on the rise in MP pollution brought on by increased PPE use, as well as the various disinfection techniques currently used to treat COVID-19 pandemic-related wastes, including steam, microwave, autoclave, and incinerator (Tagorti and Kaya 2022). Tiny plastic particles that are thought to be emerging pollutants have garnered a lot of attention in recent years. In order to create MPs (1–5 mm) and nano plastics (NPs, 1–1000 nm) from larger plastic debris, bigger plastic debris must undergo mechanical abrasion, photochemical oxidation, and biological degradation. Compared to MPs, NPs' environmental destiny, ecosystem toxicity, and possible dangers have received less research yet. With a focus on currently understudied areas such as the environmental fate in agroecosystems, migration in porous media, weathering, and toxic effects on plants, this review offers a state-of-the-art summary of current research on NPs. To better comprehend the interactions between NPs, ecosystems, and human society, an integrated framework is presented. More research should concentrate on the entire ecosystem, including freshwater, ocean, groundwater, soil, and air (Wagner and Lambert 2018). Additionally, more efforts should be made to examine the ageing and aggregation of NPs in the environmental conditions that are relevant. Future research should examine the environmental behaviour of naturally aged NPs rather than synthesised polystyrene nanobeads since organically weathered plastic debris may have unique physicochemical properties. MPs are persistent contaminants that are becoming more prevalent in the terrestrial subsurface, and there is growing evidence that they have a substantial impact on the biological and ecological processes of soils (Henry, Laitala, and Grimstad 2019). Among other things, major sources of MPs include leachate from landfills, wastewater irrigation, land spreading of sewage sludge and biowaste composts, and plastic mulching film used in horticulture fields. The combined effects of MPs and various other environmental contaminants (heavy metals, organic pollutants, and antibiotics) on soil ecosystems are another topic covered. Another important aspect of MPs research is the standardisation of methods for MPs detection, quantification, and characterisation in soils. The report finishes by outlining knowledge gaps and offering suggestions for how to prioritise future research. In recent years, plastic pollution and its consequences on the environment have drawn attention on a global scale. However, there hasn't been much focus on how plastics are regulated from a life cycle viewpoint or how regulatory gaps might be closed to reduce and eliminate environmental exposure to macro- and MP risks. In this chapter, we map European regulations starting from the point where plastic carrier bags are produced until they reach the environment. Plastics may break down into MPs or even NPs after being exposed to the environment, which will change their original properties, reactivity, and structure and complicate and diversify study. More than 80% of plastics come from land-based activities. Due to their low density, the debris frequently floats in the water's upper surface layer before reaching the sea. Debris may settle occasionally as a result of hetero-aggregation with algae, suspended solids, and detritus (Ariza-tarazona et al. 2018). Additionally, the mechanical and structural properties of MPs may be lost during their degradation by biotic and/or abiotic processes, increasing their specific surface area for physicochemical interactions with other contaminants and microbial colonisation. According to studies, MPs and NPs have the ability to absorb harmful contaminants and transport them from one location to another. These harmful chemicals may be ingested by aquatic species, which may then enter the human food chain with unknowable consequences (Auta, Emenike, and Fauziah 2017). The latent impact of MPs and NPs on aquatic life is unclear, as these plastic particles have toxic chemicals associated with their production and sorption from the environment. Most of the investigations that were published calculated the MPs and NPs sorption capacities of various contaminants and tracked their sorption behaviour (Lv et al. 2019). According to studies, important variables like the type of plastic particles, size, shape, pH of the solution, salinity, ionic strength, level of crystallinity, chemical characteristics of pollutants, age, presence of coexisting surfactants, morphology, porosity, level of weathering of plastic polymers, specific surface area, and chemical characteristics of pollutants may affect the sorption and transformation of environmental toxic pollutants by MPs and NPs. Therefore, while evaluating the function of MPs and NPs as carriers of environmentally harmful contaminants, these factors should not be disregarded. Recent research found that ultraviolet (UV) irradiation could decrease the sorption of organic pollutants onto the surface of MPs by altering their surface hydrophobicity, morphology, and chemical properties. This suggests a reduced risk of the transportation of both non-polar and polar organic contaminants. In addition, photocatalysis is a revolutionary and cutting-edge technology to degrade MPs, but it is still unknown what will happen to the by-products in the actual world. The proposed system would also benefit from cost analysis and an optimisation study when used for commercial purposes (Carr, Liu, and Tesoro 2016). Additionally, it appears that electrooxidation is a suitable alternative for converting MPs from solid to gaseous form without producing any by-products. However, it is still unclear how anode fouling and interference from other contaminants work. In order to use this technology for commercial purposes, further research is still needed to understand what happens to gaseous products. Additionally, the use of ultrasounds is a viable option to recover MPs from activated sludge; however, it is still unknown how ultrasounds will affect the development of the bacterial community, which is necessary for the subsequent digestion process that produces energy (biogas) and reduces the amount of sewage sludge that must be dumped in landfills. Additionally, the loss of mechanical and structural properties caused by the degradation of MPs by biotic and/or abiotic processes may increase their specific surface area for physicochemical interactions with other contaminants and microbial colonisation (Alvim 2020). The assumption is that the most frequent abiotic degradation pathways in the environment are thermal, photodegradation, chemical, and mechanical. In the sun, MPs underwent chemical changes prompted by UV radiation and the atmosphere's oxidative properties. However, abiotic degradation is drastically reduced in deep water or the ocean due to the lower rates of sunlight and oxidation-mediated processes. The presence of different microbial communities of heterotrophs, symbionts, and autotrophs on the surface of plastic debris or particles, however, has led to the observation of biotic breakdown in shallower waters (Wysocki and Le Billon 2019). Typically, MPs are biodegraded outside of the cell by the use of released enzymes, which causes the breaking of polymeric chains through hydrolytic processes. This process results in tiny plastic particles of various sizes and shapes in the soil and aquatic environment. For instance, a standard plastic bag's specific surface area (0.2 m^2) would increase to 2600 m^2 after being entirely broken down into NPs with an average diameter of 40 nm. In addition, due to the increased availability of moisture and oxygen, which make these molecules more susceptible to microbial contact, the molecular weight of the polymers is lowered through chain scission (Perren, Wojtasik, and Cai 2018). Due to the production of monomers and water-soluble oligomers, the mineralization process eventually starts. For instance, bio assimilation of 60% of PE-MPs was discovered during composting. The polymer degradation varies from medium to medium, which is significant. Polycaprolactone (PCL), for instance, demonstrated greater degradation extent in sandy soils than in clay soils. Moreover, in addition to their exclusive sensitivity and adaptability to the existing environment, the decomposition of polymer materials is also influenced by the types of bacteria and the amount of biomass. For instance, *Rhodococcus* strains were able to grow on the surface of oxidised PE films comprising prooxidant additives in soil (Zhang et al. 2021). Furthermore, MPs can act as carriers of biological and chemical pollutants. They can adsorb environmental pollutants due to their high roughness, low polarity, higher specific surface area, porous structure, unique morphology, and charge opposite to pollutants. The sorption of environmental pollutants on the surface of MPs can increase their potential for transmission and binder by organisms (Rajala et al. 2020). Contrarily, the adsorbed contaminants can enter our food chain when creatures consume MPs, and the bioavailability of environmentally harmful contaminants to the organisms is increased. Additionally, MPs are capable of adsorbing harmful bacteria, viruses, and algae. The development of these harmful species may encourage gene exchange across various genera and support the spread of pathogenic and drug-resistant bacteria, which could amplify the toxicological and pathological reactions to organisms (Zhu et al. 2020). During the gene transfer, this event might create new bacterial strains. Additionally, because of their high mobility and floatability, MPs may simultaneously increase antibiotic resistance genes' mobility and spread those genes to different nearby areas. Therefore, the exchange of microbes and genes causing antibiotic resistance between societies or the environment next door may have unintended environmental effects.

3.3 CURRENT TRENDS OF MPs DEGRADATION

In the current scenario, the world is trying to combat pilling up of plastic waste recognised as an environmental burden which has become a threat to humanity as well as the environment with exceeding boundaries by entering into the food chain from land to human as various studies suggest the existence of MP in human blood. During the pandemic period, around 52 billion single-use disposable face masks were made in 2020 alone, and 1.6 billion (3%) of them ended up in the oceans, where they take 40–50 centuries to completely degrade. During the natural breakdown of these masks, tons of MPs will be discharged into the oceans, where they are likely to infiltrate our food chain. In addition, other plastic wastes have exacerbated a number of environmental issues and directly endanger human life. Because of this, there is now an abundance of plastic waste in the environment, which has negative consequences on both people and the ecosystem (Zhang et al. 2021). Dumping of MPs into oceans has been practised for decades where 5.25 trillion macroplastics and MPs are dumped in the ocean hampering marine life; therefore, removal of plastic waste via different degradation techniques such as oxidation, photochemical oxidation, photocatalytic oxidation, electrochemical oxidation, and many more is an important aspect. Due to their low water solubility, complicated structure, and lack of biodegradability, plastics are challenging to break down although our mother nature has always supported us in one or another way to maintain sustainability and achieve a flourishing environment where sunlight tends to be the primary source of thermal energy which aids in the degradation of plastic at a slow pace, and therefore introducing photocatalytic degradation can speed up the process (Lee and Li 2021). The creation and implementation of creative and efficient remediation techniques are essential given the urgency and necessity of resolving this issue. Among the many new methods, photocatalytic degradation has shown considerable promise for tackling MPs pollution (Li et al. 2019).

3.3.1 Advanced Oxidation Processes

AOPs, a reliable method for removing chemicals, have recently demonstrated outstanding performance in the degradation of persistent organic contaminants in water mediums. They form different reactive oxygen species (ROS), such as hydroxyl radical ($^{\bullet}OH$, $E_0 = 2.7$ V versus normal hydrogen electrode [NHE]) for Fenton and sulphate radical ($SO_4^{\bullet}$, $E_0 = 3.1$ V versus NHE) (Duan et al. 2020). According to a number of studies, Fenton treatment can efficiently break down plastic waste into valuable products (Wang and Liu 2017; Wang, Pfleger, and Kim 2017), and sulphate radical (SR) based on AOPs (SR-AOPs) is observed as a knock-down arrangement to degrade a variety of obstinate organic pollutants in complex water environments (Wang et al. 2017). According to a paper by Kang et al. (2019), SR-AOPs have demonstrated improved catalytic breakdown capability for the degradation of cosmetic MPs, which primarily consist of polyethylene. Due to their large redox potentials, MPs are probably better at oxidising (Ren et al. 2019). These species directly prompted the breakdown of MP chains, the construction of beneficial materials, or even the complete mineralisation of MPs. The full discussion of AOPs, including direct photodegradation, electrochemical oxidation processes, and photocatalytic oxidation, focuses on how they degrade materials.

3.3.2 Photochemical Oxidation

Photodegradation plays a significant role in the disintegration of polymers into their monomeric units (Gewert et al. 2015; Ren et al. 2019). UV radiation from sunlight may result in the formation of free radicals and cause the addition of oxygen, the addition of hydroxyl groups along with cross-linking of chemical chains. This process caused the surface of the MPs to flake and shatter (Zhu et al. 2019). UV light was thought to be the main factor influencing this process. MPs are naturally degraded by sunlight in an unregulated way. Therefore, it is crucial to research the connection between MPs' ageing characteristics and ageing severity. Earlier studied the degradation process of MPs under lab-accelerated photodegradation settings and discovered that oxygen-containing groups and cracks were generated on the surfaces of MPs. However, due to the relatively low quantity of ROS in the aquatic environment, nothing was known about the effects of ROS on the ageing processes of MPs (Song et al. 2017). Additionally, MP photodecomposition in the field is incredibly sluggish, especially in watery conditions. Removal of polystyrene MPs (PS-MPs) appears to be imperative research because it is one of the plastic products that is most frequently found in aquatic settings. When used as a model, PS-MPs were aged for up to 150 days in an aquatic environment with simulated sunlight (wavelengths between 295 and 2500 nm).

They came to the conclusion that under ideal experimental conditions, photoageing of PS-MPs was discovered, elucidating the mechanism behind ROS formation and MPs photodeterioration in simulated sunlight. ROS, for example, singlet oxygen (1 O_2), $^{\bullet}OH$, H_2O_2, and superoxide radical ($O_2^{\bullet}$), were sensed in the suspension of PS-MPs because of light illumination. The extent of MP photoageing and the kinds of transitional products in this photochemical arrangement, however, were not examined in this work. Furthermore, prolonged exposure to simulated light irradiation readily increased energy use or even caused minor light pollution. According to reports, polystyrene compounds cannot be degraded by the Fenton process (Feng et al. 2011). It was described that the photo-Fenton process for the photodegradation of polystyrene had primarily used for large plastic films rather than MPs or NPs (Kemp and McIntyre 2006; Zan et al. 2006). The primary limiting element in a Fenton-based reaction is the oxidant. Furthermore, because H_2O_2 has a free radical scavenging effect, the secondary pollution brought on by iron ions leaching and significant sludge formation prevents their practical implementation (Kang et al. 2019; Zhou and Feng 2017). Moreover, the peroxy-monosulphate ion formed during photochemical oxidation is catalytically encouraged to form active radicals supported by durable carbon nanosprings, which can then be used to degrade cosmetic MPs (Kang et al. 2019). Despite the fact that this AOPs was unable to directly break down MPs into beneficial products. The non-toxic organic by-products of MP degradation may be broken down by microbes, which can then convert them into other useful products like biofuels, protein, and sugar that can be used by people in a sustainable and environmentally acceptable manner. The carbon cycle in nature can therefore be realised through this technique. These will serve as the basis for the discussion on the photocatalytic degradation of MPs on catalysts in the section following.

3.3.3 Photocatalytic Oxidation

Photocatalytic degradation refers to the degradation of pollutants by using light energy where the process is based on photo-oxidation due to UV radiation and catalytic oxidation. A semiconductor used as a photocatalyst is frequently employed to absorb light and speed up the photoreaction rate with the help of photocatalysts, resulting in the production of ROS such as hydroxyl radicals and superoxide ions when they interact with the H_2O, OH, and O_2 adsorbed on the semiconductor's surface. This leads to start the breakdown of plastic, which results in chain breakage, branching, cross-linking, and eventually total mineralisation into carbon dioxide and water. The perfect photocatalyst would be able to absorb light at room temperature, be very stable against photo-corrosion, and be non-toxic. According to Klavarioti, Mantzavinos, and Kassinos (2009), photocatalysis is also regarded as an advanced oxidation process for removing MPs. Due to its many advantageous qualities, such as its remarkable oxidation-reduction capability, high-temperature

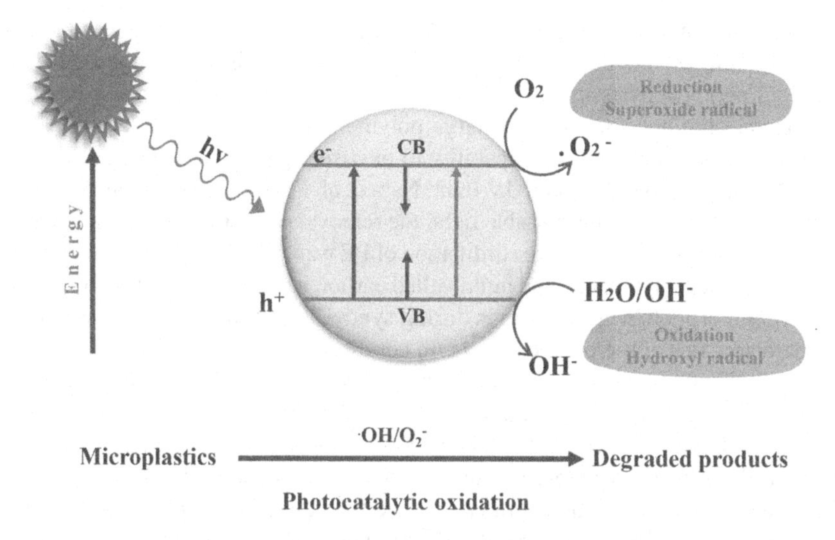

FIGURE 3.2 Photocatalytic oxidation. The conversion of the long chain of MPs to small forms of MPs by the UV light from the sun through the production of superoxide radicals that helps in the degradation of MPS, which helps in addition of hydroxyl group in the MPs.

stability, chemical steadiness, economic effectiveness, and environmental friendliness, titanium dioxide (TiO_2) is the most extensively used photocatalyst (Du et al. 2021). When the absorbed photon energy (E) exceeds the semiconductor's bandgap energy (E_g) ($E > E_g$), the electrons in the valence band (e) will move into the conduction band (CB), causing the production of positive holes (h^+) in the VB and the separation of electron-hole pairs (Nakata and Fujishima 2012). According to Nakata and Fujishima (2012), both species (e and h^+) react with OH, O_2, or H_2O to form highly ROS, which directly start the destruction processes of organic pollutants and plastic products (Uheida et al. 2021). The electrons (e) are driven from the photocatalyst's VB to CB and produce an energetic electron (e)-hole (h^+) pair when exposed to UV light with an irradiation energy that is equal to or greater than the photocatalyst's bandgap (Figure 3.2), as shown in the following equation.

$$\mathrm{TiO_2 hv} \rightarrow \mathrm{e^-} + \mathrm{h^+}$$

A part of the electrons in TiO_2 will quickly recombine with holes in femtoseconds, while the remaining have a lengthier lifetime. The quantity of photogenerated electrons and holes transferred to the TiO_2 surface for chemical processes involving oxidation and reduction is decreased by the recombination of electrons and holes as upon exposure to light, photo-induced electrons and holes may be compelled to flow via the electric field built into the semiconductors, decreasing carrier recombination. Scavengers and the insertion of trap sites, which reduce the recombination of electrons and holes in the semiconductor's bulk, are thus appealing methods to boost catalytic activity and are crucial to understand that a built-in electric field at the heterojunction interface and the behaviour of photo-induced carrier transfer are dependent on a number of variables, including semiconductor semi-conductivity, work function, and conduction band/valence band potentials (Sharma et al. 2019).

Model photocatalysts made of TiO_2-based nanoparticles were widely employed to accelerate the decomposition of large polymers and MPs in visible light, which was aptly due to their superior capability to oxidise organic contaminants. To break down PE and polystyrene under UV light, Nabi et al. (2021) used TiO_2 photocatalysts manufactured in a lab. Under visible light, the removal of LDPE was enhanced by the use of TiO_2 nanotubes, while the elimination of PE was increased by the use of copper phthalocyanine, polypyrrole, and multiwalled carbon nanotubes (Ali et al. 2016). TiO_2 underwent morphological alterations during synthesis to enhance the physicochemical characteristics of the photocatalysts. In certain investigations, materials based on zinc oxide (ZnO) were also used to degrade MPs and plastics due to their strong catalytic activity and high redox potential. ZnO, which has a bandgap similar to that of titanium dioxide, is frequently employed as a substitute for titanium dioxide. ZnO has been the subject of numerous investigations, and some findings also indicated a high rate of photocatalytic degradation. The breakdown of LDPE film employing ZnO as a photocatalyst was suggested by Tofa et al. (2019). As photocatalysts for photodegradation, iron oxide, cadmium sulphide, zinc sulphide, tungsten oxide, tin oxide, bismuth vanadate, and non-metallic carbon nitride are frequently used. Although the photocatalyst is essential for photodegradation, the degradation rate is also strongly influenced by the polymer structure, the mass of the photocatalyst, and the experimental apparatus (Du et al. 2021). One of the key benefits of photocatalytic technologies is their capacity to break down a variety of MPs types, including the frequently encountered polymers polypropylene (PP), polyethylene (PE), and polystyrene (PS).

In the process of photocatalytic degradation, the shape of photocatalysts is extremely important. The overall photocatalytic performance can be affected by changes in surface area, porosity, and active site accessibility caused by various morphologies. Understanding the process and improving photocatalytic performance depend on how the impact variables, photocatalytic mechanisms, and catalyst morphology interact. For instance, impact factors like pH, temperature, and the concentration of MPs can have an impact on the shape of the catalyst, which in turn can have an impact on the production of reactive species and the photocatalytic mechanism. By increasing improved MP adsorption, greater formation of reactive species, and effective utilisation of active sites on the catalysts, it is feasible to improve the efficacy of the photocatalytic process. It is important to consider how operating parameters affect the photocatalytic degradation of MPs. The pH of the solution, the termination rate, and the ease of producing free radicals which are necessary for releasing hydrogen atoms from polymer chains all have an impact on the photocatalytic degradation efficiency of MPs. For the process to be optimised and improved, it is essential to comprehend how these elements affect the morphology of the photocatalysts and the photocatalytic mechanism. For instance, the pH of the solution affects the production of reactive species that are involved in photocatalytic action, such as hydroxyl radicals and superoxide radicals, as well as the adsorption of MPs onto the overall surface of the photocatalyst. The rate of reaction and the production of reactive species are both impacted by temperature too, which eventually influences the overall degradation efficiency. Last but not least, the MPs concentration can change the active sites that are present on the photocatalyst surface, which in turn can affect the morphology and functionality of the photocatalysts (Martic, Tabobondung, and Gao 2022).

3.3.4 Electrochemical Oxidation

Electrochemical oxidation is another advanced oxidation process widely used for wastewater treatment where strong oxidising species are produced that break down the pollutants and produce CO_2, H_2O, and intermediates or may entirely mineralise when enough energy and a supportive electrolyte are provided to the system. It is made up of two electrodes, anode and cathode, and is divided into two types, that is, anodic oxidation and indirect cathode oxidation. Anodic oxidation (AO) is the process that is more frequently used and generally advised among the two (Du et al. 2021). Through the transmission of charges, it directly oxidises organic contaminants on an anode's surface. Another name for indirect cathode oxidation is the electro-Fenton method (EF) where hydroxyl radicals or reagents (such as hydrogen peroxide, peroxy-monosulphate, ozone, etc.) in an aqueous solution are used to indirectly oxidise contaminants (Figure 3.3).

The breakdown of hydrogen peroxide by the catalytic reaction of Fe^{2+} results in the formation of the hydroxyl radical or other ROS. By causing a redox reaction, free radicals are in charge of degrading organic contaminants. It has been demonstrated that with this method, it is possible to efficiently convert a variety of organic contaminants, including MPs, dyes, antipyretics, and antibiotics, into straightforward

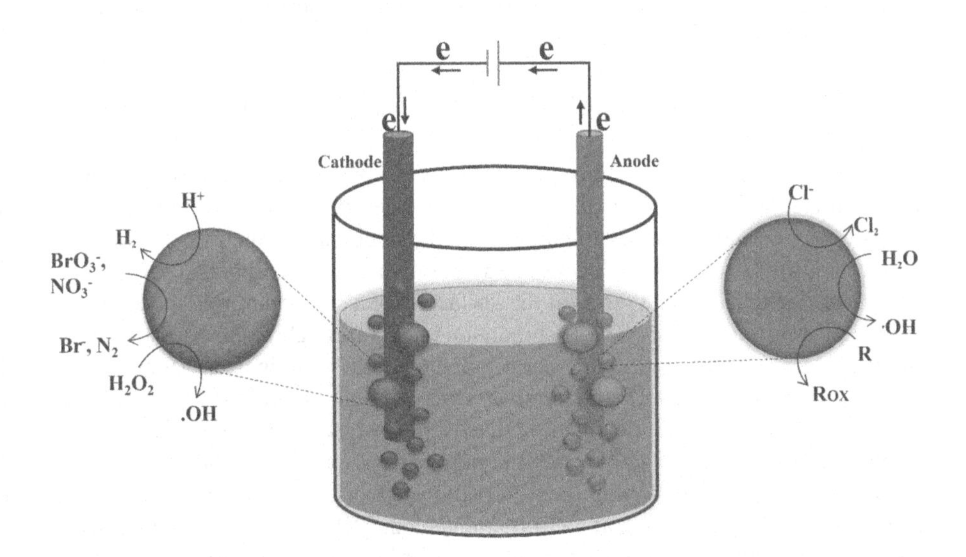

FIGURE 3.3 Electrochemical oxidation. The MPs degradation through electrochemical oxidation, the anode attracts negative charged ions, and the cathode attracts positive charged ions in the MPs, which leads to the degradation.

and non-toxic by-products like water vapour and carbon dioxide without the use of chemical agents (Sutradhar 2022). The breakdown of PVC MPs through an Electro-Fenton (EF) system that operates by a TiO_2/graphite cathode was proposed (Miao et al. 2020). The efficiency of electrochemical oxidation is affected by a number of variables, such as the anode's surface area and material, the type and concentration of the electrolyte employed, length of the degradation reaction, and current intensity (Kiendrebeogo et al. 2020).

3.4 MICROORGANISM-MEDIATED BIODEGRADATION OF MPs

Polymer biodegradation is made up of three procedures: (a) microbial adhesion to the polymer's surface, (b) the use of the polymer as a carbon source, and (c) polymer disintegration. Large polymers break down into low-molecular-weight monomers and oligomers when microorganisms adhere to their surface and break them down by secreting enzymes to gain energy for their growth. The degradation by microorganisms occurs aerobically in which microorganisms break down large organic compounds into smaller compounds by using oxygen as an electron acceptor. The by-products of this process are water and carbon dioxide.

$$\text{Carbon}\left(\text{plastic}\right) + \text{Oxygen} \rightarrow \text{Carbon dioxide} + \text{Water} + \text{Carbon residual}$$

And, anaerobically in which oxygen is not utilised for the breakdown of compounds by the action of microorganisms, and instead of oxygen, nitrate, sulphate, iron, manganese, and CO_2 act as an final electron acceptor to break down large organic compounds into smaller compounds.

$$\text{Carbon}\left(\text{plastic}\right) \rightarrow \text{Methane} + \text{Carbon dioxide} + \text{Water} + \text{Carbon residual}$$

Microbial degradation of micro-nano plastics is a practical, affordable, and clean way to remediate environmental contaminants. Biostimulation and bio-augmentation are two crucial processes to enhance the biodegradation rate of contaminants in nutrient-limited environments. MNPs' additive and hydrophobic nature and persistence of organic pollutants on the surface play a significant role in the remediation process. Microbial plastic degradation occurs through biodeterioration, biofragmentation, assimilation, and mineralisation (Figure 3.4). It increases with abiotic factors such as UV radiation and photo-oxidation. Higher molecular weight plastic polymers hinder microbial degradation due to the presence of large molecular fragments. Microorganisms depolymerise the polymers through intracellular and extracellular degradation, converting them into end products. The final consumption of these end products is biological natural attenuation. MPs are degraded by physicochemical and microbial factors but are less susceptible than other degradable materials. Microorganisms have been isolated from environmental samples to investigate the degradation behaviour of MPs.

The breakdown of MPs is a comprehensive process that combines physicochemical and microbiological degradation variables in various situations (Ammala

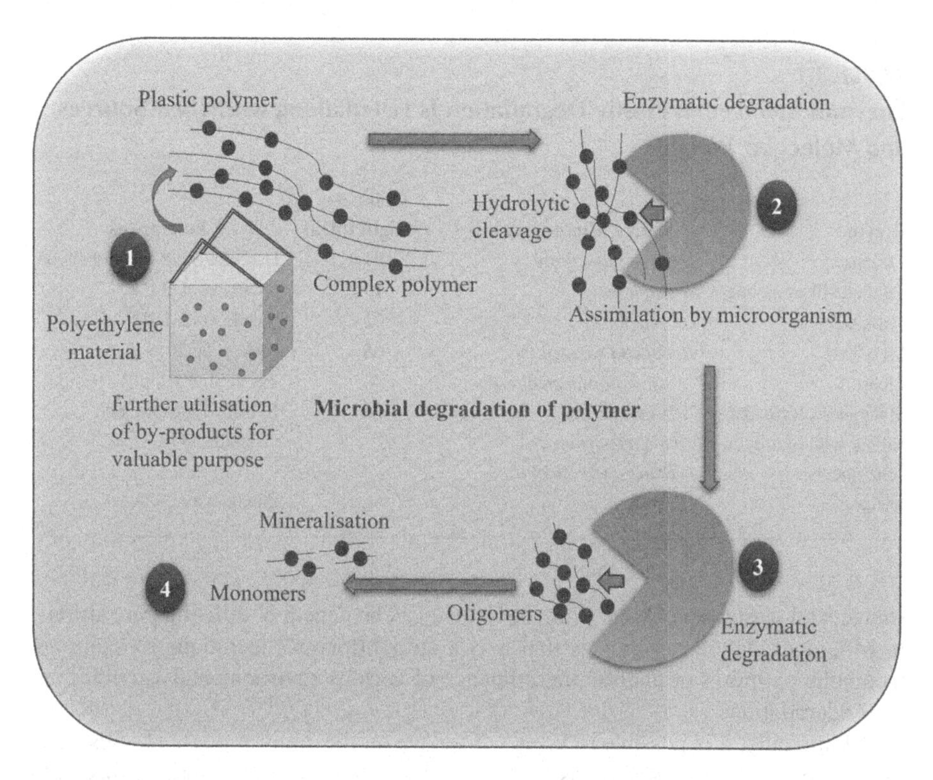

FIGURE 3.4 Microbial polymers degradation. In this process, the enzymes produced by the microbes and the long polymers degrade into smaller oligomers, dimers, and monomers, and finally mineralization takes place.

et al. 2011). The transformation of MPs has been significantly impacted by microbial degradation, despite the fact that MPs are less vulnerable to microbial assault than other degradable materials (Rujnić-Sokele and Pilipović 2017). In addition to providing support for microbial colonisation and growth, MPs also act as a carbon source, creating a novel ecological niche for microorganisms. Multiple kinds of MPs-degrading microorganisms have been isolated from diverse environmental samples in order to study the behaviour of MPs degradation. In this section, we've compiled a list of the microbes responsible for MPs degradation and divided them into four groups: pure bacterial cultures, pure fungal cultures, bacterial consortia, and biofilms. Also, MPs changes before and after degradation as well as the impact of microbes on degradation are reviewed.

Bacteria, the most prevalent group of all creatures, make up a significant portion of microorganisms. They typically inhabit soil, water, and the environment, and many species are well-known for their capacity to break down contaminants (Bakir et al. 2016). Numerous studies have looked into the use of bacteria for MPs degradation currently. Studies on the microbially mediated breakdown of MPs in lab settings have mostly employed pure bacterial cultures. Although some of these cultures came from culture collections, the majority of these cultures were isolated from sediment,

TABLE 3.1
Enzymes Involved in Plastic Degradation Is Listed along with Their Sources and Molecular Weight

Enzyme	Source Microorganism	Molecular Weight (kDa)	References
Cutinase	*Aspergillus oryzae*	21.6	Hasegawa and Machida 2005
Cutinase-like enzyme	*Paraphoma*	19.7	Suzuki et al. 2014
Cutinase	*C. magnus*	21	Suzuki et al. 2013
MHETase	*Ideonella sakaiensis*	65	Palm et al. 2019
Lipase	*Amycolatopsis mediterannei*	48	Tan et al. 2021
Manganese peroxidase along with lignolytic enzyme	*Phanerochaete chrysosporium* and *Trametes versicolor*	86	Iiyoshi, Tsutsumi, and Nishida 1998
Laccase	*E. coli*	66	Zhang et al. 2023

sludge, and wastewater by enrichment culturing. The benefit of utilising pure strains in MPs degradation research is that it is a straightforward technique to examine metabolic pathways or analyse the influence of various environmental variables on MPs degradation.

Additionally, it is possible to keep a close eye on the entire process of MPs degradation by functional bacteria as well as any changes to MPs (Janssen, Heymsfield, and Ross 2002). Bacterial isolates that can degrade MPs have been discovered more frequently recently, and their MP-degrading characteristics and effects have drawn more attention (Table 3.1). Two pure bacterial cultures were isolated from mangrove sediment and used in Auta's study for the degradation of PP MPs (Auta et al. 2017). The weight loss of PP MPs induced by *Rhodococcus* sp. strain 36 and *Bacillus* sp. strain 27 was 6.4% and 4.0%, respectively, after 40 days of incubation. The activity of the microorganisms was also seen to have caused a variety of pores and irregularities to form on the treated PP MPs' surfaces. These findings suggested that MPs could become adherent, colonised by pure cultures of bacteria isolated from the environment, and be harmed. Auta also tested two isolates, *B. cereus*, and *B. gottheilii*, for their capacity to biodegrade various MPs in order to explore the biodegradation of these microorganisms. For PE, polyethylene terephthalate (PET), and PS, the weight loss percentages of MPs particles induced by *B. cereus* were 1.6%, 6.6%, and 7.4%, respectively. Furthermore, *B. gottheilii* caused MPs weight losses of 6.2%, 3.0%, 3.6%, and 5.8% for PE, PET, PP, and PS, respectively. Interestingly, the surfaces of some MPs were observed to have become rough and possessed.

3.5 ADVANCES IN BIOTECHNOLOGY FOR MPs BIODEGRADATION

Despite recent progress in plastic biodegradation, there is still little consensus over which biotechnological advancement is more useful for MPs biodegradation. Given that the MPs are a blend of several polymers, combination or sophisticated techniques of polymer biodegradation can be optimised for the removal of MPs from certain environments. In contrast to objects with a small surface area (like a water bottle), MPs with a big surface area may have the potential for fast biodegradation. The essential component of biostimulation is the supply of nutrients to encourage the use of local microbial biodegradation. Several research show that the slow biodegradation of some polymers, including due to the smaller size at which PLA, PHB, PVC, and PE, cause their fragmentation however, they are more persistent in the biodegradation of plastic as MPs. Membrane and enzyme technology are being used to degrade micro-nano plastics in wastewater treatment plants. Due to the development of chemical sludge, traditional treatment techniques including granular activated carbon and sand filtration, coagulation-flocculation, precipitation, and sedimentation are ineffective. Membrane and enzyme technology is broadly utilised to clean water in WWTP. Membrane ultrafiltration is a proven method for getting rid of contaminants such as organic acids, microorganisms, and soluble microbial products from wastewater treatment. Membrane bioreactor technology combines traditional biological sludge treatment with a micro-ultrafiltration device to remove waste materials more effectively than traditional techniques. Even after ultrafiltration of 269 pollutants, nano-sized particles such as polyvinyl chloride, PET, polystyrene, and polyethylene are still visible. Solutions for eliminating nano-sized contaminants are being offered by membrane technology, such as the immobilisation of nano-sized elements on or within the matrix or membrane. Nanofiltration has been efficacious in desalination and elimination of boron from seawater, and nanopores have the potential to aid as fine filters.

The key difference between plastic and MPs is a certain size range. As formerly said, owing to biotic and abiotic causes, micro-nano plastics are either produced extensively from big plastic pieces or come from a variety of sources. The daily growth of MNP contaminants in the environment and the resulting toxicological effects on diverse creatures warn against increasing plastic manufacturing or usage. Designing plastics that are safe for the environment and hasten their natural disintegration is a superior strategy to eliminate MNPs in the ecosystem. In order to investigate practical ways to remove micro-nano plastics from the environment, nanotechnology may be a potential field. The interesting methods through which nanotechnology promotes plastic breakdown are many. By including nanoparticles in the microbial cultures, the biodegradation of plastics is accelerated. For instance, SiO_2 nanoparticles have been shown to have an impact on the development of bacteria that break down plastic at varying concentrations (Pathak and Navneet 2017). By lowering the lag phase and significantly increasing the pace at which bacterial strains degrade plastic in this instance, SiO_2 nanoparticles improve growth (Pathak and Navneet 2017). When present in larger quantities, fullerene 60 nanoparticles are often regarded as hazardous. However, it demonstrates a faster rate of plastic decomposition when given at a low dose. The addition of fullerene 60 nanoparticles to the bacteria's minimum medium enhances performance and causes the bacterium to break down LDPE by debilitating its hydrocarbon chain and producing different side products. According to a different study, the addition of nanoparticles, such as superparamagnetic iron oxide (SPION) and nano-barium titanate (NBT), may accelerate the breakdown of LDPE via interactions with bacterial metabolism (Kapri, Zaidi, and Goel 2010). Intriguingly, the presence of nanoparticles in bacterial culture medium encourages their development by triggering an early rise of the exponential phase, but it also improves the biodegradation effectiveness of a prospective consortium of microbes that can break down plastics (Kapri et al. 2010). These results imply the significance of interactions between nanoparticles and microorganisms. Investigating different nanoparticles for their potential to support microbial remediation methods may aid in the commercialisation of plastic breakdown.

3.6 CURRENT RESEARCH ON THE BIODEGRADATION OF MPs USING MICROBES

Recent years have seen the discovery of numerous bacterial isolates that can break down polymeric polymers (MPs). Two bacterial cultures were isolated from mangrove sediment and used for PP MPs degradation. *Bacillus* sp. strain 27 and *Rhodococcus* sp. strain 36 both caused PP MPs to lose weight after 40 days of incubation, although at different rates – 6.4% and 4.0%, respectively (Auta et al. 2017). Two bacterial isolates, *B. cereus* and *B. gottheilii*, were assayed to disintegrate different types of MPs. *B. cereus* mediated MPs particles had a weight reduction percentage of 1.6%, 6.6%, and 7.4% for PE, PET, and PS, respectively. Additionally, *B. gottheilii* induced MPs weight loss percentages were 6.2%, 3.0%, 3.6%, and 5.8% for PE, PET, PP, and PS, respectively. For a study on MPs degradation, a number of bacterial consortia have been produced via lab acclimation trials. In an investigation by Park (Park and Kim 2019), the degradation of PE MPs by a mesophilic bacterial consortium obtained from landfill sediment was analysed. Two bacterial consortia were domesticated, and the results suggested that the Souda community was more effective than the Agios community at reducing the weight of HDPE MPs (Tsiota et al. 2018). The worm *Lumbricus terrestris* was responsible for the LDPE's deterioration. The earthworm's gut microbes absorbed and degraded LDPE MPs when they were consumed, indicating that the worm's gastrointestinal tract has microbes that can break down MPs (Lwanga et al. 2017). The microbial-mediated degradation of MPs is initiated by the secretion of enzymes that catalyse the cleavage of MPs chains into monomers. Recent studies have focused on the separation and identification of enzymes in microorganisms. A novel bacterium, *Ideonella sakaiensis* 201-F6, can use PET as its major energy and carbon source (Tanasupawat et al. 2016). An *AlkB* family alkane hydroxylase was active towards PE samples with molecular weights of up to 27,000 Da (Yoon et al. 1996). These studies show that enzymes play a crucial role in the breakdown of MPs and call for more research into this area. Extensive research is required since the enzymes involved in the microbial-mediated breakdown of plastics are currently poorly known.

3.7 ENZYMES INVOLVED IN BIODEGRADATION OF PLASTIC

The recalcitrant MPs can be removed by different physical and chemical methods. But these methods have several drawbacks such as production of harmful secondary contaminants, high energy, and chemical cost, and risk to the life of manpower. Studies have shown that a variety of microbes using their plastic-degrading enzymes (PDEs) can break down plastics (Kumar et al. 2023). Enzyme-mediated biocatalytic disintegration of polymers is a useful and effective substitute for recycling and treating plastic (Zhu, Wang, and Wei 2022). The search for biocatalysts for plastic processing and an understanding of microorganisms mediated plastic degradation have spurred the development of plastic-degrading microbial strains (PDMs) and PDEs over the past several decades. The microbial degradation of MPs converts them into sugars and biofuels (Cholewinski et al. 2022). Therefore, MPs can be biodegraded to create valuable products as well as protect the environment. MP degradation involves four processes, namely, biodeterioration bio-fragmentation, assimilation, and mineralisation. Enzymes that are involved in the degradation of plastics attach to the polymer's backbone and break it down into monomers. However, this process is dependent on the chemical complexity, functional group, crystallinity, and molecular weight. *Actinobacteria* and *Proteobacteria* primarily possess the genes encoding potential polyester hydrolases. Giangeri et al. (2022) have demonstrated the role of multiple enzymes for plastic degradation such as alkane 1-monooxygenase, a feruloyl esterase, cutinase, triacylglycerol lipase, and protocatechuate 4,5-dioxygenase and medium-chain acyl-CoA dehydrogenase (Giangeri et al. 2022). Additionally, polyethylene can be depolymerised by two enzymes found in waxworm saliva.

Enzymatic procedures like oxidation, hydrolysis, and hydroxylation are responsible for the polymer chains to disintegrate into oligomers and monomers. It is believed that high-molecular-weight polymers are first broken down by extracellular enzymes released by microbes. Further, the subunits (dimers or oligomers) are assimilated by the microbial cells for mineralisation (Mohanan et al. 2020). Mineralisation is characterised by the formation of CO_2, CH_4, H_2O, N_2, and numerous other metabolic substances.

The different enzymes that degrade plastics have different catalytic centres. For instance, PET-degrading enzymes belong to serine hydrolases, which have a catalytic triad of serine, histidine, and aspartic acid. Enzymes that break down PLA have been categorised as proteases, lipases, and esterases and they exhibit comparatively higher sequence diversity. PHA depolymerase belongs to the carboxylesterase family and has a conserved histidine near the oxyanion hole in their catalytic triad of serine, histidine, and aspartic acid.

The PDE belongs to the category of cutinase and cutinase-like enzymes extracted from *Cryptococcus magnus* exhibit optimal activity at 40°C and pH 7.5. Bivalent cations also influence the catalytic activity of PDEs. For instance, the enzyme activity of cutinase is enhanced by a 2.5 mM concentration of Ca^{2+} or Mg^{2+} (Suzuki et al. 2013) (Table 3.1).

3.8 PLASTIC-DEGRADING ENZYME IN ADVERSE ENVIRONMENT

The plastic-degrading microorganisms such as *Bacillus, Pseudomonas, Aspergillus,* etc. are either thermophilic or mesophilic. They degrade plastic at moderate to high temperature (Zhang et al. 2023). Researchers are trying to incorporate plastic-degrading genes into such microorganisms that can degrade plastic at low temperature. Using cold active laccase, Zhang et al. achieved low temperature biodegradation of polyethylene MP. They express the gene of cold active laccase into the surface of *Escherichia coli* in order to degrade plastic at low temperature. They confirmed 50% remaining enzyme activity in 4 days at 15°C (Zhang et al. 2023).

Furthermore, finding new halotolerant bacteria that can break down MPs is crucial since MPs typically contaminate marine environments with high salinity. Adithama et al. (2023) describe the biological degradation of low-density polyethylene MP by halotolerant bacteria. They found a lowering crystallinity index based on X-ray diffraction technique (Adithama et al. 2023).

3.9 INCREASING THE EFFICIENCY OF PLASTIC-DEGRADING MICROBIAL ENZYMES

The enzyme involved in MP degradation can be immobilised for their better functioning and reducing the possibility of denaturation. The development of enzyme immobilisation along with genetic engineering methods has made it possible to use peroxidases, laccases, and hydrolases to remove toxins from the environment. Such immobilised enzymes accelerate microplastic degradation by photogenerated radicals based oxidation and by promoting bacterial growth (Tang et al. 2022).

Genetic engineering and cloning play important roles in enhancing the function of PDEs. For instance Shinozaki et al. isolated a gene encoding *PaE*, a gene of biodegradable PDE from *Pseudozyma antarctica*, and cloned it into *Saccharomyces cerevisiae* (Shinozaki et al. 2013). It has also been discovered that protein engineering of the PDE is effective at speeding up plastic decomposition. In order to increase the capacity of microorganisms to break down plastic, PDEs genes are also coded using synthetic biology.

3.10 KNOWLEDGE GAPS AND FUTURE RECOMMENDATIONS

Plastics are fascinating because of their longevity, yet their durability has a negative impact on the environment. The concentration of MPs will increase as plastic manufacture and distribution increase (Priya et al. 2022). MPs have an alarmingly long cycle of biodegradation, which leads to bioaccumulation and biomagnification at all trophic levels of the food chain after they enter the food chain. In addition to biological components (microorganisms), a number of physicochemical elements are also involved in the biodegradation of MPs (Kumar et al. 2023). Evaluation of biodegradation strategies is critical for incorporating microplastic waste into a circular economy Cholewinski et al. 2022. Through the use of multi-omic and culture-independent methods, new computational tools and sequencing technology, genetic engineering, and protein engineering are hastening scientific discoveries for plastic waste management. However, the use of such technology must be justified in order to prevent harm to the ecosystem and the survival of other species.

This chapter emphasised the most significant knowledge gaps regarding environmental concentrations, sources, and ecological effects of microplastic pollution in terrestrial ecosystems. The understanding of microplastic concentrations in freshwater systems is fast expanding, but it has not yet been connected to ecological implications. The quantity, make-up, and variety of microplastic particles entering the environment are all highly speculative. This section emphasises the difficult task of comprehending the dynamics and effects of MPs as an environmental pollutant, particularly as they relate to freshwater and terrestrial environments, and how data from marine studies can be used to infer or predict what might happen in these less well-studied systems. Additionally, nanomaterial studies can shed light on the behaviour and fate of particulates. In the realm of nanomaterial research, it is crucial to advance the subject of study. The defining of "microplastics" as an environmental pollutant, the creation of standardised techniques for gathering, processing, and analysing environmental samples, and the potential repercussions of this continually accumulating pollution are the most crucial details in this chapter. In order to learn more about the polymers present in terrestrial and freshwater ecosystems, it is crucial to continue developing spectroscopy methods, as well as alternatives like differential scanning calorimetry (DSC) and thermo-gravimetric analysis (TGA). These methods have already been used to identify freshwater and terrestrial nanoparticles. It's also critical to comprehend the potential consequences of this fragmentation-related pollution that keeps building up.

According to the aforementioned data in this work, certain knowledge gaps remain in the field of MPs breakdown by AOPs (direct photodegradation, photocatalytic oxidation, and electrochemical oxidation) and microbiological processes. (1) How do the suggested techniques prevent the release of hazardous intermediate products into model ecosystems while MPs are breaking down? If not, what steps are being made to stop the pollution? (2) How can AOPs methods selectively break the chemical bonds of polymer MPs to produce usable organics at a high rate? (3) What procedures are employed to extract CH_2O_2, $C_2H_4O_2$, CH_4O, and C_2H_6O, among others with high purity after the degradation process? (4) Whether any surviving MPs are environmentally damaging following the degrading processes described in this literature? (5) There has been ineffective collection and detection of MPs' biodegradation products. (6) Whether microbial consortia's ability to break down and use MPs is influenced by interactions between distinct microbes and numerous enzymes (Chen et al., 2022). Current research shows how different photocatalysts can degrade big plastics like PP, PE, and PS. Since most plastic films and particles only partially disintegrate by UV light, it appears that the active oxygen species produced under visible light are insufficient to cause chain cleavage and subsequent oxidation reactions. Given this, more research on its use in MPs deterioration is required. The majority of these proposed photocatalytic systems did not study the final degradation products using LC-MS or other detection techniques, despite the fact that they showed partial degradation. These procedures might also release VOCs. In addition, a separate issue needs to be adequately considered (Qi et al. 2017). For large-scale development, it will be necessary to create resilient catalysts that can keep their performance through several cycles. Most current research focuses on specific photocatalytic systems, which may be limited by elements like quick electron-hole recombination, inappropriate band values and placements, and slow surface reaction kinetics. These restrictions result in subpar photocatalytic degradation performance, characterised by poor efficiency and prolonged irradiation times, which prevents the development of new applications (Chen et al. 2022; Xie et al. 2023). How can the decomposition of MPs be prevented from releasing hazardous intermediate products into simulated environments? What steps are taken to stop pollution, if any, in that case?

In order to selectively break down MPs into usable and non-toxic compounds by the control of the breakage of chemical bonds, high activity of degradation and selectivity of catalysts must be analysed. For biodegradation, particular functional microorganisms must be cultivated. Additionally, to increase the efficiency of microbial-mediated MPs degradation, future studies should make use of genomes and proteomics engineering methods (Park and Kim 2019). High performance liquid chromatography-high resolution mass spectrometry (HPLC-HRMS), positron annihilation spectroscopy, Fourier transform infrared spectroscopy techniques, and free radical trapping experiments can all reveal the selective degradation process of MPs. The best extraction techniques for high purity degradation products should be investigated (Uheida et al. 2021). A toxicity study needs to be done to determine the toxicity of the intermediate compounds that are produced while MPs degrade (Zurier and Goddard 2021). It is challenging to gather degradation products from the soil or other ecosystems, in contrast to the aquatic environment. Therefore, the development of an extraction method for degradation products has to receive much more attention. And because products are too complex, accurate analytical procedures for GC-MS or HPLC-MS should be developed (Zhu et al. 2020). MPs' degradation products and effectiveness should be examined to determine the effects of the interactions between various functional microbiological organisms and numerous enzymes on the degradation of MPs. It is necessary to create standardised and precise techniques for sampling, measuring, and identifying MPs in various environmental compartments. For sample collection, separation, and analysis for every type of MP, regardless of origin, shape, size, or composition, suitable consistent models must be constructed (Akdogan and Guven 2019). Despite extensive prior studies on MPs, there is still a dearth of complete information regarding MP pollution, concentrations, and distributions in various environmental compartments. In order to give fundamental information on the pollution situation and the distribution of MPs worldwide, extensive global monitoring programmes must be implemented. Evaluation of the contributions of anthropogenic and natural activities to MP contamination in the atmosphere, soil, and water is also crucial (Belzagui et al. 2020). We know very little about how MPs move through the soil and atmosphere or what happens to them there. Consequently, it is crucial to look into this to remove from the environment by degradation and disintegration of MPs in terrestrial ecosystems, including both natural and human-made mechanisms. Further investigation is required into the vertical distribution of these MPs in water columns as well as the downstream transit of MPs via river water into the ocean. A major factor in MPs' environmental degradation is weathering. Therefore, it is vital to use accelerated weathering studies to evaluate the impacts of both short- and long-term weathering on MP characteristics. Further study is needed in the area of biodegradation to better understand how biofilms arise, how they affect MP degradation, and other related issues such as water salinity and organic matter (Lwanga et al. 2017). Microorganisms are crucial to the biodegradation of MPs. The majority of earlier research on this subject, however, has been on aerobic microbes. Future studies should concentrate on both aerobic and anaerobic microorganisms as well as the use of various molecular approaches to explore the various pathways involved in MP degradation in order to achieve robust MP degradation and remediation. Research into and promotion of the use of biodegradable plastic materials should be done in order to control the production and use of plastic materials. The knowledge base of MPs has grown over the past ten years as a result of growing scientific interest, but there are still many important concerns and problems that need to be solved. Our ability to analyse the regional and temporal patterns of these pollutants is hampered by inconsistent sample design and MP processing. Therefore, it is crucial to provide a unified and integrated sampling and processing technique for future research. Net sampling, a time-saving technique with the advantages of covering wide sampling regions and minimising the water volume of samples, is the perfect approach for determining the level of MP (items/m^3) in freshwater and marine water. Inter-study comparison has been complicated by the use of varied net aperture sizes, trawling speeds, and durations (Veerasingam et al. 2021).

3.11 CONCLUSIONS

Due to their negative effects on the environment, MP pollution has drawn a lot of attention. However, finding effective strategies for MP repair remains a formidable challenge. This review summarised the present methods for MPs degradation, including AOPs and biodegradation, the degradation mechanism, and the state of development. It also made some suggestions for future research areas based on the identified information gaps. As previously mentioned, due to their tiny size, MPs were mostly removed from the environment by regular incineration and landfills for the treatment of large plastic products. The most popular MPs remediation method at the moment is AOPs, but other methods, particularly those based on photocatalytic oxidation and electrochemical oxidation with catalysts, should be regarded seriously given the decrease in MPs' breakdown activity after numerous cycling tests. Another possible strategy to get rid of the MPs contamination is biodegradation. However, it typically takes a long period, and it is yet unclear how it degrades. Therefore, choosing the appropriate decomposition techniques for MPs treatment aimed at various MP kinds is important. This review gives readers a better knowledge of how AOPs and microbes decompose MPs and offers suggestions for improving MPs' decomposition efficiency.

REFERENCES

Adithama, R. M., I. Munifah, D. H. Y. Yanto, and A. Meryandini. 2023. "Biodegradation of Low-Density Polyethylene Microplastic by New Halotolerant Bacteria Isolated from Saline Mud in Bledug Kuwu, Indonesia." *Bioresource Technology Reports* 22: 101466. doi: 10.1016/j.biteb.2023.101466.

Akdogan, Z. and B. Guven. 2019. "Microplastics in the Environment: A Critical Review of Current Understanding and Identification of Future Research Needs." *Environmental Pollution* 254: 113011. doi: 10.1016/j.envpol.2019.113011.

Ali, S. S., I. A. Qazi, M. Arshad, Z. Khan, T. C. Voice, and C. T. Mehmood. 2016. "Photocatalytic Degradation of Low Density Polyethylene (LDPE) Films Using Titania Nanotubes." *Environmental Nanotechnology, Monitoring & Management* 5: 44–53. doi: 10.1016/j.enmm.2016.01.001.

Alvim, C. B. 2020. "Wastewater Treatment Plant as Microplastics Release Source – Quantification and Identification Techniques." *Journal of Environmental Management* 255. doi: 10.1016/j.jenvman.2019.109739.

Ammala, A., S. Bateman, K. Dean, E. Petinakis, P. Sangwan, S. Wong, Q. Yuan, L. Yu, C. Patrick, and K. H. Leong. 2011. "An Overview of Degradable and Biodegradable Polyolefins." *Progress in Polymer Science* 36(8): 1015–49. doi: 10.1016/j.progpolymsci.2010.12.002.

Ariza-tarazona, M. C., J. Francisco, V. Barbieri, E. I. Cedillo-gonzález, M. Camila, J. F. Villarreal, V. Barbieri, C. Siligardi, E. I. Cedillo-gonzález, M. C. Ariza-tarazona, J. F. Villarreal-chiu, V. Barbieri, and E. I. Cedillo-gonzález. 2018. "New strategy for microplastic degradation: Green photocatalysis using a protein-based porous N-TiO_2 semiconductor." *Ceramics International* 45 (7): 9618–24. doi: 10.1016/j.ceramint.2018.10.208.

Atugoda, T., M. Vithanage, H. Wijesekara, N. Bolan, A. K. Sarmah, M. S. Bank, S. You, and Y. Sik. 2021. "Interactions between Microplastics, Pharmaceuticals and Personal Care Products: Implications for Vector Transport." *Environment International* 149: 106367. doi: 10.1016/j.envint.2020.106367.

Auta, H. S., C. U. Emenike, and S. H. Fauziah. 2017. "Screening of *Bacillus* Strains Isolated from Mangrove Ecosystems in Peninsular Malaysia for Microplastic Degradation." *Environmental Pollution* 231: 1552–59. doi: 10.1016/j.envpol.2017.09.043.

Bakir, A., I. A. O'Connor, S. J. Rowland, A. J. Hendriks, and R. C. Thompson. 2016. "Relative Importance of Microplastics as a Pathway for the Transfer of Hydrophobic Organic Chemicals to Marine Life." *Environmental Pollution* 219: 56–65. doi: 10.1016/j.envpol.2016.09.046.

Belzagui, F., C. Gutiérrez-Bouzán, A. Álvarez-Sánchez, and M. Vilaseca. 2020. "When Size Matters – Textile Microfibers into the Environment." In: Cocca, M., et al. Proceedings of the 2nd International Conference on Microplastic Pollution in the Mediterranean Sea. ICMPMS 2019. Springer Water. Springer, Cham. doi: 10.1007/978-3-030-45909-3_13.

Bittner, G. D., M. S. Denison, C. Z. Yang, M. A. Stoner, and G. He. 2014. "Chemicals Having Estrogenic Activity Can Be Released from Some Bisphenol A-Free, Hard and Clear, Thermoplastic Resins." *Environmental Health* 13: 103. doi: 10.1186/1476-069X-13-103.

Carr, S. A., J. Liu, and A. G. Tesoro. 2016. "Transport and Fate of Microplastic Particles in Wastewater Treatment Plants." *Water Research* 91: 174–82. doi: 10.1016/j.watres.2016.01.002.

Chen, C. C., Y. Shi, Y. Zhu, J. Zeng, W. Qian, S. Zhou, J. Ma, K. Pan, Y. Jiang, and Y. Tao. 2022. *Water Research* 219: 118536. https://doi.org/10.1016/j.watres.2022.118536

Cholewinski, A., E. Dadzie, C. Sherlock, W. A. Anderson, T. C. Charles, K. Habib, S. B. Young, and B. Zhao. 2022. "A Critical Review of Microplastic Degradation and Material Flow Analysis towards a Circular Economy." *Environmental Pollution* 315: 120334. doi: 10.1016/j.envpol.2022.120334.

Cole, M., P. Lindeque, C. Halsband, and T. S. Galloway. 2011. "Microplastics as Contaminants in the Marine Environment: A Review." *Marine Pollution Bulletin* 62 (12): 2588–97. doi: 10.1016/j.marpolbul.2011.09.025.

De Souza Machado, A. A., S. Hempel, M. C. Rillig, W. Kloas, and C. Zarfl. 2017. "Microplastics as an Emerging Threat to Terrestrial Ecosystems." *Global Change Biology* 24 (4): 1405–16. doi: 10.1111/gcb.14020.

Du, H., Y. Xie, and J. Wang. 2021. "Microplastic Degradation Methods and Corresponding Degradation Mechanism: Research Status and Future Perspectives." *Journal of Hazardous Materials* 418: 126377. doi: 10.1016/j.jhazmat.2021.126377.

Duan, P., Y. Qi, S. Feng, X. Peng, W. Wang, Y. Yue, Y. Shang, Y. Li, B. Gao, and X. Xu. 2020. "Enhanced Degradation of Clothianidin in Peroxymonosulfate/Catalyst System via Core-Shell FeMn @ N-C and Phosphate Surrounding." *Applied Catalysis B: Environmental* 267: 118717. doi: 10.1016/j.apcatb.2020.118717.

Elsamahy, T., J. Sun, S. E. Elsilk, and S. S. Ali. 2023. "Biodegradation of Low-Density Polyethylene Plastic Waste by a Constructed Tri-Culture Yeast Consortium from Wood-Feeding Termite: Degradation Mechanism and Pathway." *Journal of Hazardous Materials* 448: 130944. doi: 10.1016/j.jhazmat.2023.130944.

Feng, H. M. J. C. Zheng, L. Ngai-Yu, Y. Lei, K. H. Kong, H. Q. Yu, T.-C. Lau, and M. H. W. Lam. 2011. "Photoassisted Fenton Degradation of Polystyrene." *Environmental Science & Technology* 45 (2): 744–50. doi: 10.1021/es102182g.

Gaylor, M., E. Harvey, R. C. Hale, M. O. Gaylor, E. Harvey, and R. C. Hale. 2013. "Polybrominated Diphenyl Ether (PBDE) Accumulation by Earthworms (Eisenia Fetida) Exposed to Biosolids-, Polyurethane Foam Microparticle- and Penta-BDE-Amended Soils Biosolids-, Polyurethane Foam Microparticle- and Penta-BDE-Amended Soils Department." *Environmental Science & Technology* 47 (23): 13831–39. doi: 10.1021/es403750a.

Gewert, B., M. M. Plassmann, and M. Macleod. 2015. "Environmental Science Processes & Impacts Pathways for Degradation of Plastic Polymers Floating in the Marine Environment." *Environmental Science: Processes & Impacts* 17: 1513. doi: 10.1039/c5em00207a

Geyer, R., J. R. Jambeck, and K. L. Law. 2017. "Production, Use, and Fate of All Plastics Ever Made." *Science Advances* 3 (7): e170078. doi: 10.1126/sciadv.1700782.

Giangeri, G., M. S. Morlino, N. De Bernardini, M. Ji, M. Bosaro, V. Pirillo, P. Antoniali, G. Molla, R. Raga, L. Treu, and S. Campanaro. 2022. "Preliminary Investigation of Microorganisms Potentially Involved in Microplastics Degradation Using an Integrated Metagenomic and Biochemical Approach." *Science of The Total Environment* 843: 157017. doi: 10.1016/j.scitotenv.2022.157017.

Henry, B., K. Laitala, and I. Grimstad. 2019. "Microfibres from Apparel and Home Textiles: Prospects for Including Microplastics in Environmental Sustainability Assessment." *Science of the Total Environment* 652: 483–94. doi: 10.1016/j.scitotenv.2018.10.166.

Iiyoshi, Y., Y. Tsutsumi, and T. Nishida. 1998. "Polyethylene Degradation by Lignin-Degrading Fungi and Manganese Peroxidase." *Journal of Wood Science* 44 (3): 222–29. doi: 10.1007/BF00521967.

Janssen, I., S. B. Heymsfield, and R. Ross. 2002. "Low Relative Skeletal Muscle Mass (Sarcopenia) in Older Persons Is Associated with Functional Impairment and Physical Disability." *Journal of the American Geriatrics Society* 50 (5): 889–96. doi: 10.1046/j.1532-5415.2002.50216.x.

Kang, J., L. Zhou, X. Duan, H. Sun, Z. Ao, and S. Wang. 2019. "Degradation of Cosmetic Microplastics via Functionalized Carbon Nanosprings." *Matter* 1 (3): 745–58. doi: 10.1016/j.matt.2019.06.004.

Kapri, A, M. G. H. Zaidi, and R. Goel. 2010. "Implications of SPION and NBT Nanoparticles upon In Vitro and In Situ Biodegradation of LDPE Film." *Journal of Microbiology and Biotechnology* 20 (6): 1032–41. doi: 10.4014/jmb.0912.12026

Kaushal, J., M. Khatri, and S. K. Arya. 2021. "Recent Insight into Enzymatic Degradation of Plastics Prevalent in the Environment: A Mini-Review." *Cleaner Engineering and Technology* 2: 100083. doi: 10.1016/j.clet.2021.100083.

Kemp, T. J. and R. A. McIntyre. 2006. "Influence of Transition Metal-Doped Titanium(IV) Dioxide on the Photodegradation of Polystyrene." *Polymer Degradation and Stability* 91 (12): 3010–19. doi: 10.1016/j.polymdegradstab.2006.08.005.

Kiendrebeogo, M., M. R. Karimi Estahbanati, A. Khosravanipour Mostafazadeh, P. Drogui, and R. D. Tyagi. 2020. "Treatment of Microplastics in Water by Anodic Oxidation: A Case Study for Polystyrene." *Environmental Pollution* 116168. doi: 10.1016/j.envpol. 2020.116168.

Klavarioti, M., D. Mantzavinos, and D. Kassinos. 2009. "Removal of Residual Pharmaceuticals from Aqueous Systems by Advanced Oxidation Processes." *Environment International* 35 (2): 402–17. doi: 10.1016/j.envint.2008.07.009.

Kumar, M., X. Xiong, M. He, D. C. W. Tsang, J. Gupta, E. Khan, S. Harrad, D. Hou, Y. Sik, and N. S. Bolan. 2020. "Microplastics as Pollutants in Agricultural Soils." *Environmental Pollution* 265: 114980. doi: 10.1016/j.envpol.2020.114980.

Kumar, V., N. Sharma, L. Duhan, R. Pasrija, J. Thomas, M. Umesh, S. K. Lakkaboyana, R. Andler, A. S. Vangnai, M. Vithanage, M. K. Awasthi, W. Y. Chia, P. LokeShow, and D. Barceló. 2023. "Microbial Engineering Strategies for Synthetic Microplastics Clean Up: A Review on Recent Approaches." *Environmental Toxicology and Pharmacology* 98: 104045. doi: 10.1016/j.etap.2022.104045.

Lee, Q. Y. and H. Li. 2021. "Photocatalytic Degradation of Plastic Waste: A Mini Review." *Micromachines* 12 (8): 907. doi: 10.3390/mi12080907.

Li, J., H. Liu, and J. P. Chen. 2018. "Microplastics in Freshwater Systems: A Review on Occurrence, Environmental Effects, and Methods for Microplastics Detection." *Water Research* 137: 362–74. doi: 10.1016/j.watres.2017.12.056.

Li, W., Y. Zhang, N. Wu, Z. Zhao, W. Xu, Y. Ma, and Z. Niu. 2019. "Colonization Characteristics of Bacterial Communities on Plastic Debris Influenced by Environmental Factors and Polymer Types in the Haihe Estuary of Bohai Bay, China." *Environmental Science and Technology* 53 (18): 10763–73. doi: 10.1021/acs.est.9b03659.

Lv, X., Q. Dong, Z. Zuo, Y. Liu, X. Huang, and W.-M. Wu. 2019. "Microplastics in a Municipal Wastewater Treatment Plant: Fate, Dynamic Distribution, Removal Efficiencies, and Control Strategies." *Journal of Cleaner Production* 225: 579–86. doi: 10.1016/j. jclepro.2019.03.321.

Lwanga, E. H., J. M. Vega, V. Ku Quej, J. D. Los Angeles, L. Sanchez, C. Chi, G. E. Segura, G. Henny, M. Van Der Ploeg, A. A. Koel, and G. Violette. 2017. "Field Evidence for Transfer of Plastic Debris along a Terrestrial Food Chain." *Scientific Reports* 7: 14071. doi: 10.1038/s41598-017-14588-2.

Maeda, H., Y. Yamagata, K. Abe, F. Hasegawa, M. Machida, R. Ishioka, K. Gomi, and T. Nakajima. 2005. "Purification and Characterization of a Biodegradable Plastic-Degrading Enzyme from *Aspergillus oryzae*." *Applied Microbiology and Biotechnology* 67 (6): 778–88. doi: 10.1007/s00253-004-1853-6.

Martic, S., M. Tabobondung, and S. Gao. 2022. "Emerging Electrochemical Tools for Microplastics Remediation and Sensing." 2022 (August): 1–8. doi: 10.3389/fsens.2022. 958633.

Miao, F., Y. Liu, M. Gao, X. Yu, P. Xiao, S. Wang, and X. Wang. 2020. "Degradation of Polyvinyl Chloride Microplastics via an Electro-Fenton-Like System with a Tio_2/ Graphite Cathode." *Journal of Hazardous Materials* 3: 123023. doi: 10.1016/j.jhazmat. 2020.123023.

Mohanan, N., Z. Montazer, P. K. Sharma, and D. B. Levin. 2020. "Microbial and Enzymatic Degradation of Synthetic Plastics." *Frontiers in Microbiology* 11 (November). doi: 10. 3389/fmicb.2020.580709.

Nabi, I., A. U. R., Bacha, F. Ahmad, and L. Zhang. 2021. "Application of Titanium Dioxide for the Photocatalytic Degradation of Macro- and Micro-Plastics: A Review." *Journal of Environmental Chemical Engineering* 9: 105964. doi: 10.1016/j.jece.2021.105964.

Nakata, K. and A. Fujishima. 2012. "TiO_2 Photocatalysis: Design and Applications." *J Photochem Photobiol C* 13 (3): 169–89. doi: 10.1016/j.jphotochemrev.2012.06.001

Napper, I. E. and R. C. Thompson. 2020. "Plastic Debris in the Marine Environment: History and Future Challenges." *Global Challenges* 4: 1900081. doi: 10.1002/gch2.201900081.

Palm, G. J., L. Reisky, D. Böttcher, H. Müller, E. A. P. Michels, M. C. Walczak, L. Berndt, M. S. Weiss, U. T. Bornscheuer, and G. Weber. 2019. "Structure of the Plastic-Degrading *Ideonella sakaiensis* MHETase Bound to a Substrate." *Nature Communications* 10 (1): 1–10. doi: 10.1038/s41467-019-09326-3.

Park, S. Y. and C. G. Kim. 2019. "Biodegradation of micro-polyethylene particles by bacterial colonization of a mixed microbial consortium isolated from a landfill site." *Chemosphere* 222: 527–33. doi: 10.1016/j.chemosphere.2019.01.159.

Pathak, V. M. and Navneet. 2017. "Review on the Current Status of Polymer Degradation: A Microbial Approach." *Bioresources and Bioprocessing* 4 (1). doi: 10.1186/s40643-017-0145-9.

Perren, W., A. Wojtasik, and Q. Cai. 2018. "Removal of Microbeads from Wastewater Using Electrocoagulation." doi: 10.1021/acsomega.7b02037.

Prata, J. C., J. P. Costa, I. Lopes, and A. C. Duarte. 2020. "Environmental Exposure to Microplastics: An Overview on Possible Human Health Effects." *Science of the Total Environment* 134455. doi: 10.1016/j.scitotenv.2019.134455.

Priya, K. L., K. R. Renjith, C. J. Joseph, M. S. Indu, R. Srinivas, and S. Haddout. 2022. "Fate, Transport and Degradation Pathway of Microplastics in Aquatic Environment — A Critical Review." *Regional Studies in Marine Science* 56: 102647. doi: 10.1016/j.rsma.2022.102647.

Qi, Y., S. Chen, M. Li, Q. Ding, Z. Li, J. Cui, B. Dong, F. Zhang, and C. Li. 2017. Achievement of Visible-Light-Driven Z-Scheme Overall Water Splitting Using Barium-Modified Ta_3N_5 as a H_2^- Evolving Photocatalyst. *Chemical Science* 8: 437–43. doi: 10.1039/C6SC02750D.

Rajala, K., O. Grönfors, M. Hesampour, and A. Mikola. 2020. "Removal of Microplastics from Secondary Wastewater Treatment Plant Effluent by Coagulation/Flocculation with Iron, Aluminum and Polyamine-Based Chemicals." *Water Research* 116045. doi: 10.1016/j.watres.2020.116045.

Ren, L., L. Men, Z. Zhang, F. Guan, J. Tian, B. Wang, J. Wang, Y. Zhang, and W. Zhang. 2019. "Biodegradation of Polyethylene by *Enterobacter* sp. D1 from the Guts of Wax Moth *Galleria mellonella*." *International Journal of Environmental Research and Public Health* 16 (11). doi: 10.3390/ijerph16111941.

Rujnić-Sokele, M. and A. Pilipović. 2017. "Challenges and Opportunities of Biodegradable Plastics: A Mini Review." *Waste Management and Research* 35 (2): 132–40. doi: 10.1177/0734242X16683272.

Sharma, K., V. Dutta, S. Sharma, P. Raizada, A. Hosseini-bandegharaei, P. Thakur, and P. Singh. 2019. "Recent Advances in Enhanced Photocatalytic Activity of Bismuth Oxyhalides for Efficient Photocatalysis of Organic Pollutants in Water: A Review." *Journal of Industrial and Engineering Chemistry* 78: 1–20. doi: 10.1016/j.jiec.2019.06.022.

Shinozaki, Y., T. Morita, X. H. Cao, S. Yoshida, M. Koitabashi, T. Watanabe, K. Suzuki, Y. Sameshima-Yamashita, T. Nakajima-Kambe, T. Fujii, and H. K. Kitamoto. 2013. "Biodegradable Plastic-Degrading Enzyme from *Pseudozyma antarctica*: Cloning, Sequencing, and Characterization." *Applied Microbiology and Biotechnology* 97 (7): 2951–59. doi: 10.1007/s00253-012-4188-8.

Song, Y. K., S. H. Hong, M. Jang, G. M. Han, S. W. Jung, and W. J. Shim. 2017. "Combined Effects of UV Exposure Duration and Mechanical Abrasion on Microplastic Fragmentation by Polymer Type." *Environmental Science & Technology* 51 (8): 4368–76. doi: 10.1021/acs.est.6b06155.

Sutradhar, M. 2022. "A Review of Microplastics Pollution and Its Remediation Methods: Current Scenario and Future Aspects." *Archives of Agriculture and Environmental Science* 7 (2): 288–93. doi: 10.26832/24566632.2022.0702019.

Suzuki, K., M. T. Noguchi, Y. Shinozaki, M. Koitabashi, Y. Sameshima-Yamashita, S. Yoshida, T. Fujii, and H. K. Kitamoto. 2014. "Purification, Characterization, and Cloning of the Gene for a Biodegradable Plastic-Degrading Enzyme from Paraphoma-Related Fungal Strain B47-9." *Applied Microbiology and Biotechnology* 98 (10): 4457–65. doi: 10.1007/s00253-013-5454-0.

Suzuki, K., H. Sakamoto, Y. Shinozaki, J. Tabata, T. Watanabe, A. Mochizuki, M. Koitabashi, T. Fujii, S. Tsushima, and H. K. Kitamoto. 2013. "Affinity Purification and Characterization of a Biodegradable Plastic-Degrading Enzyme from a Yeast Isolated from the Larval Midgut of a Stag Beetle, *Aegus laevicollis*." *Applied Microbiology and Biotechnology* 97 (17): 7679–88. doi: 10.1007/s00253-012-4595-x.

Tagorti, G. and B. Kaya. 2022. "Chemosphere Genotoxic Effect of Microplastics and COVID-19: The Hidden Threat." *Chemosphere* 286 (P3): 131898. doi: 10.1016/j.chemosphere.2021.131898.

Tan, Y., G. T. Henehan, G. K. Kinsella, and B. J. Ryan. 2021. "An Extracellular Lipase from *Amycolatopsis mediterannei* Is a Cutinase with Plastic Degrading Activity." *Computational and Structural Biotechnology Journal* 19: 869–79. doi: 10.1016/j.csbj.2021.01.019.

Tanasupawat, S., T. Takehana, S. Yoshida, K. Hiraga, and K. Oda. 2016. "*Ideonella sakaiensis* sp. nov., Isolated from a Microbial Consortium That Degrades Poly(ethylene Terephthalate)." *International Journal of Systematic and Evolutionary Microbiology* 66 (8): 2813–18. doi: 10.1099/ijsem.0.001058.

Tang, K. H. D., S. S. M. Lock, P.-S. Yap, K. W. Cheah, Y. H. Chan, C. L. Yiin, A. Z. E. Ku, A. C. M. Loy, B. L. F. Chin, and Y. H. Chai. 2022. "Immobilized Enzyme/Microorganism Complexes for Degradation of Microplastics: A Review of Recent Advances, Feasibility and Future Prospects." *Science of The Total Environment* 832: 154868. doi: 10.1016/j.scitotenv.2022.154868.

Tofa, T. S., L. K. Karthik, P. Swaraj, and D. Joydeep. 2019. "Visible Light Photocatalytic Degradation of Microplastic Residues with Zinc Oxide Nanorods." *Environmental Chemistry Letters* 17 (3): 1341–46. doi: 10.1007/s10311-019-00859-z.

Tsiota, P., K. Karkanorachaki, E. Syranidou, M. Franchini, and N. Kalogerakis. 2018. "Microbial Degradation of HDPE Secondary Microplastics: Preliminary Results." *Springer Water* 181–88. doi: 10.1007/978-3-319-71279-6_24.

Uheida, A., M. Abdel-Rehim, W. Hamd, and J. Dutta. 2021. "Visible Light Photocatalytic Degradation of Polypropylene Microplastics in a Continuous Water Flow System." *Journal of Hazardous Materials* 406: 124299. doi: 10.1016/j.jhazmat.2020.124299.

Veerasingam, S., M. Ranjanib, R. Venkatachalapathyb, A. Bagaevc, V. Mukhanovd, D. Litvinyukd, M. Mugilarasane, K. Gurumoorthif, L. Guganathanb,V. M. Aboobackera, and P. Vethamony. 2021. "Contributions of Fourier Transform Infraredspectroscopy in Microplastic Pollution Research: A Review." *Critical Reviews in Environmental Science and Technology* 51(22): 2681–743. doi: 10.1080/10643389.2020.1807450.

Wagner, M. and L. Scott. 2018. *Freshwater Microplastics. Emerging Environmental Contaminants?* Springer Cham. doi: 10.1007/978-3-319-61615-5.

Wang, C., B. F. Pfleger, and S. W. Kim. 2017. "Reassessing *Escherichia coli* as a Cell Factory for Biofuel Production." *Current Opinion in Biotechnology* 45: 92–103. doi: 10.1016/j.copbio.2017.02.010.

Wang, S. and X. Liu. 2017. "China's City-Level Energy-Related CO_2 Emissions: Spatiotemporal Patterns and Driving Forces." *Applied Energy* 200: 204–14. doi: 10.1016/j.apenergy.2017.05.085.

Wysocki, Ina Tessnow-von and P. Le Billon. 2019. "Plastics at Sea: Treaty Design for a Global Solution to Marine Plastic Pollution." *Environmental Science and Policy* 100: 94–104. doi: 10.1016/j.envsci.2019.06.005.

Xie, A. M. Jin, J. Zhu, Q. Zhou, L. Fu, and W. Wu. 2023. "Photocatalytic Technologies for Transformation and Degradation of Microplastics in the Environment: Current Achievements and Future Prospects." *Catalysts* 13 (5): 846. doi: 10.3390/catal13050846.

Yoon, J.-S., I.-J. Chin, M.-N. Kim, and C. Kim. 1996. "Degradation of Microbial Polyesters: A Theoretical Prediction of Molecular Weight and Polydispersity." *Macromolecules* 29 (9): 3303–7. doi: 10.1021/ma950314k.

Yuan, J., J. Ma, Y. Sun, T. Zhou, Y. Zhao, and F. Yu. 2020. "Microbial Degradation and Other Environmental Aspects of Microplastics/Plastics." *Science of the Total Environment* 715: 136968. doi: 10.1016/j.scitotenv.2020.136968.

Zan, L., S. Wang, W. Fa, Y. Hu, L. Tian, and K. Deng. 2006. "Solid-Phase Photocatalytic Degradation of Polystyrene with Modified Nano-TiO_2 Catalyst." *Polymer* 47 (24): 8155–62. doi: 10.1016/j.polymer.2006.09.023.

Zhang, A., Y. Hou, Y. Wang, Q. Wang, X. Shan, and J. Liu. 2023. "Highly Efficient Low-Temperature Biodegradation of Polyethylene Microplastics by Using Cold-Active Laccase Cell-Surface Display System." *Bioresource Technology* 382: 129164. doi: 10.1016/j.biortech.2023.129164.

Zhang, L., J. Liu, Y. Xie, S. Zhong, and P. Gao. 2021. "Occurrence and Removal of Microplastics from Wastewater Treatment Plants in a Typical Tourist City in China." *Journal of Cleaner Production* 291: 125968. doi: 10.1016/j.jclepro.2021.125968.

Zhang, X., Y. Li, D. Ouyang, J. Lei, Q. Tan, and L. Xie. 2021. "Systematical Review of Interactions between Microplastics and Microorganisms in the Soil Environment." *Journal of Hazardous Materials* 418: 126288. doi: 10.1016/j.jhazmat.2021.126288.

Zhou, X. and C. Feng. 2017. "The Impact of Environmental Regulation on Fossil Energy Consumption in China: Direct and Indirect Effects." *Journal of Cleaner Production* 142: 3174–83. doi: 10.1016/j.jclepro.2016.10.152.

Zhu, B., D. Wang, and N. Wei. 2022. "Enzyme Discovery and Engineering for Sustainable Plastic Recycling." *Trends in Biotechnology* 40 (1): 22–37. doi: 10.1016/j.tibtech.2021.02.008.

Zhu, K., H. Jia, Y. Sun, Y. Dai, C. Zhang, X. Guo, T. Wang, and L. Zhu. 2020. "Long-Term Phototransformation of Microplastics under Simulated Sunlight Irradiation in Aquatic Environments: Roles of Reactive Oxygen Species." *Water Research* 173: 115564. doi: 10.1016/j.watres.2020.115564.

Zhu, K., H. Jia, S. Zhao, T. Xia, X. Guo, T. Wang, and L. Zhu. 2019. "Formation of Environmentally Persistent Free Radicals on Microplastics under Light Irradiation." *Environmental Science & Technology* 53 (14): 8177–86. doi: 10.1021/acs.est.9b01474.

Zurier, H. and J. Goddard. 2021. "Biodegradation of microplastics in food and agriculture." *Current Opinion in Food Science* 37: 1–8. doi: 10.1016/j.cofs.2020.09.001.

4 Applications of Microbial Biosurfactants in Oil Recovery

Sombir Pannu, Lalit Mohan Pandey, and Pankaj Tiwari

4.1 INTRODUCTION

Biosurfactants are produced by biological resources like microbes, plants, and animals. These are amphiphilic compounds that adsorb at the interface of two immiscible liquids to lessen the interfacial tension (IFT) (Das & Mukherjee, 2005; Pruthi & Cameotra, 2003). In order to facilitate oil recovery, biosurfactants can lessen the IFT between oil and injected fluid and lower the surface tension (ST) of water below 25–30 mN/m. Microbial production of biosurfactants is the most common method. In this direction, potential microorganisms are isolated, screened for surface-active properties, and cultivated in the first stage. Reservoir samples (core and formation water), oil-contaminated sites, or warm water springs are rich sources for the isolation of microbes (Cooper et al., 1981; Datta et al., 2018; Sharma et al., 2019). Screening of biosurfactants for the oil recovery is done based on their surface activity measurements, and sometimes oil separating ability could be the screening criteria (Rabiei et al., 2013). The literature has explored the combined effects of two techniques that are predicted as the key mechanisms of enhanced oil recovery (EOR): wettability alteration and IFT decrease (Crescente et al., 2006; Datta et al., 2022; Hirasaki & Zhang, 2004; Zhang & Austad, 2006).

In EOR schemes, surfactants showed great potential to recover the trapped oil. Surfactants decrease the irreducible oil saturation of the rock and widen the range of recoverable oil from the reservoir (Jørgensen, 2013). However, synthetic surfactants come with the drawback of environmental contamination. To overcome this problem, biosurfactants are prioritized over synthetic surfactants (Rabiei et al., 2013). Positioning against synthetic surfactants, biosurfactants have the benefits of lower toxicity, biodegradability, and environmental friendliness (Al-Wahaibi et al., 2014). Utilizing biosurfactants and bacteria that produce biosurfactants is an alternative clean method of EOR and remediating environmental contamination (Datta et al., 2020; Pacwa-Płociniczak et al., 2011). Microbial enhanced oil recovery (MEOR) is the phrase used to describe the use of either potential bacteria or their metabolites, such as biosurfactants, in the extraction of oil (Joshi et al., 2008). A microbe can be injected exogenous or can be utilized as indigenous depending on its metabolites or requirements of the culture conditions (substrate, pH, temperature, oxygen, salinity). To be

DOI: 10.1201/9781003407683-4

compatible with reservoir conditions, microbes are cultured between 45°C and 80°C, whereas some thermophiles can withstand temperatures near 120°C (Karray et al., 2016). Similar treatments are done for the pressure and salinity to test the compatibility with reservoir environments. Besides huge advantages, the practical application of biosurfactants in the fields is limited because of limited bulk production (Banat et al., 2010). *Bacillus* sp., *Pseudomonas*, and *Rhodococcus* sp. are the most generally used microbes strains to produce *in-situ* or *ex-situ* biosurfactants (Youssef et al., 2009). *B. subtilis* strain is the most often used bacteria for biosurfactant production because of the high surface activity of the lipopeptide group (Nitschke & Pastore, 2006). *B. subtilis* strains have been reported for producing biosurfactants that can reduce ST and IFT to 25–30 mN/m and 1 mN/m, respectively (Ghojavand et al., 2008).

This chapter focuses on the applications of biosurfactants for oil recovery. To test the suitability of a biosurfactant for EOR applications, surface properties like wettability alteration and reduction in IFT are primarily screened. The stability of isolated microbes or their metabolites is explored under reservoir conditions to confirm the practical applicability for EOR. Lab-scale core flooding experiments are performed using real core samples to simulate the actual field conditions. Various researchers have performed experimental and simulation investigations to estimate the impact of biosurfactants in oil recovery. Different biosurfactant solutions have been used as an injecting fluid in core flooding experiments against different types of rock or core samples. Biosurfactants were found efficient for oil recovery, but their efficiency can be further improved by injecting them synergistically with other compounds like chemical surfactants, alkalis, or polymers (Al-Sulaimani et al., 2012; Moradi et al., 2019).

4.2 PRODUCTION AND EXTRACTION OF BIOSURFACTANTS

Biosurfactants are generally low molecular weight compounds, and biosurfactants with higher molecular compounds can be classified as bioemulsifiers. Bioemulsifiers can be used as biopolymers, as they possess polymeric properties. Bioemulsifiers are used to improve macroscopic sweep efficiency (Varjani & Upasani, 2017). Biosurfactants can be classified based on their origin and ionic type (Figure 4.1). Glycolipids, rhamnolipids, sophorolipids, mannosylerythritol, and terehalolipids are the best-suited biosurfactants that reduce surface tension the most. These are the most commonly used biosurfactants because of their very high surface activity (Banat et al., 2010).

Based on the origin or spot of production, the MEOR process is classified as *in-situ* or *ex-situ*. In *in-situ* MEOR, biosurfactants are produced by microbes (indigenous or injected) in the reservoir after consumption of nutrients supplied to them. In contrast, in the case of *ex-situ* MEOR, biosurfactants are produced by cultivating microbes in laboratories under controlled conditions, which are applied for oil recovery. Many potential microbes have been isolated for the production of biosurfactants. The properties of biosurfactants are also found to vary according to their microbial origin (Morita et al., 2016). Various potential microbial strains are used for the production (fermentation) of biosurfactants. The produced biosurfactants are

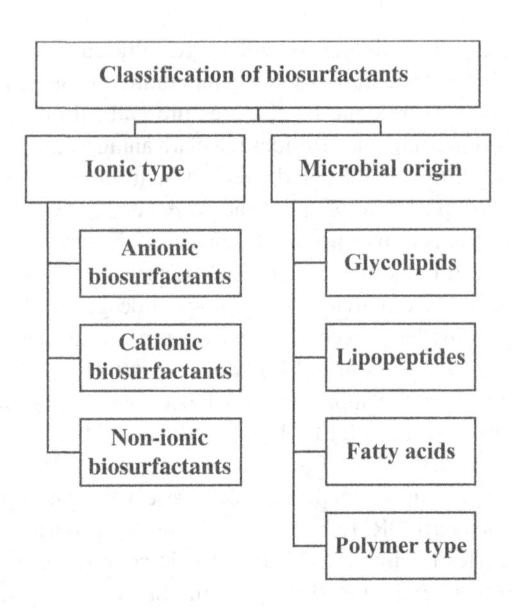

FIGURE 4.1 Classification of biosurfactants.

separated from the microbes using a centrifuge and extracted by acid precipitation and/or solvent extract.

Several analytical methods are used to characterize the produced biosurfactants. Other enhanced characterization methods include nuclear magnetic resonance (NMR), Fourier transform infrared spectroscopy (FTIR), and high-performance liquid chromatography (HPLC) (Datta et al., 2018; Mondal et al., 2017; Nitschke & Pastore, 2006; Sharma et al., 2019; Varjani & Upasani, 2017). FTIR analysis of the produced biosurfactant reveals the presence of aliphatic hydrocarbons and peptides, which verify the produced metabolites as a surfactant. Peaks near about 2800–2900 cm^{-1} indicate the existence of methyl or methylene group, and peaks at 1602 cm^{-1} confirm the N-H group. Peaks at 1250 cm^{-1} to 1700 cm^{-1} refer to the presence of C-H bending and C-O group of aliphatic chain. Certain groups like alkoxy, alkene, and aromatic can be found from 750 cm^{-1} to 1150 cm^{-1} peaks. NMR spectra were investigated for ^{1}H and ^{13}C. Peaks obtained from 14–30 ppm in ^{13}C spectra indicate the presence of sp^3, $-CH$, $-CH_2$, and $-CH_3$ aliphatic or amide groups. Peak obtained nearby 177 ppm points the carboxylic acid or amide groups. For ^{1}H spectra, aliphatic peaks are obtained in the range of 0.75–1.34 ppm (Sharma et al., 2019). The surface-active properties of the produced biosurfactants are tested in terms of critical micelle concentration (CMC), wettability, emulsification index, and IFT. The CMC value of the biosurfactants signifies the agglomeration of biosurfactants to form micelles (Batista et al., 2006). Sometimes the drop-collapse test is also conducted to differentiate between the bioemulsifiers and biosurfactants.

One of the biosurfactant's best advantages is no energy consumption like thermal processes (Elshafi, 2011), yet the economics of the production process limit their bulk production. Typically, substrate costs contribute about 30–40% of the entire production expenses. The benefits of microbial production of biosurfactants are the possible usage of cheap substrates, such as sugar, glycerol, and molasses. The by-products of agro-industry present either in solid or liquid form and rich in carbon source are compatible with the substrate for biosurfactant production (Makkar & Cameotra, 1997). Many researchers are working on cheaper raw materials like molasses, corn syrup, and agro-products for biosurfactant production (Al-Bahry et al., 2013; Joshi et al., 2008; Varjani & Upasani, 2017). The key challenge in biosurfactant production is choosing a suitable substrate that contains the right balance of nutrients, which promotes optimized cell growth and production of desired metabolites(s).

4.3 STABILITY OF BIOSURFACTANTS

For the practical oil recovery application, a biosurfactant and/or producer microbes must withstand reservoir conditions. The reservoir is generally at 70–90°C temperature and pressure higher than 3000–15000 psi depending upon the reservoir depth and geographical location. Hence, the produced biosurfactants are characterized for thermal, pH, salinity, and pressure stability, as depicted in Figure 4.2 (Phukan et al., 2019a; Phukan et al., 2019c; Saha et al., 2018). The stability tests are carried out before and after ageing (Datta et al., 2018). Thermogravimetric analysis (TGA) is performed for thermal stability. TGA estimates the relation between mass decomposition under the effect of a specific heating rate. An insignificant reduction in mass depicts the stability of the selected samples (Saha et al., 2018). Thermostability was observed by Datta et al. (2018) for the biosurfactant (lipopeptide) produced by *B. subtilis*, and negligible mass loss was observed up to 150°C. Foam stability determines

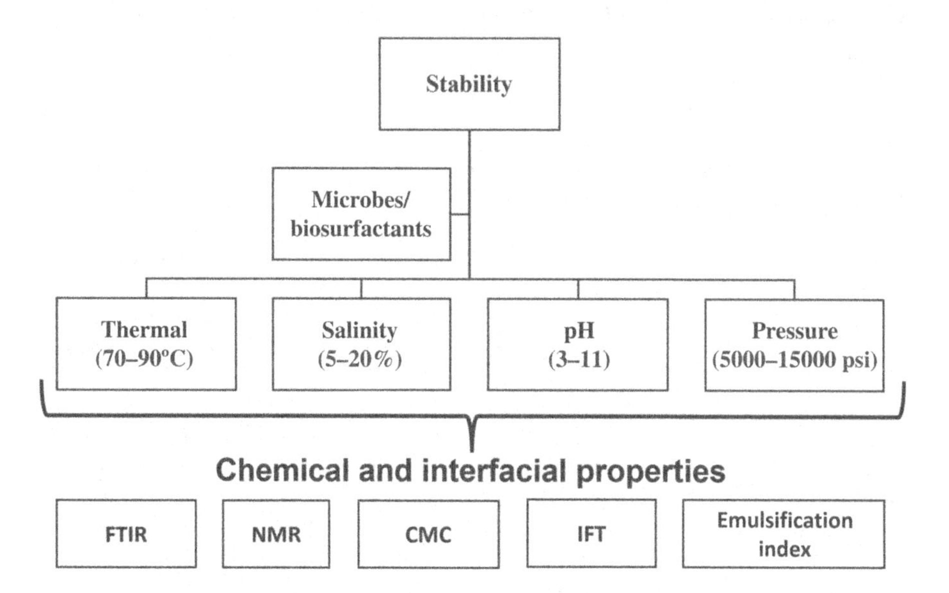

FIGURE 4.2 A schematic representation of thermal, salinity, pH, and pressure stability testing of the biosurfactants.

the foaminess or foamability of any surfactant. It includes the foam volume formed under specific conditions and the time duration for which that foam stays. The foaming ability coefficient relates to these factors, and correlation is reported by Phukan et al. (2019b). The higher the foaming ability coefficient, the greater will be the stability of the surfactant.

The functional groups of biosurfactants are characterized by FTIR analysis before and after pH, salinity, pressure, and thermal stability tests (Elshafi, 2011). Aged and non-aged samples were kept at different pH and incubated for multiple days at reservoir temperature (Al-Bahry et al., 2013; Joshi et al., 2008). For the salinity test, NaCl is added to the biosurfactant at varying concentrations of 0–15% (w/v). The produced biosurfactants have been found to be quite stable till 121°C and in the pH range of 2.5–12.0 while salt concentration up to 20% NaCl as listed in Table 4.1. The change in characteristics and functional groups of aged and non-aged samples determines the stability of the surfactant. A reservoir-compatible and stable surfactant does not show variations in physical, chemical, and interfacial properties for aged and non-aged samples (Bordoloi & Konwar, 2008; Makkar & Cameotra, 1997). Biosurfactants showed high stability when pH was varied from 6.0 to 12.0, and high surface tension and IFT were observed at pH between 2.0 and 4.0 due to precipitation (Al-Bahry et al., 2013). Distorted structure and precipitation cause a reduction in the capability of biosurfactants to reduce surface tension. Instability was observed by Gudiña et al. (2013) and others at lower pH values due to presence of negative ions (Batista et al., 2006; Ghojavand et al., 2008). *B. tequilensis* MK 729017

TABLE 4.1
Comparative Stability Studies of Biosurfactants Produced by Different Isolates

Isolate	Ageing Time (days)	Temperature (°C)	Salt (%)	pH	Surface Tension (mN/m)	Reference
B. tequilensis MK 729017	1	80	5	3.0–11.0	30 ± 2	Datta et al., 2020
P. aeruginosa MTCCC7815	5	45	NA	3.0–10.0	26.2 ± 0.5	Sharma et al., 2019
B. subtilis B30	9	121	5	6.0–12.0	26.63 ± 0.45	Al-Bahry et al., 2013
P. aeruginosa (MTCC7815)	4	100	NA	2.5–11.0	36.2 ± 0.3	Bordoloi & Konwar, 2008
P. aeruginosa (MTCC7812)	4	100	NA	2.5–11.0	38.4 ± 0.5	Bordoloi & Konwar, 2008
B. subtilis R1	9	80	7	6.0–12.0	33	Joshi et al., 2008
Bacillus HS3	9	80	7	6.0–12.0	36	Joshi et al., 2008
B. subtilis LB5a	2	100	20	–	26.6	Nitschke & Pastore, 2006
B. subtilis MTOC 2423	NA	100	NA	4.5–10.5	40	Makkar & Cameotra, 1997

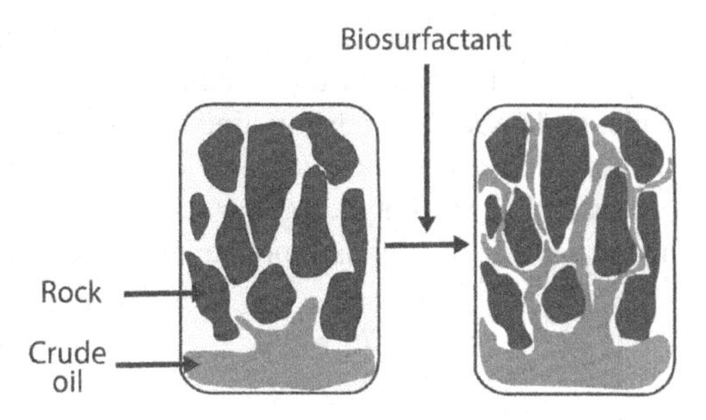

FIGURE 4.3 Mechanism of oil recovery by biosurfactants (image source: Pacwa-Płociniczak et al., 2011).

acquired from North-India reservoir soil (Assam) was aged for 24 h at 80°C by Datta et al. (2020), who also measured the ST before and after the thermal exposure. The identical ST measurements showed that the generated biosurfactants were thermally stable.

4.4 MECHANISM OF OIL RECOVERY

Biosurfactants lower the intermolecular tensions between two immiscible liquids by adsorbing at their contact. This is similar to how synthetic surfactants work, the method for reducing surface tension or IFT (Kong et al., 2017). For oil recovery utilizing biosurfactants, wettability change, and interfacial reduction are the most crucial mechanisms and governing elements. Biosurfactants improve microscopic sweep efficiency by lowering the IFT between oil and injecting fluid. This reduces the amount of residual oil saturation (Jørgensen, 2013). Biosurfactants lower the capillary forces, which increases the volume of trapped oil that may be recovered. High capillary forces cause oil to become stuck between pores. Biosurfactants bind to the oil-water contact to create emulsions. Together with the injected fluid, the formed emulsion stabilizes the desorbed oil and makes it easier to remove. Oil can travel through the pores of rock due to reduced capillary forces (Figure 4.3) (Kong et al., 2017; Pacwa-Płociniczak et al., 2011).

Another substantial oil recovery mechanism is the wettability alteration. Altered wettability correlates with the IFT and shows the wetting ability of the biosurfactants (Datta et al., 2018). Wettability alteration between rock surface and injected fluid can be visualized using the contact angle measurement. Surface-active agent wettability has been discovered to be both hydrophilic and hydrophobic (Saha et al., 2018; Singh et al., 2021).

4.5 PARAMETERS AFFECTING THE OIL RECOVERY FACTOR

An essential aspect of the successful implementation of biosurfactant flooding on any reservoir is screening appropriate surfactants. Parameters such as pH, pressure drop, wettability alteration, IFT reduction, and viscosity reduction are some of the controlling elements for improved oil recovery (Marie, 2010; Maudgalya et al., 2007).

The oil recovery factor, which can be divided into viscous force and capillary force, can be greatly influenced by all of the factors. Reducing capillary forces helps to mobilize trapped residual oil by reducing the binding forces that hold it in place. IFT must be decreased in order to retrieve the remaining oil, which increases the capillary number. The dimensionless quantity, capillary number *Nc*, is used to correlate the viscous force and the capillary forces as expressed by Eq. (4.1) (Jørgensen, 2013).

$$Nc = \frac{Fv}{Fc} = \frac{\upsilon\mu}{\sigma} \tag{4.1}$$

Here, *Fv* is the viscous forces, *Fc* is the capillary force, υ is the velocity of the displacing fluid, μ is the viscosity, and σ is the IFT between two phases. The capillary number must exceed the critical value for residual oil recovery. It can be either done by increasing mobilizing viscous force or decreasing capillary binding forces. Other parameters like porosity, permeability, and metabolites affect the *in-situ* and *ex-situ* modes of MEOR (Sarafzadeh et al., 2013). Before flooding, wettability modification and IFT reduction could be carried out to determine the bio-potential surfactants for oil extraction.

4.5.1 Surface Tension and IFT Reduction

The relationship between macroscopic sweep efficiency and microscopic sweep efficiency can be used to calculate the oil extraction factor. By infusing viscous fluid and displacing the remaining oil, macroscopical sweep efficiency might be increased. The efficiency of the microscopic sweep can be increased by lowering the remaining oil saturation (Jørgensen, 2013). Thus, the easiest and most accurate approach to evaluate surface activity is to measure surface tension and IFT. Numerous research has been carried out to find out the effect of IFT reduction on oil recovery (Afrapoli et al., 2010; Al-Sulaimani et al., 2012; Donaldson & Obeida, 1980; Pannu et al., 2022; Rabiei et al., 2013). For experiments with no incubation period, IFT reduction was dominating, and similar results were obtained at different conditions. As the incubation period increases, the dominance of wettability alteration is enhanced, and the impact of IFT reduction decreases. Therefore, the IFT reduction phenomenon is the determining factor for core flooding experiments with no incubation period.

4.5.2 Wettability Alteration

The exact mechanism behind the wettability alteration is not explicably recognized. The best-fitting processes, however, include the adherence of biosurfactants to surfaces, the formation of biofilms, and the adsorption of components from bacterial solutions (Crescente et al., 2006; Karimi et al., 2012). There are various methods to conduct the wettability alteration test. Contact angle (CA) measurement is the most generally used method to perform the wettability test. Reduction in CA demonstrates the conversion of oil-wet rock to water-wet rock, which shows a decrease in irreducible oil saturation (Afrapoli et al., 2009). Rabiei et al. (2013) conducted wettability tests and concluded that wettability is a time-consuming phenomenon and can be seen only in incubated or shut-in experiment set-ups. Afrapoli et al. (2009), Karimi et al. (2012), and Negin et al. (2017) also proved wettability alteration as the primary mechanism to enhance the efficiency of biosurfactants. Karimi et al. (2012) also observed the presence of the cells in broth helps in altering the wettability significantly, thus improving the oil extraction.

Afrapoli et al. (2009) and Crescente et al. (2006) calculated wettability indices and their impact on oil recovery from surfactant producing and non-surfactant producing bacteria. Imbibition and drainage are the wettability-driven processes that depend upon the capillary pressure. The decrease in capillary pressure during the experiment signifies the dominance of the wettability alteration mechanism. Afrapoli et al. (2009) successfully showed that the adsorption of components from bacterial solution and bacterial mass attaching to the surface of grain greatly impact the efficiency of wettability. The wettability alteration mechanism was more suitable for the core having higher adsorption or rough surface. Wettability shows the proportional relation with the shut-in period. Higher oil recovery through wettability alteration is strongly supported by the shut-in period (Bao et al., 2009).

4.5.3 Viscosity Reduction, Salinity, Pressure Drop, and pH

Viscosity reduction, salinity, pressure drop, and pH focus on various aspects of the oil recovery mechanism. Viscosity reduction improves the mobility of the oil, whereas pressure maintenance ensures the gas remains trapped in the oil, enhancing gas drive and reducing oil viscosity. Salinity and pH help in improving the favourable conditions for the microbes. Several tests have been conducted by various researchers (Crescente et al., 2006; Donaldson & Obeida, 1980; Rabiei et al., 2013; Sarafzadeh et al., 2013; Zhang & Austad, 2006) to demonstrate the impact of viscosity reduction, salinity, pressure drop, and pH. Variations in pressure can be observed during selective plugging. Pressure drop could also be beneficial for determining the selective plugging efficiency during polymer injection or biopolymer production during MEOR. Salinity and pH could also be functions of acid generation during MEOR. Change in salinity shows some effect on the oil recovery as salinity affects the ST and IFT between oil and water (Zhang & Austad, 2006). No direct effect of salinity was observed on oil recovery, whereas salinity significantly affects microbial growth. A change in IFT caused by the change in salinity eventually causes a change in the oil recovery factor (Al-Bahry et al., 2013). Similarly, pH does not affect oil recovery directly, but pH variation can cause significant changes in microbial growth and biosurfactant production (Elshafi, 2011).

4.6 EXPERIMENTAL INVESTIGATIONS OF BIOSURFACTANT-BASED OIL RECOVERY

Majority of the matured oilfields in the world are at the stage of EOR with the consideration of implementing the EOR scheme to be economically compatible (Rabiei et al., 2013). Biosurfactants ease the recovery process and help in extracting inaccessible oil (Alvarado & Manrique, 2010). Reservoir-like conditions could be formed using a sand-packed column at the laboratory scale. Parameters like porosity and permeability are decided by the size of the sand particles in the column (Gudiña et al., 2013). Biosurfactants reduce the oil binding forces and recover the trapped oil from the pore throat. The viscous forces mobilize oil droplets trapped in pores, and the correlation between viscous and ST forces is controlled by capillary number. The actual performance of a biosurfactant could only be estimated by implementing the surfactant flooding in the reservoir, but economically this is not feasible. Initially, experiments are conducted at a laboratory scale on the core/sand samples to replicate the reservoir conditions to a great extent. However, the heterogeneity of rock, damage zones, and fault surfaces are some of the factors that are difficult to reproduce (Caine et al., 1996; Treffeisen & Henk, 2020).

4.6.1 Core Flooding Studies on Sand Pack and Model Columns

Joshi et al. (2008) conducted flooding experiments on sand pack columns using biosurfactants produced from *B. licheniformis* K51, *B. subtilis* R1, *B. subtilis* 20B, and *Bacillus* strain HS3. Broth prepared from isolates was passed through the sand pack and around 25–33% of the residual oil left after brine flooding was recovered. Bordoloi and Konwar (2008) conducted several flooding experiments under laboratory conditions and observed oil recovery as a function of temperature. Selected microbes were kept in different carbon source mediums to assess their surface tension-reducing properties. They found that oil recovery from the *in-situ* biosurfactant is significantly lower than that of the *ex-situ* mode. *In-situ* cultured microbes could not survive higher than room temperature, whereas *ex-situ* biosurfactant outperformed at 90°C than 70°C. The highest recovery of 62% was obtained at 90°C, 8–12% higher than the oil recovered at 70°C. Suthar et al. (2008) conducted *ex-situ* flooding experiments on sand pack columns with different strains of the *B. licheniformis*. An additional oil recovery (AOR) of 43.1% was obtained from bioemulsifier metabolites of *B. licheniformis*. Bioemulsifiers showed higher stability and lower IFT. Bioemulsion can keep the water-oil emulsion stable and possess the properties of the surfactant-polymer mixture. The surfactant portion kept the IFT low, and the polymer portion helped in mobilizing the oil.

Gudiña et al. (2013) performed *in-situ* MEOR experiments with four different oil samples, heating oil, viscous paraffin, and two samples of heavy crude oils using *Bacillus* strains, incubated in the sand pack and the nutrients for 14 days at 40°C. AOR of 4.8% to 21.7% was obtained for different oil samples. The differences in AOR values were probably due to the viscosity and API gravity difference. Selected microbes were found to be good candidates for heavy crude oil recovery.

Fractured reservoirs are very complex to find the suitability of any method for enhanced oil recovery. Soudmand-asli et al. (2007), Zekri and Almehaideb (2003), and Zekri et al. (2003) simulated the flooding experiments on fractured samples to quantify the effect of biosurfactants on fractured reservoirs. Soudmand-asli et al. (2007) used glass micromodels of different fractured types to perform flooding with *Leuconostoc mesenteroides* and *B. subtilis* microbes. The model was incubated for four days to quantify the wettability alteration impact, and the oil recovery achieved in different cases is summarized in Table 4.2. The results demonstrate that higher oil recovery was obtained for *B. subtilis*.

TABLE 4.2
Oil Recovery Attained from Fractured Micromodel (Soudmand-asli et al., 2007)

Experiment No.	Bacteria Used	Fractured Angle Orientation	Oil Recovery (%)
1	*B. subtilis*	Non-fractured	23.1
2	*B. subtilis*	Inclined-fractured	29.9
3	*B. subtilis*	Vertical-fractured	28.8
4	*B. subtilis*	Horizontal-fractured	3.2
5	*L. mesenteroides*	Non-fractured	21.2
6	*L. mesenteroides*	Inclined-fractured	12.1
7	*L. mesenteroides*	Vertical-fractured	16.6
8	*L. mesenteroides*	Horizontal-fractured	0

4.6.2 Core Flooding Studies on the Carbonate Core Sample

Sandstone reservoirs are generally water-wet, whereas carbonate reservoirs are generally oil-wet reservoirs. Zekri et al. (2003) conducted wettability alteration tests on the carbonate rock samples. Bacteria from the *Bacillus* family were procured from the warm water springs presented locally. Beef extract and yeast were chosen as the substrate medium for the growth of microbes. The concentration of bacteria increased from 0 to 60×10^3 cells/mL at a constant salinity of 5% NaCl. It was noticed that with the elevation in bacteria concentration, CA decreases up to a specific concentration only, beyond which no changes in contact angle were observed. The formation of rings by the microbes and partition into oil and adsorption at the interface decreases the contact angle. Variation in rock type did not show any clear trend between contact angle and rock type. However, some changes were observed due to mineralogy.

Sarafzadeh et al. (2013) performed *in-situ* and *ex-situ* flooding observations on the dolomite-rich core samples. *Enterobacter cloacae*, a facultative biosurfactant producing strain, was chosen for biosurfactant production. Flooding experiments were performed in a stage-wise manner to weigh the impact of IFT and wettability alteration with the hypothesis that IFT reduction is an instantaneous phenomenon, whereas wettability alteration is a time-consuming phenomenon. In the former stage core was injected with the biosurfactant, and in the latter stage shut-in period of one week was provided. In the *in-situ* operation, the first stage flooding recovered 4% of the residual oil due to a reduction in IFT, while an additional 6.3% recovery in the second stage was due to a change in the wettability during the incubation period. A similar type of trend was also obtained for the *ex-situ* flooding, 4% in the first stage and 14% in the secondary stage. The increased recovery in the secondary stage of the *ex-situ* process was due to the improved wettability alteration and minimum residual oil saturation. In both methods of flooding, biosurfactants show the potential to recover the residual oil.

Using smart water injection and a plant-based natural surfactant called *Tribulus terrestris*, Moradi et al. (2019) carried out core flooding trials on carbonate rock. The combined impacts of IFT reduction (surfactant) and wettability alteration increased the ultimate oil extraction from 45.2% to 71.8% with the hybrid surfactant and smart water solution (smart water).

4.6.3 Core Flooding Studies on the Sandstone Core Sample

Four sandstone core samples with slight variations in physical properties were subjected to biosurfactant-based EOR experiments by Bao et al. (2009). A nutrition activator along with water was injected, and core samples were incubated for 21, 27, 36, and 48 days and AOR values were reported as 1.60%, 2.48%, 6.82%, and 9.14%, respectively. The results revealed that microbes present in the reservoir require more activation than the exogenous microbes. The synergy between indigenous and exogenous microbes depends on the reservoir conditions and air adaptation between aerobic and anaerobic microorganisms. Elshafi (2011) conducted *ex-situ* core flooding experiments using biosurfactants produced from the *B. licheniformis and B. subtilis* strains. The produced biosurfactant was flooded against the sandstone core sample of porosity 28% and permeability of 300 mD. Biosurfactant was injected in both concentrated and diluted forms. Initial water flooding recovered 38% of OOIP, and biosurfactant improved the recovery factor by 10%. Further injection of concentrated biosurfactant enhanced the recovery factor by 13% due to improved oil mobility. Excess surfactant eliminates the adsorption factor and helps in mobilizing the oil.

Al-Sulaimani et al. (2012) organized multiple core flooding experiments and evaluated the effect of synergy of synthetic surfactant and biosurfactant. The synergy of both surfactants with an appropriate mixing ratio outperformed the single surfactant results (Table 4.3). Maximum oil recovery was obtained at an equal concentration of both surfactants. The combination of both surfactants causes a change in the structure of molecules at the interface, which improves the recovery factor drastically. Rabiei et al. (2013) conducted core flooding experiments on the sandstone core samples under *in-situ* and *ex-situ* flooding conditions. Volumetric sweep efficiency was considered as the controlling factor for the oil recovery factor. For the *in-situ* mode of flooding experiment, no incubation period (quick *in-situ* flooding) and 14 days incubation resulted in oil recovery of 37.5% and 48.5% of OOIP, respectively. An identical course of action was followed for the *ex-situ* mode of flooding,

TABLE 4.3
Oil Recovery Factor at Different Mixing Ratios

	Mixing Ratio (%)		
Sl. No.	Biosurfactant	Chemical Surfactant	Recovery Factor (%)
1	100	0	23
2	0	100	30
3	75	25	46
4	50	50	50
5	25	75	39

and oil recovery of 40.5% for quick flooding and 45% for incubated flooding were marked. The synergy of both wettability alteration and reduction in IFT for the incubated *in-situ* experiment contributed to higher oil recovery.

A similar order of oil recovery has also been cited by other researchers (Joshi et al., 2008; Thomas et al., 1993; Yakimov et al., 1997; Youssef et al., 2009a). Various studies demonstrating the impact of the addition of biosurfactants on EOR in various porous media is summarized in Table 4.4. Other than laboratory experiments, some reservoir fields have been operated with biosurfactants. Table 4.5 demonstrates the case studies of the implementation of MEOR in reservoir fields.

TABLE 4.4
Biosurfactant Addition Effect on Additional Oil Recovery in Various Porous Media

Isolate	IFT (mN/m)	Core Sample	Flooding Conditions	Recovery/ AOR (%)	Reference
B. subtilis B30	3.79	Berea sandstone cores	7–9% NaCl 7 pH 60°C temperature	17–26 (light oil), 31 (heavy oil)	Al-Wahaibi et al., 2014
B. subtilis strain W19	3.28	Berea core plugs	12 ppt 60°C temperature	50	Al-Sulaimani et al., 2012
Engineered strains of *P. aeruginosa* and *Escherichia coli*	0.07	Sand-packed column	pH 52% NaCl	50	Wang et al., 1996
Pseudomonas strains	27*	Sand-packed column	7.0 pH 45°C temperature	64	Das & Mukherjee, 2005
Anaerobic enrichments from high temperature oil	–	Sand-packed column	90°C temperature	22	Banwari et al., 2005
B. mojavensis JF-2	–	Sand packed	5% NaCl	22	Maudgalya et al., 2004
Mixture strains of *Bacillus*	–	Fractured limestone cores	0–50,000 ppm salinity	6–10	Zekri et al., 2003
B. subtilis and *Pseudomonas* strains	0.052	Limestone packed column	75°C temperature	5–10	Wang et al., 2012
Anaerobic bacteria mixture	27.69*	Artificial stuff sand core flooding	5.0–7.0 pH 45°C temperature	12.1	Han et al., 2001
Thermophilic bacterial mixtures obtained from UAE water tanks	0.07	Carbonate rock	6% salinity 50°C temperature	15–20	Zekri et al., 1999

*Refers to surface tension values.

TABLE 4.5
Effect of Biosurfactant Injection on Recovery of Oil from Different Oilfields

Microbes	Oil Property	Oilfield	Impact on Oil Recovery	Reference
Indigenous microbe strain of *Bacillus* XJ2-1, *P. aeruginosa* XJ3-1	Density of 0.956 7 g/cm^3, viscosity of 19.683 mPa·s	Chunfeng Oilfield, China,	3464 tonnes of oil produced	Wang et al., 2016
Pseudomonas	Oil viscosity 145 mPa·s	Xinjiang Oilfield, China	1872 tonnes of heavy oil produced	Chai et al., 2015
B. licheniformis RS-1, *B. spizizenii* NRRL B-23049	NA	Viola limestone formation, Oklahoma	52.5 m^3 of oil produced	Youssef et al., 2013
Inherent microorganisms	NA	Providencia and Lobitos fields, Northwest Peru	19,410 and 13,907 bbl oil produced	Bybee, 2006
Anaerobic facultative microbes	Oil API gravity 29°	Vizcacheras field, Argentina	5887 m^3 of oil produced	Strappa et al., 2004
Thermophilic NG80 microbe	Dead oil density is 0.8801 g/cm^3, viscosity 6.8 mPa·s	G69 block, Northern China	13.8% additional oil recovery	Feng et al., 2007
P. aeruginosa (P-1)	NA	Daqing Oilfield, China	11.2% enhanced oil recovery	Li et al., 2002
Stimulation of indigenous bacteria	Oil API gravity 25°	Trinidadian Oil Well, Trinidad and Tobago	Increased oil recovery	Maharaj et al., 1993

4.6.4 Simulation Flooding Experiments Using Biosurfactants

Simulators hold the upper edge of greater flooding hours and can be simulated over real life-size reservoirs. Maudgalya et al. (2007) conducted experimental and simulated core flooding studies using the bacteria *B. mojavensis.* A mathematical model was prepared to replicate the flooding experiment. Mathematical equations were developed for the relations between IFT, surfactant concentration, salinity, viscosity, etc. The prepared model was initially verified against the Buckley-Leverett 1D model, and the simulated model was able to produce identical results as obtained in experiments during biosurfactant core flooding.

Kowalewski et al. (2006) tested the impact of wettability alteration and IFT reduction on recovery using biosurfactants through the reservoir simulator 'Eclipse'. Simulated core flooding was validated against the Bentheimer sandstone core flooding experiment. A sharp decrease in the IFT was reported after the injection of microbes. Oil production from experimental flooding obtained was almost identical to the simulated oil production. Change in the relative permeability was also observed, which represents the plugging in pores because of microbes.

Nielsen et al. (2010) constructed a 1D Compositional Streamline Simulator (CSL) for MEOR simulations and later modified it to 3D. The model that was created was based on the IMPEC (implicit pressure and explicit composition). The transport and reaction equations along the streamline can be solved by IMPEC. Microbes expanded, ate nutrients, and produced metabolites or surfactants as part of the chemical reactions. *Ex-situ* mode of injection was opted for, and the presence of no indigenous organism was considered. Monod expression from enzyme reaction kinetic was used for the growth rate expression.

Hosseininoosheri et al. (2016) successfully conducted the core flooding experiment in the reservoir simulator 'UTCHEM' considering chemical reactions. The primary focus was to model the kinetic equations considering all the variations that alter the growth of microbes and biosurfactant production. Hosseininoosheri et al. (2016) developed a comprehensive model to consider all the variables and replicate the core flooding experiment. Microbes were injected for 1000 days after 1000 days of water flooding. The reduction in IFT against biosurfactant concentration was observed and compared with Youssef et al. (2009). A further oil recovery of 15.4% was observed due to biosurfactants. The outcome of nutrient concentration, temperature, and biosurfactant adsorption was also observed. Salinity and pH also affected microbe growth, which ultimately affected biosurfactant production and the oil recovery factor.

4.7 CONCLUSIONS

The extensive research conducted has proven the biosurfactants to be potential candidates for EOR. Biosurfactant-based EOR has been recognized as the most assuring and progressive method to recover residual oil. Oil recovery factors can differ depending on the methods, materials, and flooding conditions. Properties like thermal stability, emulsification, surface activity, wetting, foaming, and viscosity reduction make biosurfactants a suitable candidate for oil recovery. Microscopic sweep efficiency is improved by reduced capillary forces, and high surface tensions are eliminated by reduced IFT and wettability alteration. Improvement in oil recovery using bioemulsifiers and selective plugging can also help in improving oil recovery by blocking high permeable zones or by reducing fingering. Besides all the lab-scale experiments, biosurfactant flooding has already been implemented over various reservoirs from countries like the United States, India, Malaysia, Germany, Peru, China, Australia, Russia, etc. Optimized biosurfactant production and growth from the feasible substrate would help the residual oil recovery process be more economically viable and profitable.

REFERENCES

Afrapoli, M. S., Alipour, S., & Torsaeter, O. (2010). Effect of Wettability and Interfacial Tension on Microbial Improved Oil Recovery with Rhodococcus sp 094. *SPE 129707. Improved Oil Recovery Symposium*, Tulsa, Oklahoma, USA. https://doi.org/10.2118/129707-MS

Afrapoli, M. S., Crescente, C., Alipour, S., & Torsaeter, O. (2009). The effect of bacterial solution on the wettability index and residual oil saturation in sandstone. *Journal of Petroleum Science and Engineering*, *69*(3–4), 255–260.

Al-Bahry, S. N., Al-Wahaibi, Y. M., Elshafie, A. E., Al-Bemani, A. S., Joshi, S. J., Al-Makhmari, H. S., & Al-Sulaimani, H. S. (2013). Biosurfactant production by *Bacillus subtilis* B20 using date molasses and its possible application in enhanced oil recovery. *International Biodeterioration and Biodegradation*, *81*, 141–146.

Al-Sulaimani, H., Al-Wahaibi, Y., Al-Bahry, S., Elshafle, A., Al-Bemani, A., Joshi, S., & Zargari, S. (2011). Optimization and Partial Characterization of Biosurfactants Produced by Bacillus Species and Their Potential for Ex-situ Enhanced Oil Recovery. *Society of Petroleum Engineering Journal*, 16 (03): 672–682. https://doi.org/10.2118/129228-PA

Al-Sulaimani, H., Al-Wahaibi, Y., Ai-Bahry, S., Elshafie, A., Al-Bemani, A., Joshi, S., & Ayatollahi, S. (2012). Residual-oil recovery through injection of biosurfactant, chemical surfactant, and mixtures of both under reservoir temperatures: Induced-wettability and interfacial-tension effects. *SPE Reservoir Evaluation and Engineering*, *15*(2), 210–217.

Alvarado, V., & Manrique, E. (2010). Enhanced oil recovery: An update review. *Energies*, *3*(9), 1529–1575.

Al-Wahaibi, Y., Joshi, S., Al-Bahry, S., Elshafie, A., Al-Bemani, A., & Shibulal, B. (2014). Biosurfactant production by *Bacillus subtilis* B30 and its application in enhancing oil recovery. *Colloids and Surfaces B: Biointerfaces*, *114*, 324–333.

Banat, I. M., Franzetti, A., Gandolfi, I., et al. (2010). Microbial biosurfactants production, applications and future potential. *Applied Microbiology and Biotechnology*, *87*(2), 427–444.

Banwari, L. A. L., Reedy, M. R. V., Agnihotri, A., Kumar, A., Sarbhai, M. P., Singh, M., & Misra, T. R. (2005). A process for enhanced recovery of crude oil from oil wells using novel microbial consortium. *WO2005005773A2*.

Bao, M., Kong, X., Jiang, G., Wang, X., & Li, X. (2009). Laboratory study on activating indigenous microorganisms to enhance oil recovery in Shengli Oilfield. *Journal of Petroleum Science and Engineering*, *66*(1–2), 42–46.

Batista, S. B., Mounteer, A. H., Amorim, F. R., & Tótola, M. R. (2006). Isolation and characterization of biosurfactant/bioemulsifier-producing bacteria from petroleum contaminated sites. *Bioresource Technology*, *97*(6), 868–875.

Bordoloi, N. K., & Konwar, B. K. (2008). Microbial surfactant-enhanced mineral oil recovery under laboratory conditions. *Colloids and Surfaces B: Biointerfaces*, *63*(1), 73–82.

Bybee (2006). MEOR in northwest Peru. *Journal of Petroleum Technology*, *58*(1), 48–49.

Caine, J. S., Evans, J. P., & Forster, C. B. (1996). Fault zone architecture and permeability structure. *Geology*, *24*(11), 1025–1028.

Chai, L. J., Zhang, F., She, Y. H., Banat, I. M., & Hou, D. J. (2015). Impact of a microbial-enhanced oil recovery field trial on microbial communities in a low-temperature heavy oil reservoir. *Nature Environment and Pollution Technology*, *14*(3), 455–462.

Cooper, D. G., Macdonald, C. R., Duff, S. J. B., & Kosaric, N. (1981). Enhanced production of surfactin from *Bacillus subtilis* by continuous product removal and metal cation additions. *Applied and Environmental Microbiology*, *42*(3), 408–412.

Crescente, C., Torsaeter, O., Hultmann, L., Stroem, A., Rasmussen, K., Kowalewski E. (2006). An Experimental Study of Driving Mechanisms in MIOR Processes by Using Rhodococcus sp. 094. *SPE 100033. PE/DOE Symposium on Improved Oil Recovery*, Tulsa, Oklahoma, USA. https://doi.org/10.2118/100033-MS

Das, K., & Mukherjee, A. K. (2005). Characterization of biochemical properties and biological activities of biosurfactants produced by *Pseudomonas aeruginosa* mucoid and non-mucoid strains isolated from hydrocarbon-contaminated soil samples. *Applied Microbiology and Biotechnology*, *69*(2), 192–199.

Datta, P., Pannu, S., Tiwari, P., & Pandey, L. (2022). Core Flooding Studies Using Microbial Systems. *Microbial Enhanced Oil Recovery: Principles and Potential, Green Energy and Technology*, 221–241. Springer Nature, ISBN. 9789811654657

Datta, P., Tiwari, P., & Pandey, L. M. (2018). Isolation and characterization of biosurfactant producing and oil degrading *Bacillus subtilis* MG495086 from formation water of Assam oil reservoir and its suitability for enhanced oil recovery. *Bioresource Technology, 270*, 439–448.

Datta, P., Tiwari, P., & Pandey, L. M. (2020). Oil washing proficiency of biosurfactant produced by isolated *Bacillus tequilensis* MK 729017 from Assam reservoir soil. *Journal of Petroleum Science and Engineering, 195*, 107612.

Donaldson, E. C., & Obeida, T. (1980). Ch. R-14 enhanced oil recovery at simulated reservoir conditions. In *Developments in Petroleum Science*, 31, 227–245. https://doi.org/10.1016/S0376-7361(09)70162-1

Feng, L., Wang, W., Cheng, J., et al. (2007). Genome and proteome of long-chain alkane degrading *Geobacillus thermodenitrificans* NG80-2 isolated from a deep-subsurface oil reservoir. *Proceedings of the National Academy of Sciences of the United States of America, 104*(13), 5602–5607.

Ghojavand, H., Vahabzadeh, F., Roayaei, E., & Shahraki A K. (2008). Production and properties of a biosurfactant obtained from a member of the *Bacillus* subtilis group (PTCC 1696). *Journal of Colloid and Interface Science, 324*, 172–176. https://doi.org/10.1016/j.jcis.2008.05.001

Gudiña, E. J., Pereira, J. F. B., Costa, R., Coutinho, J. A. P., Teixeira, J. A., & Rodrigues, L. R. (2013). Biosurfactant-producing and oil-degrading *Bacillus subtilis* strains enhance oil recovery in laboratory sand-pack columns. *Journal of Hazardous Materials, 261*, 106–113.

Han, P., Sun, F., & Shi, M. (2001). Microbial EOR laboratory studies on the microorganisms using petroleum hydrocarbon as a sole carbon source. *SPE-72128. SPE Asia Pacific Improved Oil Recovery Conference*, Kuala Lumpur, Malaysia.

Hirasaki, G. J., & Zhang, D. L. (2004). Surface chemistry of oil recovery from fractured, oil-wet, carbonate formations. *SPE Journal, 9*(2), 151–162.

Hosseininoosheri, P., Lashgari, H. R., & Sepehrnoori, K. (2016). A novel method to model and characterize *in-situ* bio-surfactant production in microbial enhanced oil recovery. *Fuel, 183*, 501–511.

Jørgensen, K. (2013). *Implementation of a Surfactant Model in MRST with Basis in Schlumberger's Eclipse*. PhD Thesis, Norwegian University of Science and Technology.

Joshi, S., Bharucha, C., Jha, S., Yadav, S., Nerurkar, A., & Desai, A. J. (2008). Biosurfactant production using molasses and whey under thermophilic conditions. *Bioresource Technology, 99*(1), 195–199.

Karimi, M., Mahmoodi, M., Niazi, A., Al-Wahaibi, Y., & Ayatollahi, S. (2012). Investigating wettability alteration during MEOR process, a micro/macro scale analysis. *Colloids and Surfaces B: Biointerfaces, 95*, 129–136.

Karray, F., Mezghani, M., Mhiri, N., Djelassi, B., & Sayadi, S. (2016). Scale-down studies of membrane bioreactor degrading anionic surfactants wastewater: Isolation of new anionic-surfactant degrading bacteria. *International Biodeterioration and Biodegradation, 114*, 14–23.

Kong, L., Saar, K. L., Jacquat, R., et al. (2017). Mechanism of biosurfactant adsorption to oil/water interfaces from millisecond scale tensiometry measurements. *Interface Focus, 7*(6). 20170013. https://doi.org/10.1098/rsfs.2017.0013

Kowalewski, E., Rueslåtten, I., Steen, K. H., Bødtker, G., & Torsæter, O. (2006). Microbial improved oil recovery-bacterial induced wettability and interfacial tension effects on oil production. *Journal of Petroleum Science and Engineering, 52*(1–4), 275–286.

Kreuter, J. (1996). Nanoparticles and microparticles for drug and vaccine delivery. *Journal of Anatomy, 189*(*Pt 3*), 503–505.

Li, Q., Kang, C., Wang, H., Liu, C., & Zhang, C. (2002). Application of microbial enhanced oil recovery technique to Daqing Oilfield. *Biochemical Engineering Journal, 11*(2–3), 197–199.

Maharaj, U., May, M., & Imbert, M. P. (1993). The application of microbial enhanced oil recovery to Trinidadian Oil Wells. *Developments in Petroleum Science,* 39, 245–263. https://doi.org/10.1016/S0376-7361(09)70065-2

Makkar, R. S., & Cameotra, S. S. (1997). Biosurfactant production by a thermophilic bacillus subtilis strain. *Journal of Industrial Microbiology and Biotechnology, 18*(1), 37–42.

Maudgalya, S., Corp, A. P., Knapp, R. M., & Mcinerney, M. J. (2007). SPE 106978 Microbial Enhanced – Oil – Recovery Technologies: A Review of the Past, Present, and Future. *SPE 106978. Production and Operations Symposium*, Oklahoma City, Oklahoma, USA. https://doi.org/10.2118/106978-MS

Maudgalya, S., McInerney, M. J., Knapp, R. M., Nagle, D. P., & Folmsbee, M. J. (2004). Development of bio-surfactant based microbial enhanced oil recovery procedure. *Proceedings – SPE Symposium on Improved Oil Recovery*, April 2004, 1–6.

Mondal, M. H., Sarkar, A., Maiti, T. K., & Saha, B. (2017). Microbial assisted (pseudomonas sp.) production of novel bio-surfactant rhamnolipids and its characterisation by different spectral studies. *Journal of Molecular Liquids, 242*, 873–878.

Moradi, S., Akbar, A., Bachari, Z., & Mahmoodi, H. (2019). Journal of petroleum science and engineering combination of a new natural surfactant and smart water injection for enhanced oil recovery in carbonate rock: Synergic impacts of active ions and natural surfactant concentration. *Journal of Petroleum Science and Engineering, 176*(January), 1–10.

Morita, T., Fukuoka, T., Imura, T., & Kitamoto, D. (2016). Glycolipid Biosurfactants. Reference Module in Chemistry, Molecular Sciences and Chemical Engineering. https://doi.org/10.1016/B978-0-12-409547-2.11565-3

Negin, C., Ali, S., & Xie, Q. (2017). Most common surfactants employed in chemical enhanced oil recovery. *Petroleum, 3*(2), 197–211.

Nielsen, S. M. (2010). Microbial Enhanced Oil Recovery – Advanced Reservoir Simulation. PhD Thesis, Technical University of Denmark.

Nielsen, S. M., Jessen, K., Shapiro, A. A., Michelsen, M. L., & Stenby, E. H. (2010). Microbial Enhanced Oil Recovery: 3D Streamline Simulation with Gravity Effects. *SPE-131048-MS. SPE EUROPEC/EAGE Annual Conference and Exhibition*, Barcelona, Spain. https://doi.org/10.2118/131048-MS

Nitschke, M., & Pastore, G. M. (2006). Production and properties of a surfactant obtained from *Bacillus subtilis* grown on cassava wastewater. *Bioresource Technology, 97*(2), 336–341.

Pacwa-Płociniczak, M., Płaza, G. A., Piotrowska-Seget, Z., & Cameotra, S. S. (2011). Environmental applications of biosurfactants: Recent advances. *International Journal of Molecular Sciences, 12*, 633–654.

Pannu, S., Phukan, R., & Tiwari, P. (2022). Experimental and simulation study of surfactant flooding using a combined surfactant system for enhanced oil recovery. *Petroleum Science and Technology, 40*, 2907–2924.

Phukan, R., Gogoi, S. B., & Tiwari, P. (2019a). Alkaline-surfactant-alternated-gas/CO_2 flooding: Effects of key parameters. *Journal of Petroleum Science and Engineering, 173*(September 2018), 547–557.

Phukan, R., Gogoi, S. B., & Tiwari, P. (2019b). Enhanced oil recovery by alkaline-surfactant-alternated-gas/CO_2 flooding. *Journal of Petroleum Exploration and Production Technology, 9*(1), 247–260.

Phukan, R., Gogoi, S. B., Tiwari, P., & Vadhan, R. S. (2019c). Optimization of immiscible alkaline-surfactant-alternated-gas/CO_2 flooding in an upper Assam oilfield.

SPE-195262. SPE Western Regional Meeting Proceedings, San Jose, California, USA. https://doi.org/10.2118/195262-MS

Pruthi, V., & Cameotra, S. S. (2003). Effect of nutrients on optimal production of biosurfactants by *Pseudomonas putida* – A Gujarat oil field isolate. *Journal of Surfactants and Detergents*, *6*(1), 65–68.

Rabiei, A., Sharifinik, M., Niazi, A., Hashemi, A., & Ayatollahi, S. (2013). Core flooding tests to investigate the effects of IFT reduction and wettability alteration on oil recovery during MEOR process in an Iranian oil reservoir. *Applied Microbiology and Biotechnology*, *97*(13), 5979–5991.

Saha, R., Uppaluri, R. V. S., & Tiwari, P. (2018). Effects of interfacial tension, oil layer break time, emulsification and wettability alteration on oil recovery for carbonate reservoirs. *Colloids and Surfaces A: Physicochemical and Engineering Aspects*, *559*(September), 92–103.

Sarafzadeh, P., Zeinolabedini, A., Ravanbakhsh, M., Niazi, A., & Ayatollahi, S. (2013). *Enterobacter cloacae* as biosurfactant producing bacterium: Differentiating its effects on interfacial tension and wettability alteration mechanisms for oil recovery during MEOR process. *Colloids and Surfaces B: Biointerfaces*, *105*, 223–229.

Sharma, S., Datta, P., Kumar, B., Tiwari, P., & Pandey, L. M. (2019). Production of novel rhamnolipids via biodegradation of waste cooking oil using *Pseudomonas aeruginosa* MTCC7815. *Biodegradation*, *30*(4), 301–312.

Singh, D., Roy, S., Jagat, H., & Phirani, J. (2021). Solid-fluid interfacial area measurement for wettability quantification in multiphase flow through porous media. *Chemical Engineering Science*, *231*, 116250.

Soudmand-asli, A., Ayatollahi, S. S., Mohabatkar, H., Zareie, M., & Shariatpanahi, S. F. (2007). The in situ microbial enhanced oil recovery in fractured porous media. *Journal of Petroleum Science and Engineering*, *58*(1–2), 161–172.

Strappa, L. A., De Lucia, J. P., Maure, M. A., & Lopez Llopiz, M. L. (2004). A novel and successful MEOR pilot project in a strong water-drive reservoir Vizcacheras Field, Argentina. *SPE-89456. SPE/DOE Symposium on Improved Oil Recovery*, Tulsa, Oklahoma, USA 1–24. https://doi.org/10.2118/89456-MS

Suthar, H., Hingurao, K., Desai, A., & Nerurkar, A. (2008). Evaluation of bioemulsifier mediated microbial enhanced oil recovery using sand pack column. *Journal of Microbiological Methods*, *75*(2), 225–230.

Thomas, C. P., Duvall, M. L., Robertson, E. P., Barrett, K. B., & Bala, G. A. (1993). Surfactant-Based EOR Mediated by Naturally Occurring Microorganisms. *SPE-22844. SPE Reservoir Engineering (Society of Petroleum Engineers)*, (United States), 285–291. https://doi.org/10.2118/22844-PA

Treffeisen, T., & Henk, A. (2020). Representation of faults in reservoir-scale geomechanical finite element models – A comparison of different modelling approaches. *Journal of Structural Geology*, *131*(November 2019), 103931.

Varjani, S. J., & Upasani, V. N. (2017). Critical review on biosurfactant analysis, purification and characterization using rhamnolipid as a model biosurfactant. *Bioresource Technology*, *232*, 389–397.

Wang, L. Y., Duan, R. Y., Liu, J. F., Yang, S. Z., Yang, J. D., & Mu, B. Z. (2012). Molecular analysis of the microbial community structures in water-flooding petroleum reservoirs with different temperatures. *Biogeosciences*, *9*(11), 4645–4659.

Wang, X., Yang, Y., & Xi, W. (2016). Microbial enhanced oil recovery of oil-water transitional zone in thin-shallow extra heavy oil reservoirs: A case study of Chunfeng Oilfield in western margin of Junggar Basin, NW China. *Petroleum Exploration and Development*, *43*(4), 689–694.

Yakimov, M. M., Amro, M. M., Bock, M., Fredrickson, H. L., Kessel, D. G., & Timmis, K. N. (1997). The potential of Bacillus *l*icheniformis strains for in situ enhanced oil recovery.

Journal of Petroleum Science and Engineering, *18*, 147–160. https://doi.org/10.1016/S0920-4105(97)00015-6

Youssef, N., Elshahed, M. S., & McInerney, M. J. (2009). Chapter 6 Microbial Processes in Oil Fields. Culprits, Problems, and Opportunities. In *Advances in Applied Microbiology* (1st ed., Vol. 66). Elsevier. https://doi.org/10.1016/S0065-2164(08)00806-X

Youssef, N., Simpson, D. R., McInerney, M. J., & Duncan, K. E. (2013). *In-situ* lipopeptide biosurfactant production by *Bacillus* strains correlates with improved oil recovery in two oil wells approaching their economic limit of production. *International Biodeterioration and Biodegradation*, *81*, 127–132.

Zekri, A. Y., & Almehaideb, R. (2003). Microbial and waterflooding of fractured carbonate rocks: An experimental approach. *Petroleum Science and Technology*, *21*(1–2), 315–331.

Zekri, A. Y., Almehaideb, R. A., & Chaalal, O. (1999). Project of increasing oil recovery from UAE reservoirs using bacteria flooding. *SPE-56827. SPE Annual Technical Conference and Exhibition*, Houston, Texas, USA. https://doi.org/10.2118/56827-MS

Zekri, A. Y., Ghannam, M. T., & Almehaideb, R. A. (2003). Carbonate Rocks Wettability Changes Induced by Microbial Solution. *SPE - Asia Pacific Oil and Gas Conference*, 482–491.

Zhang, P., & Austad, T. (2006). Wettability and oil recovery from carbonates: Effects of temperature and potential determining ions. *Colloids and Surfaces A: Physicochemical and Engineering Aspects*, *279*(1–3), 179–187.

5 Omics Strategies Targeting Microbes with Microplastic Detection and Biodegradation Properties

Edwin Hualpa-Cutipa, Andi Solórzano Acosta, Yadira Karolay Ravelo Machari, Fiorella Gomez Barrientos, Jorge Johnny Huayllacayan Mallqui, Fiorella Maité Arquíñego-Zárate, Andrea León Chacón, and Milagros Estefani Alfaro Cancino

5.1 INTRODUCTION

Microplastics (MPs) are plastic fragments that are less than 5 mm in size (Vivekanand, Mohapatra and Tyagi, 2021). These ubiquitous pollutants degrade gradually due to their long residence time, high stability, and fragmentation capacity (Padervand et al., 2020). Microplastics can be categorized into two main types: primary and secondary. Primary microplastics refer to microbeads that are intentionally added to cosmetics and personal care products and are not formed by erosion or degradation processes. In contrast, secondary microplastics are derived from the fragmentation of larger plastic materials (He et al., 2019). The entry of MPs into the ecosystem occurs throughout the manufacturing process of plastic products and its amount in the environment varies in the range of 10–1000 μm (Pabortsava and Lampitt, 2020). Air serves as the main pathway for suspended MPs, carrying them from their source to their final destination (Rist et al., 2018), Depending on the parameters influencing the air, these MPs can end up in either aquatic environments or inhabited areas, this can lead to potential negative impacts on both ecosystems and human health (Huang et al., 2021). Wastewater treatment plants are the main source of aquatic MPs pollution, especially because many of them are close to oceans or seas (Murphy et al., 2016). In addition, the presence of plastics in oceans that are eroding or degrading increases the concentration of MP. Rivers also transport plastics as they flow

DOI: 10.1201/9781003407683-5

(Vivekanand et al., 2021). In the case of lakes, the lack of significant dynamic flow causes sedimentation and accumulation of MP (Huerta Lwanga et al., 2017).

Microplastics, once concentrated in bodies of water, become part of the diet of marine fauna, thus forming part of the food chain. In the case of fish, the amount of microplastics they consume depends on their type of diet. In general, carnivorous fish tend to ingest less plastic compared to omnivorous fish. In addition, the latter have a low excretion capacity (Zhang et al., 2021). Invertebrates living near beaches are also affected, as they consume large quantities of white plastic. Despite knowledge of this problem, there are few studies on the impact of this pollution on the marine environment (Rey, Franklin and Rey, 2021).

The human body has the capacity to eliminate certain fragments of microplastics through the excretory tract. However, when microplastics enter the gastrointestinal tract or respiratory tract, they can reach the lymphatic and blood systems, resulting in accumulation in the organs. This accumulation can lead to inflammation as a consequence (Vivekanand et al., 2021). Therefore, the size of microplastics determines their ability to pass through the organism. Those with a size of approximately 20 μm have limited access to organs, unlike those of 10 μm, which can enter sensitive organs by crossing cell membranes and the blood-brain barrier. It could be determined that the detrimental impact on the organism depends on the physical and chemical properties of microplastics as they are composed of a variety of polymers (Campanale, Savino and Uricchio, 2020).

Due to the prolonged stay of microplastics in the organism and their capacity to accumulate and cause toxicity, it is of vital importance to implement efficient strategies to prevent these contaminants from affecting both the ecosystem and human beings. So far, a variety of microorganisms capable of degrading polymers present in plastics through enzymatic reactions have been recorded. To further discover new strains with this degradation capability, an omicron approach is required. Omicron approaches will provide genetic information that will allow investigation and identification of different microbial species with the necessary capacity to degrade polymers present in microplastics (Wani et al., 2023).

5.2 PROPERTIES AND DISPERSION ROUTES OF PLASTICS AND MICROPLASTICS IN ENVIRONMENTAL MATRICES

Plastic is a crucial material in the everyday existence of humanity. Long-chain polymers are mostly made of carbon, oxygen, hydrogen, silicon, and chloride. These elements are derived from natural gas, petroleum, and coal (Shah et al., 2008). Its importance lies in the improvements it has brought to people's daily lives, thanks to properties such as durability, excellent mechanical properties, weather resistance, and long service life (Ahmed et al., 2021). Plastics are utilized across a diverse range of industries, fulfilling crucial functions in areas such as packaging, automotive, aquaculture, fisheries, biomedicine, shipping, agriculture, building and construction, telecommunications, furniture, transportation, plumbing, personal care products, textiles, and apparel, among numerous others (Ogunola, Onada and Falaye, 2018).

From 1950 to the present, there has been an enormous growth in plastic production, which continues to increase on a large scale. An example of this growth is that in the year 1950, 1.7 million tons of plastic were produced, while in 2019 global production reached 361 million tons (Auta, Emenike and Fauziah, 2017). However, this increase in production has also led to a significant accumulation of plastics in the environment, due to their resistance to chemical and thermal degradation. Although preventive measures, such as recycling of these materials, have been implemented, they have not achieved the desired success. A minimal portion of plastic garbage has undergone the recycling process, with the majority of these contaminants ultimately being released into the environment, particularly the aquatic ecosystem. It is estimated that approximately 10% of all plastics produced end up in the marine environment. Plastic pollution has been found to be widely distributed throughout the world's oceans, as evidenced by multiple studies. Its existence and dispersion have been proven not just in the water column but also in both populated and rural regions of the Earth (Hughes et al., 2021).

Contamination of the aquatic environment by microplastics is a matter of concern; these particles are less than 5 mm in size (Thompson et al., 2009). They are usually generated and dispersed in the environment as by-products of larger plastic particles in photooxidative and mechanical processes (Robin et al., 2020), which are also called secondary MP. In contrast, plastics that are deliberately produced for a specific purpose are referred to as primary microplastics (Hidalgo-Ruz et al., 2012). The aforementioned items are primarily employed in personal care merchandise, including but not limited to deodorants, sunscreens, body washes, nail polishes, hair products, and other cosmetic-related items (Hernandez, Yousefi and Tufenkji, 2017).

The issue develops because of the escalating prevalence of MPs in diverse environmental constituents, a development that has elicited significant apprehension due to their minuscule dimensions and the substantial health hazards they entail. In light of this worry, a multitude of research has been undertaken in the past decade to ascertain the prevalence and distribution of such materials (Upadhyay and Bajpai, 2021). These studies have revealed that MPs have the ability to migrate across borders, making them a threat to all ecosystems. However, there are currently no systematic studies that comprehensively analyze the fate and transport of these particles, considering different environmental systems and necessary mitigation measures (Ahmed et al., 2021).

Little is known about the transboundary migration of microplastics in all environmental matrices. This migration originates from various sources and acts in concert, forming a multidimensional stressor group. The interdependence of these sources requires detailed scientific analysis (Upadhyay and Bajpai, 2021). In addition, the specific properties and characteristics of microplastics also influence their distribution and transfer to the environment. The shape and size of microplastics are some of these characteristics. In fact, these characteristics allow their classification, which facilitates the identification of the source and transfer pathway, as well as the degradation state and fate of microplastics. In turn, these characteristics determine their interaction with organisms and the impact they generate in ecosystems (Diniz, 2018).

The size of microplastics plays a crucial role in their interaction with biotic components and represents a long-term threat to the ecosystem. According to several studies, a relationship has been established indicating that smaller microplastics are more prone to sorption processes, which increases the possibility of adsorption of toxic contaminants on a larger surface area (Filella, 2015). In addition to size, another important characteristic of microplastics is their shape, which exhibits a wide diversity, such as spherical, granular, foamy, fibrous, flaky, bead-like, film-like, and fragments. The form in which they are manufactured serves a specific purpose. For example, primary microplastics are often in the form of microbeads found in products such as facial cleansers and scrubs (Chubarenko et al., 2016).

Macroplastics undergo natural degradation and fragmentation. During this process, a variety of mechanical forces act on the surface of the plastic, resulting in the formation of secondary microplastics of rough texture and irregular shape. The contamination of these microparticles occurs in different matrices and one of these matrices is groundwater contaminated by landfill leachate. This contamination is of particular concern due to its association with chemicals and pathogens (Rose et al., 2023). The eco-toxicological effect that occurs depends on several factors, such as the volume and toxicity of the MP present in the leachate, as well as the permeability and direction of groundwater flow. Among these factors, groundwater flow is crucial because here it moves slowly and steadily through the permeable soil strata (Ahmed et al., 2021). If a landfill contaminates groundwater with MP, a pollution plume will form and the aquatic sources in that plume will also be affected, which in turn leads to groundwater contamination. On the other hand, the dispersion of these materials from the surface of landfills into the aquatic environment and other matrices occurs through atmospheric transport. This is possible because the MPs are lightweight (O'Brien et al., 2023). Likewise, there is a variety of MPs with low buoyancy, which are carried by the wind from the landfill surface and can be deposited in aquatic bodies. Therefore, recent studies claim that wind is the main dispersal pathway in lakes, lagoons, and seas, also transporting mixtures of light plastic particles (Rezaei et al., 2019). Even, wind can transport MPs from the ground surface where solid waste landfills are located and deposit them in environments far from the initial site (Horton and Dixon, 2018).

Regarding pollution in the air, little research is available, except for the study by Abbasi (2019) where he mentioned the dispersion and concentration of microcaps, which are a type of plastic and a pollutant component. Unfortunately, no further studies have been conducted to investigate the relationship between air emissions and the concentrations of microcaps in the air. These studies are crucial as they would help identify and quantify airborne microcaps more accurately, enabling the development of effective strategies to reduce their impact on air quality.

The dispersion of microplastics on land is due to being the place where they are consumed and manufactured. Among the causes leading to their release, direct littering and inefficient waste management can be mentioned (Lechner and Ramler, 2015). Another means that influences their dispersion in the terrestrial environment are modern agricultural practices, which make use of a large amount of plastics, such as ropes, mulches, and bale wrappers. These materials, when discarded, can degrade and form secondary microplastics in that environment (Nizzetto, Futter and Langas, 2016; Hurley and Nizzetto, 2018). If microplastics are not transported and distributed from this environment to marine environments, they could accumulate massively on land. However, there are currently not enough studies to support this claim, so it is not appropriate to establish a direct connection between microplastic accumulation and specific environmental characteristics or human activities.

Therefore, in the face of the wide dispersion of MPs and the scarce research in this regard, it is crucial to invest and carry out studies in this field. These studies should focus primarily on analyzing associations between MPs, both primary and secondary, as well as examining environmental interactions that will allow us to better understand the transport mechanisms of these pollutants (Horton and Dixon, 2018). By studying the behavior of marine pollutants (i.e., microplastics) and the influence of river flow, we can gain a better understanding of how these factors impact ecosystems. This knowledge is crucial for designing effective strategies to mitigate the negative effects.

5.3 QUANTIFICATION AND IDENTIFICATION OF MICROPLASTICS IN THE ENVIRONMENT AND LIVING ORGANISMS

The U.S. National Oceanic and Atmospheric Administration (NOAA) has expanded the definition of MPs to include any fragment smaller than 5 mm. It is crucial to note that different types of plastic have varying densities. Plastic bags float because their density is considerably lower than that of seawater. However, other types of plastic, such as polyethylene terephthalate (PET) and polyester fibers used in clothing, tend to sink.

Aldana, Enríquez and Castillo (2022) conducted a study on the Caribbean's pink snail (*Strombus gigas*) to quantify the quantity of microplastic components. The snails were chosen due to their indigenous status and significant population. The study followed ethical guidelines to preserve endangered species and prevent harm. The snails were collected and analyzed, and the results showed that the entire Caribbean is contaminated by microplastics. 100% of the snails analyzed present them in their excreta. The study highlights the need to modify our activities to reduce ocean contamination and health risks. Additionally, plastic production from oil contributes to climate change. To sustain the planet, it is crucial to avoid single-use plastics and instead use eco-friendly alternatives like thermos and bags.

MPs analysis in water samples consists of five steps: sampling, separation, cleanup, identification, and confirmation. In the sampling step, nets, sieves, or pumps can be used to collect water samples, but the uneven distribution of MPs makes direct recovery difficult (Lee & Chae, 2021). Separation can be achieved by oxidation of organic material or by density separation, although further studies are needed to establish effective protocols. Cleaning is performed to remove organic matter and avoid confusion with other particles. According to Schymanski et al. (2021) for MPs identification, techniques such as visual inspection, infrared or Raman spectroscopy, thermogravimetry, and gas chromatography can be used.

It is important to consider that identification by selective sampling involves directly collecting visible plastic particles, i.e., those between 1 and 5 mm in size. However, this technique is simple and may have limitations when detecting microplastics, especially when they are mixed with other debris or lack distinctive shapes. The provenance of the samples should be taken into account when applying collection methods

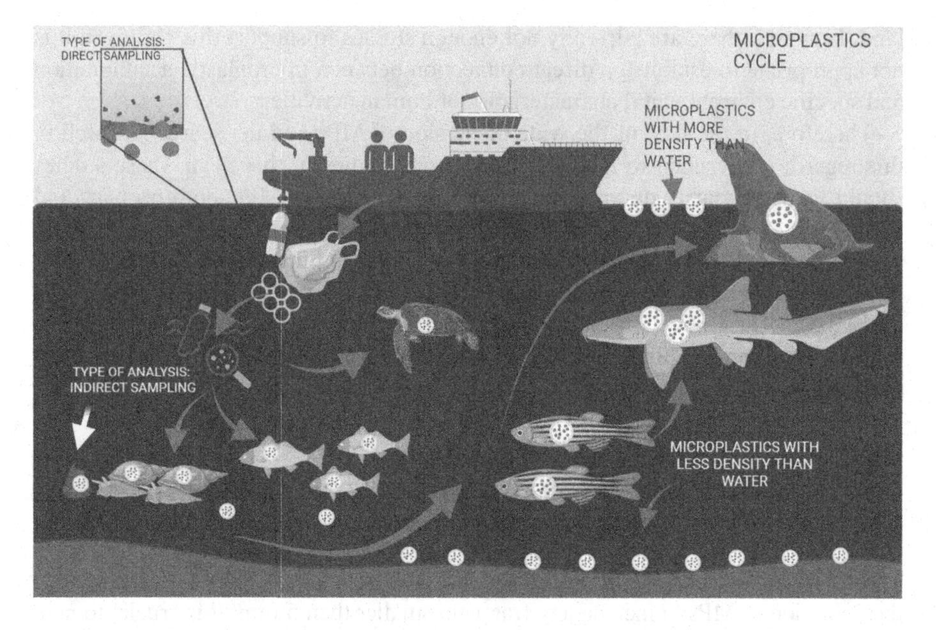

FIGURE 5.1 Microplastics cycle and marine distribution in aquatic ecosystems.

because information on the vertical distribution and fate of MPs in seawater is still limited. This is because the MPs have different densities, being found at different vertical levels (Eo et al., 2022). This type of sampling could introduce a bias in the quantification and identification of MP, as mentioned in the collection section. Studies reported by Purca and Henostroza, 2017 found Microscopic synthetic fibers in the sediments of high tidal areas, as well as in the water column and estuaries. Nine types of polymers have been identified: acrylic, alkali, poly(ethylene-propylene), polyamide, polyester, nylon, polyethylene, Poly(methyl acrylate) (PMA), and polyvinyl.

The cycle and dispersion of microplastics in marine ecosystems, considering the size and density of the particulate material, are depicted in Figure 5.1. Furthermore, it is important to consider the influence of these contaminants on the local wildlife and flora.

5.4 PHYSICOCHEMICAL STRATEGIES FOR THE MANAGEMENT OF MICROPLASTICS

Plastic is a durable, synthetic material that is resistant to both physical and chemical degradation. This characteristic has become a serious environmental problem, as plastic waste accumulates uncontrollably in various terrestrial and aquatic habitats, and even in smaller ecosystems. Over time, this waste fragments into smaller and smaller pieces (Castañeta et al., 2020).

As to the European Chemicals Agency (ECHA), the term "MP" refers to solid plastic particles that are composed of blends of polymers and functional components. Moreover, these formations might occur either deliberately or inadvertently.

An illustrative instance of the latter phenomenon can be observed in the process of abrasion, but in the case of the former, they are incorporated into various items with the intention of attaining specific characteristics (Salthammer, 2022). The main potential sources of MPs include synthetic polymers, natural polymers such as biopolymers (e.g., silk, starch, rubber, among others), and modified polymers, such as semi-synthetics, e.g., nitrocellulose and vulcanized rubber (García Diez, 2009).

MP is classified as primary and secondary. The former are intentionally manufactured for specific applications, such as pellets. They are used in cleaning products, cosmetics, paints, toothpaste, among others (Fadare et al., 2020). These sources contribute to the increase in MP and are also used in medical applications and are even included in some medicines. For example, monobutyl phthalate and dimethyl phthalate have been found at a combined concentration of 16,868 ng/mL in Asacol, a drug used for the treatment of mild to moderate ulcerative colitis. In addition, there are other drugs that include these types of compounds in their composition as excipients, flavorings, or coating (Chung et al., 2019).

On the other hand, secondary MPs is produced by the fragmentation or selective degradation of macroplastics exposed to external factors. This degradation can occur by physical and chemical phenomena, such as photodegradation, biological phenomena, such as biodegradation by bacteria and fungi, and mechanical degradation (synergistic effect between wind and wave action) (Castañeta et al., 2020). Therefore, MPs have become an emerging potential threat because they act as vectors by absorbing heavy metals, pathogens, and other pollutants. On the other hand, fish and other aquatic animals ingest microplastics and, finally, people ingest them at the tertiary level of the food chain. This can block the digestive tracts and eventually shorten people's lives (Dey, Uddin and Jamal, 2021).

The elimination of MPs is an urgent environmental problem due to the hazardous effects it has on the environment and on water sources used for human consumption. To address this problem, various physicochemical methods have been employed, such as coagulation, sedimentation, sand filtration, and clarification (Bhatt et al., 2021).

One of the practical methods used to remove MPs is the sand filter, which captures it by adhesion. In a study conducted (Zhao et al., 2021), it was found that the sand filtration system used had relatively low MP removal. However, an alternative approach for the removal of MPs involves the utilization of a fast sand filter that employs distinct layers to facilitate their separation. The presence of layers, such as silica sand, facilitates the interaction between MPs and these layers, resulting in the separation of MPs from the water (Hidayaturrahman and Lee, 2019). An alternative approach for the removal of MPs involves the utilization of chemical coagulants. This method operates by inducing a destabilization of the surface charge of MPs, leading to the formation of flocs during the coagulation process (Bhatt et al., 2021). In the investigation conducted by Skaf et al., various doses of alum were employed to eliminate the aforementioned substances from simulated drinking water (Skaf et al., 2020).

Research conducted by (Zhou et al., 2021) reported the link between the removal rate and the pH of the medium, where the findings of the study provide confirmation that the removal of MPs is more efficient in alkaline media as compared to acidic media.

Dissolved air flotation is a physicochemical method used primarily to exclude insoluble substances. It consists of dissolving air in water, which produces bubbles that adhere to the surface of the insoluble substances and are then removed (Bhatt et al., 2021). This method shows high efficiency, although its application and performance can be affected by certain parameters. An investigation was conducted on MP removal by positive modification of this method, achieving removal of 32–38% of MP (Wang et al., 2021). The primary sedimentation method has shown good performance in MP removal, reaching up to 70%. Basically, it consists of removing MP by surface sedimentation and skimming (Murphy et al., 2016).

According to García-Segura et al. (2017) electrocoagulation is based on the use of electrically generated metal electrons to form coagulants and, it has been investigated that this method can remove more than 90% of MPs, indicating good performance. For example a study found a maximum MPs removal of 99.24% was achieved at pH 7.5 (Perren, Wojtasik and Cai, 2018). In addition, an alternative approach was described by Grbic et al. (2019) known as the magnetic procedure. This method involves the synthesis of hydrophobic iron nanoparticles, which possess magnetic properties. Therefore, the nanoparticles are capable of interacting with and effectively separating MPs from water. The efficacy of this approach in the mitigation of MPs in polyethylene and polystyrene has been demonstrated. The selection of iron nanoparticles was based on their notable adsorption capability, convenient synthesis process, cost-effectiveness, and magnetic properties, with particular emphasis on their substantial surface area (Bhatt et al., 2021).

Figure 5.2 illustrates the source of microplastics resulting from spontaneous fragmentation and the detrimental effects associated with their build-up in various

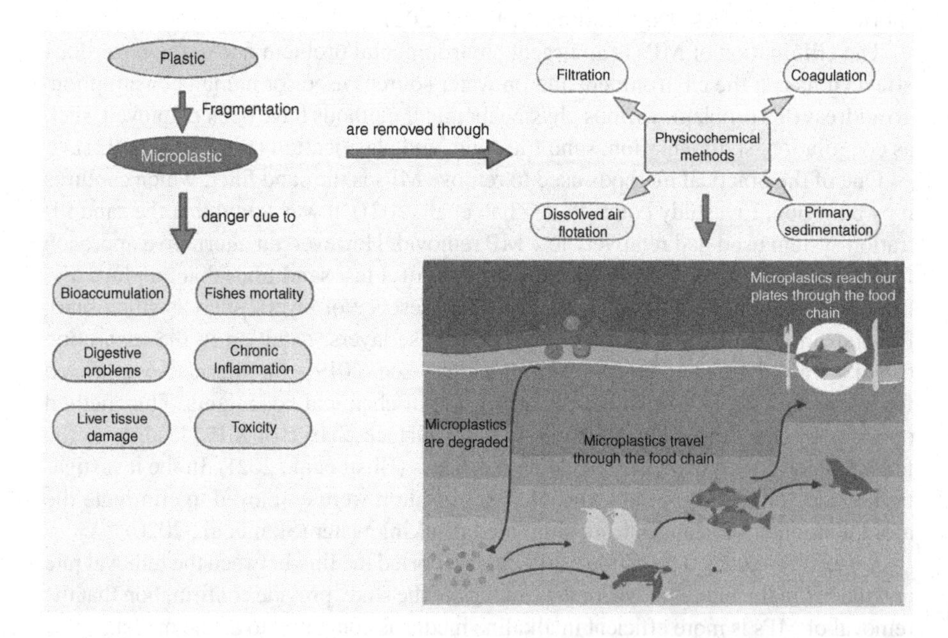

FIGURE 5.2 The adverse effects resulting from the utilization of MPs and physicochemical techniques for handling.

organs and tissues of organisms. Also highlights primary methodologies employed for the elimination of these contaminants by physicochemical approaches.

5.5 BIOLOGICAL STRATEGIES FOR DETECTION AND MANAGEMENT OF MICROPLASTIC

According to Magalhães et al. (2020) current research to detect MP using biological strategies is scarce due to their physicochemical properties, such as low water solubility and low charge density, resulting in low efficiency biological systems Also, Dey et al. (2021) states the biological detection of MPs detection is based on using microorganisms. However, some of these strategies may be more effective than others depending on the complexity of the sample and the characteristics of the MP. There are several biological strategies for the detection of MP. The following are some of them.

5.6 USING EXTRACELLULAR POLYMERIC CYANOBACTERIAL SUBSTANCES

The study conducted by Gongi et al. (2022) explores the use of cyanobacterial extracellular polymeric substances (EPS) as a material for sensor coatings, specifically for detecting microplastics. Using a sensitive membrane applied over a gold electrode, the EPS was analyzed using electrochemical impedance spectroscopy to identify four types of microplastics. The surface characteristics of the impedimetric sensor were examined using Scanning Electron Microscopy, Fourier Transform Infrared Spectroscopy, and X-ray Spectroscopy. The EPS-based sensor had high sensitivity and a low detection limit of 10-11 M (molar concentration), making it a potential approach for quantifying low amounts of microplastics.

Biosensors: These tools are scientific instruments that employ living organisms to identify and ascertain the existence of compounds within a given environment. Biosensors can be designed to detect microplastics and emit a signal when they are detected.

Engineered peptide biosensors: Peptides possess amino acid sequences that exhibit binding affinity toward types of plastics, hence facilitating the precise and discerning capture of polystyrene (PS) and polypropylene (PP) microplastics. According to study by Woo et al. (2022) peptides that bind to PS and PP (PSBPs and PPBPs) revealed that PSBP and PPBP exhibited binding affinity toward both oxidized and unoxidized MPs. The study focused on the detection of microplastics in water due to environmental contamination. introducing a new method for using peptides to identify these plastics. Polystyrene and polypropylene were chosen as plastics in the water. The binding affinity of hydrophobic peptides was assessed for each type of plastic. The peptide biosensors successfully detected microplastics in small animals' intestinal extracts. The study suggests that plastic-binding peptides could enhance microplastic detection efficiency and potentially aid in the separation of microplastics from seawater.

5.7 BIOLOGICAL STRATEGIES FOR MICROPLASTICS MANAGEMENT

These are based on the use of biological processes to degrade or remove MP. These strategies can be more sustainable and environmentally friendly than other MP removal techniques that use chemical and physical processes. There are several biological strategies for MP management that can be applied to both water and soil. Some of them are presented below.

5.7.1 Bioremediation

According to Qiu et al. (2022), the use of microorganisms such as fungi, bacteria, and algae to regulate MPs is currently being investigated, however this approach is novel, and more research is required to corroborate its efficacy (Yogalakshmi and Singh, 2020). According to Chattopadhyay (2022), *Pseudomonas*, *Arthrobacter*, and *Bacillus* are the main bacterial groups used for the biodegradation of different types of plastics such as polyethylene, polypropylene, polyurethane, and polystyrene. These microbial decompose MPs after days, weeks, or months, depending on the chemical composition and structure of the polymer. Fungi, on the other hand, have great potential to degrade polymeric contaminants. For example, the genera *Fusarium, Humicola*, and *Aspergillus* degrade polyhydroxyalkanoate, PET, and polyesters over a long period of time, generally more than 100 days, because they secrete hydrolase enzymes, which generate a long period of MPs degradation (Sánchez, 2020).

5.7.2 Membrane Technology

A revolutionary technological approach is now being developed for the remediation of wastewater polluted with MPs. This innovative method involves the creation of a dynamic membrane that functions as a protective barrier, preventing the embedding of MPs inside the membrane that supports the wastewater (Figure 5.3). According to

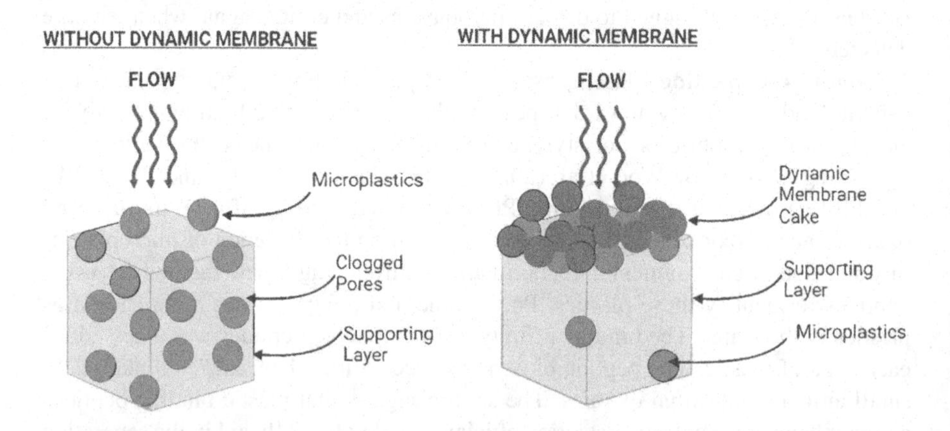

FIGURE 5.3 Membrane technology for MPs detection.

Krishnan et al. (2023) due to the selectivity of the membrane, other substances are retained by the pores. Then, the pores must have a permeable material to hold and support the membrane on the surface, as well as be robust enough to withstand the forces required in the process. In addition, this technology allows significant energy savings when compared to more elaborate processes such as microfiltration or ultrafiltration. Due to its simplicity, stability, and operational flexibility, it has advantages over other known methods, suggesting application on an industrial scale.

5.7.3 Genetically Modified Microorganisms

Genetic engineering allows us to take advantage of the metabolic pathways of the microorganisms to increase their biodegradation potential, gene editing tools that allow us to modify or express specific genes through the use of transcription activator type effector nuclease, zinc finger proteins, etc. (Dasgupta et al., 2020). Initially the appropriate gene for biodegradation is selected by identifying which types of enzymes enhance MPs biodegradation, the recombinant DNA fragment is introduced immediately into the host cell (Miri et al., 2022). According to Anand et al. (2023) the enzymes that are encoded by genes, such as dehalogenase, esterase, and laccase, are most associated with degrading microplastics, for example, *Streptomyces albogriseolus* has been identified to encode three CRISPR sequences, where the enzyme associated with polyethylene degradation is oxygenase.

5.8 MICROBIAL BIODIVERSITY ASSOCIATED WITH MICROPLASTICS: MICROBIAL COLONIZING STRATEGIES

Microbial biodegradation of MP is a complex process, where microorganisms naturally convert complex polymeric molecules into monomeric molecules and enzymes. Some MPs due to their chemical structure are more difficult to degrade, and aspects such as hydrophobicity and crystallinity make MPs less susceptible to degradation (Lu et al., 2023). The process involves in the first instance the colonization of the MPs by microorganisms, where they produce enzymes that bind to the polymer; here the polymer surface is modified, and hydrolysis of the polymer chain occurs. The extracellular enzymes produced such as lipase, esterase, laccase, and lignin peroxidase improve the hydrophilicity of the MPs by converting their functional groups into alcohols or carbonyl groups. In the second scenario, the process of bioassimilation involves the penetration of MPs through the semipermeable membrane. These MPs are then transported to the cytoplasm where they undergo degradation by intracellular enzymes. According to Yuan et al. (2022), Miloloža et al. (2022), and Mahmud et al. (2022) the mechanism and mode of transport of the monomeric units to the cell cytoplasm are influenced by membrane proteins, such as porins, which aid in the mobilization of MPs. The final step of this process involves the biodegradation of MPs into highly oxidized metabolites, including H_2O, CO_2, N_2, and CH_4. Finally, the bacteria degrade MPs more efficiently due to their proliferation and adaptation to the environmental conditions and for the degradation of hydrolyzable and non-hydrolyzable polymer, they use metabolic pathways and enzymes.

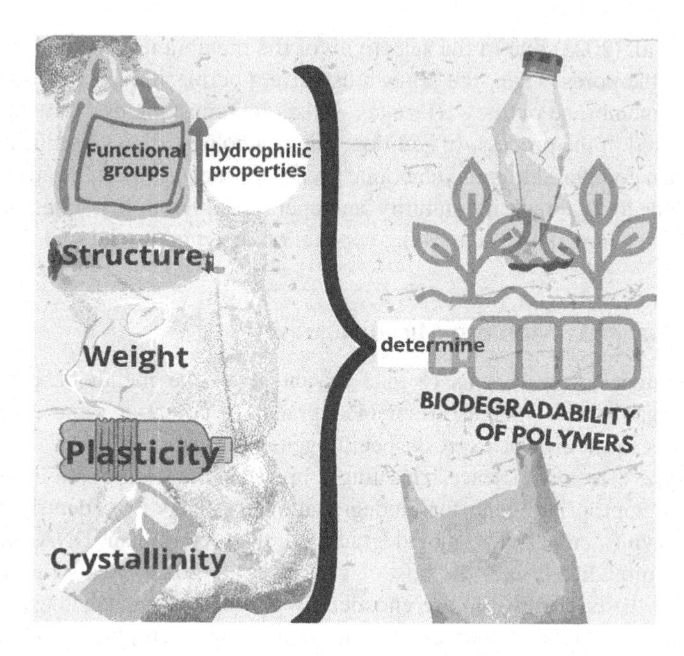

FIGURE 5.4 Characteristics of the biopolymer that determine its biodegradability.

Fungi are well adapted to living with MPs because they are rapidly nourished and thus have the potential to degrade MPs by adhering to and utilizing their metabolic capacity. The fungal cells release hydrophobic proteins, which facilitate the use of polymers as a carbon and energy source. The biodegradability of polymers is mainly determined by different physicochemical characteristics, such as the presence of functional groups that increase the hydrophilic properties, structure, weight, plasticity, number of amorphous areas, and crystallinity of the polymer, including polymer hardening, structure, plasticity, molecular structure, and presence of fragment groups (Shabbir et al., 2020). Biodegradation depends not only on the type of plastic and the type of microorganisms but also on environmental conditions such as temperature and humidity (Figure 5.4). For their study, Pires Costa, Vaz de Miranda and Pinto (2022) collected samples from various polluted environments and conducted laboratory experiments to examine the relationship between bacteria, alkalinity, and plastic biodegradation. The goal of the analysis was to determine which material has a higher degradation capability and to find out how well two types of plastic break down: LDPE, which comes from oil, and a "biodegradable" plastic bag made of branched polyethylene terephthalate and inoculum number 18, which comes from the tunicate *Didemnum* sp. Understanding the degradation capabilities of these plastics is important for assessing their environmental impact and finding more sustainable alternatives.

The researchers De Villalobos, Costa and Marín-Beltrán (2022) observed that the bacterial community in the infected samples showed a significant increase in pH, particularly in the case of LDPE. This suggests that bacteria may require an alkaline environment to thrive and promote the biodegradation of plastic polymers; however, something to consider regarding the types of bacteria under investigation is that some bacteria show different tolerance to pH changes by resisting high and low pH changes (Islami, Tazkiaturrizki and Rinanti, 2019). The research conducted by Amaral-Zettler, Zettler and Mincer (2020) revealed that bacteria belonging to the *Cobetia*, *Pseudoalteromonas*, and *Ruegeria* genera exhibited a significant preference for LDPE particles in comparison to the surrounding broth. Conversely, the *Arcobacter* and *Vibrio* groups were shown to be more prevalent in the broth.

This first process of adhesion or colonization of microorganisms is crucial to continue with biodeterioration, as it reduces the resistance and durability of the plastic, allowing its microbial action (Tang et al., 2022). In addition, it is very important that the plastic is bioaccessible, i.e., that the chemical structure of its surface is colonizable. Consequently, the incorporation of hydrophilic functional groups onto the surface of plastic polymers is frequently employed to facilitate the adherence of microbes to the substrate. The promotion of adhesion and acceleration of deterioration occurs when plastic surfaces exhibit hydrophilicity. Hence, the hydrophobicity of plastic polymers is reduced as a result of the occurrence and development of polar functional groups, which may be attributed to environmental conditions or pretreatment methods such as exposure to UV radiation or the introduction of enzymes with deep hydrophobic sites (Chamas et al., 2020). For example, carboxylesterases, which also belong to the α/β-hydrolase family, are serine esterases and have a deep hydrophobic substrate binding site. They are highly effective against PET-type oligomers; therefore, they are used to remove low molecular weight products of PET hydrolysis by polyester hydrolases (cutinases and lipases). Their use is important because TfCa carboxylesterase enzyme is able to remove hydrophilic products from highly crystalline PET polymers (Barth et al., 2016).

Significant deterioration of PET films was seen as a result of the continuous hydrolysis of the intermediates generated by MHET and BHET, using immobilized TfCa carboxylesterase. Consequently, the process of hydrolyzing MHET to TPA resulted in a reduction in the relative abundance of MHET among the overall population of breakdown products. Based on the observation of Barth et al. (2016) the immobilization of TfCa can extract a limited quantity of MHET, BHET, and TPA directly from PET films, it can be inferred that the carboxylesterase presents in the two-enzyme system, namely, TfCut2 and LC-cutinases employed in the investigation, primarily catalyzes the hydrolysis of PET degradation intermediates.

In a comparative study of bacterial colonization by polyethylene and polyethylene terephthalate MPs in freshwater environments such as lake water and tap water conducted in China by 16S rRNA amplification sequencing, it was found that different MPs are associated with different species during biofilm growth and succession. This hints that there are different bacterial survival strategies of MPs biofilms. The structure of MPs biofilms incubated in lake water sample was more robust under environmental stress, while bacteria of MP biofilms incubated in tap water samples interacted more cooperatively (Song et al., 2022).

MPs in marine and freshwater environments are polyethylene, polypropylene, etc. and take the form of fibers, granules, and fragments. Microorganisms such as bacteria, archaea, and eukaryotes form biofilms that adhere and embed themselves in the extracellular matrix of polymeric compounds. The selection of microbial biofilm components by the MP can occur through various potential mechanisms. First, the presence of a solid surface offers a suitable environment for microbial attachment. Second, plastic material contains novel organic polymers, additives, and adsorbed contaminants that can serve as a carbon source for microbial metabolism. Third, secondary members of the microbial biofilm can adhere to the polysaccharides present in the biofilm or to the primary colonizers of the plastic (McCormick, Hoellein and Londres, 2016).

Puglisi et al. conducted a study examining the selective bacterial colonization processes on polyethylene waste samples. The samples were distinguished by various colors, yet all shared the common characteristic of being composed of polyethylene. The researchers employed infrared (IR) and differential scanning calorimetry (DSC) analyses to determine varying degrees of degradation. Additionally, Fourier-transform Raman spectroscopy and X-ray fluorescence techniques were utilized to evaluate the extent of degradation and the presence of pigments in the samples. There are different varieties of plastics and each variety of these plastics harbors different bacterial communities according to several analytical studies performed on polyethylene waste at the terrestrial level. According to Puglisi et al. (2019), among the microorganisms associated with MP, they have found the case of the bacterium *Ideonella sakaiensis* which is associated with PET type plastic waste, whereby, most studies on communities of microorganisms associated with MPs were conducted in aquatic environments. Also, they reported the presence of pathogenic bacteria such as *Vibrio* species and its pathogen *Vibrio parahaemolyticus* harboring in polyethylene, polypropylene, and polystyrene MP of aquatic environments.

The aforementioned research provides confirmation of the documented presence of potentially harmful bacteria in marine MPs and emphasizes the pressing requirement for comprehensive biogeographical investigations of marine MPs, as stated by Kirstein, Kirmizi and Wichels (2016).

The link between MPs of varying colors and the colonization preference of bacteria in the aquatic environment is depicted in Figure 5.5.

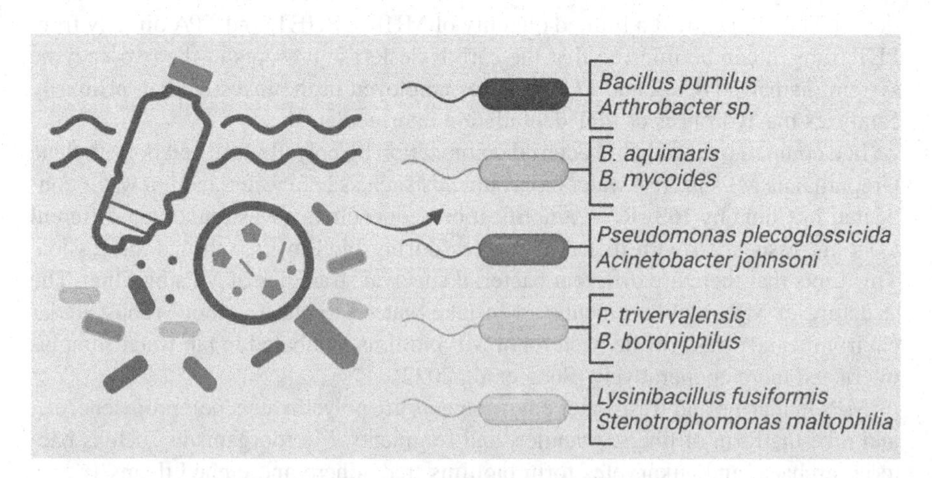

FIGURE 5.5 Potential correlation between the color of plastic and the microorganisms that reside on its surface.

5.9 MICROBIAL BIODEGRADATION OF MICROPLASTICS

Review by Kumar et al. (2023), explores microplastics, tiny plastic particles used in everyday items, their sources, distribution, biodegradation, and factors influencing their biodegradation. It also explores microbial biotechnology techniques, such as PET-ase and MHET-ase, and the challenges related to plastic removal using current information. According to Maleki Rad, Moghimi and Azin (2022), some MPs due to their chemical structure are more difficult to degrade, and aspects such as hydrophobicity and crystallinity make MPs less susceptible to degradation.

5.9.1 Degradation Process

Degradation process involves the colonization of the MPs by microorganisms, where they produce enzymes that bind to the polymer. Here the polymer surface is modified, forming a biofilm, and hydrolysis of the polymer chain occurs. The hydrophilicity of the microbial polymers can be enhanced by the enzymatic production of extracellular enzymes, including lipase, esterase, laccase, and lignin peroxidase (Kumar et al., 2023). According to Yuan et al. (2020), these enzymes facilitate the conversion of functional groups within the microbial polymer into alcohols or carbonyl groups, therefore improving its hydrophilic properties (;). In the process of bioassimilation according Miloloža et al. (2022), MPs infiltrate the semipermeable membrane, which is situated within the cytoplasm. Once inside, the MPs undergo degradation by intracellular enzymes. The transportation of the individual units of the MPs into the cytoplasm is influenced by membrane proteins, specifically porins, which facilitate the mobilization of the MPs (). Finally, during the last stage, the process of MPs biodegradation leads to the formation of extensively oxidized metabolites, including H_2O, CO_2, N_2, and CH_4 (Mahmud et al., 2022).

There are several stages or biological processes included in the biodegradation of particulate matter (PMs), wherein each phase possesses a unique characteristic that facilitates the complete breakdown of minute pollutants (microplastics) into mineral components. Multiple biological pathways are implicated in these phenomena.

The process of microbial biodegradation is characterized by its potential for sluggish rates, which are influenced by several circumstances. These elements encompass the concentration and size of the microbial population, the chemical composition and structure of the material being degraded, the prevailing environmental conditions, the pH level, and the presence of chemical additives, among other variables.

5.10 MICROORGANISMS THAT DEGRADE MICROPLASTICS

In the biodegradation of MPs, there are two types: aerobic and anaerobic respiration. The aerobic condition uses oxygen as an electron acceptor, decomposing complex organic compounds into simple ones and finally producing CO_2 and H_2O. Sulfate, nitrate, and iron groups take the place of electron acceptors in the second condition. In the last condition, complex polymers are degraded into smaller units, producing CH_2, CO_2, and H_2O as final products. In this process, the overall goal is to break down organic matter and release energy. (Srikanth et al., 2022).

TABLE 5.1
Microbial Strains with MP Biodegradation Capability

Microorganisms	Degradable MPs	Enzymes Produced	References
Pseudomonas sp	Polyethylene succinate	Esterase	Tribedi et al., 2012
Lysinibacillus xylanilyticus	LDPE	–	Esmaeili et al., 2013
Bacillus sp.	LDPE	Catalase	Thakur, 2012
Achromobacter xylosoxidans	PVC	–	Saeed, Iqbal and Deeba, 2022
Microbulbifer hydrolyticus	LDPE	Laccase	Li et al., 2020
Cladosporium halotolerans	HDPE	Oxidoreductase	Di Napoli et al., 2022
Bjerkandera adusta	HDPE	Laccase	Kang et al., 2019
Phanerochaete chrysosporium	PLA and PS polymer	–	Wu et al., 2023
Penicillium spp. *Aspergillus* sp.	PE, PP, and PS	–	Oliveira et al., 2020
Aspergillus flavus	Polyethylene	Oxidases	Zhang et al., 2020

HDPE, high-density polyethylene; LDPE, low-density polyethylene; PE, polyethylene; PLA, polylactic acid or polylactide; PP, polypropylene; PS, polystyrene; PVC, polyvinyl chloride.

Fungi are well adapted to living with microplastics as they are rapidly nourished and thus have the potential to degrade MPs by attaching to and utilizing their metabolic capacity, through cellular secretion of hydrophobic proteins that allow them to utilize the polymers as a source of carbon and energy (Rogers et al., 2020). On the other hand, the bacteria degrade MPs more efficiently due to their proliferation and adaptation to environmental conditions that allow them the degradation of hydrolyzable and non-hydrolyzable polymers, where they use metabolic pathways and enzymes (Miloloža et al., 2022).

Table 5.1 categorizes many microorganisms that have exhibited the capacity to generate enzymes possessing biodegradative attributes toward diverse microplastics derived from various origins.

5.11 OMICS APPROACHES FOR BIOPROSPECTION MICROBIAL AND THEIR BIOMOLECULES FOR DETECTION AND BIODEGRADATION OF MICROPLASTICS

Microplastics (MPs) pollution in our ecosystem negatively affects the environment and living organisms (Viljakainen and Hug, 2021). In the face of poor recycling practices, plastic is accumulating more and more every day in places such as landfills or water bodies, highlighting the urgency of addressing this issue. According to this author, the rate at which microplastics break down depends on factors like temperature, humidity, pH, the type of microorganism present, and the specific enzyme involved in the process. For example, the bacterium *Ideonella sakaiensis*, a type of microorganism, was found to break down and absorb PET (polyethylene terephthalate) using a two-enzyme system.

Microbial bioprospecting is a process that involves searching for and identifying microorganisms with specific traits that can be useful for various applications. In the case of recycling PET, researchers (De Jesus & Alkendi, 2023) have used microbial bioprospecting to find microorganisms that can use two enzymes to break down and absorb PET. This approach is significant because it provides a potential solution for the recycling of PET waste, which is a major environmental concern due to its long decomposition time and harmful effects on ecosystems.

The identification of enzymes and microorganisms capable of degrading plastic is carried out by means of clear zone tests, weight loss measurements, visual observation, and measurement of CO_2 evaluation. However, this type of technique has limitations (Viljakainen and Hug, 2021).

With the speed of DNA sequencing now ten times faster, and the development of computational biology techniques for annotating sequence data, microbial lineages can now be queried without the need for culturing (Viljakainen and Hug, 2021).

Omics approaches, including transcriptomics and proteomics, have revolutionized the field of microbial bioprospecting. These high-throughput technologies allow researchers to analyze the entire set of genes (transcriptome) or proteins (proteome) of a microorganism, providing a comprehensive view of its genetic material and protein expression. This level of analysis enables scientists to identify key genes and proteins involved in important biological processes, such as the degradation of microplastics (MPs), and understand how these microorganisms adapt and respond to their environment (Hosseini et al., 2022). This knowledge can inform strategies for mitigating the impact of microplastics on ecosystems and human health. For better detection, it is recommended to combine proteomics with metagenomics and transcriptomics. By comparing the data from these approaches, we can identify the genes responsible for protein production and gain a more comprehensive understanding of the mechanisms involved in degrading microplastics (MPs) (Hosseini et al., 2022).

5.12 LIMITATION AND FUTURE PERSPECTIVES

Plastics present in the environment serve as novel surfaces for the establishment of microbial communities, resulting in the identification of enzymes and microbes capable of breaking down plastics. According to Viljakainen and Hug (2021), the progress in the field of plastic microbiology, propelled by computational microbiology, is speeding up the process of making new findings using methods that do not rely on traditional culture techniques and by analyzing several types of biological data. This is leading to the identification of specific areas for protein engineering and enhancing the way plastic waste is managed.

However, the disadvantage of the metagenomic method is that it is not always possible to determine the taxon origin of a plastic-degrading enzyme (PDE). Despite major advances, much of the metagenome often fails to cluster (Tully, Graham and Heidelberg, 2018) or assembled genomes are produced in poor quality metagenomes

(<50% completed). However, according to Parks et al., (2017) sequencing and computational techniques have progressed, enabling the retrieval of microbial genomes from metagenomes without the need for culture. As a result, 7,903 bacterial and archaeal genomes have been reconstructed from more than 1,500 public metagenomes, leading to a 30% increase in phylogenetic diversity. The data also indicates that the diversity of the Patescibacteria superphylum differs depending on the protein marker sets used. Therefore, by reconstructing the genomes of environmental microbes using metagenomics, researchers can study the metabolic capabilities of resistant microbes. This knowledge is crucial for understanding how these microbes contribute to human health issues and environmental problems, allowing researchers to develop effective strategies for mitigating their impact.

Despite the initial utilization of metagenomic screens by researchers for the purpose of identifying novel PDEs and plastic-degrading microorganisms (PDMs), more advanced techniques like functional metagenomics, metatranscriptomics, and metaproteomics have not yet been widely embraced for the investigation of plastic-microbe interactions. These methodologies necessitate collaboration among researchers with specialized knowledge in plastic degradation, environmental biology, and computational biology, in addition to their inherent complexity and high cost (Viljakainen and Hug, 2021).

Microbial communities have the inherent ability to efficiently degrade many contaminants, including plastic compounds. While the pace of decomposition may not be enough for widespread use, several hydrolytic enzymes derived from diverse microbes have demonstrated the ability to destroy various types of plastics through distinct mechanisms. Nevertheless, there is currently a dearth of research focused on the discovery and characterization of enzymes related to the breakdown kinetics and substrate specificity of various polymers. Hence, it is important to do research on microorganisms capable of thriving in many environments, including stressful situations. This is because these microbes may be manipulated to utilize carbon polymers as an energy source for their growth.

Efficient catabolic pathways and molecular engineering techniques can enable the production of a wide range of valuable products from plastic waste. For example, certain plastics can be broken down into monomers that can be used as building blocks for the synthesis of new polymers with desired properties. These polymers can then be used in various industries, such as packaging, construction, and textiles. Additionally, plastic waste can be converted into biofuels or other chemical compounds that can be used as renewable energy sources or raw materials for the production of pharmaceuticals and other high-value chemicals (Kumari and Chaudhary, 2020). Furthermore, the utilization of plastic waste in these alternative ways can contribute to reducing environmental pollution and promoting a more sustainable future.

Most of the research referring to the microbial colonization of plastic trash has been undertaken in aquatic ecosystems, with a primary focus on comparing various kinds of plastics. Zettler et al. (2013) conducted a significant study wherein they conducted a comparative analysis of polyethylene and polypropylene wastes. The study revealed notable distinctions in the bacterial communities inhabiting these wastes, as compared to those found in seawater. Specifically, the bacterial communities exhibited dissimilar structures and lower levels of diversity. Furthermore, the study highlighted distinct variations in the bacterial species colonizing the two types of plastic materials. To the best of our understanding, a comprehensive analysis comparing various forms of polyethylene in both aquatic and terrestrial settings has not been conducted.

REFERENCES

Abbasi, S., Keshavarzi, B., Moore, F., Turner, A., Kelly, F.J., Dominguez, A.O. and Jaafarzadeh, N. (2019). Distribution and potential health impacts of microplastics and microrubbers in air and street dusts from Asaluyeh County, Iran. *Environmental Pollution*, *244*, 153–164. Doi: 10.1016/j.envpol.2018.10.039.

Ahmed, M.B., Rahman, Md. S., Alom, J., Hasan, M.D.S., Johir, M.A.H., Mondal, M.I.H., Lee, D.-Y., Park, J., Zhou, J.L. and Yoon, M.-H. (2021). Microplastic particles in the aquatic environment: A systematic review. *Science of The Total Environment*, *775*, p. 145793. Doi: 10.1016/j.scitotenv.2021.145793.

Aldana, D., Enríquez, M. and Castillo, V. (2022). El Caribe y su contaminación por microplásticos. *Revista Ciencia*, *73*(2), pp. 8–13. https://www.revistaciencia.amc.edu.mx/images/revista/73_2/PDF/Ciencia_73-2.pdf

Ali Shah, A., Hasan, F., Hameed, A. and Ahmed, S. (2008). Biological degradation of plastics: A comprehensive review. *Biotechnology Advances*, *26*(3), pp. 246–265. Doi: 10.1016/j.biotechadv.2007.12.005.

Amaral-Zettler, L.A., Zettler, E.R. and Mincer, T.J. (2020). Ecology of the plastisphere. *Nature Reviews Microbiology*, *18*(3), Article 3. Doi: 10.1038/s41579-019-0308-0.

Anand, U., Dey, S., Bontempi, E., et al. (2023). Biotechnological methods to remove microplastics: A review. *Environmental Chemistry Letters*, pp. 1–24. Doi: 10.1007/s10311-022-01552-4.

Auta, H.S., Emenike, C.U. and Fauziah, S.H. (2017). Distribution and importance of microplastics in the marine environment: A review of the sources, fate, effects, and potential solutions. *Environment International*, *102*, pp. 165–176. Doi: 10.1016/j.envint.2017.02.013.

Barth, M., Honak, A., Oeser, T., Wei, R., Belisário-Ferrari, M.R., Then, J., Schmidt, J. and Zimmermann, W. (2016). A dual enzyme system composed of a polyester hydrolase and a carboxylesterase enhances the biocatalytic degradation of polyethylene terephthalate films. *Biotechnology Journal*, *11*(8), pp. 1082–1087. Available at: https://pubmed.ncbi.nlm.nih.gov/27214855/ [Accessed 18 Mar. 2023].

Bhatt, P., Pathak, V.M., Bagheri, A.R. and Bilal, M. (2021). Microplastic contaminants in the aqueous environment, fate, toxicity consequences, and remediation strategies. *Environmental Research*, *200*, p. 111762.

Campanale, C., Massarelli, C., Savino, I., Locaputo, V. and Uricchio, V.F. (2020). A detailed review study on potential effects of microplastics and additives of concern on human health. *International Journal of Environmental Research and Public Health*, *17*(4), p. 1212. Doi: 10.3390/ijerph17041212.

Castañeta, G., Gutiérrez, A.F., Nacaratte, F. and Manzano, C.A. (2020). Microplastics: A contaminant that grows in all environmental areas, its characteristics and possible risks to public health from exposure. *Revest Bolivian de Química*, *37*(3), pp. 142–157.

Chamas, A., Moon, H., Zheng, J., Qiu, Y., Tabassum, T., Jang, J.H., Abu-Omar, M., Scott, S.L. and Suh, S. (2020). Degradation rates of plastics in the environment. *ACS Sustainable Chemistry & Engineering*, *8*(9), pp. 3494–3511. Doi: 10.1021/acssuschemeng.9b06635.

Chattopadhyay, I. (2022). Role of microbiome and biofilm in environmental plastic degradation. *Biocatalysis and Agricultural Biotechnology*, *39*, 102263. Doi: 10.1016/j.bcab.2021.102263.

Chubarenko, I., Bagaev, A., Zobkov, M. and Esiukova, E. (2016). On some physical and dynamical properties of microplastic particles in marine environment. *Marine Pollution Bulletin*, *108*(1–2), pp. 105–112. Doi: 10.1016/j.marpolbul.2016.04.048.

Chung, B.Y., Choi, S.M., Roh, T.H., Lim, D.S., Ahn, M.Y., Kim, Y.J., Kim, H.S. and Lee, B.M. (2019). Risk assessment of phthalates in pharmaceuticals. *Journal of Toxicology and Environmental Health, Part A*, *82*(5), pp. 351–360.

De Jesus, R., & Alkendi, R. (2023). A minireview on the bioremediative potential of microbial enzymes as solution to emerging microplastic pollution. *Frontiers in Microbiology*, *13*, p. 1066133. Doi: 10.3389/fmicb.2022.1066133

De Villalobos, N.F., Costa, M.C. and Marín-Beltrán, I. (2022). A community of marine bacteria with potential to biodegrade petroleum-based and biobased microplastics. *Marine Pollution Bulletin*, *185*, p. 114251. Available at: https://www.sciencedirect.com/science/article/pii/S0025326X2200933X [Accessed 18 Mar. 2023].

Dey, T.K., Uddin, M.E. and Jamal, M (2021). Detection and removal of microplastics in wastewater: Evolution and impact. *Environmental Science and Pollution Research International*, *28*(14), pp. 16925–16947. Doi: 10.1007/s11356-021-12943-5.

Diez, S.G. (2009). Referencias históricas y evolución de los plásticos. *Revista Iberoamericana de polímeros*, *10*(1), pp. 71–80.

Di Napoli, D., Silvestri, B., Giusy Castagliuolo, Carpentieri, A., Luciani, G., Antimo Di Maro, Sorbo, S., Pezzella, A., Zanfardino, A. and Varcamonti, M. (2022). High density polyethylene (HDPE) biodegradation by the fungus Cladosporium halotolerans. *FEMS Microbiology Ecology*, *99*(2). Doi: 10.1093/femsec/fiac148.

Diniz, C. (2018). Micro- and nanoplastics in the environment: Research and policymaking. *Current Opinion in Environmental Science & Health*, *1*, pp. 12–16. Doi: 10.1016/j.coesh.2017.11.002.

Eo, S., Hong, S.H., Song, Y.K., Han, G.M., Seo, S., Park, Y.-G. and Shim, W.J. (2022). Underwater hidden microplastic hotspots: Historical ocean dumping sites. *Water Research*, *216*, p. 118254. Doi: 10.1016/j.watres.2022.118254

Esmaeili, A., Pourbabaee, A.A., Alikhani, H.A., Shabani, F. and Esmaeili, E. (2013). Biodegradation of Low-Density Polyethylene (LDPE) by Mixed Culture of Lysinibacillus xylanilyticus and Aspergillus niger in Soil. *PLOS ONE*, *8*(9), p. e71720. Doi: 10.1371/journal.pone.0071720.

Fadare, O.O., Wan, B., Guo, L.H. and Zhao, L. (2020). Microplastics from consumer plastic food containers: Are we consuming it? *Chemosphere*, *253*, p. 126787.

Filella, M. (2015). Questions of size and numbers in environmental research on microplastics: methodological and conceptual aspects. *Environmental Chemistry (Collingwood, Vic.)*, *12*(5), pp. 527–538.

García Díez, S. (2009). Referencias históricas y evolución de los plásticos. *Revista Iberoamericana de Polímeros*, *10*(1), pp. 71–80. https://dialnet.unirioja.es/servlet/articulo?codigo=3694957

García-Segura, S., Eiband, M.M.S., de Melo, J.V. and Martínez-Huitle, C.A. (2017). Electrocoagulation and advanced electrocoagulation processes: A general review about the fundamentals, emerging applications and its association with other technologies. *Journal of Electroanalytical Chemistry*, *801*, pp. 267–299.

Gongi, W., Touzi, H., Sadly, I., et al. (2022). A novel impedimetric sensor based on cyanobacterial extracellular polymeric substances for microplastics detection. *Journal of Polymers and the Environment*, *30*(11), pp. 4738–4748. Doi: 10.1007/s10924-022-02555-6.

Grbic, J., Nguyen, B., Guo, E., et al. (2019). Magnetic extraction of microplastics from environmental Samples. *Environmental Science and Technology Letters*, *6*(2), pp. 68–72. Doi: 10.1021/acs.estlett.8b00671.

He, P., Chen, L., Shao, L., Zhang, H. and Lü, F. (2019). Municipal solid waste (MSW) landfill: A source of microplastics? – Evidence of microplastics in landfill leachate. *Water Research*, *159*, pp. 38–45. Doi: 10.1016/j.watres.2019.04.060.

Hernandez, L.M., Yousefi, N. and Tufenkji, N. (2017). Are There Nanoplastics in Your Personal Care Products? *Environmental Science & Technology Letters*, *4*(7), pp. 280–285. Doi: 10.1021/acs.estlett.7b00187.

Hidalgo-Ruz, V., Gutow. L., Thompson, R.C. and Thiel, M. (2012). Microplastics in the Marine Environment: A Review of the Methods Used for Identification and Quantification. *Environmental Science & Technology*, *46*(6), pp. 3060–3075. Doi: 10.1021/es2031505.

Hidayaturrahman, H. and Lee, T.G. (2019). A study on characteristics of microplastic in wastewater of South Korea: Identification, quantification, and fate of microplastics during treatment process. *Marine Pollution Bulletin*, *146*, pp. 696–702.

Horton, A.A. and Dixon, S.J. (2018). Microplastics: An introduction to environmental transport processes. *Wiley Interdisciplinary Reviews: Water*, *5*(2), p. e1268. Doi: 10.1002/WAT2.1268.

Hosseini, H., Al-Jabri, H.M., Moheimani, N.R., et al. (2022). Marine microbial bioprospecting: Exploitation of marine biodiversity towards biotechnological applications-a review. *Journal of Basic Microbiology*, *62*(9), pp. 1030–1043. Doi: 10.1002/jobm.202100504.

Huang, D., Tao, J., Cheng, M., Deng, R., Chen, S., Yin, L. and Li, R. (2021). Microplastics and nanoplastics in the environment: Macroscopic transport and effects on creatures. *Journal of Hazardous Materials*, *407*, p. 124399. Doi: 10.1016/j.jhazmat.2020.124399.

Huerta Lwanga, E., Gertsen, H., Gooren, H., Peters, P., Salánki, T., van der Ploeg, M., Besseling, E., Koelmans, A.A. and Geissen, V. (2017). Incorporation of microplastics from litter into burrows of *Lumbricus terrestris*. *Environmental Pollution*, *220*, pp. 523–531. Doi: 10.1016/j.envpol.2016.09.096.

Hughes, M., Clapper, H., Burgess, R.M. and Ho, K.T. (2021). Human and ecological health effects of nanoplastics: May not be a tiny problem. *Current Opinion in Toxicology*, *28*, pp. 43–48. Doi: 10.1016/j.cotox.2021.09.004.

Hurley, R. and Nizzetto, L. (2018). Fate and occurrence of micro(nano)plastics in soils: Knowledge gaps and possible risks. *Current Opinion in Environmental Science & Health*, *1*, pp. 6–11. Doi: 10.1016/j.coesh.2017.10.006.

Islami, A.N., Tazkiaturrizki, T. and Rinanti, A. (2019). The effect of pH-temperature on plastic allowance for low-density polyethylene (LDPE) by *Thiobacillus* sp. and *Clostridium* sp. *Journal of Physics: Conference Series*, *1402*(3), p. 033003. Doi: 10.1088/1742-6596/1402/3/033003 [Accessed 18 Mar. 2023].

Kang, B.R., Kim, S.B., Song, H.A. and Lee, T.K. (2019). Accelerating the Biodegradation of High-Density Polyethylene (HDPE) Using *Bjerkandera adusta* TBB-03 and Lignocellulose Substrates. *Microorganisms*, *7*(9), Article 9. Doi: 10.3390/microorganisms7090304.

Kirstein, I, Kirmizi, S and Wichels, A. (2016). Dangerous hitchhikers? Evidence for potentially pathogenic Vibrio spp. on microplastic particles. *Marine Environmental Research*, *120*, pp. 1–8. Doi: 10.1016/j.marenvres.2016.07.004.

Krishnan, R.Y., Manikandan S., Subbaiya R., et al. (2023). Recent approaches and advanced wastewater treatment technologies for mitigating emerging microplastics contamination–A critical review. *The Science of the Total Environment*, *858*(Pt 1), p. 159681. Doi: 10.1016/j.scitotenv.2022.159681.

Kumar, V., Sharma, N., Duhan, L., et al. (2023). Microbial engineering strategies for synthetic microplastics clean up: A review on recent approaches. *Environmental Toxicology and Pharmacology*, *98*(104045), p. 104045. Doi: 10.1016/j.etap.2022.104045.

Kumari, A. and Chaudhary, D.R. (2020). Engineered microbes and evolving plastic bioremediation technology. *Bioremediation of Pollutants*, pp. 417–443. Available at: https://www.sciencedirect.com/science/article/pii/B9780128190258000211 [Accessed 24 Mar. 2023].

Lechner, A. and Ramler, D. (2015). The discharge of certain amounts of industrial microplastic from a production plant into the river Danube is permitted by the Austrian legislation. *Environmental Pollution*, *200*, pp. 159–160. Doi: 10.1016/j.envpol.2015.02.019.

Lee, J. and Chae, K.-J. (2021). A systematic protocol of microplastics analysis from their identification to quantification in water environment: A comprehensive review. *Journal of Hazardous Materials, 403*, p. 124049. 10.1016/j.jhazmat.2020.124049.

Li, Z., Wei, R., Gao, M., Ren, Y., Yu, B., Nie, K., Xu, H. and Liu, L. (2020). Biodegradation of low-density polyethylene by *Microbulbifer hydrolyticus* IRE-31. *Journal of Environmental Management, 263*, p. 110402. 10.1016/j.jenvman.2020.110402.

Magalhães, S., Alves, L., Medronho, B., et al. (2020). Microplastics in ecosystems: From current trends to bio-based removal strategies. *Molecules (Basel, Switzerland), 25*(17), p. 3954. Doi: 10.3390/molecules25173954.

Mahmud, A., Wasif, M.M., Roy, H., et al. (2022). Aquatic microplastic pollution control strategies: Sustainable degradation techniques, resource recovery, and recommendations for Bangladesh. *Water, 14*(23), p. 3968. Doi: 10.3390/w14233968.

Maleki Rad, M., Moghimi, H. and Azin, E (2022). Biodegradation of thermo-oxidative pretreated low-density polyethylene (LDPE) and polyvinyl chloride (PVC) microplastics by Achromobacter denitrificans Ebl13. *Marine Pollution Bulletin, 181*(113830), p. 113830. Doi: 10.1016/j.marpolbul.2022.113830.

McCormick, A., Hoellein, T and Londres, M. (2016). Microplastic in surface waters of urban rivers: Concentration, sources, and associated bacterial assemblages. *Ecosphere, 7*(11), e01556. Doi: 10.1002/ecs2.1556.

Miloloža, M., Cvetnić, M., Grgić, D.K., et al. (2022). Biotreatment strategies for the removal of microplastics from freshwater systems. A review. *Environmental Chemistry Letters, 20*(2), pp. 1377–1402. Doi: 10.1007/s10311-021-01370-0.

Miri, S., Saini, R., Davoodi, S.M., et al. (2022). Biodegradation of microplastics: Better late than never. *Chemosphere, 286*(Pt 1), p. 131670. Doi: 10.1016/j.chemosphere.2021.131670.

Murphy, F., Ewins, C., Carbonnier, F. and Quinn, B. (2016). Wastewater treatment works (WwTW) as a source of microplastics in the aquatic environment. *Environmental Science & Technology, 50*(11), pp. 5800–5808. Doi: 10.1021/acs.est.5b05416.

Nizzetto, L., Futter, M. and Langaas, S. (2016). Are agricultural soils dumps for microplastics of urban origin? *Environmental Science & Technology, 50*(20), pp. 10777–10779. Doi: 10.1021/acs.est.6b04140

O'Brien, S., Rauert, C., Ribeiro, F., Okoffo, E.D., Burrows, S.D., O'Brien, J.W., Wang, X., Wright, S.L. and Thomas, K.V. (2023). There's something in the air: A review of sources, prevalence and behaviour of microplastics in the atmosphere. *Science of The Total Environment, 874*, p. 162193. Doi: 10.1016/j.scitotenv.2023.162193.

Ogunola, O.S., Onada, O.A. and Falaye, A.E. (2018). Mitigation measures to avert the impacts of plastics and microplastics in the marine environment (a review). *Environmental Science and Pollution Research, 25*(10), pp. 9293–9310. Doi: 10.1007/s11356-018-1499-z.

Oliveira, T.A. de, Barbosa, R., Mesquita, A.B.S., Ferreira, J.H.L., Carvalho, L.H. de and Alves, T.S. (2020). Fungal degradation of reprocessed PP/PBAT/thermoplastic starch blends. *Journal of Materials Research and Technology, 9*(2), pp. 2338–2349. Doi: 10.1016/j.jmrt.2019.12.065

Pabortsava, K. and Lampitt, R.S. (2020). High concentrations of plastic hidden beneath the surface of the Atlantic Ocean. *Nature Communications, 11*(1). Doi: 10.1038/s41467-020-17932-9.

Padervand, M., Lichtfouse, E., Robert, D. and Wang, C. (2020). Removal of microplastics from the environment. A review. *Environmental Chemistry Letters, 18*(3), pp. 807–828. Doi: 10.1007/s10311-020-00983-1.

Parks, D.H., Rinke, C., Chuvochina, M., Chaumeil, P.-A., Woodcroft, B.J., Evans, P.N., Hugenholtz, P. and Tyson, G.W. (2017). Recovery of nearly 8,000 metagenome-assembled genomes substantially expands the tree of life. *Nature Microbiology, 2*(11), pp. 1533–1542. Available at: https://www.nature.com/articles/s41564-017-0012-7 [Accessed 24 Mar. 2023].

Perren, W., Wojtasik, A. and Cai, Q. (2018). Removal of microbeads from wastewater using electrocoagulation. *ACS Omega*, *3*(3), pp. 3357–3364.

Pires Costa, L., Vaz de Miranda, D.M. and Pinto, J.C. (2022). Critical evaluation of life cycle assessment analyses of plastic waste pyrolysis. *ACS Sustainable Chemistry & Engineering*, *10*(12), pp. 3799–807.

Puglisi, E., Romaniello, F., Galletti, S., Boccaleri, E., Frache, A. and Cocconcelli, P.S. (2019). Selective bacterial colonization processes on polyethylene waste samples in an abandoned landfill site. *Scientific Reports*, *9*(1), Article 1. Doi: 10.1038/s41598-019-50740-w.

Purca, S. and Henostroza, A. (2017). Presencia de microplásticos en cuatro playas arenosas de Perú. *Revista Peruana de Biología*, *24*(1), pp. 101–106. Doi: 10.15381/rpb.v24i1.12724

Rey, S.F., Franklin, J. and Rey, S.J. (2021). Microplastic pollution on island beaches, Oahu, Hawai'i. *PLOS ONE*, *16*(2), p. e0247224. Doi: 10.1371/journal.pone.0247224.

Rezaei, M., Riksen, M.J.P.M., Sirjani, E., Sameni, A. and Geissen, V. (2019). Wind erosion as a driver for transport of light density microplastics. *Science of the Total Environment*, *669*, pp. 273–281. Doi: 10.1016/J.SCITOTENV.2019.02.382

Rist, S., Carney Almroth, B., Hartmann, N.B. and Karlsson, T.M. (2018). A critical perspective on early communications concerning human health aspects of microplastics. *Science of The Total Environment*, *626*, pp. 720–726. Doi: 10.1016/j.scitotenv.2018.01.092.

Robin, R.S., Karthik, R., Ramesh, R., Ganguly, D.K., Anandavelu, I. and Mugilarasan, M. (2020). Holistic assessment of microplastics in various coastal environmental matrices, southwest coast of India. *Science of the Total Environment*, *703*, pp. 134947–134947. Doi: 10.1016/j.scitotenv.2019.134947.

Rogers, K.L., Carreres-Calabuig, J.A., Gorokhova, E., et al. (2020). Micro-by-micro interactions: How microorganisms influence the fate of marine microplastics. *Limnology and Oceanography Letters*, *5*(1), pp. 18–36. Doi: 10.1002/lol2.10136.

Rose, P.K., Jain, M., Kataria, N., Sahoo, P.K., Garg, V.K. and Yadav, A. (2023). Microplastics in multimedia environment: A systematic review on its fate, transport, quantification, health risk, and remedial measures. *Groundwater for Sustainable Development*, *20*, pp. 100889–100889. Doi: 10.1016/j.gsd.2022.100889.

Saeed, S., Iqbal, A. and Deeba, F. (2022). Biodegradation study of Polyethylene and PVC using naturally occurring plastic degrading microbes. *Archives of Microbiology*, *204*(8). Doi: 10.1007/s00203-022-03081-8.

Salthammer, T. (2022). Microplastics and their additives in the indoor environment. *Angewandte Chemie International Edition*, *61*(32), p. e202205713.

Sánchez, C. (2020). Fungal potential for the degradation of petroleum-based polymers: An overview of macro- and microplastics biodegradation. *Biotechnology Advances*, *40*, p. 107501. Doi: 10.1016/j.biotechadv.2019.107501.

Schymanski, D., Oßmann, B.E., Benismail, N., Boukerma, K., Dallmann, G., von der Esch, E., Fischer, D., Fischer, F., Gilliland, D., Glas, K., Hofmann, T., Käppler, A., Lacorte, S., Marco, J., Rakwe, M.E., Weisser, J., Witzig, C., Zumbülte, N. and Ivleva, N.P. (2021). Analysis of microplastics in drinking water and other clean water samples with micro-Raman and micro-infrared spectroscopy: Minimum requirements and best practice guidelines. *Analytical and Bioanalytical Chemistry*, *413*(24), pp. 5969–5994. Doi: 10.1007/s00216-021-03498-y.

Shabbir, S., Faheem, M., Ali, N., Kerr, P.G., Wang, L.-F., Kuppusamy, S. and Li, Y. (2020). Periphytic biofilm: An innovative approach for biodegradation of microplastics. *Science of The Total Environment*, *717*, p. 137064. Available at: https://www.sciencedirect.com/science/article/abs/pii/S004896972030574X?via%3Dihub [Accessed 18 Mar. 2023].

Skaf, D.W., Punzi, V.L., Rolle, J.T. and Kleinberg, K.A. (2020). Removal of micron-sized microplastic particles from simulated drinking water via alum coagulation. *Chemical Engineering Journal*, *386*, p. 123807.

Song, Y., Zhang, B., Zou, L., Xu, F., Wang, Y., Xin, S., Wang, Y., Zhang, H., Ding, N., & Wang, R. (2022). Comparative Analysis of Selective Bacterial Colonization by

Polyethylene and Polyethylene Terephthalate Microplastics. *Frontiers in Microbiology, 13.* Doi: 10.3389/fmicb.2022.836052.

Srikanth, M., Sandeep, T.S.R.S., Sucharitha, K., et al. (2022). Biodegradation of plastic polymers by fungi: A brief review, *Bioresources and Bioprocessing, 9*(1). Doi: 10.1186/s40643-022-00532-4.

Tang, K.H.D., Lock, S.S.M., Yap, P.-S., Cheah, K.W., Chan, Y.H., Yiin, C.L., Ku, A.Z.E., Loy, A.C.M., Chin, B.L.F. and Chai, Y.H. (2022). Immobilized enzyme/microorganism complexes for degradation of microplastics: A review of recent advances, feasibility and future prospects. *Science of The Total Environment, 832*, p. 154868. Available at: https://www.sciencedirect.com/science/article/abs/pii/S0048969722019611 [Accessed 18 Mar. 2023].

Thakur, P. (2012). Screening of plastic degrading bacteria from dumped soil area [MSc]. http://ethesis.nitrkl.ac.in/3141/

Thompson, R.C., Moore, C.J., vom Saal, F.S. and Swan, S.H. (2009). Plastics, the environment and human health: Current consensus and future trends. *Philosophical Transactions of the Royal Society B: Biological Sciences, 364*(1526), pp. 2153–2166. Doi: 10.1098/rstb.2009.0053.

Tribedi, P., Sarkar, S., Mukherjee, K. and Sil, A.K. (2012). Isolation of a novel Pseudomonas sp from soil that can efficiently degrade polyethylene succinate. *Environmental Science and Pollution Research, 19*(6), pp. 2115–2124. Doi: 10.1007/s11356-011-0711-1.

Tully, B., Graham, E. and Heidelberg, J. (2018). The reconstruction of 2,631 draft metagenome-assembled genomes from the global oceans. *Scientific Data, 5*, p. 170203. Available at: https://www.nature.com/articles/sdata2017203.pdf [Accessed 24 Mar. 2023].

Tursi, A., Baratta, M., Easton, T., et al. (2022). Microplastics in aquatic systems, a comprehensive review: Origination, accumulation, impact, and removal technologies. *RSC Advances, 12*(44), pp. 28318–28340. Doi: 10.1039/d2ra04713f.

Upadhyay, K. and Bajpai, S. (2021). Microplastics in landfills: A comprehensive review on occurrence, characteristics and pathways to the aquatic environment. *Nature Environment and Pollution Technology, 20*(5). Doi: 10.46488/NEPT.2021.v20i05.009.

Viljakainen, V.R. and Hug, L.A. (2021). New approaches for the characterization of plastic-associated microbial communities and the discovery of plastic-degrading microorganisms and enzymes. *Computational and Structural Biotechnology Journal, 19*, pp. 6191–6200. Doi: 10.1016/j.csbj.2021.11.023.

Vivekanand, A.C., Mohapatra, S. and Tyagi, V.K. (2021). Microplastics in aquatic environment: Challenges and perspectives. *Chemosphere, 282*, p. 131151. Doi: 10.1016/j.chemosphere.2021.131151.

Wang, Y., Li, Y.N., Tian, L., Ju, L. and Liu, Y. (2021). The removal efficiency and mechanism of microplastic enhancement by positive modification dissolved air flotation. *Water Environment Research, 93*(5), pp. 693–702.

Wani, A.K., Akhtar, N., Naqash, N., Rahayu, F., Djajadi, D., Chopra, C., Singh, R., Mulla, S.I., Sher, F. and Américo-Pinheiro, J.H.P. (2023). Discovering untapped microbial communities through metagenomics for microplastic remediation: Recent advances, challenges, and way forward. *Environmental Science and Pollution Research.* Doi: 10.1007/s11356-023-25192-5.

Woo, H., Kang, S.H., Kwon, Y., et al. (2022). Sensitive and specific capture of polystyrene and polypropylene microplastics using engineered peptide biosensors. *RSC Advances, 12*(13), pp. 7680–7688. Doi: 10.1039/d1ra08701k.

Wu, F., Guo, Z., Cui, K., Dong, D., Yang, X., Li, J., Wu, Z., Li, L., Dai, Y., & Pan, T. (2023). Insights into characteristics of white rot fungus during environmental plastics adhesion and degradation mechanism of plastics. *Journal of Hazardous Materials*, *448*, p. 130878. 10.1016/j.jhazmat.2023.130878.

Yogalakshmi, K.N. and Singh, S. (2020). "Plastic waste: Environmental hazards, its biodegradation, and challenges," in *Bioremediation of Industrial Waste for Environmental Safety*. Singapore: Springer Singapore, pp. 99–133.

Yuan, J., Ma, J., Sun, Y., et al. (2020). Microbial degradation and other environmental aspects of microplastics/plastics. *The Science of the Total Environment*, *715*(136968), p. 136968. Doi: 10.1016/j.scitotenv.2020.136968.

Zettler, E.R., Mincer, T.J. and Amaral-Zettler, L. (2013). Life in the "Plastisphere": Microbial communities on plastic marine debris. *Environmental Science & Technology*, *47*(13), pp. 7137–7146. Doi: 10.1021/es401288x.

Zhang, C., Wang, J., Zhou, A., Ye, Q., Feng, Y., Wang, Z., Wang, S., Xu, G. and Zou, J. (2021). Species-specific effect of microplastics on fish embryos and observation of toxicity kinetics in larvae. *Journal of Hazardous Materials*, *403*, p. 123948. Doi: 10.1016/j.jhazmat.2020.123948.

Zhang, J., Gao, D., Li, Q., Zhao, Y., Li, L., Lin, H., Bi, Q. and Zhao, Y. (2020). Biodegradation of polyethylene microplastic particles by the fungus *Aspergillus flavus* from the guts of wax moth *Galleria mellonella*. *Science of the Total Environment*, *704*, p. 135931. 10.1016/j.scitotenv.2019.135931.

Zhao, L., Su, C., Liu, W., et al. (2020). Exposure to polyamide 66 microplastic leads to effects performance and microbial community structure of aerobic granular sludge. *Ecotoxicology and Environmental Safety*, *190*, p. 110070. Doi: 10.1016/J.ECOENV.2019.110070.

Zhao, S., Wang, T., Zhu, L., Xu, P., Wang, X., Gao, L. and Li, D. (2021). Corrigendum to 'Analysis of suspended microplastics in the Changjiang Estuary: Implications for riverine plastic load to the ocean' [Water Res. 161 (2019) 560–569/48672]. *Water Research*, *195*, p. 116987. Doi: 10.1016/j.watres.2021.116987.

6 Application of Bioinformatics

Integrated Computational Tools in Green Technologies

M. Nithya Kruthi, Sebika Panja, Abhishek Kumar, Abhishek Sengupta, and Priyanka Narad

6.1 INTRODUCTION

Paulien Hogeweg and Ben Hesper first introduced the term "bioinformatics" in 1970, referring to the study of information processes in biological systems. Bioinformatics represents a critical intersection of biology, computer science, and information technology, according to the National Centre for Biotechnology Information (Fredj Tekaia of the Institut Pasteur further refines this definition, citing bioinformatics as the application of mathematical, statistical, and computational methodologies to solve biological problems using DNA and amino acid sequence data (Hogeweg 2011). Bioinformatics encompasses the management and computational analysis of an array of biological information, spanning from genes, genomes, and proteins to cells, ecological systems, medical data, and even extending to areas such as robotics and artificial intelligence.

The advent of bioinformatics has significantly transformed the field of genomics. Following the sequencing of the first complete microbial genome, that of *Haemophilus influenzae* in 1995, numerous microbial genomes have since been sequenced and deposited in GenBank. Bioinformatics promises immense potential for the agricultural sector, particularly with the sequencing of animal and plant genomes. Bioinformatic techniques facilitate the identification and functional delineation of genes within these genomes. This novel genetic knowledge could stimulate the development of superior crop varieties characterized by enhanced resilience to drought, disease, and pests (Zenda et al. 2021).

6.1.1 Green Technologies

Green technologies, also known as sustainable or clean technologies, are innovative solutions designed to minimize negative environmental impacts and promote the efficient use of resources. These technologies aim to address the challenges posed by climate change, pollution (Du, Cheng, and Yao 2021), resource depletion, and other ecological concerns. One of the key areas of focus for green technologies is renewable energy. These technologies harness energy (Bartlett 2005) from natural sources

DOI: 10.1201/9781003407683-6

that are constantly replenished, such as sunlight, wind, water, and geothermal heat. Solar power, for instance, utilizes photovoltaic cells to convert sunlight into electricity, while wind turbines generate electricity from wind energy.

Another aspect of green technologies is energy (Xiong, Ma, and Ji 2019) efficiency. These technologies aim to reduce energy consumption and optimize energy use in various sectors. Appliances that are energy efficient, including light emitting diode (LED) lighting solutions, function to mitigate energy wastage and diminish overall electricity consumption. In the construction industry, green technologies promote sustainable building practices. Energy-efficient building designs, such as passive solar architecture and proper insulation, minimize energy consumption for heating, cooling, and lighting. Additionally, technologies like rainwater harvesting systems and greywater recycling enable efficient water management in buildings (Wanjiru and Xia 2018). Agriculture and food production also benefit from green technologies. Green technologies extend to various other sectors as well. In the manufacturing industry, sustainable manufacturing processes and materials reduce waste generation, energy consumption, and environmental pollution.

The adoption and advancement of green technologies require supportive policies, research and development, and public awareness. Governments across the world are putting in place rules and rewards to promote the use of green technologies, such as feed-in tariffs for renewable energy generation and tax credits for energy-efficient upgrades (Xu and Ma 2021). Research institutions and private companies invest in developing innovative green technologies, exploring new materials, and improving existing processes. Public awareness campaigns and education programs are essential in informing individuals about green technologies. Also known as sustainable or clean technologies, these innovative solutions are designed to minimize negative environmental impacts and promote the efficient use of resources. Green technologies also play a crucial role in waste management and recycling. Recycling technologies facilitate the recovery and reuse of materials from waste streams, reducing the need for raw materials and minimizing waste disposal. In the construction industry, green technologies promote sustainable building practices. Energy-efficient building designs, such as passive solar architecture and proper insulation, minimize energy consumption for heating, cooling, and lighting. Additionally, technologies like rainwater harvesting systems and greywater recycling enable efficient water management in buildings (Chel and Kaushik 2018). Green technologies extend to various other sectors as well. In the manufacturing industry, sustainable manufacturing processes and materials reduce waste generation, energy consumption, and environmental pollution. Advanced recycling technologies enable the recovery of valuable resources from electronic waste, promoting a circular economy (Srivastav et al. 2023).

6.2 HARNESSING BIOINFORMATICS FOR SUSTAINABLE GREEN TECHNOLOGIES

Bioinformatics is a multidisciplinary field that combines biology, computer science, and information technology to analyze biological data, unravel complex biological processes, and develop innovative solutions. It plays a crucial role in advancing green technologies by providing valuable insights, optimizing processes, and facilitating sustainable practices. Let's explore the importance of bioinformatics in the context of green technologies.

1. **Genomics and Proteomics:** In the field of green technologies, genomics and proteomics data can be utilized to identify genes and proteins responsible for desirable traits, such as drought resistance, disease resistance, and enhanced photosynthesis.
2. **Metagenomics:** Metagenomics involves the study of genetic material collected directly from environmental samples, allowing researchers to explore the microbial communities and their functional potential (Zhang et al. 2021). In the context of green technologies, metagenomics helps identify microbial enzymes and metabolic pathways that can be harnessed for various applications.
3. **Systems Biology:** Bioinformatics plays a vital role in systems biology, which aims to understand biological systems as a whole by integrating data from various sources. By analyzing complex networks of genes, proteins, and metabolites, bioinformatics helps uncover interactions and regulatory mechanisms within biological systems. This knowledge is invaluable for optimizing processes in green technologies. For instance, in biofuel production, bioinformatics analysis of metabolic networks can guide the engineering of microorganisms to enhance fuel production efficiency (Liang, Zhou, and Xu 2016).
4. **Molecular Modeling and Drug Discovery:** Bioinformatics is instrumental in facilitating molecular modeling and simulations, key processes that significantly influence the advancement in therapeutics discovery and the creation of eco-friendly chemical compounds. By utilizing computational techniques (Aminpour, Montemagno, and Tuszynski 2019), researchers can predict the properties, behavior, and interactions of molecules.
5. **Synthetic Biology:** Synthetic biology involves designing and constructing biological systems for useful applications. Bioinformatics tools and algorithms facilitate the design and assembly of genetic components, enabling the creation of synthetic organisms with desired properties. In green technologies (Carbonell et al. 2016), synthetic biology can be employed to develop microorganisms capable of producing biofuels, biodegradable plastics, or other sustainable materials. The flowchart can showcase the different stages of bioinformatics analysis, such as data collection, data integration, data analysis, and application development (Figure 6.1). A simplified flowchart that showcases the different stages of bioinformatics analysis highlight the application of integrated computational tools in green technologies.

6.3 INTEGRATED COMPUTATIONAL TOOLS

Integration of bioinformatics in green technologies not only enhances the understanding of biological processes but also accelerates the development and optimization of sustainable solutions. By leveraging the power of computational analysis, data integration, and modeling, bioinformatics contributes to the advancement of renewable energy, sustainable agriculture, environmental conservation, and more.

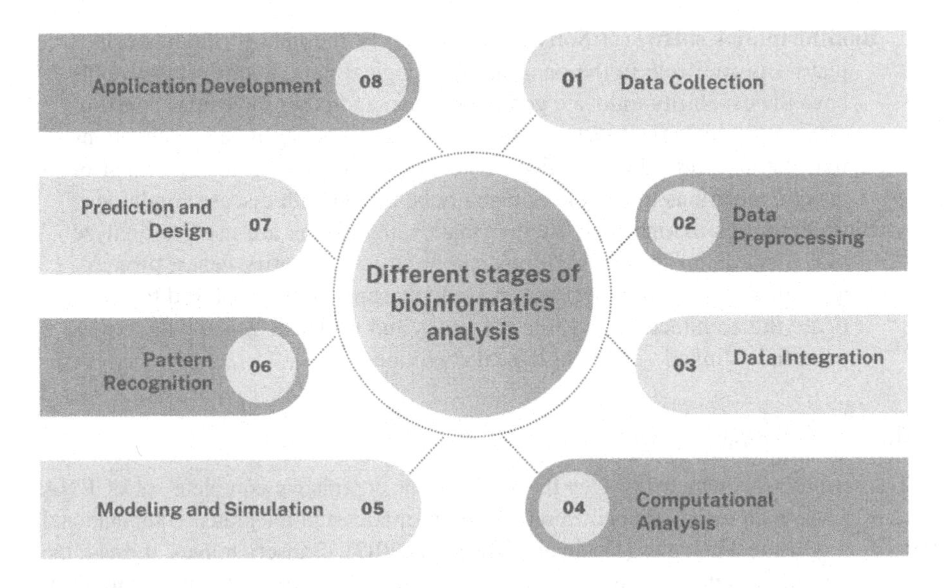

FIGURE 6.1 Different stages of bioinformatics analysis.

6.3.1 Genome Analysis

Genome analysis is a subfield of bioinformatics that focuses on the analysis of genetic information. Genome analysis serves as a comprehensive tool to discover genes, proteins, and various other biomolecules (Bayat 2002). It enables in-depth investigation into the architecture and roles of genes and further aids in the identification of genetic alterations that might be linked to pathological conditions. Genome analysis has a wide range of applications in green technologies. To illustrate, genome analysis can be employed to

- Identify genes involved in the production of biofuels, such as ethanol and biodiesel.
- Identify genes involved in the degradation of pollutants.
- Identify genes involved in plant growth and development.
- Determine the genetic components contributing to plant resilience against pest infestations and pathogenic diseases.

As advancements in genome analysis persist, it is plausible to anticipate an escalating proliferation of its applications, reinforcing its position as an indispensable instrument in the future progression of green technologies. Here are some of the integrated computational tools that are used in genome analysis:

Bioinformatics databases: Bioinformatics databases store biological data, such as genome sequences, protein sequences, and gene annotations. These databases are essential for genome analysis, as they provide researchers with the data they need to study genes and proteins.

Bioinformatics software: Software employed in the field of bioinformatics plays a pivotal role in the analysis of biological data. Such software tools have the capability to locate genes, proteins, and other molecular constituents within biological systems. Further, these tools facilitate investigations into the structure and function of genes, as well as the identification of genetic mutations that could potentially be linked with disease conditions.

Bioinformatics algorithms: Bioinformatics algorithms are used to analyze biological data. These algorithms can be used to identify genes, proteins, and other biological molecules. These tools can also be employed to investigate the architecture and roles of genes and to detect genetic alterations potentially linked with pathological conditions.

6.3.2 Transcriptomics

Transcriptomics, which involves the study of an organism's complete set of RNA transcripts, is an essential application of bioinformatics and integrated computational tools in green technologies (Dolan and coworkers 2017). Transcriptomics, through the examination of gene expression patterns and RNA molecules, delivers meaningful understandings of the functional constituents of genomes, the mechanisms of regulation, and the proceedings at the cellular level. Let's explore how transcriptomics contributes to sustainable practices in areas such as agriculture, environmental conservation, and renewable energy, with some examples.

1. **Crop Improvement:** Transcriptomics is a critical component in the advancement of crop plants for sustainable agriculture. Through the utilization of high-throughput sequencing techniques and bioinformatics software, scientists are able to investigate the transcriptomes of diverse plant tissues in response to varying environmental conditions, stressors, and developmental phases. This analysis facilitates the identification of essential genes and pathways associated with crucial physiological processes.

   Example: Transcriptomics analysis of rice plants subjected to drought stress can identify genes involved in water use efficiency and drought tolerance (Baldoni 2022).
2. **Environmental Monitoring and Conservation:** Transcriptomics aids in monitoring and conserving natural ecosystems. By analyzing the gene expression profiles of indicator species or environmental samples, researchers can assess the health of ecosystems, detect environmental disturbances, and identify potential indicators of pollution or habitat degradation.
3. **Microbial Ecology:** Transcriptomics enables the study of microbial communities and their functional potentials in various environments. By analyzing the transcriptomes of microbial species or metatranscriptomics from environmental samples, researchers can uncover the interactions and ecological roles of microorganisms.

   Example: The analysis of transcriptomics in soil microbial communities has the potential to unveil the genes and metabolic pathways associated with the degradation of organic matter, nutrient cycling, and interactions between plants and microbes. This valuable knowledge can serve as a roadmap for the creation of microbial-driven approaches aimed at promoting sustainable soil fertility management and augmenting crop productivity (McCarren et al. 2010).

4. **Biofuel Production:** Transcriptomics analysis contributes to the development of sustainable biofuel production processes. By studying the gene expression profiles of microorganisms or plants involved in biofuel production, researchers can identify genes and regulatory networks responsible for efficient conversion of biomass to biofuels.

   Example: Transcriptomics analysis of algae species used for biofuel production can identify genes involved in lipid biosynthesis, carbon fixation, and stress response. This knowledge can guide genetic engineering approaches to develop high-yielding algae strains for sustainable biofuel production.

   Plant-Microbe Interactions: Transcriptomics analysis helps in understanding the intricate interactions between plants and beneficial microbes. By studying the gene expression profiles of both plants and associated microbes, researchers can unravel the molecular mechanisms underlying symbiotic or mutualistic relationships. Transcriptomics aids in identifying genes involved in nutrient exchange, disease resistance, and stress tolerance, leading to the development of sustainable agricultural practices.

   Example: Transcriptomics analysis of legume-rhizobia interactions can identify genes involved in nitrogen fixation and nutrient exchange. This work is useful to develop microbial inoculants for sustainable nitrogen management and reduced fertilizer dependency.

Transcriptomics, combined with bioinformatics and integrated computational tools, plays a vital role in advancing green technologies. By providing a comprehensive understanding of gene expression patterns and regulatory networks, transcriptomics enables the development of sustainable practices, conservation strategies, and the design of innovative solutions for a greener and more environmentally friendly future.

6.3.3 Proteomics

Proteomics, the study of proteins and their functions within a biological system, is another vital application of bioinformatics and integrated computational tools in green technologies. By leveraging computational algorithms, data analysis techniques, and advanced bioinformatics tools, proteomics contributes to various sustainable practices in areas such as agriculture, environmental conservation (Lacerda and Reardon 2009), and renewable energy. Let's explore some examples of how proteomics analysis benefits green technologies.

1. **Proteomics in Environmental Conservation:** Proteomics analysis contributes to biodiversity conservation efforts and the management of endangered species. By studying the proteomes of rare or endangered species,

researchers can gain insights into their physiological adaptations, stress responses, and unique characteristics.

2. **Sustainable Aquaculture:** Proteomics analysis plays a significant role in improving aquaculture practices by enhancing fish health, nutrition, and production efficiency. By studying the proteomes of aquaculture species, researchers can identify proteins associated with growth, immunity, and stress responses.
3. **Protein Engineering and Biocatalysis:** Proteomics analysis, combined with computational tools, facilitates protein engineering and the design of biocatalysts for green technologies. By studying the structure-function relationships of proteins and enzymes, researchers can identify key amino acid residues and domains responsible for specific activities or properties. Computational algorithms aid in predicting protein structures, modeling protein-protein interactions, and designing mutant enzymes with desired characteristics (Woodley 2018).
4. **Sustainable Forestry and Wood Utilization:** Proteomics analysis contributes to understanding the molecular mechanisms of wood formation, decay, and chemical composition. By studying the proteomes of trees and wood-degrading organisms, researchers can identify enzymes involved in lignocellulosic degradation, lignin modification, and cellulose synthesis. This knowledge supports the development of sustainable forestry practices, including the selection of trees with desirable wood properties and the optimization of wood processing for bioenergy, pulp, and paper industries (Hernandez-Morcillo et al. 2022).
5. **Sustainable Waste Management and Bioplastics:** Proteomics analysis contributes to the development of microbial-based systems for waste management and bioplastic production. By studying the proteomes of microorganisms involved in waste degradation or bioplastic synthesis, researchers can identify enzymes and metabolic pathways for efficient waste conversion and biopolymer production.

The integration of proteomics and bioinformatics in green technologies revolutionizes our understanding of protein functions and opens doors to innovative solutions for sustainable practices. By unraveling the complexities of proteins and their interactions, proteomics analysis provides valuable insights for optimizing agricultural productivity, developing bioenergy solutions, mitigating environmental pollution, and promoting conservation efforts.

6.3.4 Metabolomics

Metabolomics, a branch of bioinformatics and integrated computational tools, plays a significant role in green technologies. It involves the comprehensive analysis of metabolites, the small molecules involved in cellular metabolism, in various biological systems. By combining analytical techniques with computational algorithms, metabolomics enables the identification, quantification, and analysis of metabolites, providing valuable insights into biochemical pathways, metabolic networks, and the overall metabolic status of organisms. Let's explore how metabolomics contributes to sustainable practices in areas such as agriculture, environmental monitoring, and renewable energy.

1. **Industrial Biotechnology:** Metabolomics facilitates the optimization of bioproduction processes in the biofuel, pharmaceutical, and chemical industries.
   Example: Metabolomics analysis of microbial strains used for biofuel production (Ramamurthy et al. 2021) can identify key metabolites associated with high yields and increased production rates, guiding the engineering of more efficient and sustainable biofuel production processes.
2. **Nutritional Science:** Metabolomics provides insights into the complex interactions between diet, metabolism, and human health. By analyzing metabolite profiles in biological samples, such as blood, urine, or tissues, bioinformaticians can identify metabolic signatures associated with specific dietary patterns, nutrient metabolism, or disease conditions.
   Example: Metabolomics analysis of human blood samples can identify metabolites associated with the consumption of specific food groups (Guasch-Ferré, Bhupathiraju, and Hu 2018), enabling the development of personalized dietary interventions for improved health outcomes and disease prevention.
3. **Bioenergy and Renewable Resources:** Metabolomics analysis contributes to the development of sustainable bioenergy production systems and the utilization of renewable resources.
   Example: Metabolomics analysis of lignocellulosic biomass degradation by microbial communities can identify key metabolic intermediates and enzymes involved in the efficient conversion of biomass into biofuels (Vélez-Mercado et al. 2021), facilitating the development of sustainable and economically viable bioenergy processes.

6.4 BIOINFORMATICS IN MICROBIAL COMMUNITIES

In this section, we aim to delve into comprehensive analysis and comprehension of the microbial compositions in a particular system, possibly related to agriculture, biotechnology, or environmental science. By analyzing the microbial compositions and diversity/similarity, researchers can gain insights into the complex interactions and metabolic networks between different microorganisms. There have been recent events that led to remarkable advancement in high-throughput sequencing technologies and methods that has led to substantial progress in analyzing metagenomic data derived from microbial communities (Ortiz-Estrada et al., 2019). One compelling evidence of this progress is a discernible rise in the number of articles in scholarly journals that discuss metagenomics and pyrosequencing, particularly in areas such as waste treatment, bioenergy, and bioremediation. The production of massive content of genomic data from a variety of microorganisms present in a sample is made possible by these sequencing techniques, such as pyrosequencing in the span period from 2008 to 2021 (Figure 6.2).

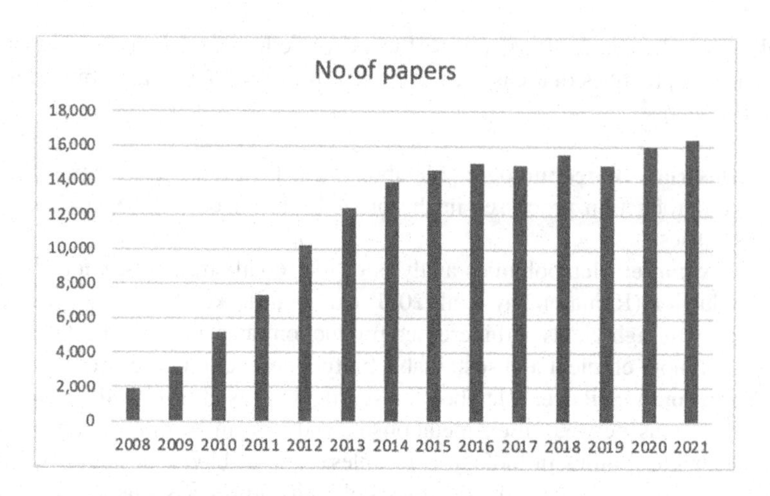

FIGURE 6.2 Numbers of papers published during the tenure of 2008–2021.

The availability of high-resolution sequencing technology has transformed the field of metagenomics by enabling researchers to analyze complex microbial communities in unprecedented detail (Laskar et al. 2018). These sequencing techniques, like pyrosequencing, enable the production of enormous amounts of genomic data from a variety of microorganisms present in a sample. Scientists can learn by collecting such vast amounts of data, more about the genetic makeup and functional potential of microbial communities, which will help them understand their roles in a variety of applications, such as waste treatment (Antwi et al. 2017), bioenergy production (Nie et al. 2017), and bioremediation (Kuppusamy et al. 2016). The enormous amount of metagenomic data generated has simultaneously become easier for researchers to process and analyze due to the falling costs of bioinformatics technologies and tools.

In 2008, the results of the first metagenomics examination of the microbial populations in biogas digesters were reported (Hassa et al. 2018). This groundbreaking research heralded the start of a new era in the comprehension of and utilization of the potential of microbial communities to produce biogas. Over time, the development of effective software tools, improved standard operating procedures, and improved gene sequence analysis technologies helped metagenomics analysis advance (Faust 2021).

In the "2030 Framework for Climate and Energy," the European Council established objectives for exploiting renewable energy and lowering greenhouse gas emissions in 2014 (Fekete et al. 2021). Innovative strategies were needed to reduce greenhouse gas emissions by 40% from 1990 levels and use renewable energy at least at a rate of 27% by 2030. One such strategy, microbial community metagenomics analysis, has emerged, underscoring its significance in addressing these environmental challenges (Rose et al. 2016). Our understanding of microbial compositions and network-based correlation analysis has substantially improved because of the advent of bioinformatics tools such as MG-RAST, MetaVelvet, and Genovo (Kumar Awasthi et al. 2020). Researchers have been able to analyze sizable metagenomic datasets and gain insightful knowledge by using these bioinformatics tools. For instance, the popular metagenomic analysis pipeline MG-RAST offers a thorough platform for characterizing microbial communities and investigating their potential functionalities (Liu et al. 2021). Moreover, the integration of data mining, artificial intelligence, and statistical approaches within these bioinformatics tools has enhanced our ability to extract meaningful information from complex metagenomic datasets. Researchers can learn more about the ecological functions of various microbial taxa and their contributions to important processes like nutrient cycling, carbon fixation, and biodegradation by identifying these correlations (Gill, Lee, and McGuire 2017). Numerous disciplines, including environmental science, agriculture, biotechnology, and human health, will be significantly impacted by this understanding (Gupta et al. 2017).

6.4.1 Genomics: 16S rRNA

The identification of the 16S rRNA gene sequence, which is found in mostly all microorganisms and contains both conserved and variable sections (Fuks et al. 2018), was a significant advance in genomics. The best marker for analyzing phylogenetic relationships between organisms in microbial communities has emerged like the 16S rRNA gene (Rashid et al. 2020). The hypervariable areas can be selectively amplified and sequenced, allowing for a thorough evaluation of microbial diversity.

6.4.2 Next-Generation Technologies

Sanger sequencing was created in 1977 as a result of the groundbreaking work of Fred Sanger and Alan R. Coulson (John Ogbe et al. 2016). With gradual advancements, this technique transformed DNA sequencing for the next three decades. In contrast to Sanger sequencing, which might produce read lengths of 700–1000 base pairs, NGS typically produces shorter read lengths, ranging from 30 to 500 base pairs (Dapprich et al. 2016). Six basic techniques are frequently used in modern NGS research. These methods include fragmenting the target DNA, joining adapter molecules to the fragments to speed up sequencing reactions, amplifying the fragments through cloning techniques, sequencing the DNA fragments themselves, gathering the raw data produced during the sequencing process, and finally base calling, which transforms the raw data into readable DNA sequences (Slatko, Gardner, and Ausubel 2018).

6.4.3 Pyrosequencing

Pyrosequencing, a rapid DNA sequencing technique that enables the real-time determination of DNA sequence by the detection of bioluminometric signals, was developed as a result of this fundamental notion (Li 2021). The first commercially accessible next-generation sequencing (NGS) device, the 454 devices, uses pyrosequencing as its core technique (Besser et al., 2018). The detection of pyrophosphate molecules, which are generated during DNA synthesis, is the basis for this sequencing technique. Every time a nucleotide is incorporated, a series of enzymatic processes including DNA polymerase, luciferase, apyrase, and ATP sulfurylase begin, which eventually produce visible light. Various prerequisite tasks must be completed before the actual pyrosequencing analysis. The template DNA is initially immobilized by being bound to magnetic beads. The target DNA fragments are subsequently amplified using an emulsion polymerase chain reaction (PCR). Following the deposition of the DNA fragments on the magnetic beads onto a specially created plate, the enzymatically generated light reactions are carefully quantified. Microfluidics technology is used during the sequencing run to ensure efficient handling and regulated input of nucleotides and enzymes at predefined intervals (Suea-Ngam et al. 2020). The greatly improved sequencing capacity of pyrosequencing is one of its outstanding benefits. A single DNA sample can produce thousands of sequences in a matter of hours once the complementary DNA strand has been synthesized. The technique is made faster and more efficient by performing the sequencing process simultaneously for each fragment. Long read lengths and high-resolution genomic data have made it possible to conduct in-depth analyses of microbial relationships through PacBio sequencing (Frank et al. 2018). Real-time monitoring of microbes communities (Ciuffreda, Rodríguez-Pérez, and Flores 2021) using nanopore sequencing has enabled on-site identification of possible pathogens or pollution indicators in water quality evaluation, permitting quick reactions to environmental risks. Hence, this NGS technology enables the investigation of functional genes and metabolic pathways within microbial communities.

6.5 SUSTAINABLE WASTE TREATMENT

The need to control and lessen the effects of waste on the environment has led to a long history of evolution in waste treatment technologies. The present waste treatment uses physical, chemical, and biological procedures to remove contaminants and prevent tainted sources. Bioremediation techniques have been developed more recently because of an increasing focus on environmentally friendly waste management strategies (Quintella et al., 2019). Utilizing microorganisms' inherent capabilities to break down contaminants and clean up contaminated settings is known as bioremediation. Anaerobic digestion, composting, and biofiltration are the three processes explained in bioremediation in this chapter (Figures 6.3 and 6.4).

6.5.1 Anaerobic Digestion

Utilizing the natural process of microbial decomposition in the absence of oxygen, i.e., anaerobic digestion (AD), plays a vital role in the treatment of garbage. Organic waste products, such as agricultural wastes, food waste, and wastewater sludge, are broken down by AD systems through a sequence of biological interactions (Abbas et al. 2021). The waste is transformed through this process into useful by-products, principally digestate and biogas. In AD, hydrolytic bacteria first hydrolyze complex organic molecules into simpler ones. After that, these substances undergo additional metabolism known as acidogenesis and acetogenesis, which produces organic acids and volatile fatty acids. Biogas is created when methanogenic bacteria finally transform these intermediate products into methane and carbon dioxide. Methane makes up the majority of biogas, which is a renewable source of energy that can be used for electricity generation, heating, or as a vehicle fuel (Ullah Khan et al. 2017). AD has

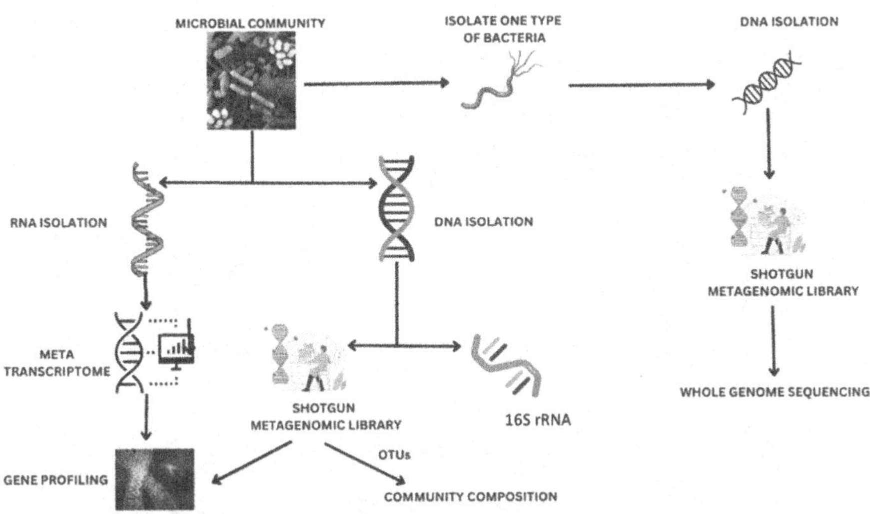

FIGURE 6.3 Various molecular methods from microbial samples.

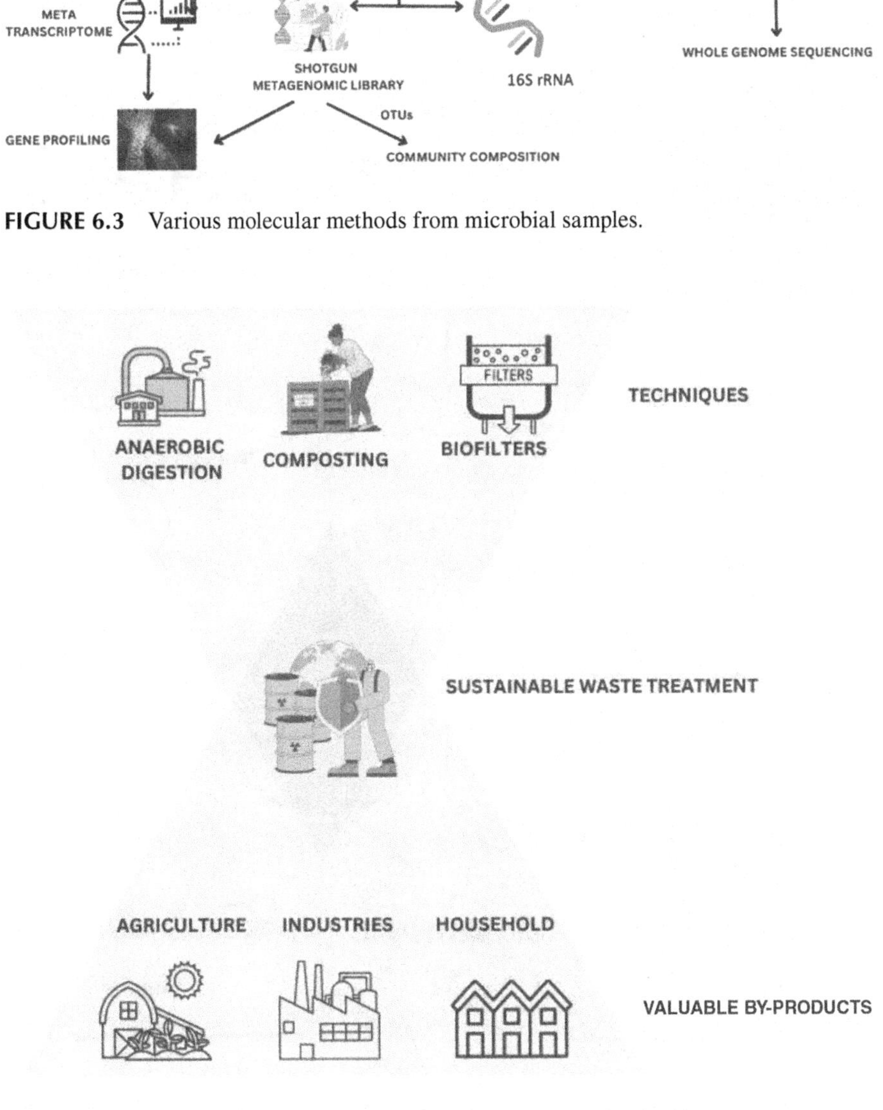

FIGURE 6.4 Sustainable waste treatment (techniques and by-products).

environmental advantages in addition to being a useful waste treatment technique. By capturing and using methane, a strong greenhouse gas, as a renewable energy source, it reduces greenhouse gas emissions. Additionally, AD lowers the volume of organic waste, reducing the requirement for landfilling and the concerns of environmental degradation that come with it (Figure 6.5).

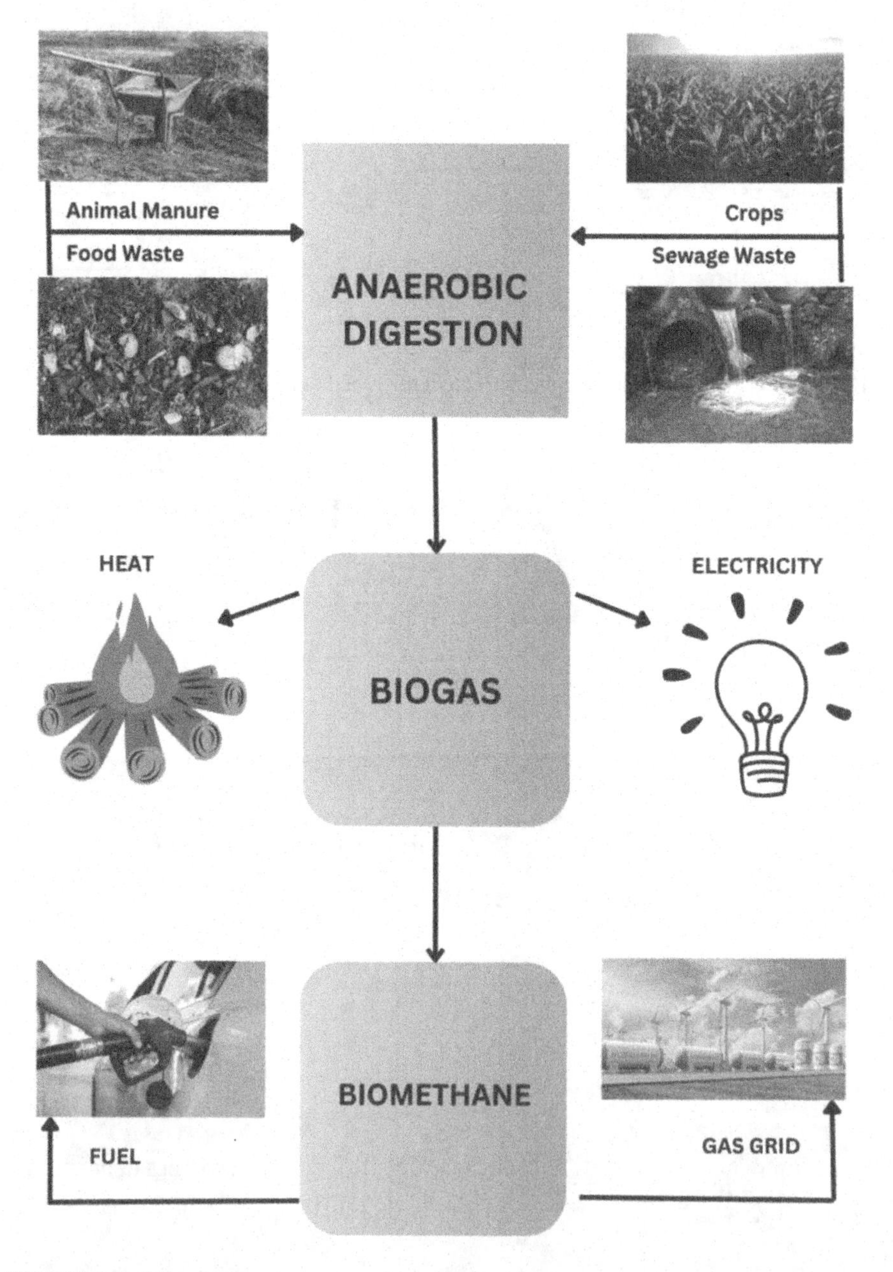

FIGURE 6.5 Bioenergy.

6.5.2 Composting

Waste products are gathered and heaped in composting containers. The decomposition process is started by the microbial community, which may already exist or be introduced through inoculation. Bacteria play a key part in the first breakdown, best demonstrated by taxa like *Bacillus*, *Pseudomonas*, and *Actinobacteria* (Kertesz and Thai 2018). The resulting compost is a valuable resource that may be used to enrich the soil, strengthen it, increase its capacity to store water, and encourage healthy plant growth.

6.5.3 Biofiltration

A sustainable waste treatment technique called biofiltration uses live organisms' metabolic processes, such as those of bacteria, fungi, and algae, to effectively remove pollutants and toxins from various environmental matrices like soil, water, and air. The microbes in this biofilter use the contaminants as a source of nutrients, allowing them to break them down into harmless by-products like water and carbon dioxide.

Researchers use bioinformatics to examine 16S rRNA gene sequences obtained through pyrosequencing to pinpoint the bacterial species that are prevalent in AD, composting, and biofiltration systems. This characterization makes it possible to assess the composition, variety, and abundance of bacterial communities involved in these processes. Additionally, 16S rRNA analysis aids in identifying phylogenetic connections across bacterial species, revealing details about their evolutionary ties and functional roles. Various species have been identified as significant contributors to pollutant degradation in biofiltration systems.

6.6 BIOENERGY

Utilizing microorganisms to produce bioenergy entails taking advantage of their metabolic capacity to transform organic matter into useful energy in the form of biofuels or other valuable products. This procedure includes three unique stages: choosing the substrate, growing the bacteria, and recovering the finished product (Kumar et al. 2018). The organic molecules in the substrates must be metabolizable by the chosen bacteria. The grown microorganisms use their particular metabolic pathways during the metabolic conversion step to convert the organic matter contained in the substrate into the necessary bioenergy products. Depending on the particular microorganism and the desired end product, different biochemical pathways may be used throughout this metabolic conversion process (Figure 6.6).

6.6.1 Anaerobic Digestion

High-throughput sequencing and other molecular methods have made it possible to analyze microbial communities and expose the dynamics of those populations during anaerobic digestion. This information makes it easier to create plans for microbial consortium optimization and process stability enhancement. The NGS technologies make it easier to categorize microbial communities, estimate their diversity, and

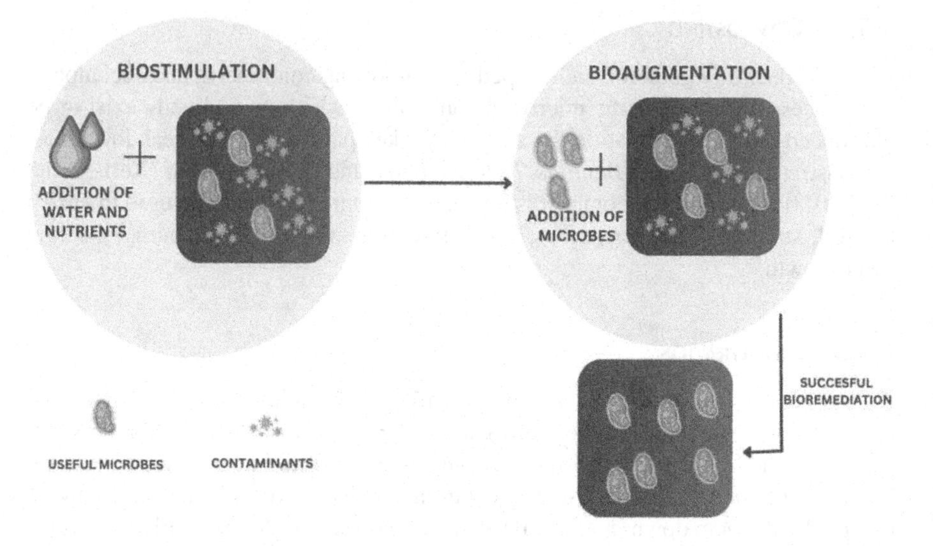

FIGURE 6.6 Bioremediation.

conduct comparative analysis using software programs like MOTHUR and QIIME (Quantitative Insights Into Microbial Ecology) (Straub et al. 2020).

6.6.2 Microbial Fuel Cell

A type of bioenergy technology known as microbial fuel cells (MFCs) uses microorganisms' metabolic pathways in order to generate electricity.

Bioinformatics methods, like 16S rRNA sequencing, pyrosequencing, metagenomics, transcriptomics, and proteomics, have significantly contributed to the scientific understanding and enhancement of bacteria in MFCs for bioenergy production. Based on sequence similarity, operational taxonomic units (OTUs) representing various bacterial species or groupings can be created using bioinformatics analysis of pyrosequencing date (Mysara et al., 2017). Through this information, researchers gather knowledge about the mechanisms and interactions between bacteria and the electrode surface, which can be used to enhance the performance of MFCs.

6.7 BIOREMEDIATION

Microbial activity-based bioremediation has become a practical and long-term method for cleaning up a variety of environmental contaminants. Metagenomic and metatranscriptomic analyses used in genomic research have shown particular genetic components and metabolic routes involved in how various pollutants break down. This information makes it easier to comprehend the dynamics of microbial communities and makes it possible to develop specialized bioremediation techniques (Ojuederie and Babalola 2017). Optimizing microbial metabolic activity and enhancing the effectiveness of bioremediation are the goals of genetic changes (Figure 6.7).

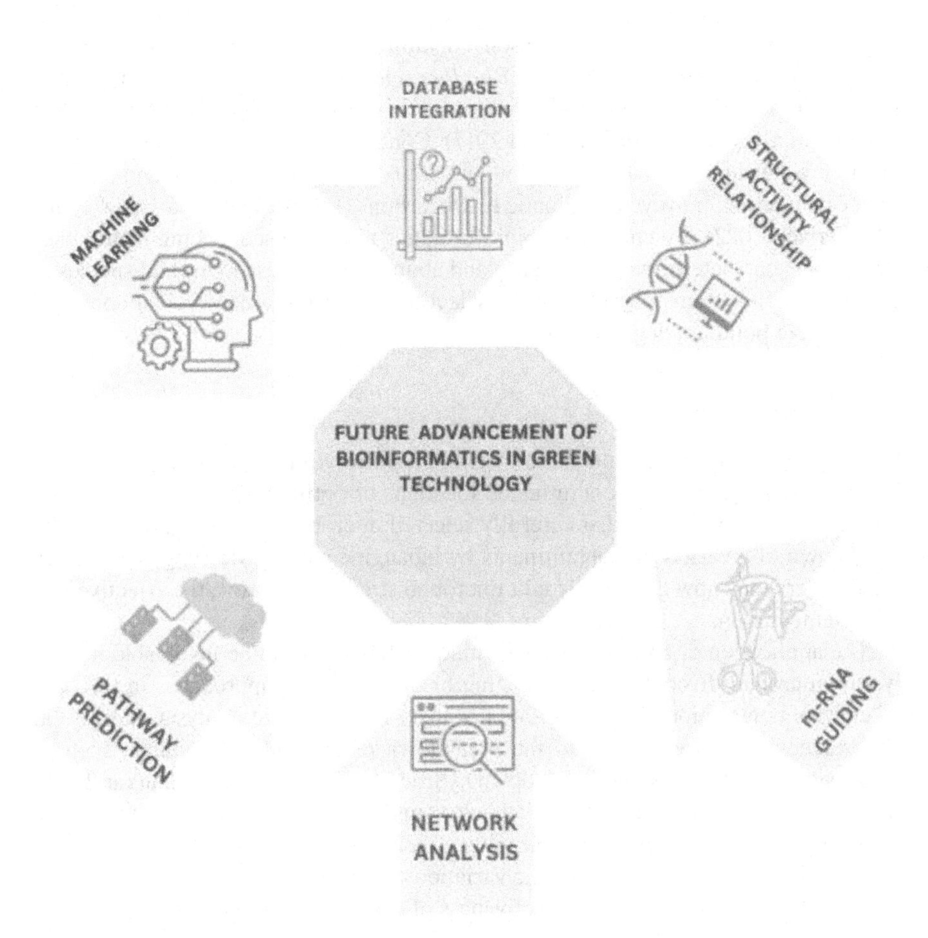

FIGURE 6.7 Future advancement of bioinformatics in green technology.

6.7.1 Biostimulation

The process of employing microbes in biostimulation as bioremediation entails using them to break down or eliminate contaminants from contaminated areas (Kebede et al. 2021). In the process of biostimulation, the addition of particular nutrients promotes the growth of microorganisms that may break down the desired contaminants. These nutrients provide the microbial population with energy and building blocks, fostering their metabolic activities and speeding up the rate of disintegration. Metagenomics and metatranscriptomics methods reveal the genetic potential and gene expression patterns of bacteria discovered in polluted environments. This information aids in choosing the best microbial consortia for biostimulation and improving their functionality.

Bioinformatics analysis has discovered microbial species with pertinent capabilities for use in biostimulation projects that target particular contaminants, such as petroleum hydrocarbons. *Pseudomonas putida* is well-known for its ability to degrade a wide range of hydrocarbon contaminants, including petroleum-based compounds and aromatic pollutants. *Rhodococcus erythropolis* is often utilized in bioremediation efforts for hydrocarbon-contaminated sites, especially for degrading long-chain alkanes (Koshlaf and Ball 2017). Contrarily, *Bacillus subtilis* is a multipurpose bacterium that can break down a variety of organic contaminants, including pesticides and polycyclic aromatic hydrocarbons (PAHs) (Espinosa-Ortiz, Rene, and Gerlach 2022). By employing bioinformatics tools and sequencing techniques, researchers can determine the presence and abundance of these microbial species in biostimulation systems, facilitating the selection of appropriate microbial consortia for effective pollutant degradation.

6.7.2 Bioaugmentation

Microorganism-based bioremediation techniques involve either screening and isolating microorganisms from contaminated locations or optimizing microbial communities. Studies have shown that carefully selected microbial strains can enhance the breakdown of a variety of contaminants by bioaugmentation (Mawang et al. 2021). Researchers may now genetically edit microbial strains to improve the effectiveness of bioremediation.

The application of bioinformatics techniques has shown to be invaluable in analyzing microbial diversity and optimizing bioaugmentation approaches in the field of environmental bioremediation. For instance, metagenomic analysis may reveal the existence of genes linked to the breakdown of particular pollutants, such as those encoding enzymes that break down hydrocarbons (Thakur and Shankar 2022). Scientists can identify the genetic changes underlying enhanced biodegradation skills by contrasting the genomes of highly effective degrading organisms with those of less efficient species. These genetic variations can be used to enhance bioaugmentation techniques and boost the effectiveness of environmental remediation.

6.8 FUTURE DIRECTIONS AND CHALLENGES

The field of bioinformatics has immense potential to contribute to sustainable green technologies.

6.8.1 Emerging Technologies

Microbe communities play a critical role in green technology, and next-generation sequencing (NGS) technologies have transformed the field of green technology. NGS technologies have improved from the first-generation sequencing method like Sanger sequencing, yet they still have drawbacks. For the advancement of sustainable green technology, new sequencing methods that can produce lengthy reads with high precision and at a lower cost are essential.

As the microbial communities are complex, diverse, and dynamic, microbial metagenomics pose various difficulties like pre-treatment, assembly, binning, and annotation. For instance, a study to analyze the microbial populations found in activated sludge samples from wastewater treatment plants used T-RFLP, FISH, and qPCR in conjunction with metagenomics-based bioinformatics methods (Kumar et al. 2020).

There is always an opportunity to improve current bioinformatics tools and data processing techniques. One such tactic is machine learning (Kornej et al. 2020), which comprises teaching algorithms to recognize patterns and make predictions based on big datasets. Another approach is network analysis, which simulates interactions between microbial species or microbes and their environment. Network analysis (Schmidt et al. 2019) can assist us in comprehending the functioning of microbial communities and the ecological interactions among different species. Additionally, the assessment of chemical safety in environmental management is made possible by bioinformatics techniques. The development of numerous applications has been aided by integrating bioinformatics with chemical databases and toxicity prediction (Zhao et al. 2020). Bioinformatics tool MetaRouter (Singh et al. 2021) manages and mines heterogeneous biodegradation-related data to facilitate bioremediation research. Researchers can find patterns, correlations, and trends in the datasets using MetaRouter's sophisticated data mining tools. A deeper understanding of biodegradation processes, identifying prospective biodegradation pathways, and discovering novel enzymes or microbial populations engaged in pollutant breakdown can all be accomplished using this information.

6.8.2 Data Integration and Analysis

To efficiently arrange and retrieve data, MetaRouter makes use of complex indexing systems, such as the SRS indexing system (Shekhar, Godheja, and Modi 2020). This indexing system enables speedy and precise searches, ensuring that researchers can find pertinent material within the database with ease. A strong and secure framework for data storage is also provided by the use of PostgreSQL (Dahunsi et al. 2021). The system helps researchers find hidden relationships, produce hypotheses, and decide on bioremediation solutions by combining various datasets and offering data integration and analysis tools.

To advance green technology, bioinformatics and wet lab methods must be integrated. With its computer tools and algorithms, bioinformatics makes it possible to analyze and interpret vast amounts of biological data. Wet lab methods, on the other hand, use data generation and experimentation to deliver crucial biological knowledge. Utilizing the full potential of bioinformatics in the design of effective bioremediation and climate-resilient crop production systems requires collaboration between these two fields. The limitations and constraints of each omics platform should be taken into account by the new data integration techniques.

Another important aspect to consider is eliminating discrepancies and ensuring the quality of the provided data, which can be facilitated by the creation of defined methods for data processing and analysis. Researchers in the field of sustainable green technologies must collaborate and share data to advance the discipline. The creation of platforms and databases for data exchange would accelerate research progress and facilitate the development of new technologies.

6.8.3 Ethical and Legal Considerations

The new data integration methods should take into account the constraints and limitations of each omics platform (Graw et al. 2021). Each strategy has drawbacks in terms of the number of samples, technical variables, batch effects, sequencing depths, statistical power, sample preparation, and a wide range of additional variables.

The ability of a study to identify significant effects when they exist is referred to as statistical power. The capacity to recognize significant relationships or trends in the context of green technology may be constrained by a lack of statistical power. A key element influencing bioinformatics studies' precision and generalizability is the sample size. Insufficient sample sizes cause skewed or unreliable results; obtaining large and diverse sample sizes might be difficult for studies in green technology, especially when working with rare or endangered species or in particular settings.

Data integration might be difficult as a result of these variables' potential to produce systematic biases or disparities among datasets. Standardization and normalizing procedures are used to lessen these restrictions. Numerous computational techniques, like ComBat (Choudhury 2021) and Surrogate Variable Analysis (SVA) (Chen et al. 2017), were created to manage batch impacts. These techniques seek to eliminate or correct batch effects, enabling precise omics data potential for using bioinformatics to address environmental issues, and open the door to a greener, more sustainable future.

6.9 CONCLUSION

6.9.1 Implications for Sustainable Green Technologies

The application of bioinformatics and integrated computational tools in green technologies has significant implications for the development of sustainable practices. These tools can potentially revolutionize various sectors, from energy production and waste management to agriculture and biodiversity conservation. Biofuel production stands to benefit immensely from bioinformatics. The potential to genetically modify organisms for efficient biofuel production holds the promise of a renewable and carbon-neutral source of energy. Bioinformatics can help identify key enzymes and metabolic pathways in these organisms, increasing biofuel yield, reducing cost, and making biofuels a more viable alternative to fossil fuels.

The implications of bioinformatics for agriculture are vast. Genomic studies can lead to the development of more robust and resilient crops, reducing dependency on pesticides and fertilizers and promoting sustainable farming practices. This genetic modification could lead to higher crop yields, enhancing food security and promoting a more sustainable agricultural system. Genetic modification, while promising, raises ethical and safety concerns that need to be addressed. Additionally, the handling of large-scale biological data poses privacy and security challenges.

6.9.2 Summary of Key Points

Bioinformatics, a multidisciplinary field that combines biology, information technology, and computer science, has demonstrated immense potential in various sectors, including green technologies. The application of bioinformatics in green technology is aimed at developing sustainable, eco-friendly solutions that cause minimal harm to the environment.

A crucial application of bioinformatics in green technologies has been observed in the realm of biofuel production. The combination of genomic data and computational tools allows researchers to identify and modify organisms capable of producing biofuels more efficiently. These modified organisms could potentially convert biomass into biofuel more efficiently than conventional methods, leading to higher yields and a reduced carbon footprint. Another significant area where bioinformatics shows promise is in waste management. By understanding the genetic makeup of bacteria and other microorganisms, it's possible to engineer species that can degrade waste more efficiently, reducing the volume of waste that ends up in landfills and oceans. Similarly, the study of microbial genomes can aid in the development of novel bioremediation strategies, utilizing organisms to eliminate or neutralize pollutants from a contaminated site.

In conclusion, the integration of bioinformatics with green technologies offers a path toward a more sustainable and environmentally friendly future. The amalgamation of computational tools and biological data enables us to better understand and manipulate our world in a way that aligns with the principles of sustainability. As we continue to develop and refine these tools, the possibilities for creating a more sustainable world continue to grow.

REFERENCES

Abbas, Y, S Yun, Z Wang, Y Zhang, X Zhang, and K Wang. 2021. Recent advances in bio-based carbon materials for anaerobic digestion: A review. Renew Sust Energ Rev. doi: 10.1016/j.rser.2020.110378.

Aminpour, M, C Montemagno, and JA Tuszynski. 2019. An overview of molecular modeling for drug discovery with specific illustrative examples of applications. Molecules. 24(9):1693. doi: 10.3390/molecules24091693.

Antwi, P, J Li, P Opoku Boadi, J Meng, E Shi, C Xue, Y Zhang, and F Ayivi. 2017. Functional bacterial and archaeal diversity revealed by 16S rRNA gene pyrosequencing during potato starch processing wastewater treatment in an UASB. Bioresour Technol. 235: 348–357. doi: 10.1016/j.biortech.2017.03.141.

Awasthi, MK, B Ravindran, S Sarsaiya, et al. 2020. Metagenomics for taxonomy profiling: Tools and approaches. Bioengineered. doi: 10.1080/21655979.2020.1736238.

Baldoni, E. 2022. Improving drought tolerance: Can comparative transcriptomics support strategic rice breeding? Plant Stress. 3:100058.

Bartlett, A. 2005. Farmer Field Schools to promote Integrated Pest Management in Asia: the FAO Experience.

Bayat, A. 2002. Science, medicine, and the future: Bioinformatics. BMJ. 324(7344):1018–1022. doi: 10.1136/bmj.324.7344.1018.

Carbonell, P, A Currin, AJ Jervis, NJ Rattray, N Swainston, C Yan, E Takano, and R Breitling. 2016. Bioinformatics for the synthetic biology of natural products: Integrating across the design-build-test cycle. Nat Prod Rep. 33(8):925–932. doi: 10.1039/c6np00018e.

Chel, A, and G Kaushik. 2018. Renewable energy technologies for sustainable development of energy efficient building. Alex Eng J. 57(2):655–669.

Chen, J, E Behnam, J Huang, MF Moffatt, DJ Schaid, L Liang, and X Lin. 2017. Fast and robust adjustment of cell mixtures in epigenome-wide association studies with SmartSVA. BMC Genom. 18(1). doi: 10.1186/s12864-017-3808-1.

Choudhury, A. 2021. The role of machine learning algorithms in materials science: A state of art review on industry 4.0. Arch Comput Methods Eng. 28(5):3361–3381. doi: 10.1007/s11831-020-09503-4.

Ciuffreda, L, H Rodríguez-Pérez, and C Flores. 2021. Nanopore sequencing and its application to the study of microbial communities. Comput Struct Biotechnol J. doi: 10.1016/j.csbj.2021.02.020.

Dahunsi, FM, AJ Joseph, OA Sarumi, and OO Obe. 2021. Database management system for mobile crowdsourcing applications. doi: 10.4314/njt.v40i4.18.

Dapprich, J, D Ferriola, K Mackiewicz, et al. 2016. The next generation of target capture technologies – large DNA fragment enrichment and sequencing determines regional genomic variation of high complexity. BMC Genom. 17(1). doi: 10.1186/s12864-016-2836-6.

Du, K, Y Cheng, and X Yao. 2021. Environmental regulation, green technology innovation, and industrial structure upgrading: The road to the green transformation of Chinese cities. Energ Econ. 98:105247. doi: 10.1016/j.eneco.2021.105247.

Espinosa-Ortiz, EJ, ER Rene, and R Gerlach. 2022. Potential use of fungal-bacterial co-cultures for the removal of organic pollutants. Crit Rev Biotechnol. doi: 10.1080/07388551.2021.1940831.

Faust, K. 2021. Open challenges for microbial network construction and analysis. ISME J. doi: 10.1038/s41396-021-01027-4.

Fekete, H, T Kuramochi, M Roelfsema, et al. 2021. A review of successful climate change mitigation policies in major emitting economies and the potential of global replication. Renew Sust Energ Rev. 137(March). doi: 10.1016/j.rser.2020.110602.

Frank, J, S Lücker, RHAM Vossen, MSM Jetten, RJ Hall, HJM Op Den Camp, and SY Anvar. 2018. Resolving the complete genome of Kuenenia stuttgartiensis from a membrane bioreactor enrichment using single-molecule real-time sequencing. Sci Rep. 8(1). doi: 10.1038/s41598-018-23053-7.

Fuks, G, M Elgart, A Amir, A Zeisel, PJ Turnbaugh, Y Soen, and N Shental. 2018. Combining 16S rRNA gene variable regions enables high-resolution microbial community profiling. Microbiome. 6(1). doi: 10.1186/s40168-017-0396-x.

Gill, AS, A Lee, and KL McGuire. 2017. Phylogenetic and functional diversity of total (DNA) and expressed (RNA) bacterial communities in urban green infrastructure bioswale soils. Appl Environ Microbiol. 83(16). doi: 10.1128/AEM.00287-17.

Graw, S, K Chappell, CL Washam, A Gies, J Bird, MS Robeson, and SD Byrum. 2021. Multi-omics data integration considerations and study design for biological systems and disease. Mol Omics. doi: 10.1039/d0mo00041h.

Guasch-Ferré, M, SN Bhupathiraju, and FB Hu. 2018. Use of metabolomics in improving assessment of dietary intake. Clin Chem. 64(1):82–98. doi: 10.1373/clinchem.2017.272344.

Gupta, V, M Sengupta, J Prakash, and BC Tripathy. 2017. An Introduction to Biotechnology. In Basic and Applied Aspects of Biotechnology, 1–21. Singapore: Springer Singapore. doi: 10.1007/978-981-10-0875-7_1.

Hassa, J, I Maus, S Off, A Pühler, P Scherer, M Klocke, and A Schlüter. 2018. Metagenome, metatranscriptome, and metaproteome approaches unraveled compositions and functional relationships of microbial communities residing in biogas plants. Appl Microbiol Biotechnol. doi: 10.1007/s00253-018-8976-7.

Hernandez-Morcillo, M, M Torralba, T Baiges, A Bernasconi, G Bottaro, S Brogaard, F Bussola, E Díaz-Varela, D Geneletti, CM Grossmann, J Kister, M Klingler, L Loft, M Lovric, C Mann, N Pipart, JV Roces-Díaz, S Sorge, M Tiebel, L Tyrväinen, E Varela, G Winkel, and T Plieninger. 2022. Scanning the solutions for the sustainable supply of forest ecosystem services in Europe. Sustain Sci. 17(5):2013–2029. doi: 10.1007/s11625-022-01111-4.

Hogeweg, P. 2011. The roots of bioinformatics in theoretical biology. PLoS Comput Biol. 7(3):e1002021. doi: 10.1371/journal.pcbi.1002021.

John Ogbe, R, DO Ochalefu, and OB Olaniru. 2016. Bioinformatics advances in genomics – A review. Int J Curr Pharm Rev Res. 8. https://www.researchgate.net/publication/304648097.

Kebede, G, T Tafese, EM Abda, M Kamaraj, and F Assefa. 2021. Factors influencing the bacterial bioremediation of hydrocarbon contaminants in the soil: Mechanisms and impacts. J Chem. doi: 10.1155/2021/9823362.

Kertesz, MA, and M Thai. 2018. Compost bacteria and fungi that influence growth and development of Agaricus bisporus and other commercial mushrooms. Appl Microbiol Biotechnol. doi: 10.1007/s00253-018-8777-z.

Kornej, J, CS Börschel, EJ Benjamin, and RB Schnabel. 2020. Epidemiology of atrial fibrillation in the 21st century: Novel methods and new insights. Circ Res. doi: 10.1161/CIRCRESAHA.120.316340.

Koshlaf, E, and AS Ball. 2017. Soil bioremediation approaches for petroleum hydrocarbon polluted environments. AIMS Microbiol. doi: 10.3934/MICROBIOL.2017.1.25.

Kumar, V, R Kothari, VV Pathak, and SK Tyagi. 2018. Optimization of substrate concentration for sustainable biohydrogen production and kinetics from sugarcane molasses: Experimental and economical assessment. Waste Biomass Valori. 9(2):273–281. doi: 10.1007/s12649-016-9760-5.

Kumar, V, I Shekhar Thakur, A Kumar Singh, and MP Shah. 2020. Application of Metagenomics in Remediation of Contaminated Sites and Environmental Restoration. In Emerging Technologies in Environmental Bioremediation, 197–232. Elsevier. doi: 10.1016/B978-0-12-819860-5.00008-0.

Kuppusamy, S, P Thavamani, M Megharaj, K Venkateswarlu, YB Lee, and R Naidu. 2016. Pyrosequencing analysis of bacterial diversity in soils contaminated long-term with PAHs and heavy metals: Implications to bioremediation. J Hazardous Mater. 317(November):169–179. doi: 10.1016/j.jhazmat.2016.05.066.

Lacerda, CMR, and KF Reardon. 2009. Environmental proteomics: Applications of proteome profiling in environmental microbiology and biotechnology. Brief Funct Genomics. 8:75–87.

Laskar, F, S Das Purkayastha, A Sen, MK Bhattacharya, and BB Misra. 2018. Diversity of methanogenic archaea in freshwater sediments of lacustrine ecosystems. J Basic Microbiol. doi: 10.1002/jobm.201700341.

Li, Y. 2021. Modern epigenetics methods in biological research. Methods. doi: 10.1016/j.ymeth.2020.06.022.

Liang, M, X Zhou, and C Xu. 2016. Systems biology in biofuel. Phys Sci Rev. 1(11):20160047.

Liu, Z, A Ma, E Mathé, M Merling, Q Ma, and B Liu. 2021. Network analyses in microbiome based on high-throughput multi-omics data. Brief Bioinform. doi: 10.1093/bib/bbaa005.

Lowe, R, N Shirley, M Bleackley, S Dolan, and T Shafee. 2017. Transcriptomics technologies. PLoS Comput Biol. 13(5):e1005457. doi: 10.1371/journal.pcbi.1005457.

Mawang, CI, AS Azman, ASM Fuad, and M Ahamad. 2021. Actinobacteria: An eco-friendly and promising technology for the bioaugmentation of contaminants. Biotechnol Rep. doi: 10.1016/j.btre.2021.e00679.

McCarren, J, JW Becker, DJ Repeta, Y Shi, CR Young, RR Malmstrom, SW Chisholm, and EF DeLong. 2010. Microbial community transcriptomes reveal microbes and metabolic pathways associated with dissolved organic matter turnover in the sea. Proc Natl Acad Sci USA. 107(38):16420–16427. doi: 10.1073/pnas.1010732107.

Nie, Y, R Chen, X Tian, and YY Li. 2017. Impact of water characteristics on the bioenergy recovery from sewage treatment by anaerobic membrane bioreactor via a comprehensive study on the response of microbial community and methanogenic activity. Energy. 139:459–467. doi: 10.1016/j.energy.2017.07.168.

Ojuederie, OB, and OO Babalola. 2017. Microbial and plant-assisted bioremediation of heavy metal polluted environments: A review. Int J Environ Res Public Health. doi: 10.3390/ijerph14121504.

Ramamurthy, PC, S Singh, D Kapoor, P Parihar, J Samuel, R Prasad, A Kumar, and J Singh. 2021. Microbial biotechnological approaches: Renewable bioprocessing for the future energy systems. Microb Cell Fact. 20(1):55. doi: 10.1186/s12934-021-01547-w.

Rashid, Z, SMH Gilani, A Ashraf, S Zehra, A Azhar, KA Al-Ghanim, F Al-Misned, S Mahboob, and S Galani. 2020. Benchmark taxonomic classification of chicken gut bacteria based on 16S rRNA gene profiling in correlation with various feeding strategies. J King Saud Univ Sci. 32(1):1034–1041. doi: 10.1016/j.jksus.2019.09.013.

Rose, R, B Constantinides, A Tapinos, DL Robertson, and M Prosperi. 2016. Challenges in the analysis of viral metagenomes. Virus Evol. 2(2). doi: 10.1093/ve/vew022.

Schmidt, JE, AD Kent, VL Brisson, and M Gaudin. 2019. Agricultural management and plant selection interactively affect rhizosphere microbial community structure and nitrogen cycling. Microbiome 7(1). doi: 10.1186/s40168-019-0756-9.

Shekhar, SK, J Godheja, and DR Modi. 2020. Molecular Technologies for Assessment of Bioremediation and Characterization of Microbial Communities at Pollutant-Contaminated Sites. In Bioremediation of Industrial Waste for Environmental Safety, 437–474. Springer Singapore. doi: 10.1007/978-981-13-3426-9_18.

Singh, AK, M Bilal, HMN Iqbal, and A Raj. 2021. Trends in predictive biodegradation for sustainable mitigation of environmental pollutants: Recent progress and future outlook. Sci Total Environ. doi: 10.1016/j.scitotenv.2020.144561.

Slatko, BE, AF Gardner, and FM Ausubel. 2018. Overview of next-generation sequencing technologies. Curr Protoc Mol Biol 122(1). doi: 10.1002/cpmb.59.

Srivastav, AL, Markandeya, N. Patel, M Pandey, AK Pandey, AK Dubey, A Kumar, AK Bhardwaj, and VK Chaudhary. 2023. Concepts of circular economy for sustainable management of electronic wastes: Challenges and management options. Environ Sci Pollut Res Int. 30:48654–48675. doi: 10.1007/s11356-023-26052-y.

Straub, D, N Blackwell, A Langarica-Fuentes, A Peltzer, S Nahnsen, and S Kleindienst. 2020. Interpretations of environmental microbial community studies are biased by the selected 16S rRNA (gene) amplicon sequencing pipeline. Front Microbiol 11(October). doi: 10.3389/fmicb.2020.550420.

Suea-Ngam, A, L Bezinge, B Mateescu, PD Howes, AJ Demello, and DA Richards. 2020. Enzyme-assisted nucleic acid detection for infectious disease diagnostics: Moving toward the point-of-care. ACS Sens. 5(9): 2701–2723. doi: 10.1021/acssensors.0c01488.

Thakur, C, and J Shankar. 2022. Bioinformatics Integration to Biomass Waste Biodegradation and Valorization. In: Verma, P (ed). Enzymes in the Valorization of Waste. CRC Press, Boca Raton.

Xu, T, and J Ma. 2021. Feed-in tariff or tax-rebate regulation? Dynamic decision model for the solar photovoltaic supply chain. Appl Math Model. 89(2):1106–1123.

Ullah Khan, I, MHD Othman, H Hashim, T Matsuura, AF Ismail, M Rezaei-DashtArzhandi, and IW Azelee. 2017. Biogas as a renewable energy fuel – A review of biogas upgrading, utilisation and storage. Energy Convers Manag. doi: 10.1016/j.enconman.2017.08.035.

Vélez-Mercado, MI, AG Talavera-Caro, KM Escobedo-Uribe, S Sánchez-Muñoz, MP Luévanos-Escareño, F Hernández-Terán, A Alvarado, and N Balagurusamy. 2021. Bioconversion of lignocellulosic biomass into value added products under anaerobic conditions: Insight into proteomic studies. Int J Mol Sci. 22(22):12249. doi: 10.3390/ijms222212249.

Wanjiru, E, and X Xia. 2018. Sustainable energy-water management for residential houses with optimal integrated grey and rain water recycling. J Clean Prod. 170:1151–1166.

Woodley JM. 2018. Integrating protein engineering with process design for biocatalysis. Philos Trans A Math Phys Eng Sci. 376(2110):20170062. doi: 10.1098/rsta.2017.0062.

Xiong, S, X Ma, and J Ji. 2019. The impact of industrial structure efficiency on provincial industrial energy efficiency in China. J Clean Prod. 215:952–962. doi: 10.1016/j.jclepro.2019.01.095

Yang-Yang Gao, Wei Zhao, Yuan-Qin Huang, Vinit Kumar, Xiao Zhang, and Ge-Fei Hao. 2024. In silico environmental risk assessment improves efficiency for pesticide safety management, Science of The Total Environment, Volume 908, 167878, ISSN 0048-9697, https://doi.org/10.1016/j.scitotenv.2023.167878.

Zenda, T, S Liu, A Dong, J Li, Y Wang, X Liu, N Wang, and H Duan. 2021. Omics-facilitated crop improvement for climate resilience and superior nutritive value. Front Plant Sci. 12:774994. doi: 10.3389/fpls.2021.774994.

Zhang, L, F Chen, Z Zeng, M Xu, F Sun, L Yang, X Bi, Y Lin, Y Gao, H Hao, W Yi, M Li, and Y Xie. 2021. Advances in metagenomics and its application in environmental microorganisms. Front Microbiol. 12:766364. doi: 10.3389/fmicb.2021.766364.

7 Exploring the Next-Generation Polymers

Synthesis and Emerging Applications of UHMW Protein Polymers

Swati Srivastava and Saurabh Sudha Dhiman

7.1 INTRODUCTION

In contemporary times, the ubiquity of the term "polymer" is an exemplification of its wide-ranging application. It can be defined as a macromolecule composed of numerous monomer units imparting specific properties and functionalities. A seminal example of a polymer is benzene (C_6H_6), comprising three acetylene (C_2H_2) molecules formed by cyclic polymerization (Lutz, 1961). Likewise, in 1869, John Wesley Hyatt pioneered the synthesis of the first synthetic polymer, plastic. This substance was created by treating cotton-derived cellulose with cellulose and emerged as a viable substitute for ivory, ushering in the material versatility era (Cobb & Brooks, 2017). The notable contributions of Hermann Staudinger, Wallace Carothers, and Herman Mark have been pivotal in propelling the polymer science field. Herman Mark made a significant stride in elucidating the concept of macromolecules. Hermann Staudinger, on the other hand, formulated the concept of macromolecules which proved to be instrumental in unraveling the polymer molecular architecture. Notably, Wallace Carothers stands out for his pioneering role in the creation of the first commercially available polymer, nylon.

Swiftly, plastic permeated every aspect of human existence, potentially eclipsing all other man-made material in terms of production volume. From rudimentary packaging to the pivotal biomedical industry, plastic has become pervasive and, simultaneously, fueling concerns related to land and marine pollution (Geyer et al., 2017). Following the advent of plastic, the quest for novel material development accelerated giving rise to materials such as Bakelite, unsaturated polyesters, polyvinyl chloride, styrene, polymethyl methacrylate, nylon 6, and many others (Feldman, 2008). Simultaneously, a plethora of other polymers were meticulously designed and synthesized. For instance, artificial fiber known as viscose made their debut, although natural fibers, such as silk and wool, were incorporated into the textile

 DOI: 10.1201/9781003407683-7

industry way before. Following the commercialization of these synthetic fibers, progress in the fiber domain was introduced by the introduction of polyvinyl alcohol (PVA) fiber, polyacrylonitrile (PAN), and poly(m-phenylene isophthalamide) (MPD) (Feldman, 2008). Furthermore, synthetic rubber entered the commercial arena, paving the path for extensive utilization of hydrocarbons for rubber production (Feldman, 2008). From an alternate perspective, these endeavors were an attempt to circumvent the dependence on naturally occurring polymers that had prevailed. The transformative impact of polymers on human civilization remains unparalleled until a thorough assessment of their environmental repercussions becomes imperative. Plastic emerged as a primary contributor to multifarious forms of pollution. Synthetic polymers derived from petroleum, augmented with multiple stabilizers, are also identified as culprits for releasing harmful toxins to the environment, eventually increasing the environmental impact of polymers (Amarakoon et al., 2022; Ragitha & Edison, 2022).

The comprehensive classification of polymers encompasses three categories: natural polymers, synthetic polymers, and semi-synthetic polymers. Synthetic polymers are further classified based on biodegradability distinguishing as biodegradable and non-biodegradable polymers (Prajapati et al., 2019). This classification is based on the polymer's propensity for biodegradation, a crucial distinction in the polymer landscape. Biodegradable polymers can be subclassified based on their origin, distinguishing between synthetic and natural sources. Synthetic biodegradable polymers, e.g., polyglycolic acid (PGA), polylactic acid (PLA), poly(lactic-co-glycolic acid) (PLGA), polycaprolactone (PCL), and others, originate from petroleum-derived resources rendering them non-renewable. On the contrary, natural biodegradable polymers, e.g., chitin, cellulose, PHA, PHB, collagen, gelatin, and others, are sourced from biological resources endowing them with complete renewability (Vroman & Tighzert, 2009). The contemporary emphasis on natural biodegradable polymers, majorly polysaccharide and polypeptide polymers, stems from their capability of replacing environmentally detrimental alternatives. Biopolymers consisting of carbon, oxygen, and nitrogen backbone are inherently conducive to microbial degradation into benign by-products such as carbon dioxide and water, therefore perpetuating a natural recycling process. Notably, biopolymers demonstrate flexibility and serve as a smart responder to environmental stress and enzyme activity. This responsiveness is attributed to the constant manipulation of their structure in response to these stimuli (Gheorghita et al., 2021). Genetic engineering and recombinant DNA technology have played a significant role in elevating polymer production capabilities. The incorporation of genetically modified PHA and PHB into the polymer segment is considered to be an advancement in the replacement of traditional plastic (Getachew & Woldesenbet, 2016; Sehgal & Gupta, 2020). The imperative for biodegradable plastics has prompted research efforts focused on leveraging naturally occurring microbes capable of producing PHA and PHB. Minimal genetic modification in these microorganisms leads to bulk production with tailored functional groups enhancing the sustainability with the market (Sehgal & Gupta, 2020). Bioplastics is not the limit to the horizon of polymer science; concurrent research is underway for producing highly specialized protein polymers and polysaccharide polymers, characterized by robust tensile properties, e.g., compared to archetypical counterparts.

Protein polymers are naturally occurring, but are often limited in availability, thereby necessitating the implementation of genetic modification technology in microbial host cells for their production. These polymers are crafted through genetic modifications within the microbial genome, allowing tailored designs to meet specific requirements. These genetically enhanced polymers synthesized using a microbial host system represent the next-generation polymers and primarily include protein polymers which are in high demand (Ghosh et al., 2021; Moradali & Rehm, 2020; Ramezaniaghdam et al., 2022; Werten et al., 2019). A notable inclusion in the next-generation polymers is the ultra-high molecular weight (UHMW) protein polymers. The UHMW proteins are characterized by their high molecular weight, extensive amino acid chain length (>500 amino acids), and presence of repetitive sequences endowed with highly specific characteristics; e.g., the highly repetitive PVEK segment of the titin protein is responsible for elasticity and force development (Ding et al., 2011; Linke et al., 1998). Among UHMW proteins, a human structural protein known as titin protein, coded by the *TTN* gene, has garnered significant attention. Titin is considered to be a promising protein polymer due to its distinctive physical properties, characterized by exceptional tensile strength, high elasticity, and efficient energy utilization (Schuetz et al., 2015). These UHMW protein polymers offer advantages over conventional polymers, as they are biocompatible, easily degradable, renewable, and, with optimization, amenable for bulk production. However, the production of these polymers poses formidable challenges including post-translational modification, extraction, and purification requirements. Proteins such as titin are naturally found in vertebrates and present difficulties in expression within microbial systems due to various factors, including microorganism metabolic capacity and genetic burden associated with the production of a specific biomolecule (Bowen et al., 2021). The recent advancements in computational biology and genetic engineering have facilitated in determining factors such as transformation rate and genetic stability, increasing the feasibility to achieve successful expression and synthesis of designed UHMW proteins (De Frates et al., 2018; Gupta & Nayak, 2015; Xue et al., 2019). The focus of the next-generation polymers is their wide-ranging applicability across industries, from the prosthetic industry and drug delivery systems due to their biocompatibility and tensile strength to their implementation in production of temperature-sensitive textiles.

These next-generation polymers are expected to mitigate the drawbacks associated with conventional polymers, specifically in case of environmental impact, and provide user-friendly material. This chapter aims to provide a detailed overview of polymers, microbial production of polymers by applying genetic engineering and computational biology, next-generation polymer synthesis, challenges, and expected applications in the future. The overarching objective is to enhance understanding of the current trajectories of polymer science with a growing emphasis on assessing environmental impacts.

7.2 HISTORY OF POLYMER

Polymers have been part of human life since long before a complete understanding of science existed (Burford, 2019). Rubber, a natural polymer, was discovered in the early 1700s and subsequently brought into industrialization. Following this came into the picture plastic, the first synthetic polymer in the 1800s. It was initially developed as an ivory substitute but soon found wide-ranging applications (Arias & van Dijk, 2019). The introduction of Bakelite in 1907 is considered as the starting point of the polymer age, and soon plastic, the first synthetic polymer, became an integral part of our lives. Its usage evolved from insulating radar cabling during the Second World War to consumer products, including plastic bags, containers, and even artificial hip and knee replacements (Geyer, 2020). However, despite all the advantages, there is a detrimental environmental impact of plastic over time. Nevertheless, proliferation of synthetic polymers is unceasing (Amarakoon et al., 2022; Thompson et al., 2009). Synthetic polymers consist of multiple monomer units exhibiting a wide range of properties in terms of mechanical, thermal, and even degradation characteristics (Prieto & Guelcher, 2014). These polymers were developed in the last six decades, becoming an irreplaceable part of the material world. The evolution and shift of polymers are represented in Figure 7.1. Synthetic polymers are derived from petroleum oil or engineered to meet specific requirements. They have gained such popularity that nearly half of the chemical industries are conducting research on them and are expected to continue with each step toward advancement. A few well-known and approved polymers are polyesters, polyethylene, nylon, Teflon, and saran (McKeen, 2017; McQuarrie et al., 2011).

In response to the environmental concerns and continued demand for polymers, the research shifted direction toward bio-based polymers, which were eco-friendlier and compatible. Although the replacement of the synthetic polymer may not be feasible, the success of biopolymers in various application areas cannot be disapproved (Babaremu et al., 2023). The introduction of PLA was transformative as it serves as a replacement for plastic due to its mechanical and strength properties similar

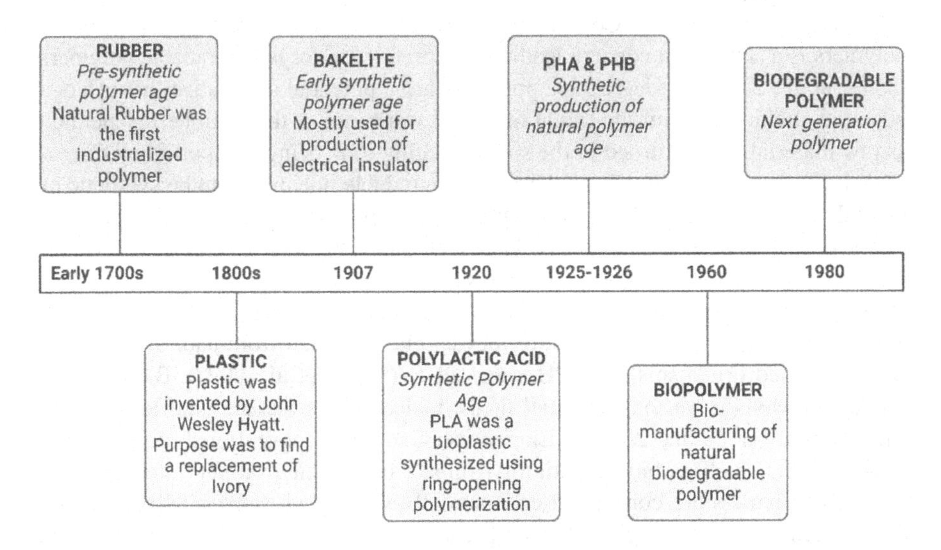

FIGURE 7.1 Overview of timeline representing the introduction of different polymers and their specific characteristics.

to polystyrene (Parker et al., 2011). Modified PLA exhibits properties similar to polyethylene, making it suitable for diverse applications. For instance, they are utilized for the production of single-use laboratory consumables (Freeland et al., 2022; Vroman & Tighzert, 2009). In the mid-1920s, polyhydroxyalkanoates (PHA) and polyhydroxybutyrate (PHB) were also discovered, expanding the realm of bioplastic (Cobb & Brooks, 2017; Sehgal & Gupta, 2020).

In the 1980s, biodegradable synthetic polymers were developed to address environmental concerns, while not affecting the regular requirements. Another key addition is the protein-based polymers, which are poised to transform the current scenario. These polymers fall under the category of natural biodegradable polymers as they are synthesized in vitro with the application of multiple recombinant technologies in biological systems. These protein polymers fall under biopolymers (Baranwal et al., 2022a).

7.3 BIOPOLYMER

The classification of polymers is based on various factors due to the multitude of polymers currently in use and under development. Polymers consist of chemically bonded monomer units, transformed into functional materials (Kfoury et al., 2013). Broadly, polymers are categorized based on the following factors: (a) origin, (b) structure, (c) polymerization, (d) molecular force, (e) monomer unit, and (f) thermal behavior (Kfoury et al., 2013; Okolie et al., 2023; Patel et al., 2023). Natural polymers can be further divided into three subcategories based on their chemical structure: (a) polypeptides/protein polymers, (b) polysaccharides/polycarbonates, and (c) polyesters (Aravamudhan et al., 2014). Natural polymers originate from biological sources and are extracted from plants and animals. Examples of natural polymers include wool from sheep, silk from silkworms, cellulose from plants, and protein from variable sources (Caillol, 2021). Concerning environmental impact, the degradation of the polymers is a significant concern leading to a preference for biodegradable polymers (Samir et al., 2022). Biodegradable polymers are a potential alternative that offers a potential solution to limit the threat of global warming, as they can be regenerated as raw material, and returned to the soil, enriching it by being composted by microorganisms (Luyt & Malik, 2018). These biodegradable polymers can be synthetic or natural, e.g., PGA, PLA, PBS, protein polymers, cellulose, etc. Figure 7.2 represents the classification of biodegradable polymers. One such section of the polymer is natural biodegradable polymers or biopolymers (Vroman & Tighzert, 2009).

Biopolymers are either extracted from biomass, e.g., fiber, soy protein, collagen, casein, and cross-linked lipid, or they can be extracted from natural or genetically modified organisms, e.g., PHA and PHB (Samir et al., 2022). Biopolymers find applications in various essential fields, including the medical and food industries (Baranwal et al., 2022a). Biopolymers surpass conventional polymers in many aspects, such as biocompatibility, impact on the environment, and applications. Biopolymers are considered economically viable and user-friendly materials due to their biological origin. The comprehensive application of the biopolymer is extended from medical to military to cosmetics as well (Gowthaman et al., 2021). Considering the wide-ranging application, biopolymers have a foothold in various

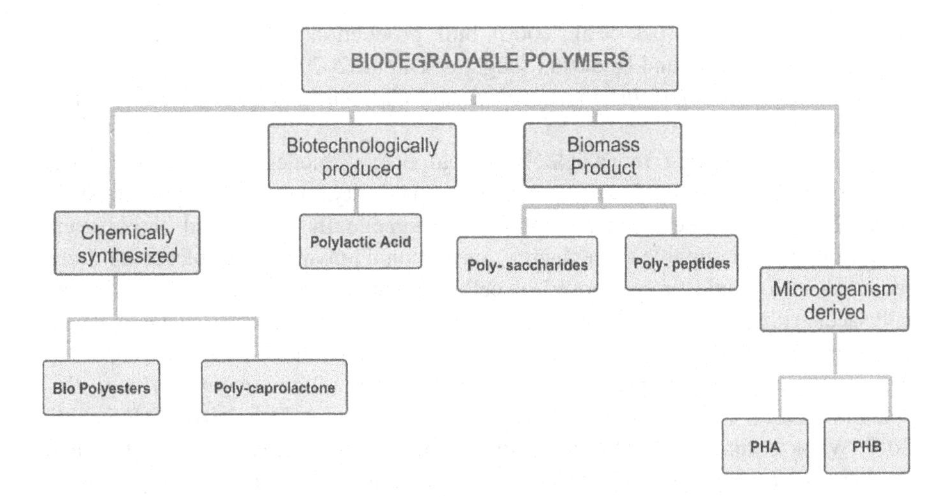

FIGURE 7.2 Classification of biodegradable polymers based on their origin.

industrial segments, including food packaging, drug delivery, and medical devices to agriculture and construction. These polymers have their application in pharmaceuticals in the form of microcapsules and hydrogels. In agriculture, they are expected to enhance water recovery and act as soil conditioners, whereas in the food packaging industry, they are utilized for edible film packaging and emulsifiers. Another key application of these biocompatible polymers is in medical devices which include implants and prosthetics (Baranwal et al., 2022b).

7.4 NEXT-GENERATION POLYMERS

Concerns regarding the development of biopolymers that can be degraded and recycled have led to the development of diverse polymers, e.g., UHMW protein polymers, PHA, PHB, and PLA. Next-generation polymers have drawn researchers' colossal interest due to their intriguing thermal, chemical, and structural properties. Conventional polymers have the disadvantage of their detrimental effects on the environment and replacing them with advanced biopolymers has become a significant objective. The new generation of polymers is designed to compete with the latter in terms of their properties, quantity, and cost-effectiveness. All biopolymers, including protein-based polymers and polysaccharide polymers, which are genetically modified, can be considered as next-generation polymers (Halley et al., 2011). The incorporation of genetic engineering and recombinant technology to biologically synthesize biopolymers has eased it for researchers to replicate proteins and polysaccharides, which exhibit significantly similar properties and can potentially replace conventional polymers. Genetic engineering has substantially contributed to establishing new ways of synthesizing biopolymers (Gupta et al., 2021). Heterologous genetic engineering, the expression of genes in other organisms, has potentially increased the likelihood of producing modified biopolymers. A few examples are the production of spider silk using *Pichia pastoris* and *Escherichia coli* (Fahnestock

& Irwin, 1997; Fahnestock et al., 2000), bulk production of PHA by multiple species of archaebacteria and bacteria (Sehgal & Gupta, 2020), and bacterial cellulose production by *Agrobacterium tumefaciens* and *Gluconacetobacter xylinus* (Lahiri et al., 2021).

Properties including renewable biological origin, biodegradability, structural integrity, thermal physical properties, biocompatibility, and eco-energy efficiency make these biopolymers consummate. These genetically engineered biopolymers are designed to be suitable for replacing conventional polymers; e.g., plastic has been replaced with bioplastic composed of cellulose (Steven et al., 2022). Protein and polysaccharide polymers, which are cost-effective and regenerative, make excellent biopolymers with high-end applications in the biomedical field. Drug delivery using these polymers is highly preferred because of their low immunogenicity; e.g., chitosan is used for efficient protein delivery and collagen for drug delivery (Jao et al., 2017; Wang et al., 2020). Multiple studies reported the successful production and implementation of biopolymers in various segments of society, and currently UHMW protein, e.g., titin, has gained popularity among researchers. UHMW protein polymers are a new addition to the section of next-generation polymers with enhanced properties such as tensile strength and elasticity. UHMW proteins are large proteins with long repeated chains of amino acids and specific structures providing meritorious mechanical strength, making them superior to the currently available polymers.

UHMW protein polymers are captivating due to their mechanical strength, which can be utilized to produce hydrogel and fibers. One such protein found in vertebrates is titin which is being explored rigorously. Titin protein is one of the largest proteins found in humans, with a molecular weight of ~4 MDa, and is composed of ~34,000 amino acids (Eldemire et al., 2021; Granzier et al., 2005; Labeit et al., 1997). Titin is a potential next-generation polymer due to its elasticity, high tensile strength, and biological origin. In molecular terms, the protein is primarily encoded by the TTN gene where titin serves as an anchoring molecule that extends from the Z-line to the M-line (Bang et al., 2001) Within this extension, there are several distinct regions; starting from the I-band, which contributes to the protein's elasticity, there are PEVK regions which are responsible for the passive elasticity, and finally there is the A-band, a combination of thick and thin filament governing the muscle contraction. The macromonomer of titin protein consists of folded globular domains which, upon stretching, release energy, lengthen the protein, and generate the mechanical force required for the functioning of muscles (Mártonfalvi & Kellermayer, 2014; Wang et al., 2019). This tandem modular structure is the main feature that makes it a potential polymer with high-end applications. These distinctive characteristics and structural segments render it an appealing candidate for a robust biomaterial.

7.5 PRODUCTION OF UHMW PROTEIN POLYMER

Protein structures include a sequence of amino acids bonded by peptide bonds. Experimental attempts to produce protein were initiated in the early 1900s with the successful chemical synthesis of a dipeptide, glycyl glycine (Nilsson et al., 2005). If not specified for biopolymers, proteins generally have a vital role in pharmaceuticals, resulting in the requirement of their synthesis since the feasibility of their isolation from their natural source is low. Apart from that, multiple protein polymers are required in bulk quantity, e.g., silk and wool. This demands the artificial production of these polymers to meet the industrial scale requirement. Production of polypeptides or protein polymers has two defined methodologies, i.e., chemical synthesis and biomanufacturing.

7.6 CHEMICAL SYNTHESIS

Chemical synthesis of the protein polymers focuses on the formation of amide bonds required to form amino acid sequences followed by litigation to form polymer. A conventional technique for peptide production involves the addition of a free amine to activate the carboxyl group in the presence of a coupling reagent, forming a peptide bond upon condensation (James & Numat, 2013). Another well-accepted technique for chemical synthesis is the solid-phase method. The solid phase method for peptide synthesis utilizes an insoluble support, e.g., resins, to which an amino acid is attached via its carboxyl group facilitating coupling with another amino acid which has an activated and protected functional group. In this reaction, the protected amino group is replaced with the consecutive amino acid added to the coupling reaction (James & Numat, 2013). The solid-phase method is automated with the implementation of resins as the immobilization surface. An amino acid with α-amino protecting group along with a side chain is immobilized on the resin. In addition to soluble amino acids, the α-amino protecting group is deprotected, and the active amino group couples with the activated carboxyl group. The deprotection of the α-amino protecting group and coupling is continued till the sequence is achieved, followed by the cleaving of the sequence from the resin (Coin et al., 2007; Fields, 2001; Nilsson et al., 2005). Synthesis of the peptide is followed by ligation, which is done via native chemical ligation, α-ketoacid-hydroxylamine ligation, serine/threonine ligation, and cysteine/penicillamine ligation (Tan et al., 2020). Advantages of the automated protein synthesis technology include the chemically defined structure of the obtained protein polymer and possible alteration of the function or properties of the protein by addition/deletion of any functional group. A crucial disadvantage of this methodology is its limitation to producing protein at a small scale. Chemical synthesis of small protein polymers is now in regular practice. However, the synthesis of large protein polymers remains challenging both at laboratory scale and industry due to their complexity and size. Refolding of the synthetic peptides is required to obtain functional protein, as only linear peptide sequences are formed (Nilsson et al., 2005; Tan et al., 2020). The drawback of this methodology implies its incompetence in producing UHMW protein polymers.

7.7 BIOMANUFACTURING

Overcoming these challenges, biological synthesis plays a vital role in protein synthesis. The biological synthesis of protein involves advanced technologies, genetic engineering, recombinant technology, and protein engineering to produce the desired protein. The biosynthesis of protein requires genetic modification of host cells by incorporating the gene that encodes for the protein of interest. The biosynthesis of the protein polymer unfolds through a series of steps, commencing with the alteration of the host cell through introduction of the gene encoding the target protein. Subsequent stages involve the careful selection of transformants by employing selection plates, e.g., ampicillin and kanamycin, the cultivation of the transformed cells within a bioreactor to facilitate protein production. Ultimately, the expressed protein is obtained either by harvesting cells in the case of intracellular production or by employing separation techniques to extract it from the medium in instances of extracellular secretion. This orchestrated process encapsulates the journey from genetic transformation to protein extraction in the realm of biosynthesis. Protein engineering plays a significant role in manipulating the protein sequence as per the requirements. Alteration in the amino acid sequence can lead to a change in the structure of the protein, which ultimately affects the properties of the protein as well (Ferrari & Cappello, 1997). This technology is only applicable for small and medium-sized proteins, but in the case of UHMW proteins, where the number of repetitive amino acid sequences is higher, the production of multiple peptides of both repetitive and non-repetitive sequences is required by means of a similar protocol. These peptides require polymerization to embellish into a functional UHMW protein. Polymerization of biosynthesized UHMW protein can be done by cross-linking of protein monomer using side chain chemistry, for example, tag reaction, enzyme-mediated protein conjugation, and small ubiquitin-like modifier enzyme-mediated conjugation. Another polymerization method is protein backbone ligation via the formation of new peptide bonds, which is done with the help of split-inteins or transpeptidase, e.g., sortase. Biosynthesis of UHMW protein gives the advantage of upscaled production and no post-translational modification such as refolding of the protein. The drawback of the explained method in the case of UHMW protein synthesis are genetic instability, the need for codon optimization, and translational difficulties (Jeon et al., 2023). Besides these drawbacks, the biological synthesis method is preferred over chemical synthesis for UHMW proteins (Nothling et al., 2020).

7.8 BIOMANUFACTURING CHALLENGES

UHMW protein polymers have gained sufficient attention because their mechanical and thermal properties prove their significance in industrial applications. However, their production remains challenging, as plentiful factors, from the host cell to the purification process, are yet to be optimized. One of the crucial challenges researchers face in biological synthesis is genetic instability and low translational efficiency due to the limited genome of the host cells. The development of understanding the structure and sequence of the protein is a key aspect of UHMW protein polymer production. Microbial host cells have a certain size limit for the protein they are capable of producing, and if there is any information above that limit, it becomes a burden for the host cell leading to the truncation of that particular gene (Bowen et al., 2021). Currently, the concept of expression of the fragmented protein is new without any explained protocol. Since the number of repetitive sequences is high, the requirement of expression of both repetitive and non-repetitive polypeptide sequences is a concern that is both tedious and tricky. Fragmentation of the protein leads to a reduction of the weight of the protein sequence, but those fragments are somewhat uneven as the fragmentation is done on the basis of the repetitive and non-repetitive sequences. The lack of a defined protocol for high-yield production remains challenging as optimization is still in process (James & Numat, 2013). An appropriate host cell system that can evenly express the protein is not available. Due to the size of the protein, synthesis of full-length protein is not possible in the currently available host cells. The identification of the repetitive as well as non-repetitive sequences in the large protein sequence is difficult. There are higher chances of misinterpretation, which could lead to the production of non-functional proteins. Also, the uneven sequences lead to variable expressions of the sequences, as some sequences are suitable for microbial production while some are large enough to get truncated (Tang & Chilkoti, 2016). Another key hurdle in production is the structural instability of the produced protein by various host cells since certain host cells do not have the post-translational modification of the produced protein. For example, *E. coli* does not have this function, whereas it is a key feature of *P. pastoris* (Viswanath, 2021).

Furthermore, biological processes that can ease the protein's expression are yet to be defined. Along with this, the procedure of polymerization capable of providing the full-length protein with both structural and functional properties intact is still to be optimized. Technologies and modifications required for increasing the yield of the recombinant protein polymer to match the industrial scale are still uncovered (Bowen, 2019). The identification of linker sequences that can be applied to enhance the polymerization process must be scrutinized. Overcoming these challenges requires further exploration and solutions to the knowledge gaps in this section (Bowen, 2019).

7.9 CURRENT AND FUTURE APPLICATIONS

The increasing demand for a greener solution for every material has led to the exploration of green polymers. Designed and synthesized biopolymers have been brought into industrialization for their efficient deliverables and properties. A common example is a synthesis of protein polymer fibers, silk, and wool in the textile industries (Gheorghita et al., 2021). In a similar manner, protein polymers and UHMW protein polymers have been explored for their application in diverse industrial segments and research areas (Figure 7.3). These application areas have been discussed in the following sections.

7.10 APPLICATION IN BIOMEDICAL

The biocompatibility and biodegradability of the protein polymers have promoted the replacement of conventional polymers in the biomedical segments (Baranwal et al., 2022b; Biswas et al., 2022). The application of these polymers encompasses utilization in tissue engineering, wound healing, drug delivery, and biosensors. For instance, arginine-modified films have represented efficient cell viability, fibronectin has been applied for wound healing since they support cell migration required for damage repair, and protein polymer-antibody conjugates are applied for streamlined and non-toxic drug delivery (Goncalves et al., 2022; Mir et al., 2018; Sengupta & Prasad, 2018). Similarly, the utilization of UHMW protein polymers within these

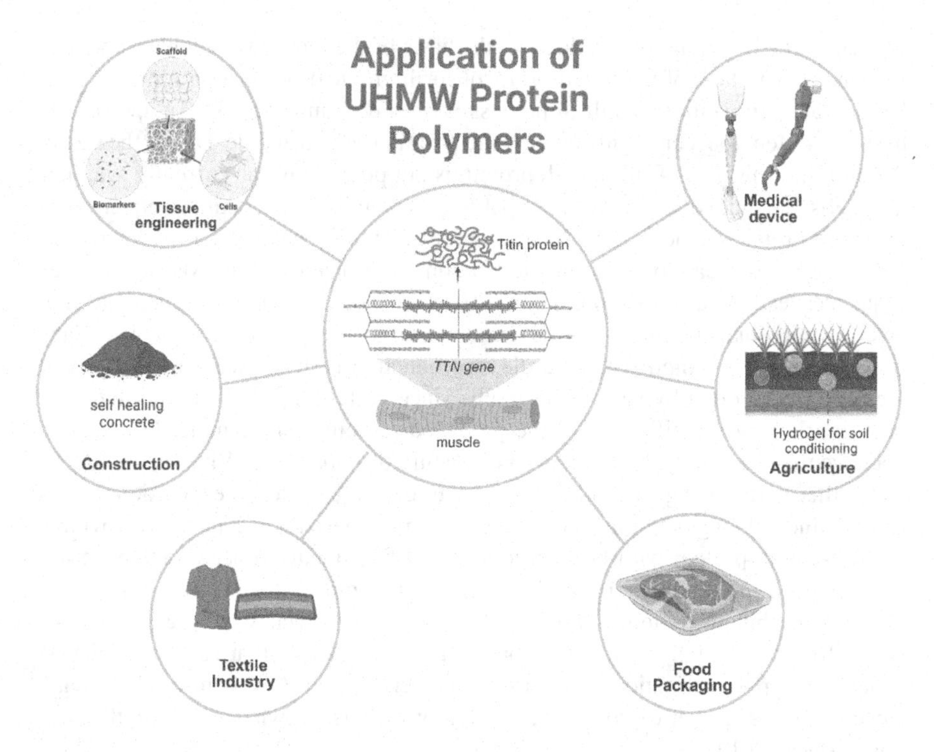

FIGURE 7.3 The overview of potential research areas and the applications of the UHMW protein as a transformative biomaterial (Ma et al., 2023).

segments is in process since the properties in these polymers are enhanced and more specific.

Their application has also been investigated in medical device segments. Currently, the application of protein and UHMW protein polymers is being explored in the case of implants and prosthetics, due to the high-end applications in tissue engineering and wound healing. Owing to the enhanced properties of being non-toxic, high mechanical strength, and elasticity, these polymers qualify for biomaterial which is suitable for production of the implants and prosthetics (Rabe et al., 2020).

7.11 APPLICATION IN AGRICULTURE

One of the key reasons for the introduction of protein polymers in the field of agriculture is the chemical and ecological changes due to anthropogenic activities. Climate change, water pollution, and chemical fertilizer have been observed to hamper the soil. The properties of protein polymers which vary based on structural heterogenicity, thermal activity, and hydrophilic behavior have been utilized for several agricultural products. For example, whey protein, biowaste of cheese industry, has found its application in multiple segments of agriculture. The whey protein is an essential polymer utilized for encapsulation of beneficial and soil-friendly bacteria which aids in controlling pests and protecting plants from diseases due to infestation. Apart from the whey protein, poultry feathers are a rich source of keratin, thereby are utilized for soil conditioning as they are rich in nitrogen (Chang et al., 2021; Joardar & Rahman, 2018; Saberi Riseh et al., 2023).

7.12 APPLICATION IN FOOD PACKAGING

In an attempt to make food packaging products more environmentally friendly and safe for consumers, the utilization of protein polymers was initiated. Mechanical strength, gas barrier, non-water solubility, flexibility, and other handful of properties were considered the key parameters for edible protein films and coats. Both plant and animal proteins were demonstrated to be effective biomaterials for this application. Corn zein films are utilized for packaging of dry fruits as they are not water soluble and grease resistant. Similarly, gelatin films with antioxidants are used for protection of frozen meat. Other protein polymers that have been applied in the food packaging segments are keratin, egg albumin, milk protein, casein, and whey (Cuq et al., 1998; Shah et al., 2023).

7.13 APPLICATION IN TEXTILE

Protein polymers have a major advantage in their properties over conventional polymers as they are biocompatible and highly extensible with flexible stiffness. They are considered a suitable biomaterial for the production of clothing, sutures, bandages, and filters. Casein fibers are developed as an alternative to wool while seaweed fibers are being researched for production of gloves, socks, bedding, and filters with antimicrobial effects. Spider silk, collagen, and fibronectin are commonly used in the textile industry whereas hagfish slime polymers are also being researched for their application in the clothing industry (Deravi et al., n.d.; Karthik & Rathinamoorthy, 2018). Another commercially available protein polymer is represented by Azlon fibers, characterized as biofibers derived from regenerated protein fiber. Sourced from proteins found in soy, peanuts, corn, and milk, these biofibers are specifically employed in the textile industry. These fibers possess moisture absorbing, UV protection, and antistatic properties (Karthik & Rathinamoorthy, 2018).

7.14 APPLICATION IN COSMETIC AND CONSTRUCTION

Multiple protein polymers are utilized for developing cosmetic products. Collagen is one such example that has been used in cosmetic products and injections as they are used for repairing dermatological defects such as acne scars and wrinkles. Keratin is popular for its use in hair treatment products. Shell material which is common in cosmetics includes proteins like gelatin and soy (Alves et al., 2020; Mitura et al., 2020).

The application of admixture and additives in construction has been upgraded over the past years (Section 7.9). Biopolymers are added to the admixtures in order to strengthen the properties of concrete. One such protein polymer which has been used in the admixtures is casein. The casein admixture usage exhibited increased flexural strength of the cement sand (Brzyski et al., 2021).

7.15 CONCLUSIONS

Progression from the synthetic polymer to a greener solution led to the establishment of biopolymers. Exploration and successful implementation of the biopolymers accelerated the progress toward an environmentally friendly solution to the objective of replacing conventional polymers. Efficient application of protein polymers in multiple industries such as biomedical, construction, agriculture, and cosmetics preceded the further forage for more potential protein polymers. UHMW protein polymers are the new research interest because of their enhanced mechanical and elastic properties as well as their biocompatible and non-toxic characteristics. These next-generation polymers have the potential to be versatile and efficient biomaterials overpowering both the conventional and biopolymers. The advent of genome-editing and phenotypic characterization techniques enables the biomanufacturing of UHMW protein-derived polymers. Overall, biological synthesis to produce the UHMW protein is suitable in terms of their advantages of post-translation modification and undisturbed protein structure. Despite the standardization of operational and extraction strategies, several knowledge gaps pertaining to the in situ polymerization and functionalization of UHMW proteins remain underexplored.

REFERENCES

Alves, T. F. R., Morsink, M., Batain, F., Chaud, M. V., Almeida, T., Fernandes, D. A., da Silva, C. F., Souto, E. B., & Severino, P. (2020). Applications of natural, semi-synthetic, and synthetic polymers in cosmetic formulations. *Cosmetics*, *7*(4), 75. https://doi.org/10.3390/cosmetics7040075

Amarakoon, M., Alenezi, H., Homer-Vanniasinkam, S., & Edirisinghe, M. (2022). Environmental impact of polymer fiber manufacture. *Macromolecular Materials and Engineering*, *307*(11), 2200356. https://doi.org/10.1002/MAME.202200356

Aravamudhan, A., Ramos, D. M., Nada, A. A., & Kumbar, S. G. (2014). Natural polymers: Polysaccharides and their derivatives for biomedical applications. In *Natural and Synthetic Biomedical Polymers* (pp. 67–89). https://doi.org/10.1016/B978-0-12-396983-5.00004-1

Arias, M., & van Dijk, P. J. (2019). What is natural rubber and why are we searching for new sources? *Frontiers for Young Minds*, *7*. https://doi.org/10.3389/frym.2019.00100

Babaremu, K., Oladijo, O. P., & Akinlabi, E. (2023). Biopolymers: A suitable replacement for plastics in product packaging. *Advanced Industrial and Engineering Polymer Research*. https://doi.org/10.1016/J.AIEPR.2023.01.001

Bang, M.-L., Centner, T., Fornoff, F., Geach, A. J., Gotthardt, M., McNabb, M., Witt, C. C., Labeit, D., Gregorio, C. C., Granzier, H., & Labeit, S. (2001). The complete gene sequence of titin, expression of an unusual ≈700-kDa titin isoform, and its interaction with obscurin identify a novel Z-line to i-band linking system. *Circulation Research*, *89*(11), 1065–1072. https://doi.org/10.1161/hh2301.100981

Baranwal, J., Barse, B., Fais, A., Delogu, G. L., & Kumar, A. (2022). Biopolymer: A Sustainable Material for Food and Medical Applications. *Polymers*, *14*(5). https://doi.org/10.3390/POLYM14050983

Biswas, M. C., Jony, B., Nandy, P. K., Chowdhury, R. A., Halder, S., Kumar, D., Ramakrishna, S., Hassan, M., Ahsan, M. A., Hoque, M. E., & Imam, M. A. (2022). Recent advancement of biopolymers and their potential biomedical applications. *Journal of Polymers and the Environment*, *30*(1), 51–74. https://doi.org/10.1007/s10924-021-02199-y

Bowen, C. H. (2019). A Platform for Microbial Production of Ultra-High Molecular Weight Protein-Based Materials. *McKelvey School of Engineering Theses & Dissertations*. https://doi.org/10.7936/wt7f-dr15

Bowen, C. H., Sargent, C. J., Wang, A., Zhu, Y., Chang, X., Li, J., Mu, X., Galazka, J. M., Jun, Y.-S., Keten, S., & Zhang, F. (2021). Microbial production of megadalton titin yields fibers with advantageous mechanical properties. *Nature Communications*, *12*(1), 5182. https://doi.org/10.1038/s41467-021-25360-6

Brzyski, P., Suchorab, Z., & Łagód, G. (2021). The influence of casein protein admixture on Pore size distribution and mechanical properties of lime-metakaolin paste. *Buildings*, *11*(11), 530. https://doi.org/10.3390/BUILDINGS11110530

Burford, R. (2019). Polymers: A historical perspective. *Journal & Proceedings of the Royal Society of New South Wales*, *152*, 242–250.

Caillol, S. (2021). Special issue "Natural polymers and biopolymers II." *Molecules*, *26*(1). https://doi.org/10.3390/MOLECULES26010112

Chang, L., Xu, L., Liu, Y., & Qiu, D. (2021). Superabsorbent polymers used for agricultural water retention. *Polymer Testing*, *94*, 107021. https://doi.org/10.1016/j.polymertesting.2020.107021

Cobb, K. C., & Brooks, M. (2017). The age of plastic: Ingenuity and responsibility. https://doi.org/10.5479/si.19492367.7

Coin, I., Beyermann, M., & Bienert, M. (2007). Solid-phase peptide synthesis: From standard procedures to the synthesis of difficult sequences. *Nature Protocols*, *2*(12), 3247–3256. https://doi.org/10.1038/nprot.2007.454

Cuq, B., Gontard, N., & Guilbert, S. (1998). Proteins as agricultural polymers for packaging production. *Cereal Chemistry*, *75*(1), 1–9. https://doi.org/10.1094/CCHEM.1998.75.1.1

De Frates, K., Markiewicz, T., Gallo, P., Rack, A., Weyhmiller, A., Jarmusik, B., & Hu, X. (2018). Protein polymer-based nanoparticles: Fabrication and medical applications. *International Journal of Molecular Sciences*, *19*(6). https://doi.org/10.3390/IJMS19061717

Deravi, C., Golecki, H. M., & Kit Parker, K. Protein-based textiles: Bio-inspired and bio-derived materials for medical and non-medical applications terms of use share your story. *Journal of Chemical and Biological Interfaces*, *1*(1), 25–34. https://doi.org/10.1166/jcbi.2013.1009

Ding, S., Wang, X., & Barron, A. E. (2011). Gene Libraries open up. *Nature Materials*, *10*(2), 83–84. https://doi.org/10.1038/nmat2955

Eldemire, R., Tharp, C. A., Taylor, M. R. G., Sbaizero, O., & Mestroni, L. (2021). The sarcomeric spring protein titin: Biophysical properties, molecular mechanisms, and genetic mutations associated with heart failure and cardiomyopathy. *Current Cardiology Reports*, *23*(9), 121. https://doi.org/10.1007/S11886-021-01550-Y

Fahnestock, S. R., & Irwin, S. L. (1997). Synthetic spider dragline silk proteins and their production in *Escherichia coli*. *Applied Microbiology and Biotechnology*, *47*(1), 23–32. https://doi.org/10.1007/s002530050883

Fahnestock, S. R., Yao, Z., & Bedzyk, L. A. (2000). Microbial production of spider silk proteins. *Reviews in Molecular Biotechnology*, *74*(2), 105–119. https://doi.org/10.1016/S1389-0352(00)00008-8

Feldman, D. (2008). Polymer history. *Designed Monomers and Polymers*, *11*(1), 1–15. https://doi.org/10.1163/156855508X292383

Ferrari, F. A., & Cappello, J. (1997). Biosynthesis of protein polymers. In *Protein-Based Materials* (pp. 37–60). Birkhäuser Boston. https://doi.org/10.1007/978-1-4612-4094-5_2

Fields, G. B. (2001). Introduction to peptide synthesis. *Current Protocols in Protein Science*, *26*(1), 18.1.1–18.1.9. https://doi.org/10.1002/0471140864.PS1801S26

Freeland, B., McCarthy, E., Balakrishnan, R., Fahy, S., Boland, A., Rochfort, K. D., Dabros, M., Marti, R., Kelleher, S. M., & Gaughran, J. (2022). A review of polylactic acid as a replacement material for single-use laboratory components. *Materials*, *15*(9), 2989. https://doi.org/10.3390/ma15092989

Getachew, A., & Woldesenbet, F. (2016). Production of biodegradable plastic by polyhydroxybutyrate (PHB) accumulating bacteria using low cost agricultural waste material. *BMC Research Notes*, *9*(1), 509. https://doi.org/10.1186/s13104-016-2321-y

Geyer, R. (2020). A brief history of plastics. In *Mare Plasticum – The Plastic Sea* (pp. 31–47). Springer International Publishing. https://doi.org/10.1007/978-3-030-38945-1_2

Geyer, R., Jambeck, J. R., & Law, K. L. (2017). Production, use, and fate of all plastics ever made. *Science Advances*, *3*(7). https://doi.org/10.1126/SCIADV.1700782

Gheorghita, R., Anchidin-Norocel, L., Filip, R., Dimian, M., & Covasa, M. (2021). Applications of biopolymers for drugs and probiotics delivery. *Polymers*, *13*(16). https://doi.org/10.3390/POLYM13162729

Ghosh, S., Lahiri, D., Nag, M., Dey, A., Sarkar, T., Pathak, S. K., Atan Edinur, H., Pati, S., & Ray, R. R. (2021). Bacterial biopolymer: Its role in pathogenesis to effective biomaterials. *Polymers*, *13*(8), 1242. https://doi.org/10.3390/polym13081242

Goncalves, A. G., Hartzell, E. J., Sullivan, M. O., & Chen, W. (2022). Recombinant protein polymer-antibody conjugates for applications in nanotechnology and biomedicine. *Advanced Drug Delivery Reviews*, *191*, 114570. https://doi.org/10.1016/j.addr.2022.114570

Gowthaman, N. S. K., Lim, H. N., Sreeraj, T. R., Amalraj, A., & Gopi, S. (2021). Advantages of biopolymers over synthetic polymers: Social, economic, and environmental aspects. In *Biopolymers and Their Industrial Applications* (pp. 351–372). https://doi.org/10.1016/B978-0-12-819240-5.00015-8

Granzier, H., Wu, Y., Siegfried, L., & LeWinter, M. (2005). Titin: Physiological function and role in cardiomyopathy and failure. *Heart Failure Reviews*, *10*(3), 211–223. https://doi.org/10.1007/s10741-005-5251-7

Gupta, S., Chaubey, K. K., Khandelwal, V., Sharma, T., & Singh, S. V. (2021). Genetic engineering approaches for high-end application of biopolymers: Advances and future prospects. *Microbial Polymers*, 619–630. https://doi.org/10.1007/978-981-16-0045-6_24

Gupta, P., & Nayak, K. K. (2015). Characteristics of protein-based biopolymer and its application. *Polymer Engineering & Science*, *55*(3), 485–498. https://doi.org/10.1002/PEN.23928

Halley, P. J., Dorgan, J. R., Halley, P. J., & Dorgan, J. R. (2011). Next-generation biopolymers: Advanced functionality and improved sustainability. *MRS Bulletin*, *36*(9), 687–691. https://doi.org/10.1557/MRS.2011.180

James, P., & Numat, K. (2013). Polymerization of peptide polymers for biomaterial applications. In *Polymer Science*. InTech. https://doi.org/10.5772/46141

Jao, D., Xue, Y., Medina, J., & Hu, X. (2017). Protein-based drug-delivery materials. *Materials*, *10*(5). https://doi.org/10.3390/MA10050517

Jeon, J., Subramani, S. V., Lee, K. Z., Jiang, B., & Zhang, F. (2023). Microbial synthesis of high-molecular-weight, highly repetitive protein polymers. *International Journal of Molecular Sciences*, *24*(7), 6416. https://doi.org/10.3390/ijms24076416

Joardar, J. C., & Rahman, M. M. (2018). Poultry feather waste management and effects on plant growth. *International Journal of Recycling of Organic Waste in Agriculture*, *7*(3), 183–188. https://doi.org/10.1007/s40093-018-0204-z

Karthik, T., & Rathinamoorthy, R. (2018). Sustainable biopolymers in textiles: An overview. In *Handbook of Ecomaterials* (pp. 1–27). Springer International Publishing. https://doi.org/10.1007/978-3-319-48281-1_53-1

Kfoury, G., Raquez, J. M., Hassouna, F., Odent, J., Toniazzo, V., Ruch, D., & Dubois, P. (2013). Recent advances in high performance poly(lactide): From "green" plasticization to super-tough materials via (reactive) compounding. *Frontiers in Chemistry*, 1. https://doi.org/10.3389/FCHEM.2013.00032

Labeit, S., Kolmerer, B., & Linke, W. A. (1997). The giant protein titin. *Circulation Research*, *80*(2), 290–294. https://doi.org/10.1161/01.RES.80.2.290

Lahiri, D., Nag, M., Dutta, B., Dey, A., Sarkar, T., Pati, S., Edinur, H. A., Kari, Z. A., Noor, N. H. M., & Ray, R. R. (2021). Bacterial cellulose: Production, characterization, and application as antimicrobial agent. *International Journal of Molecular Sciences*, *22*(23), 12984. https://doi.org/10.3390/IJMS222312984

Linke, W. A., Ivemeyer, M., Mundel, P., Stockmeier, M. R., & Kolmerer, B. (1998). Nature of PEVK-titin elasticity in skeletal muscle. *Proceedings of the National Academy of Sciences of the United States of America*, *95*(14), 8052. https://doi.org/10.1073/PNAS.95.14.8052

Lutz, E. F. (1961). The cyclic trimerization of acetylenes over a Ziegler catalyst. *Journal of the American Chemical Society*, *83*(11), 2551–2554. https://doi.org/10.1021/ja01472a029

Luyt, A. S., & Malik, S. S. (2018). Can biodegradable plastics solve plastic solid waste accumulation? In *Plastics to Energy: Fuel, Chemicals, and Sustainability Implications* (pp. 403–423). William Andrew Applied Science Publishers. https://doi.org/10.1016/B978-0-12-813140-4.00016-9

Ma, Q., Ma, S., Liu, J., Pei, Y., Tang, K., Qiu, J., Wan, J., Zheng, X., & Zhang, J. (2023). Preparation and application of natural protein polymer-based Pickering emulsions. *E-Polymers*, *23*(1). https://doi.org/10.1515/epoly-2023-0001

Mártonfalvi, Z., & Kellermayer, M. (2014). Individual globular domains and domain unfolding visualized in overstretched titin molecules with atomic force microscopy. *PLOS ONE*, *9*(1), e85847. https://doi.org/10.1371/JOURNAL.PONE.0085847

McKeen, L. W. (2017). Polyolefins, polyvinyls, and acrylics. *Permeability Properties of Plastics and Elastomers*, 157–207. https://doi.org/10.1016/B978-0-323-50859-9.00009-9

Mir, M., Ali, M. N., Barakullah, A., Gulzar, A., Arshad, M., Fatima, S., & Asad, M. (2018). Synthetic polymeric biomaterials for wound healing: A review. *Progress in Biomaterials*, *7*(1), 1. https://doi.org/10.1007/S40204-018-0083-4

Mitura, S., Sionkowska, A., & Jaiswal, A. (2020). Biopolymers for hydrogels in cosmetics: Review. *Journal of Materials Science: Materials in Medicine*, *31*(6), 50. https://doi.org/10.1007/s10856-020-06390-w

Moradali, M. F., & Rehm, B. H. A. (2020). Bacterial biopolymers: From pathogenesis to advanced materials. *Nature Reviews Microbiology*, *18*(4), 195–210. https://doi.org/10.1038/s41579-019-0313-3

Nilsson, B. L., Soellner, M. B., & Raines, R. T. (2005). Chemical synthesis of proteins. *Annual Review of Biophysics and Biomolecular Structure*, *34*(1), 91–118. https://doi.org/10.1146/annurev.biophys.34.040204.144700

Nothling, M. D., Fu, Q., Reyhani, A., Allison-Logan, S., Jung, K., Zhu, J., Kamigaito, M., Boyer, C., & Qiao, G. G. (2020). Progress and perspectives beyond traditional RAFT polymerization. *Advanced Science*, *7*(20), 2001656–2001656. https://doi.org/10.1002/ADVS.202001656

Okolie, O., Kumar, A., Edwards, C., Lawton, L. A., Oke, A., McDonald, S., Thakur, V. K., & Njuguna, J. (2023). Bio-based sustainable polymers and materials: From processing to biodegradation. *Journal of Composites Science*, 7(6), 213. https://doi.org/10.3390/JCS7060213

Parker, K., Garancher, J.-P., Shah, S., & Fernyhough, A. (2011). Expanded polylactic acid – An eco-friendly alternative to polystyrene foam. *Journal of Cellular Plastics*, *47*(3), 233–243. https://doi.org/10.1177/0021955X11404833

Patel, G. M., Shah, V., Bhaliya, J., Pathan, P., & Nikita, K. M. (2023). Polymer-based nanomaterials: an introduction. *Smart Polymer Nanocomposites: Design, Synthesis, Functionalization, Properties, and Applications*, 27–59. https://doi.org/10.1016/B978-0-323-91611-0.00018-9

Prajapati, S. K., Jain, A., Jain, A., & Jain, S. (2019). Biodegradable polymers and constructs: A novel approach in drug delivery. *European Polymer Journal*, *120*, 109191. https://doi.org/10.1016/j.eurpolymj.2019.08.018

Prieto, E. M., & Guelcher, S. A. (2014). Tailoring properties of polymeric biomedical foams. In *Biomedical Foams for Tissue Engineering Applications* (pp. 129–162). Elsevier. https://doi.org/10.1533/9780857097033.1.129

Rabe, R., Hempel, U., Martocq, L., Keppler, J. K., Aveyard, J., & Douglas, T. E. L. (2020). Dairy-inspired coatings for bone implants from whey protein isolate-derived self-assembled fibrils. *International Journal of Molecular Sciences*, *21*(15), 5544. https://doi.org/10.3390/ijms21155544

Ragitha, V. M, & Edison, L. K. (2022). Safety issues, environmental impacts, and health effects of biopolymers. *Handbook of Biopolymers* (pp. 1–27). Springer, Singapore. https://doi.org/10.1007/978-981-16-6603-2_54-1

Ramezaniaghdam, M., Nahdi, N. D., & Reski, R. (2022). Recombinant spider silk: Promises and bottlenecks. *Frontiers in Bioengineering and Biotechnology*, *10*. https://doi.org/10.3389/fbioe.2022.835637

Saberi Riseh, R., Gholizadeh Vazvani, M., Hassanisaadi, M., Thakur, V. K., & Kennedy, J. F. (2023). Use of whey protein as a natural polymer for the encapsulation of plant biocontrol bacteria: A review. *International Journal of Biological Macromolecules*, *234*, 123708. https://doi.org/10.1016/j.ijbiomac.2023.123708

Samir, A., Ashour, F. H., Hakim, A. A. A., & Bassyouni, M. (2022). Recent advances in biodegradable polymers for sustainable applications. *NPJ Materials Degradation*, *6*(1), 1–28. https://doi.org/10.1038/s41529-022-00277-7

Schuetz, J.-H., Wentao, P., & Vana, P. (2015). Titin-mimicking polycyclic polymers with shape regeneration and healing properties. *Polymer Chemistry*, *6*(10), 1714–1726. https://doi.org/10.1039/C4PY01458H

Sehgal, R., & Gupta, R. (2020). Polyhydroxyalkanoate and its efficient production: An eco-Friendly approach towards development. *3 Biotech*, *10*(12), 549. https://doi.org/10.1007/S13205-020-02550-5

Sengupta, P., & Prasad, B. L. V. (2018). Surface modification of polymers for tissue engineering applications: Arginine acts as a sticky protein equivalent for viable cell accommodation. *ACS Omega*, *3*(4), 4242–4251. https://doi.org/10.1021/ACSOMEGA.8B00215

Shah, Y. A., Bhatia, S., Al-Harrasi, A., Afzaal, M., Saeed, F., Anwer, M. K., Khan, M. R., Jawad, M., Akram, N., & Faisal, Z. (2023). Mechanical properties of protein-based food packaging materials. *Polymers*, *15*(7), 1724. https://doi.org/10.3390/polym15071724

Steven, S., Fauza, A. N., Mardiyati, Y., Santosa, S. P., & Shoimah, S. M. (2022). Facile preparation of cellulose bioplastic from *Cladophora* sp. algae via hydrogel method. *Polymers*, *14*(21), 4699. https://doi.org/10.3390/POLYM14214699

Tan, Y., Wu, H., Wei, T., & Li, X. (2020). Chemical protein synthesis: Advances, challenges, and outlooks. *Journal of the American Chemical Society*, *142*(48), 20288–20298. https://doi.org/10.1021/JACS.0C09664

Tang, N. C., & Chilkoti, A. (2016). Combinatorial codon scrambling enables scalable gene synthesis and amplification of repetitive proteins. *Nature Materials*, *15*(4), 419–424. https://doi.org/10.1038/nmat4521

Thompson, R. C., Moore, C. J., Saal, F. S. V., & Swan, S. H. (2009). Plastics, the environment and human health: Current consensus and future trends. *Philosophical Transactions of the Royal Society B: Biological Sciences*, *364*(1526), 2153. https://doi.org/10.1098/RSTB.2009.0053

Viswanath, V. (2021). A short communication on *Pichia pastoris* vs. *E. coli*: Efficient expression system. *Annals of Proteomics and Bioinformatics*, *5*(1), 049–050. https://doi.org/10.29328/journal.apb.1001016

Vroman, I., & Tighzert, L. (2009). Biodegradable polymers. *Materials*, *2*(2), 307–344. https://doi.org/10.3390/ma2020307

Wang, R., Li, J., Li, X., Guo, J., Liu, J., & Li, H. (2019). Engineering protein polymers of ultra-high molecular weight *via* supramolecular polymerization: Towards mimicking the giant muscle protein titin. *Chemical Science*, *10*(40), 9277–9284. https://doi.org/10.1039/C9SC02128K

Wang, K., Liu, M., & Mo, R. (2020). Polysaccharide-based biomaterials for protein delivery. *Medicine in Drug Discovery*, *7*, 100031. https://doi.org/10.1016/J.MEDIDD.2020.100031

Werten, M. W. T., Eggink, G., Cohen Stuart, M. A., & de Wolf, F. A. (2019). Production of protein-based polymers in *Pichia pastoris*. *Biotechnology Advances*, *37*(5), 642–666. https://doi.org/10.1016/j.biotechadv.2019.03.012

Xue, Y., Lofland, S., & Hu, X. (2019). Thermal conductivity of protein-based materials: A review. *Polymers*, *11*(3). https://doi.org/10.3390/POLYM11030456

8 Flexible Fungal Materials
An Industrial Perspective

Shreya Kapoor, Nidhi Verma*, Harsh Charak,*
Kaushiki Dutta, Aishani Gupta,
and Vandana Gupta

8.1 INTRODUCTION

The incorporation of living things in nanotechnology and material science with the aim to achieve the controlled development of materials from biological resources is enticing major research efforts (Fratzl and Barth, 2009; Meyers et al., 2013; Niemeyer, 2001). The key factors driving materials-related study towards polymeric materials made from natural sources are fossil fuel depletion (Ceseracciu et al., 2015; Epicoco, 2016; Shalwan and Yousif, 2013) and global environmental issues related to degradation difficulties of synthetic plastics and derived materials (Derraik, 2002; Webb et al., 2013). Meanwhile, mycelial-based biomaterials or extended fungal materials also known as FFMs could act as an alternative to plastics, leather, textiles, and other composites and are created from waste. Their utilisation could represent a fundamental shift in the way we now conduct the production. Additionally, after their useful lives are over, the fungal products can be recycled to create new materials. These can be used as feed or fertiliser, or to enhance the structure of the soil (Grimm and Wösten, 2018; Meyer et al., 2020). There are different fermentation-based methods to produce extended and pure fungal biomaterials or FFMs, which include liquid state-submerged, liquid state-surface, and solid-state fermentation (reviewed in Peeters et al., 2023). The mycelium of the fungus is made up of a network of tiny filaments that range in diameter from 1 to 30 μM and extend from one spore into every crevice of the substrate (Fricker et al., 2007; Islam et al., 2017). Due to its low energy requirement for development, lack of production of by-products, and wide range of potential applications, mycelium has recently attracted increased interest in both academic and commercial investigations (Holt et al., 2012; Jones et al., 2017; Nawawi et al., 2020; Gandia et al., 2021; Pelletier et al., 2013). Proteins, chitin, and glucans are among the various layers that make up each mycelium filament (Haneef et al., 2017). Chitin, a polysaccharide abundant in fungal cell walls, and is one of the primary components of mycelium; it is a naturally occurring polymer that is widely distributed in both the exoskeletons of crustaceans and fungi. It provides exceptional mechanical strength, flexibility, and biodegradability, making it an ideal alternative to

*Both the authors have contributed equally to the manuscript and both are the first authors.

 DOI: 10.1201/9781003407683-8

conventional materials derived from fossil fuels (Jones et al., 2020a). Chitin has been utilised to create nonwoven fabrics and gels which communicate with the open tissue around a wound to promote healing, necessitating research to investigate the multi-scale structures (Jayakumar et al., 2011; Muzzarelli, 2012; Muzzarelli et al., 2007). In research on wound dressing, both fungal and crustacean chitin are used; however, there are notable differences in their structure, characteristics, and processing (Jones et al., 2020a). Mycelium-based sandwich composites (MBSC) and mycelium-based foam (MBF) are two distinct mycelium-based composite materials that have been researched and developed (Girometta et al., 2019). MBF is produced by uniformly growing fungus in small particles of agricultural waste (Appels et al., 2019). As the mycelium structure expands, fibres are produced that connect particles of agricultural waste to create a porous material (Bartnicki-Garcia, 1968; Jiang et al., 2017; Karana et al., 2018). As agricultural wastes and mycelium mix to create a sandwich structure with better bending rigidity, MBSC adds natural fibre fabric as the upper and bottom layers in addition to the central core (Jiang et al., 2017). In contrast to expanded polystyrene, or EPS, foams, MBF, and MBSC as the 'mycelium bricks' or 'panels' have both demonstrated durability, are lightweight, and offer environmental benefits in packaging, building insulating material, and interior design (Girometta et al., 2019; Holt et al., 2012; Jones et al., 2017, 2018; Xing et al., 2018). Thermal insulation, water absorption, and fire safety properties of FFMs are also discussed before (Robertson, 2020). Moreover, the first bioengineered viable fungal material which shows self-healing property with minimum intervention has been reported recently (Elsacker et al., 2023). This chapter showcases different types, salient features, potential applications, advantages, and examples of FFMs in detail which will be a value additions to mycotechnology for a greener future. Also, different types of FFMs, their key features, source organisms, applications, commercial products, and 'bench to market' projects are listed (Table 8.1).

8.2 TYPES OF FLEXIBLE FUNGAL MATERIAL

Mycelium-based materials, commonly referred to as flexible fungal materials, are generated from the mycelium, i.e. the vegetative structure of filamentous fungi. Mycelium is the subterranean root of fungus consisting of a thin network of thread-like branches and structural polymers, viz. glycoproteins, beta-glucan, and chitin. It secretes enzymes like laccase and peroxidase that facilitate the partial digestion of lignocellulosic materials (Fairus et al., 2022). Utilising the capacity of the fungi to consume and convert cellulose substrates into natural composites, one may develop a diverse range of materials by cultivating the fungi on a variety of substrates (Collet, 2017). Depending upon their structural features and composition, FFMs are of different types as discussed below.

8.2.1 Pure Mycelium Foams

Mycelium foams are exclusively derived from fungal biomass, predominantly aerial and explorative hyphae. These are cost-effectively and easily generated within

TABLE 8.1
Different Types of FFMs, Their Key Features, Source Organisms, Applications, and Well Recognised Commercial Products/Projects

FFM Type	Key Features	Source Organism	Applications	Commercial Product
Pure mycelium foams	Flexible, low density, high porosity, high surface area	*Fusarium graminearum, Ganoderma carnosum, Ganoderma curtisii, Lentinus crinitus*	(a) Leather-like materials (b) Meat (c) Skincare	(a) Mylo™ (b) Atlast™ (c) Mycoflex™
Mycelium composites	Acoustic potential, high tensile strength, low thermal conductivity	*Fusarium oxysporum, Ganoderma applanatum, Ganoderma resinaceum, Pleurotus ostreatus*	(a) Building material (b) Packaging (c) Insulation	(a) Tallinn Architecture Biennale Installation (b) MycoComposite™
Mycelium textiles	Lightweight, high breathability, seamless fabrication	*Ganoderma lucidum, Schizophyllum commune, Agaricus bisporus, P. ostreatus*	Clothing, accessories, and footwear	MycoTex, MycoWorks™
Paper-like materials	Fire resistance, healing potential, solvent stability, cell adhesion, tunable wettability	*Grifola frondosa, Hypsizygus marmoreus, Trametes multicolour, Tricholoma terreum*	Wound dressing, nanopapers, filtration systems	–

14 days by employing liquid- or solid-state fermentation (van den Brandhof & Wösten, 2022). *Fusarium graminearum and Ganoderma carnosum* are a few of the commonly used fungal species for the synthesis of this material. Generally, these mycelial mats are heated at temperatures exceeding 60°C before processing to suppress biological activities. The absence of lignocellulosic material imparts low density, i.e. 30–50 kg/m^3, rendering them suitable for fabrication of light-load shipping materials as a substitute for polystyrene or polyurethane (Jones et al., 2020b). These exhibit a lower tensile strength and compressive modulus ranging 0.1–0.3 MPa and 0.6–2.0 MPa, respectively. Furthermore, the inherent porous nature facilitates the development of cellular scaffolds with profound applications in the biomedical and food sectors (Gandia et al., 2021).

8.2.2 Paper-Like Material

The potential of fungal mycelium for the fabrication of paper-like materials is attributed to their high fibre content and structural resemblance shared by fungal chitin and plant-derived cellulose. These are rich in structural (chitin and beta-glucan) and non-structural components (cytoplasm lipids) (King & Watling, 1997) and can be manufactured using liquid-state fermentation or homogenised basidiocarps. Depending upon the parent fungal source used, the fabricated papers can be hydrophilic or hydrophobic. The former can be produced using fungal basidiocarps while the latter can be developed using culture grown fungal biomass (Gandia et al., 2021). Contrary to conventional wood-pulp derived papers, these bring forth additional perks like wound healing potential and heavy metal chelation. Furthermore, these offer higher tensile strength, i.e. 0.7–9.6 MPa, solvent stability, fire resistance, and tunable wettability. Lastly, chitin-β-glucan complexes extracted from fungi when subjected to hot pressing yield nanopapers. By varying the chitin to β-glucan ratio and fibril dimensions, nanopapers with diverse properties can be obtained; e.g. higher beta-glucan content yields opaque elastomeric materials while higher chitin yields tough materials exhibiting strong tensile properties (Nawawi et al., 2020).

8.2.3 Fungal Composite and Hybrid

Mycelium-based composites (MBCs) are generated by bonding fungal mycelium with lignocellulosic biodegradable materials, viz. hemp shives, different types of straw, and sawdust. The fungal mycelium for fabrication of MBCs is predominantly derived from brown-rot and white-rot fungi, e.g. *Pleurotus ostreatus* and *Ganoderma lucidum*. Fungal mycelium functions as a binding agent, holding together the substrate particles to form a solid composite (Udayanga & Miriyagalla, 2021). In contrast to other mycelium products, MBCs are easy to manufacture and are corrosion-resistant. MBCs possess environmentally favourable traits, such as less energy consumption during manufacture, no waste creation, and easy recyclability. Owing to their exceptional properties including low density (0.10–0.39 g/cm^3), low thermal conductivity (0.10–0.18 W/(m K)), excellent sound absorption, fireproof qualities and fire resistance, MBCs are recognised as an emerging class of widely adopted biomaterials with conceivable applications in the domains of automotive manufacturing, home furnishings, construction, structural insulative, and soundproof panels (Fairus et al., 2022). Furthermore, the acoustic absorbance properties attributed to airflow resistance possessed by these materials make them appropriate for the development of soundproof panels. Additionally, these can be used in shoe insoles, insulating lofts, soft furnishings, and skincare products (Gandia et al., 2021). These are being commercialised as coffee-tables, bowls, and lampshades in Europe and the USA. Numerous on-going scientific and commercial projects aim to exploit and further enhance the properties of MBCs (Appels et al., 2019). Limitations such as low tensile strength, hygroscopicity, and propensity towards biocorrosion are evident in these materials, necessitating further improvement to ensure their utmost reliability (Sydor et al., 2022; Vandelook et al., 2021).

8.2.4 Fungal Leathers/Amadou/Mycoleathers/Vegan Leather

These are obtained by subjecting the naturally occurring basidiocarps on tree trunks (*Fomes fomentarius* and similar fungi), etc. in wilds and/or basidiocarps and mycelial mats obtained from cultured fungi. Various physical and chemical modifications are done for basidiocarps including pounding the basidiocarps flat, boiling or soaking in a solution of potassium nitrate. While treatment of mycelial mats include but is not limited to dehydration, protein denaturation, densification, and texturing (Gandia et al., 2021). Solid-state fermentation is most suitable for fabrication of leather-like materials because of low water, energy, and sterility requirements (Jones et al., 2020b). The texture, appearance, and mechanical properties including tensile strength and density of the resulting material significantly bear a striking resemblance to bovine leather. The tuning of mycelial development and leather processing affects tensile strength and tear strength. Furthermore, treatment with 20% polyethylene glycol has been reported to improve mechanical properties (Raman et al., 2022). These materials can potentially replace animal-based leather and derived products. Mylo™ is a well-recognised commercially available mycelium-based leather.

8.2.5 Fungal Textiles

Mycelium may be treated to produce fibres that can be used to make textiles. A research project called 'Mycelium Textiles' reported the possibility of transforming fungal mycelia into biodegradable textiles by utilising agar, coffee dregs, and natural fibres like soya bean, organic cotton, and hemp as nutritive substrates. The interaction of the growing mycelium with these materials leads to the emergence of an array of textures, surface patterns, and surface tensions. With the variation in supporting components utilised, different fabrics were obtained. Tartan mycelium, characterised by the presence of rusty brown stripes on a white background, was developed using unbleached hemp strips that were arranged in a check pattern on mould pre-filled with coffee dregs. Besides, growing mycelium on broken or worn-out lace and cloth coated with coffee dregs yields mycelium lace and mycelium velvet, respectively. Lastly, the fungi grown directly on the substrate led to the emergence of self-patterns resembling floral designs yielding a mycelium rubber (Collet, 2017; Rathinamoorthy et al., 2023).

8.3 FEATURES AND ADVANTAGES OF FFMs

8.3.1 Flexibility and Stretchability

Flexibility is the material's capacity to bend or compress easily without getting deformed or ruptured when subjected to stretching forces. The flexibility of mycelium-based materials is attributed to their porous structure and the intermolecular interaction between the mycelium and substrate particles. Higher porosity and fewer intermolecular interactions provide large air gaps, thus more flexibility (Yang et al., 2017). Hence it can be concluded that pure mycelium foams lacking additional materials are more flexible than mycelium composites. Other parameters like substrate type, nutrient profile, and manufacturing conditions also affect the flexibility of the material. For instance, cotton-fibre-based myco-composites exhibited more flexibility than straw-based composites (Jones et al., 2020b). Glycerol is a commonly used plasticiser to increase flexibility of the material as it reduces intermolecular interaction (Appels et al., 2019). Flexibility is desired in mycelium-based leathers, textiles, and foams, but not in packaging or construction materials (Vandelook et al., 2021).

8.3.2 Strength and Durability

The overall mechanical strength of the mycelium materials is an important parameter to determine their suitability for various applications. Compressive strength, tensile strength, and flexural strength are the direct measures of materials' mechanical properties. The materials' ability to resist deformation when being compressed, i.e. pulled together, is compressive strength. Lignin-rich substrates like sawdust, mycelium of the rapidly growing fungal species, e.g. *Pleurotus ostreatus*, and short cultivation time increase the compressive strength of the mycelial materials (Alemu et al., 2022; Ghazvinian & Gürsoy, 2022). Mycelium-based material fabricated by growing *Trametes versicolour* on substrates like paddy straw and sawdust demonstrate a good compressive strength of 347 kPa (Alemu et al., 2022). Tensile strength is the material's resistance towards tension, i.e. a force pulling it apart. It can be enhanced by utilising chitin-rich mycelia, cellulose-rich substrate, and hot pressing. The strength conferred by chitin is attributed to its rigid structure that is formed as a result of hydrogen bonding (Chan et al., 2021). Multi-layered mycelial material can exhibit tensile strength ranging from 200 to 1200 kPa (Haneef et al., 2017). In contrast to compressive strength, tensile strength is not influenced by substrate and fungal species (Girometta et al., 2019). Lastly, flexural strength is the capacity to resist bend or deformation in the presence of stretching forces. A small specimen size and pressing can significantly increase the flexural strength. A direct correlation has been found between the mechanical strength and the density of the material (Jones et al., 2020b).

Thermal stability, fire resistance, and other exceptional properties make FFMs highly durable (Gandia et al., 2021). The structural components of mycelium specifically lignin, hydrobin, and chitosan provide thermal stability and fire resistance to mycelium-based composites (Yang et al., 2017). Tear resistance is another important parameter influencing the durability of materials. Mycelium-based materials have been reported to have lower tear resistance than synthetic counterparts (Gandia et al., 2021).

8.3.3 Breathability and Moisture Content

The porous nature of mycelium composite significantly contributes to its moisture content regulation and breathability. Mycelium composites have a strong affinity towards water. It is attributed to hydroxyl groups of the cellulose and partly to the hydrophilic porous mycelium that facilitates water uptake by capillary action (Girometta et al., 2019). It has been suggested that substrate type and species significantly affect the composite's ability to absorb water (Haneef et al., 2017). For instance, sawdust-derived composite exhibited lower water absorption than composite developed using coffee husk (Bitting et al., 2022). Mycelium composites designed using *Pleurotus* sp. grown on grain fibres have been reported to absorb up to 278% water within 24 h. Due to their composition and capacity to absorb water, these mycelium-based materials may serve as a means for growing plants; however, high water absorption by composites limits their applicability in domains like construction, as it may lead to destructive problems like leaking walls or roof cavities (Alemu et al., 2022). Hot pressing, air drying, and bio-based coatings, such as polyfurfuryl alcohol resin, can reduce the affinity of the composites towards the water. Additionally, the substrates with small particle sizes provide less porosity and increased density, ultimately leading to low water absorption ability (Aiduang et al., 2022).

8.3.4 Acoustic Properties

Mycelium materials exhibit absorption of low frequency sound waves (<1500 Hz) that can be attributed to its highly porous and fibrous nature. At 1000 Hz, mycelium materials have been reported to exhibit 70–75% absorption (Aiduang et al., 2022). Porosity governs the entry of sound waves into the materials, while fibrous structure offers resistance to acoustic waves, converting their mechanical motion into heat energy, thus reducing their amplitude. Thin fibres provide tortuous paths and greater airflow resistance, thus better acoustic absorption (Jones et al., 2020c). The density of the material significantly affects the absorbed frequency, with dense structures favouring the absorption of high frequencies >2000 Hz. The exceptional acoustic properties of the mycelium materials make them highly suitable for the development of soundproof architectural systems. The perpetual road noise for mycelium composites has been reported to be lower, i.e. 45.5–60 dB, than that of conventional absorbers, e.g. plywood (65 dB). Substrate fillers like rice straw, switchgrass, flax shive, and sorghum fibre have been well-recognised for conferring excellent acoustic properties to the mycelium materials. The mycelium materials to be utilised as acoustic absorbers should not be compressed as compression of the materials significantly reduces its thickness, thus negatively impacting the acoustic properties (Girometta et al., 2019).

8.4 MANUFACTURING PROCESSES OF FLEXIBLE FUNGAL MATERIALS

8.4.1 Cultivation and Growth of Fungal Species

The cultivation of the selected fungal species begins with the isolation of pure culture followed by spawn production, substrate selection (lignocellulosic) and sterilisation (Houette et al., 2022). A spawn serves as an inoculum to facilitate substrate colonisation and is prepared by placing the pure culture onto a sterilised grain filled into glass bottles. It is then incubated in a controlled environment with appropriate temperature (25–30°C) and humidity (60–65%) for a specific period of time. The incubation period can range from 6 to 20 days depending upon the desired qualities of the end product, fungal species, the substrate used, and the growth conditions provided (Alemu et al., 2022). Furthermore, during this period, the CO_2 concentrations (low) and light conditions (dark) are strictly maintained to promote anamorphism and suppress the formation of fruiting bodies that might interfere with other resources (Attias et al., 2020).

Based on the desired quality of the end product, solid-state fermentation or liquid state fermentation can be employed to provide a suitable growth environment. The former is preferred for mycelium-based leathers and foams, while the latter is preferred for paper-like materials. While SSF significantly minimises energy and water requirements, and provide high-volume productivity, though it involves intricate upscaling procedures. SSF systems are more useful than LSF systems in terms of low-tech settings, low capital costs, simplicity, and a lower danger of contamination (Gandia et al., 2021).

8.4.2 Material Formulation and Shaping Techniques

The specific desired shapes are obtained by growing the mycelia in a mould. The harvested mycelium mats are dehydrated to terminate further growth (Attias et al., 2020). Dehydration generally results in the formation of closed-cell foam because of the presence of microscopic air cavities formed by the virtue of evaporation of water molecules. Besides, the mycelium mats are deactivated by heating above 60°C (Alemu et al., 2022). Following this, these materials are subjected to several physical and chemical changes like baking, compacting, or coating to fabricate materials with tailored properties (Manan et al., 2021). For instance, in order to obtain wound dressing materials, the paper-like mycelium is subjected to deacetylation that results in the formation of chitosan membranes with enhanced hemostatic and antibacterial properties and ameliorated cell proliferation and adhesion. Deacetylation is usually performed by hot pressing followed by mild alkaline treatment. Pressing is performed to increase the material density and rigidity while decreasing its porosity (Gandia et al., 2021). It can either be done at low temperatures (cold pressing) or high temperatures (hot pressing) depending upon the features of the fungal species being used and the desired properties of the end product. The latter is reported to affect the thermal stability of the material (Manan et al., 2021).

8.4.3 Post-Processing

Solution-based post-processing has been suggested to augment the inherent properties of the mycelium materials. This generally employs organic solutions, organic solvents like polyphenols in combination with salts, or a cocktail of different organic solvents. To illustrate, treatment with calcium chloride solution and polyphenols significantly enhances materials' elastic and mechanical properties, including tensile strength. Additionally, it improves the shelf life and durability of the precursor materials. Mechanical properties in mycelium leather can be improved by coating it with polylactic acid (PLA) (Gandia et al., 2021). PLA dissolved in water is applied onto the mycelium materials, forming a protective water and abrasion-resistant layer with reduced combustion propensity. Furthermore, coconut oil, carnauba wax or beeswax, and cellulose derived from bacterial species can be used for the purpose of coating (Vandelook et al., 2021). The steps involved in the process of production of FFMs are depicted in the Figure 8.1.

8.5 APPLICATIONS OF FLEXIBLE FUNGAL MATERIALS

FFMs possess exceptional attributes making them indispensable for a wide array of applications across a diverse range of industries, including but not exclusively

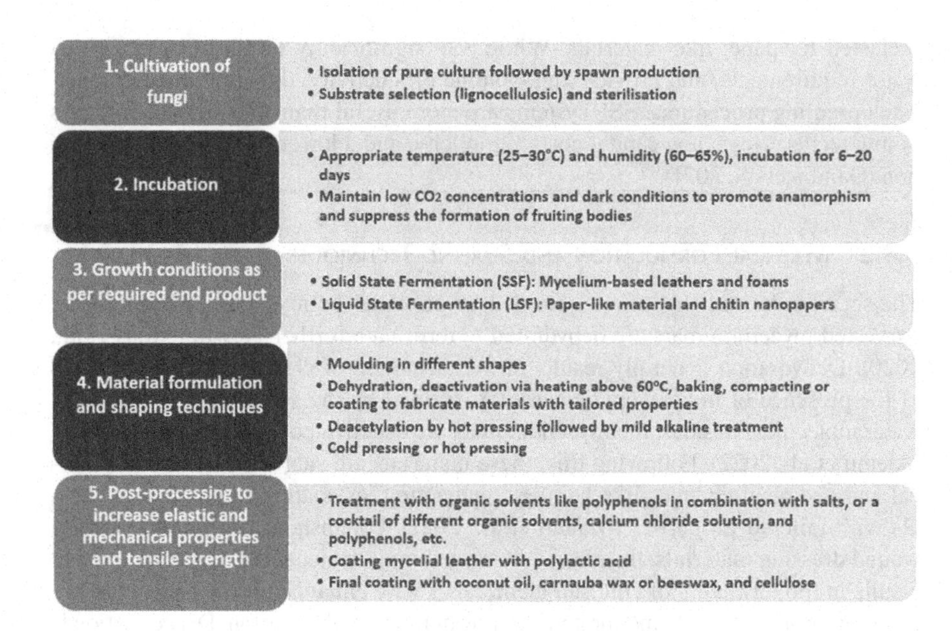

FIGURE 8.1 Steps in production of flexible fungal materials (FFMs).

biomedical, paper, packaging, cosmetic, textile, and building industries. FFMs can be combined with bioplastics for designing household and furniture products including lamps, chairs, vases, cups, and pots (Aiduang et al., 2022). Furthermore, these can be utilised in the development of biodegradable crematory and funeral products. Living cocoon, burial suit, and burial shroud are some of the well-marketed mycelium-based crematory products. Mycelium-based composites could potentially be utilised in the construction industry owing to their strong acoustic absorption, low thermal conductivity, mechanical strength, and fire resistance nature (Manan et al., 2021). Additionally, MBCs can be used in agricultural industries as soil conditioners (Fairus et al., 2022). Fungal mycelium can be used to fabricate fibrous membranes with pore sizes ranging from 10 to 50 μm in diameter (Vaišis et al., 2023) that can potentially be used in air and water filtration systems.

8.5.1 Mycelial Produce as 'Meat-Alternative'

The fibrous-protein-rich structure of the aerial fungal mycelium presents a texture and nutritional profile similar to that of traditional meat and thus could be used as a vegan substitute for animal-based meat. The first fungal-based meat alternative, popularly known as 'Quorn', was derived from *Fusarium venenatum* (Vandelook et al., 2021).

8.5.2 FFMs in Textile Industry

The production of different biomaterials from fungus through technical process is called bio-fabrication. Companies have successfully come up with fungal-based leather substitutes called 'MuSkin' (Khurana, 2021). MuSkin is a sustainable leather alternative and is derived from a fine network of threads that can grow within a period of 10 days requiring just water for growth as it derives its nutrients from soil (Khurana, 2021). When considered in the fashion and apparel industry, mycelium is the key component in constructing FFMs e.g., fabrics, it is the root-like equivalent of plant structures with a thread-like network which helps in providing basic firmness and texture same as that imparted by cotton fibres. Recently, industries have come up with construction of bioengineered mycelium produce which gives rise to firm and tough textile material (Nawawi et al., 2020). 'Bolt Threads' a US-based start-up, has used mycelial leather to successfully craft designer products such as ladies bag. MOGU is an enterprise manufacturing fungal mycelium-based materials and products.

These engineered mycelial cells form three-dimensional structures, densely interweaved and are known as 'Fine Mycelium'. They provide raw materials for apparels/textiles, that have almost the same resistance, firmness, and potential as animal leather-based garments (MycoWorks™). Most mushroom leather is made up of compressed solid foams produced from mycelium and lacks the bulky feel of typical animal-based leather used in the manufacturing of textiles.

Other positive aspects of using mycelium as textile/clothing materials is that, it is carbon negative and takes up any dye and thus results in the production of clothing material in a variety of colours (Khurana, 2021). Garments can also be easily assembled in various forms, shapes, sizes, and designs without the need for cutting and sewing using body-based modelling, thus reducing any excess waste produced. Apart from this, it also is a very eco-friendly option as the textile industry consumes high energy, lots of water, and mainly animal produce, which raises environmental and ethical issues, so replacing it with mushroom fibres is a greener approach (Khurana, 2021).

A group of scientists, reported that the growth of *Rhizopus delemar* on breadcrumbs provides fibres of chitin and chitosan. They then removed the lipids, proteins, and other constituents and retained the fibrous jelly-like substance that could be spun into yarn and used for clothing and textiles. For improvement in strength and shine, bio-based binders and glycerol were also used. To create sturdy material, layers of these individual fungal sheets were combined and crafted into the required product (Haneef et al., 2017). Basidiocarp of fungi can also play a significant role in crafting textiles. Trama, the inner, fleshy region of the basidiocarp, is used to obtain fine 'felt' like material for making garments (Gandia et al., 2021). This is done by manually extricating the fleshy part from the interior of the basidiocarp, either from fresh or soaked sample, then finally beating it with a wooden club to separate the fibre pieces. Detangle them and construct various fabric pieces (Gandia et al., 2021). Large pieces of trama, usually 20–30 cm in length, are used for making objects like hats while smaller ones are used in shaping scarves, bags, tablecloths, cleaning clothes, etc. (Gandia et al., 2021).

8.5.3 FFMs as Packaging Materials

Further, mycelium from fungus can also be used to create and design packaging materials. Fungal mycelium is a substitute for synthetic packing materials with its greatest advantage being biodegradable, environmental-friendly, and recyclable. Additional advantages include the low weight of mycelial-based packaging materials, high insulation, flame resistance, remarkable absorbance properties, great elasticity, and softening textures, making it an appreciable substitute (Majib et al., 2023). Fungal strains used in the manufacturing of such packing materials utilise corn stalks and wheat straw as substrates, thus helping in management of agricultural waste products. Feedstock obtained from agricultural waste like hemp, straw, and rice husk are also used as the substrate (Majib et al., 2023). Mycelium-based packaging material also has versatility in design and shapes. A three-dimensional reusable printed mould is used and water, agricultural wastes, and mycelium are added to it, resulting in the production of packaging material. This process occurs as a result of chitin binding to the waste into a fibre-like mass (Majib et al., 2023).

Usually mushrooms like white oyster mushrooms (*Pleurotus florida*) and yellow oyster mushrooms (*Pleurotus citrinopileatus*) are grown on rice husk, sawdust, sugarcane bagasse, teak leaves, etc. to produce natural packing material (Majib et al., 2023). The use of mycelium-based composites for packaging material has the advantage of getting composted in soil in less than a month and also releases nutrients in the soil enriching the depleted soil. This reduces the amount of waste generated making it an eco-friendly option (Patel, 2023). Mycelium grown on substrates like sawdust and coir pith can replace polystyrenes used as cushioning material in packaging goods. However, it is still being compared on its usage against synthetic packing materials in terms of availability and cost (MacArthur, 2017).

8.5.4 Biomedical and Healthcare Applications

The fruiting body of fungus shows anti-microbial, anti-ageing, and hypocholesterolemic properties (Gunawardena et al., 2014). The use of fungal mycelium in biomedical and healthcare systems is also reported where a curcumin-based mycelial wound healing patch is developed, which is demonstrated to have great strength and firmness (Khamrai et al., 2018). This patch is effective in the sustained release of curcumin from the mycelial matrix at the site of the injury which is suitable and relevant for healing of the wound and aids in cell proliferation during the healing process (Mohanty et al., 2012) and has anti-inflammatory, anti-tumour, and antioxidant properties (Aggarwal and Sung, 2009). This mycelial biomass is isolated from the fungus *Phanerochaete chrysosporium* which shows considerable antimicrobial activity against both gram-positive and gram-negative bacteria (Manan et al., 2021).

Moreover, triterpenoid compounds (TCs) show effectiveness in fighting dementia and cancer. The TCs obtained from mycelium help greatly in lowering the proliferation of PC-3, HT-29, HepG2, HeLa, and HepG2 cell lines (Wang et al., 2020). Fungal mycelium has also been reported to show shielding effects on the liver, kidney, and brain (Li et al., 2019; Xu et al., 2017). Also, chitosan, the deacetylated derivative of chitin, is widely utilised for medical applications like creating sutures, bandages, and scaffolds. It also serves as a perfect contender for drug delivery systems (Svensson et al., 2022). Moreover, trama, when heated and allowed to cool, yields a soft, thin stretchable material that can be used as a hemostatic agent in dressing wounds as it is a highly absorptive material (Gandia et al., 2021).

8.5.5 In Building Materials and Construction

Another use of fungal composites is in interior design and construction of furniture. The fungal strain *Fomes fomentarius* is mainly used in the process. First, the fungus is mixed with unused wood chips, where they grow and form a vast network of fibres. This myelinated fungus can then be given any shape to form sturdy furniture according to one's need after heat killing. A great advantage is that the fungal composite used can be moulded into any shape and can naturally stay together as a compact material. The mycelium-based composites cultivated on straw and hemp fibres act as a natural insulator providing frictional resistance and removing the extra energy in the form of heat due to their low density and thermal conductivity (Collet and Pretot, 2014). Biofoams constructed from mycelium also serve as excellent insulators and absorbers and thus are used in building materials and construction (Yang et al., 2017). Mycelium itself can absorb frequencies of up to 1500 Hz and thus replaces corks used in ceiling tiles and even can reduce sound pollution (Manan et al., 2021). The power of mycelium (especially its gluing effect) arising out of dense interlinked hyphae was first described by a Japanese scientist in 1980 (reviewed by Mojumdar et al., 2021). The construction sector witnessed the construction of the 'mycelium tower', which is 40 feet tall and made entirely out of mycelium bricks and is the tallest structure ever to be made up of fungal material. Some other notable architectural structures like the Hy-fi tower, the Mycotecture alpha, and Mycotree are designed from fungal mycelial bio-composites (Ashok et al., 2018; Mojumdar et al., 2021). To construct Hy-fi tower and Mycotecture alpha, more than a dozen meter tall, intersecting cylinders supporting thousands of mycelium blocks were used (Mojumdar et al., 2021). Mycelium materials having their own customisable properties can be used in place of foams, plastics, etc. for furnishings, woodwork, flooring material, etc. Mycelial materials having low thermal conductivity, high absorption, and fire safety can be used to replace traditional construction materials (Jones et al., 2020c). Figure 8.2 summarises the types of FFMs and their starting material.

8.5.6 In Cosmetics

The beauty industry is in constant search for natural ingredients to enhance the quality of their products. Mushrooms produce a lot of useful compounds like polysaccharides, polyphenolics, terpenoids, vitamins, and several others, which show positive effects on hair and skin. These compounds produce outstanding anti-ageing, antiwrinkle, and skin whitening outcomes. For example, the Shiitake species shows its use as a skin exfoliator with amazing skin brightening effects due to renewal of skin cells (Manan et al., 2021).

Mushroom-derived products also act as an effective moisturiser like that seen in *Tremella* mushroom; the derivative of it, i.e. carboxymethylated polysaccharide, retains up to 70% moisture for up to 4 days. Derivative from fungi *Pleurotus cornucopiae* proved to be effective against atopic dermatitis. Fungi also secrete many enzymes and antioxidants which help prevent ageing and formation of wrinkles by disrupting the body's collagen and elastin and maintaining the body's repair system (Manan et al., 2021).

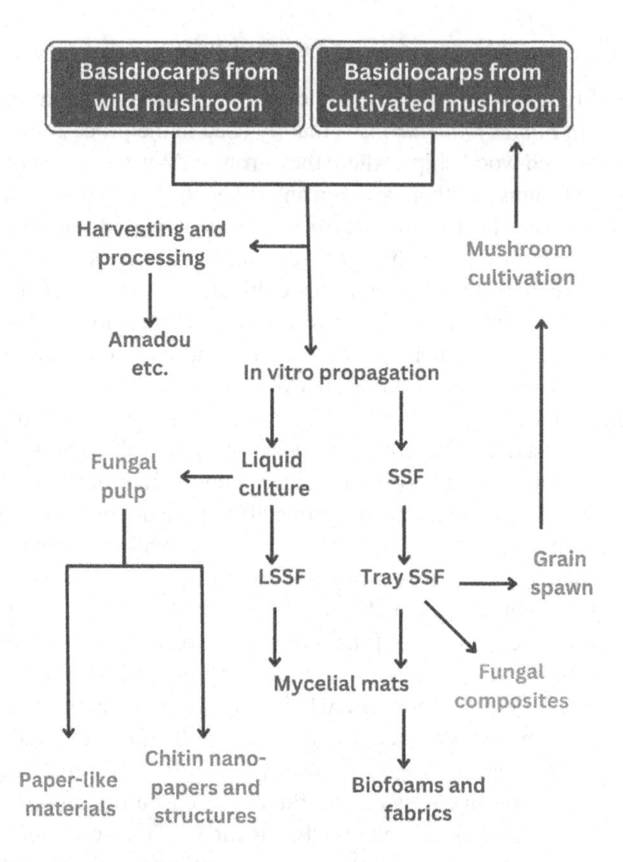

FIGURE 8.2 Types of fungal flexible material starting cultures.

8.6 ADVANTAGES, SUSTAINABILITY, AND ENVIRONMENTAL BENEFITS

The greatest and biggest advantage of FFMs is that it is biodegradable and environmental friendly. FFMs used in making paper, textiles, packaging material, biomedical area and interior designing and furniture are the 'greener' (eco-friendly) substitutes. Fungal-based biomaterials appear as promising avenues for revolutionising the manufacturing landscape, offering cost-effective and environmental friendly substitutes to their conventional counterparts (Niego et al., 2023; Robertson, 2020).

The problems and damage caused by animal-based leathers are even more than plastic or other synthetic ones because of deforestation and methane gas emissions which are linked to the livestock raised for obtaining both meat and leather. Livestock itself can raise the greenhouse emissions by a rate of 15% (Mycoworks) which can be overcome largely by fungal-based proteins and materials. Fungal mycelium-based composites used in construction of packaging material are another added advantage. With 2.8 million tons of plastic being generated each year and only 10% of it being recycled, the use of fungus-generated packaging material involving lignocellulosic compounds forms an essential part of biodegradable polymer and thus benefits the environment (Majib et al., 2023). FFMs are a sustainable alternative in fashion industries as the waste generated including toxic dyes and hazardous and carcinogenic materials disrupts the food chain and reduces the rate of photosynthesis (Pedroso-Roussado, 2023). The use of fungal materials on large and industrial scales must see that substrate availability is placed locally, i.e. nearby to the industries, to prevent an increase in the carbon footprint, which occurs when substrate needs to be brought from distant places resulting in global warming (Alemu et al., 2022).

Harnessing the potential of fungi to use agricultural and forestry by-products as nutritive substrates mitigates the production cost and reduce the reliance on non-renewable resources. FFMs simultaneously facilitate the valorisation of organic waste that poses a significant environmental threat and is expected to reach 2.2 billion tons in two years (Aiduang et al., 2022; Manan et al., 2021). Owing to their circular nature, FFMs substantially minimise carbon footprint and energy consumption contributing to overall environmental sustainability (Manan et al., 2021). This is supported by a comparative life cycle assessment of packaging material synthesised from mycelium inserts and expanded polystyrene. By employing methods like ReCiPe 2016, CED, and IPCC GWP100, the study revealed that the former exhibits low embodied energy and carbon (Enarevba & Haapala, 2023). The high adaptability of fungal species diversifies the usable feedstock and allows for the synthesis of materials with tuneable properties (Gandia et al., 2021; Manan et al., 2021; Robertson, 2020).

8.7 REAL-WORLD PROJECTS AND INNOVATIONS

The utilisation of FFM extends beyond the realm of theoretical concepts and has gained momentum in diverse markets, demonstrating tangible real-world applications. Eco-friendly leather with distinct physiochemical and structural properties has been fabricated by several companies, viz. Ecovative (Forager™), MycoWorks (Sylvania™ Reishi™), MycoTech (Mylea), and Bolt threads (Mylo™). These offer a sustainable, vegan substitute for conventional leather. Numerous well-recognised brands like Hermès, Stella McCartney, and Adidas are embracing this paradigm shift, utilising myco-based leather for the fabrication of consumer goods (Vandelook et al., 2021). Recently, NEFFA designed MycoTex fabric with the potential to serve as a substrate for clothing, accessories, and footwears. Over that, corporates like Dell and Ford motor company have unveiled the incorporation of mycelium-based foam, e.g. MycoFlex™ by Ecovative, as packaging materials (for shipping the servers) (Rajendran, 2022). Additionally, the FFMs and automotive parts like dashboards can substitute vinyl and polystyrene in furniture, e.g. Myco-Mensa, a filamentous mycelium composite table and MYCOsella, stools and chairs based on MBCs. IKEA, a Sweden-based multinational company supplying furniture and home furnishings, commits to MycoComposite as an alternative to Styrofoam for packaging. Lastly, well-established myco-based architectural projects are known; e.g. Hy-Fi, a 12 m tall edifice utilising mycelium bricks, and Tallinn Architecture Biennale

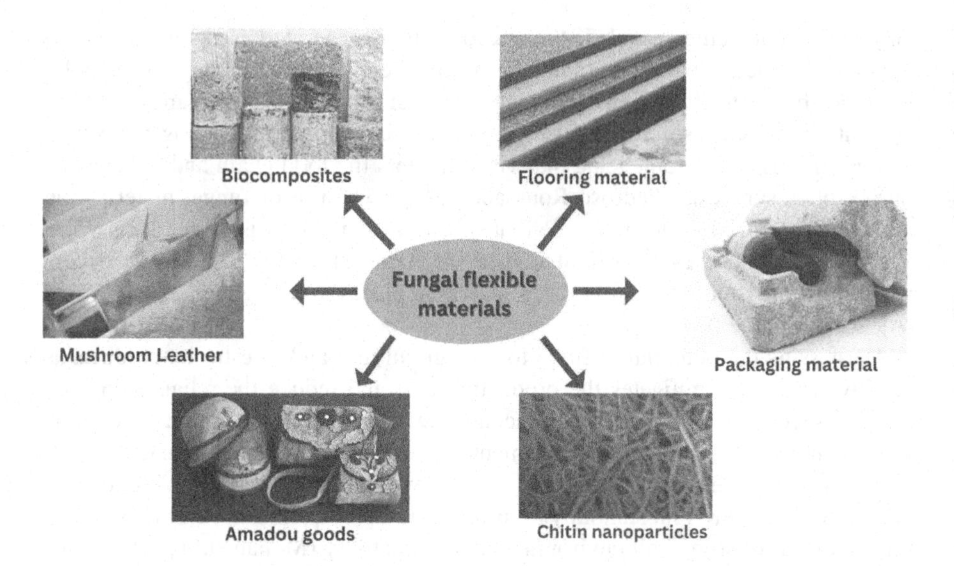

FIGURE 8.3 Commercially available products derived from fungal flexible materials.

Installation employing waste material from the timber industry. Other projects include MycoTree, The Growing Pavilion, and The Circular Garden (Bitting et al., 2022). By 2030, the monetary value of mycelium-based biomaterials is anticipated to reach USD 5.49 billion, thereby substantially contributing to global economic benefits (Almpani-Lekka et al., 2021; Niego et al., 2023). Some of the commercially available products derived from fungi are depicted in Figure 8.3.

8.8 CHALLENGES

Despite the significant advantages offered by FFMs and advancements in manufacturing technologies, the myco-based industry encounters diverse complications, with cost-effective upscaling being the primary concern (Cerimi et al., 2019; Jones et al., 2020a, 2020c). In the realm of large-scale production, the reproducibility of the material is an important attribute. Nonetheless, biological stochasticity can give rise to disparities in the composition and behaviour of the FFM, even in the presence of identical conditions. Furthermore, the scarcity of suitable infrastructure supporting the development of standardised protocols can present obstacles in ensuring the reproducibility of the material (Bitting et al., 2022; Manan et al., 2021). With bulk production, the risk of contamination rises tremendously, further complicating the upscaling. Besides, there is a dearth of literature on the durability of the FFM over prolonged periods. The flip side of biodegradability is reduced shelf life of the material. In the presence of moisture or enzyme, there might be a substantial decline in mechanical, physical, and structural properties of the materials. Lastly, myco-based materials encounter limited consumer acceptance owing to familiarity and trust in the conventional materials, perceived efficacy, and awareness (Niego et al., 2023).

8.9 SAFETY CONCERNS AND REGULATIONS

The fungal species utilized for the development of biomaterials could be pathogenic to humans, animals, or plants. Basidiomycetes, generally used for the synthesis of FFMs, are opportunistic human pathogens majorly associated with respiratory issues. Some of these fungal species are highly destructive for plantations. For instance, *Ganoderma* sp. is responsible for causing stem rots in coconut palms. In the case of FFMs in direct contact with human skin, certain individuals might develop allergic reactions in response to mycotoxins breaching the skin barrier (van den Brandhof & Wösten 2022). Furthermore, fungal species might secrete volatile compounds like alcohols and ketones responsible for attracting insects (Morath, 2012). Taking into consideration the safety concerns associated with myco-based materials, it is recommended to screen fungal species on the basis of risk assessment so as to avoid the utilisation of invasive or pathogenic fungi for the development of biomaterials. In situations where living mycelium is incorporated into the final product, pathogenic fungi should be strictly avoided. Moreover, biosafety levels should be identified for working with specific fungal species. Generally, BSL-1 is considered suitable for fungi. In compliance with the Nogya Protocol, enforced to protect the monetary benefits of the countries from their genetic resources, the use of local fungal species is suggested. This will additionally restrain the spread of fungal species in the natural environment. Alternatively, fungi can be grown with lithium chloride or other glycogen synthase kinase-3 (GSK-3) inhibitors to repress fruiting and thus invasion (van den Brandhof & Wösten2022).

8.10 IMPROVEMENTS

Strain improvement via genetic manipulation holds the potential to optimise desirable properties and reduce pathogenicity and attack by insects as revealed by a multitude of studies. In *Ganoderma* sp., overexpression of BGS1 and BGS2 genes encoding for beta-glucan synthase facilitated a 135–165% increase in beta-glucan content, significantly enhancing material strength and health-promoting potential (Mirończuk-Chodakowska et al., 2021). Similarly, a mutant strain of *Schizophyllum commune* with defective hydrophobin expression produced mycelium materials with increased density and mechanical strength (Vandelook et al., 2021). Co-cultivation with melanin-producing bacterial species provides FFM with increased resistance to ultraviolet radiation (Vandelook et al., 2021). Similarly, co-cultivation with recombinant *Streptomyces natalensis* producing natamycin suppresses contaminants like *Trichoderma*. Besides, a coating with polylactic acid or oil and mild alkaline treatments can significantly reduce the weathering effect to improve mechanical properties and tensile strength of FFMs, respectively (Alemu et al., 2022; Gandia et al., 2021). Furthermore, coating with a mixture of guayule resin and vegetable oil confers protection against insects and termites (van den Brandhof & Wösten2022).

8.11 FUNGI AND THE FUTURE

Given the tuneable properties of FFMs, on-going research, and advancements, the future of fungal-based technology holds immense potential. It is highly likely that a number of fungal species are not yet fully explored (Vandelook et al., 2021). With a cumulative effect of technological advancements and an interdisciplinary approach involving omics analysis, the dynamics of the relation among fungal species, substrate, fermentation set-up, and material biology can be understood. It will assist not only in identifying new species but also in standardising the parameters for the fabrication of optimised products. Furthermore, improvements in physiochemical properties unlock the potential of FFMs in designing electronic sensing devices, radioprotective shields for humans, and space boots (Shunk et al., 2022; Vandelook et al., 2021). These advanced studies and the value addition of fungi in the global economy need substantial research inputs for eco-friendly translational outcomes.

8.12 SUMMARY AND CONCLUSION

Mycelium growth offers a unique and cost-effective bio-fabrication method for recycling agricultural waste and by-products into sustainable biomaterials. Due to their low density, low thermal conductivity, and porosity, mycelium-based materials have better thermal and insulation properties compared to synthetic foams and conventional wood fibres. In addition, these materials are hard and highly fire resistant, making them a viable alternative to common building materials. Additionally, mycelium-based materials are lightweight, biodegradable, and offer an environmentally friendly alternative to conventional petrochemical industry-based packaging materials. Fungi is also a natural source for various bio-active compounds; this ability along with the tensile strength of mycelial compounds makes it usable to synthesise FFMs with potential applications in the medical, pharmaceutical, cosmetic, and construction industries. To date, only a countable number of fungi have been studied for their value addition contribution to materials science. The interdisciplinary approach combined with the omics approach, deeper insight into the molecular mechanisms of hyphal synthesis and material biology warrants further elucidations, making more mycelial-based bio-products with diverse applications.

Last but not least, fungi cast a revolutionary and eco-friendly manufacturing platform that offers a transformative solution to the challenges of waste management and the creation of FFMs. We can create a green economy, where waste materials are transformed into valuable resources by exploring the natural intelligence of fungi. The large-scale development and mechanical adoption of FFM's potential to revolutionise various industries will be a milestone for a better, greener, and more sustainable future.

REFERENCES

Aggarwal, B. B., & Sung, B. (2009). Pharmacological basis for the role of curcumin in chronic diseases: An age-old spice with modern targets. *Trends in Pharmacological Sciences*, *30*(2), 85–94. https://doi.org/10.1016/j.tips.2008.11.002

Aiduang, W., Chanthaluck, A., Kumla, J., Jatuwong, K., Srinuanpan, S., Waroonkun, T., Oranratmanee, R., Lumyong, S., & Suwannarach, N. (2022). Amazing fungi for eco-Friendly composite materials: A comprehensive review. *Journal of Fungi*, *8*(8), 842. https://doi.org/10.3390/jof8080842

Alemu, D., Tafesse, M., & Mondal, A. K. (2022). Mycelium-based composite: The future sustainable biomaterial. *International Journal of Biomaterials*, *2022*, 1–12. https://doi.org/10.1155/2022/8401528

Almpani-Lekka, D., Pfeiffer, S., Schmidts, C., & Seo, S. (2021). A review on architecture with fungal biomaterials: The desired and the feasible. *Fungal Biology and Biotechnology*, *8*(1), 17. https://doi.org/10.1186/s40694-021-00124-5

Appels, F. V. W., Camere, S., Montalti, M., Karana, E., Jansen, K. M. B., Dijksterhuis, J., Krijgsheld, P., & Wösten, H. A. B. (2019). Fabrication factors influencing mechanical, moisture- and water-related properties of mycelium-based composites. *Materials & Design*, *161*, 64–71. https://doi.org/10.1016/j.matdes.2018.11.027

Ashok, A., Abhijith, R., & Rejeesh, C. R. (2018). Material characterization of starch derived bio degradable plastics and its mechanical property estimation. *Materials Today: Proceedings*, *5*(1), 2163–2170.

Attias, N., Danai, O., Abitbol, T., Tarazi, E., Ezov, N., Pereman, I., & Grobman, Y. J. (2020). Mycelium bio-composites in industrial design and architecture: Comparative review and experimental analysis. *Journal of Cleaner Production*, *246*, 119037. https://doi.org/10.1016/j.jclepro.2019.119037

Bartnicki-Garcia, S. (1968). Cell wall chemistry, morphogenesis, and taxonomy of fungi. *Annual Review of Microbiology*, 22, 87–108. https://doi.org/10.1146/annurev.mi.22.100168.000511

Bitting, S., Derme, T., Lee, J., Van Mele, T., Dillenburger, B., & Block, P. (2022). Challenges and opportunities in scaling up architectural applications of mycelium-based materials with digital fabrication. *Biomimetics*, *7*(2), 44. https://doi.org/10.3390/biomimetics7020044

Cerimi, K., Akkaya, K. C., Pohl, C., Schmidt, B., & Neubauer, P. (2019). Fungi as source for new bio-based materials: A patent review. *Fungal Biology and Biotechnology*, *6*(1), 17. https://doi.org/10.1186/s40694-019-0080-y

Ceseracciu, L., Heredia-Guerrero, J. A., Dante, S., Athanassiou, A., & Bayer, I. S. (2015). Robust and biodegradable elastomers based on corn starch and polydimethylsiloxane (PDMS). *ACS Applied Materials & Interfaces* *7*(6), 3742–3753. https://doi.org/10.1021/am508515z

Chan, X. Y., Saeidi, N., Javadian, A., Hebel, D. E., & Gupta, M. (2021). Mechanical properties of dense mycelium-bound composites under accelerated tropical weathering conditions. *Scientific Reports*, *11*(1), 22112. https://doi.org/10.1038/s41598-021-01598-4

Collet, C. (2017). Grow-made textiles. In alive active adaptive: proceedings of EKSIG2017, *International Conference on Experiential Knowledge and Emerging Materials*, eds. E. G. Karana, N. Nimkulrat, K. Niedderer, and S. Camere, 24–37, Delft, The Netherlands, June 19–20.

Collet, F., & Pretot, S. (2014). Thermal conductivity of hemp concretes: Variation with formulation, density and water content. *Construction and Building Materials*, *65*, 612–619.

Derraik, J. G. B. (2002). The pollution of the marine environment by plastic debris: A review. *Marine Pollution Bulletin*, *44*, 842–852, https://doi.org/10.1016/s0025-326x(02)00220-5

Elsacker, E., Zhang, M., & Dade-Robertson, M. (2023). Fungal engineered living materials: The viability of pure mycelium materials with self-healing functionalities. *Advanced Functional Materials*, *33*(29). https://doi.org/10.1002/adfm.202301875

Enarevba, D. R., & Haapala, K. R. (2023). A comparative life cycle assessment of expanded polystyrene and mycelium packaging box inserts. *Procedia CIRP*, *116*, 654–659. https://doi.org/10.1016/j.procir.2023.02.110

Epicoco, M. (2016). Patterns of innovation and organizational demography in emerging sustainable fields: An analysis of the chemical sector. *Research Policy*, *45*, 427–441. https://doi.org/10.1016/j.respol.2015.10.013

Fairus, M. J. B. M., Bahrin, E. K., Arbaain, E. N. N., & Ramli, N. (2022). Mycelium-based composite: A way forward for renewable material. *Journal of Sustainability Science and Management*, *17*, 271–280. https://doi.org/10.46754/jssm.2022.01.018

Fratzl, P., & Barth, F. G. (2009). Biomaterial systems for mechanosensing and actuation. *Nature*, 462, 442–448. https://doi.org/10.1038/nature08603

Fricker, M., Boddy, L., & Bebber, D. (2007). "Network organisation of mycelial fungi". In *Biology of the Fungal Cell* (309–330). https://doi.org/10.1007/978-3-540-70618-2_13

Gandia, A., van den Brandhof, J. G., Appels, F. V. W., & Jones, M. P. (2021). Flexible fungal materials: Shaping the future. *Trends in Biotechnology*, *39*(12), 1321–1331. https://doi.org/10.1016/j.tibtech.2021.03.002

Ghazvinian, A., & Gürsoy, B. (2022). Mycelium-based composite graded materials: Assessing the effects of time and substrate mixture on mechanical properties. *Biomimetics*, *7*(2), 48. https://doi.org/10.3390/biomimetics7020048

Girometta, C., Picco, A. M., Baiguera, R. M., Dondi, D., Babbini, S., Cartabia, M., Pellegrini, M., & Savino, E. (2019). Physico-mechanical and thermodynamic properties of mycelium-based biocomposites: A review. *Sustainability*, *11*(1), 281. https://doi.org/10.3390/su11010281

Grimm, D., & Wösten, H. A. B. (2018). Mushroom cultivation in the circular economy. *Applied Microbiology and Biotechnology*, *102*, 7795–7803. https://doi.org/10.1007/s00253-018-9226-8

Gunawardena, D., Bennett, L., Shanmugam, K., et al. (2014). Anti-inflammatory effects of five commercially available mushroom species determined in lipopolysaccharide and interferon-γ activated murine macrophages. *Food Chemistry*, *148*, 92–96. https://doi.org/10.1016/j.foodchem.2013.10.015

Haneef, M., Ceseracciu, L., Canale, C., Bayer, I. S., Heredia-Guerrero, J. A., & Athanassiou, A. (2017). Advanced materials from fungal mycelium: Fabrication and tuning of physical properties. *Scientific Reports*, *7*, 41292. https://doi.org/10.1038/srep41292

Holt, G. A., McIntyre, G., Flagg, D., Bayer, E., Wanjura, J. D., & Pelletier, M. G. (2012). Fungal mycelium and cotton plant materials in the manufacture of biodegradable molded packaging material: Evaluation study of select blends of cotton byproducts. *Journal of Biobased Materials and Bioenergy*, *6*, 431–439. https://doi.org/10.1166/jbmb.2012.1241

Houette, T., Maurer, C., Niewiarowski, R., & Gruber, P. (2022). Growth and mechanical characterization of mycelium-based composites towards future bioremediation and food production in the material manufacturing cycle. *Biomimetics*, *7*(3), 103. https://doi.org/10.3390/biomimetics7030103

Islam, M. R., Tudryn, G., Bucinell, R., Schadler, L., & Picu, R. C. (2017). Morphology and mechanics of fungal mycelium. *Scientific Reports*, *7*, 13070. https://doi.org/10.1038/s41598-017-13295-2

Jayakumar, R., Prabaharan, M., Sudheesh Kumar, P. T., Nair S. V., Furuike, T., & Tamura, H. (2011). Biomedical Engineering, Trends in Materials Science. Edited by Anthony N. Laskovski, 1–26. https://doi.org/10.5772/13509

Jiang, L., Walczyk, D., McIntyre, G., Bucinell, R., & Tudryn, G. (2017). Manufacturing of biocomposite sandwich structures using mycelium-bound cores and preforms. *Journal of Manufacturing Process*, *28*, 50–59. https://doi.org/10.1016/j.jmapro.2017.04.029

Jones, M., Bhat, T., Huynh, T., Kandare, E., Yuen, R., Wang, C. H., & Juhn, S. (2018). Waste-derived low-cost mycelium composite construction materials with improved fire safety. *Fire and Materials*, *42*, 816–825. https://doi.org/10.1002/fam.2637

Jones, M., Huynh, T., Dekiwadia, C., Daver, F., & John, S. (2017). Mycelium composites: A review of engineering characteristics and growth kinetics. *Journal of Bionanoscience*, 11, 241–257. https://doi.org/10.1166/jbns.2017.1440

Jones, M., Kujundzic, M., John, S., & Bismarck, A. (2020a). Crab vs. mushroom: A review of crustacean and fungal chitin in wound treatment. *Marine Drugs*, *18*, 64. https://doi.org/10.3390/md18010064

Jones, M., Gandia, A., John, S., & Bismarck, A. (2020b). Leather-like material biofabrication using fungi. *Nature Sustainability*, *4*(1), 9–16. https://doi.org/10.1038/s41893-020-00606-1

Jones, M., Mautner, A., Luenco, S., Bismarck, A., & John, S. (2020c). Engineered mycelium composite construction materials from fungal biorefineries: A critical review. *Materials & Design*, *187*, 108397. https://doi.org/10.1016/j.matdes.2019.108397

Karana, E., Blauwhoff, D., Hultink, E. J., & Camere, S. (2018). When the material grows: A case study on designing (with) mycelium-based materials. *International Journal of Design*, *12*, 119–136.

Khamrai, M., Banerjee, S. L., & Kundu, P. P. (2018). A sustainable production method of mycelium biomass using an isolated fungal strain *Phanerochaete chrysosporium* (accession no: KY593186): Its exploitation in wound healing patch formation. *Biocatalysis and Agricultural Biotechnology*, *16*, 548–557.

Khurana, A. (2021). Biodegradable textile from fungus-mycelium textile. *Descriptio*, *3*(1).

King, A., & Watling, R. (1997). Paper made from bracket fungi. *Mycologist*, *11*(2), 52–54. https://doi.org/10.1016/S0269-915X(97)80033-X

Li, H., Zhao, H., Gao, Z., et al. (2019). The antioxidant and anti-aging effects of acetylated mycelia polysaccharides from *Pleurotus djamor*. *Molecules*, *24*(15), 2698.

MacArthur, E. (2017). The new plastics economy: Rethinking the future of plastics & catalysing action. Ellen MacArthur Foundation: Cowes, UK, 68. https://www.weforum.org/publications/the-new-plastics-economy-rethinking-the-future-of-plastics/

Majib, N. M., Sam, S. T., Yaacob, N. D., Rohaizad, N. M., & Tan, W. K. (2023). Characterization of fungal foams from edible mushrooms using different agricultural wastes as substrates for packaging material. *Polymers*, *15*(4), 873.

Manan, S., Ullah, M. W., Ul-Islam, M., Atta, O. M., & Yang, G. (2021). Synthesis and applications of fungal mycelium-based advanced functional materials. *Journal of Bioresources and Bioproducts*, *6*(1), 1–10. https://doi.org/10.1016/j.jobab.2021.01.001

Meyers, M. A., McKittrick, J., & Chen, P. Y. (2013). Structural biological materials: Critical mechanics-materials connections. *Science*, *339*, 773–779. https://doi.org/10.1126/science.1220854

Meyer, V., Basenko, E. Y., Benz, J. P., et al. (2020). Growing A circular economy with fungal biotechnology: A white paper. *Fungal Biology and Biotechnology*, *7*, 1–23. https://doi.org/10.1186/s40694-020-00095-z

Mirończuk-Chodakowska, I., Kujawowicz, K., & Witkowska, A. M. (2021). Beta-glucans from fungi: Biological and health-promoting potential in the COVID-19 pandemic era. *Nutrients*, *13*(11), 3960. https://doi.org/10.3390/nu13113960

Mohanty, C., Das, M., & Sahoo, S. K. (2012). Sustained wound healing activity of curcumin loaded oleic acid based polymeric bandage in a rat model. *Molecular Pharmaceutics*, *9*(10), 2801–2811. https://doi.org/10.1021/mp300075u

Mojumdar, A., Behera, H. T., & Ray, L. (2021). Mushroom Mycelia-Based Material: An Environmental Friendly Alternative to Synthetic Packaging. In: Vaishnav, A., Choudhary, D. K. (eds) *Microbial Polymers*. Springer, Singapore. pp. 131–141. https://doi.org/10.1007/978-981-16-0045-6_6

Morath, S. U., Hung, R., & Bennett, J. W. (2012). Fungal volatile organic compounds: A review with emphasis on their biotechnological potential. *Fungal Biology Reviews*, *26*(2-3), 73–83. https://doi.org/10.1016/j.fbr.2012.07.001

Muzzarelli, R. A. A. (2012). Nanochitins and nanochitosans, paving the way to eco-Friendly and energy-saving exploitation of marine resources. *Polymer Science: A Comprehensive. Reference*, *10*, 153–164. https://doi.org/10.1016/B978-0-444-53349-4.00257-0

Muzzarelli, R. A. A., Morganti, P., Morganti, G., et al. (2007). Chitin nanofibrils/chitosan glycolate composites as wound medicaments. *Carbohydrate Polymers*, *70*, 274–284. https://doi.org/10.1016/j.carbpol.2007.04.008

Nawawi, W. M., Jones, M. P., Kontturi, E., Mautner, A., & Bismarck, A. (2020). Plastic to elastic: Fungi-derived composite nanopapers with tunable tensile properties. *Composites Science and Technology*, *198*, 108327. https://doi.org/10.1016/j.compscitech.2020.108327

Niego, A. G. T., Lambert, C., Mortimer, P., Thongklang, N., Rapior, S., Grosse, M., Schrey, H., Charria-Girón, E., Walker, A., Hyde, K. D., & Stadler, M. (2023). The contribution of fungi to the global economy. *Fungal Diversity*. https://doi.org/10.1007/s13225-023-00520-9

Niemeyer, C. M. (2001). Nanoparticles, proteins, and nucleic acids: Biotechnology meets materials science. *Angewandte Chemie International Edition*, *40*, 4128–4158. https://doi.org/10.1002/1521-3773(20011119)40:22<4128::aid-anie4128>3.0.co;2-s

Patel, S. (2023). Four promising uses for mycelium. https://blog.mdpi.com/2023/02/24/four-uses-for-mycelium/ (accessed on 17 July 2023).

Pedroso-Roussado, C. (2023). The fashion industry needs microbiology: Opportunities and challenges. *mSphere*, *8*(2), e00681–22. https://doi.org/10.1128/msphere.00681-22

Peeters, E., Salueña Martin, J., & Vandelook, S. (2023). Growing sustainable materials from filamentous fungi. *The Biochemist (Lond)*, *45*(3), 8–13. https://doi.org/10.1042/bio_2023_120

Pelletier, M. G., Holt, G. A., Wanjura, J. D., Bayer, E., & McIntyre, G. (2013). An evaluation study of mycelium based acoustic absorbers grown on agricultural by-product substrates. *Industrial Crops and Products*, *51*, 480–485. https://doi.org/10.1016/j.indcrop.2013.09.008

Rajendran, R. C. (2022). Packaging applications of fungal mycelium-based biodegradable composites. In *Fungal Biopolymers and Biocomposites* (pp. 189–208). Springer Nature Singapore. https://doi.org/10.1007/978-981-19-1000-5_11

Raman, J., Kim, D.-S., Kim, H.-S., Oh, D.-S., & Shin, H.-J. (2022). Mycofabrication of mycelium-based leather from brown-rot fungi. *Journal of Fungi*, *8*(3), 317. https://doi.org/10.3390/jof8030317

Rathinamoorthy, R., Sharmila Bharathi, T., Snehaa, M., & Swetha, C. (2023). Mycelium as sustainable textile material – Review on recent research and future prospective. *International Journal of Clothing Science and Technology*, *35*(3), 454–476. https://doi.org/10.1108/IJCST-01-2022-0003

Robertson, O. (2020). Fungal Future: A review of mycelium biocomposites as an ecological alternative insulation material. DS 101: *Proceedings of NordDesign 2020*, Lyngby, Denmark, 12–14 August 2020, pp. 1–13. https://doi.org/10.35199/NORDDESIGN2020.18

Shalwan, A., & Yousif, B. F. (2013). In state of art: Mechanical and tribological behaviour of polymeric composites based on natural fibres. *Materials & Design*, 48, 14–24. https://doi.org/10.1016/j.matdes.2012.07.014

Shunk, G. K., Gomez, X. R., Kern, C., & Averesch, N. J. H. (2022). Growth of the radiotrophic fungus *Cladosporium sphaerospermum* aboard the International Space Station and effects of ionizing radiation. Biorxiv. https://doi.org/10.1101/2020.07.16.205534

Svensson, S. E., Oliveira, A. O., Adolfsson, K. H., Heinmaa, I., Root, A., Kondori, N., Ferreira, J. A., Hakkarainen, M., & Zamani, A. (2022). Turning food waste to antibacterial and biocompatible fungal chitin/chitosan monofilaments. *International Journal of Biological Macromolecules*, *209*, 618–630.

Sydor, M., Cofta, G., Doczekalska, B., & Bonenberg, A. (2022). Fungi in mycelium-based composites: Usage and recommendations. *Materials*, *15*(18), 6283. https://doi.org/10.3390/ma15186283

Udayanga, D., & Miriyagalla, S. D. (2021). Fungal mycelium-based biocomposites: An emerging source of renewable materials. In *Microbial Technology for Sustainable Environment* (pp. 529–550). Springer Singapore. https://doi.org/10.1007/978-981-16-3840-4_27

Vaišis, V., Chlebnikovas, A., & Jasevičius, R. (2023). Numerical study of the flow of pollutants during air purification, taking into account the use of eco-Friendly material for the filter—Mycelium. *Applied Sciences, 13*(3), 1703. https://doi.org/10.3390/app13031703

van den Brandhof, J. G., & Wösten, H. A. B. (2022). Risk assessment of fungal materials. *Fungal Biology and Biotechnology, 9*(1), 3. https://doi.org/10.1186/s40694-022-00134-x

Vandelook, S., Elsacker, E., Van Wylick, A., De Laet, L., & Peeters, E. (2021). Current state and future prospects of pure mycelium materials. *Fungal Biology and Biotechnology, 8*(1), 20. https://doi.org/10.1186/s40694-021-00128-1

Wang, Z., Wang, H., Kang, Z., Wu, Y., Xing, Y., & Yang, Y. (2020). Antioxidant and antitumour activity of triterpenoid compounds isolated from Morchella mycelium. *Archives of Microbiology, 202*, 1677–1685. https://doi.org/10.1007/s00203-020-01876-1

Webb, H. K., Arnott, J., Crawford, R. J., & Ivanova, E. P. (2013). Plastic degradation and its environmental implications with special reference to poly(ethylene terephthalate). *Polymers, 5*, 1–18. https://doi.org/10.3390/polym5010001

Xing, Y., Brewer, M., El-Gharabawy, H., Griffith, G., & Jones, P. (2018). Growing and testing mycelium bricks as building insulation materials. *IOP Conference Series: Earth and Environmental Science, 121*, 022032. https://doi.org/10.1088/1755-1315/121/2/022032

Xu, N., Gao, Z., Zhang, J., et al. (2017). Hepatoprotection of enzymatic-extractable mycelia zinc polysaccharides by *Pleurotus eryngii* var. *tuoliensis*. *Carbohydrate Polymers, 157*, 196–206. https://doi.org/10.1016/j.carbpol.2016.09.082

Yang, Z., Zhang, F., Still, B., White, M., & Amstislavski, P. (2017). Physical and mechanical properties of fungal mycelium-based biofoam. *Journal of Materials in Civil Engineering, 29*(7), 04017030. https://doi.org/10.1061/(ASCE)MT.1943-5533.0001866

9 Fungal Biocontrol Agents to Control and Management of Diseases in Crops

Tarun Kumar Patel

9.1 INTRODUCTION

Plant diseases are caused by a wide range of pathogens, including fungi, bacteria, viruses, and nematodes including contamination with mycotoxin (Abdulkhair & Alghuthaymi 2016; Agrios 2005; Shankar et al. 2018). *Aspergillus flavus* and *A. parasiticus* produce aflatoxins and contaminate leading crop products (Tiwari & Shankar 2018), thus impacting human health (Shankar 2021), and various strategies are adopted to control them including biocontrol methods (Atehnkeng et al. 2008; Bansal et al. 2023; Mauro et al. 2015). These pathogens can infect various parts of the plant, including the roots, stems, leaves, and fruits, and can cause a range of symptoms such as wilting, yellowing, stunting, and deformities (Pandit et al. 2022).

Crop production is significantly influenced by plant diseases as they can cause yield losses, reduced quality of produce, and increased production costs due to the use of pesticides and other control methods (Ngoune Liliane & Shelton Charles 2020; Richard et al. 2022; Rizzo et al. 2021). In severe cases, plant diseases can even lead to complete crop failure, resulting in economic losses for farmers and food shortages for consumers.

Plant diseases are also a major threat to global food security, especially in developing countries where crop losses due to diseases can be particularly devastating (Bebber et al. 2013). Apart from their immediate influence on crop production, plant diseases can lead to secondary consequences for the environment and human health. For example, the use of synthetic pesticides to control plant diseases can lead to pollution of water and soil, as well as potential health hazards for farmers and consumers (Tudi et al. 2021).

Therefore, the development of effective and sustainable methods for managing plant diseases is essential for ensuring global food security and protecting the environment and human health.

9.1.1 The Need for Alternative Control Methods to Synthetic Pesticides

9.1.1.1 Environmental Concerns

Synthetic pesticides are often made from harmful chemicals that can cause pollution and harm to the environment. These chemicals can persist in the soil and water,

DOI: 10.1201/9781003407683-9

contaminating them and harming non-target organisms, including humans (Krupke et al. 2012; Pathak et al. 2022).

9.1.1.2 Resistance

The extended utilization of synthetic pesticides has resulted in the emergence of resistance within pest populations, making them less effective over time. This has led to an increased reliance on stronger and more toxic pesticides, which can have even increased detrimental impacts on both the environment and human well-being (Siddiqui et al. 2023).

9.1.1.3 Public Health

Synthetic pesticides can pose a significant risk to human health, especially for farmworkers and consumers who are exposed to them. These chemicals have been linked to various health problems, including cancer, neurological disorders, and reproductive issues (Damalas & Eleftherohorinos 2011; Nicolopoulou-Stamati et al. 2016).

9.1.1.4 Economic Concerns

Synthetic pesticides can be expensive, especially for small-scale farmers. Using alternative methods can be more cost-effective and sustainable in the long run (Ayilara et al. 2023; Garcia 2020).

9.1.2 Fungal Biocontrol Agents and Their Advantages

Fungal biocontrol agents are natural enemies of plant pathogens, which are used to control and manage plant diseases in crops. Their general mechanism of life cycle is depicted in Figure 9.1. These agents can be classified into three groups based on their modes of action: antagonistic fungi, endophytic fungi, and mycoparasitic fungi.

Antagonistic fungi are those that directly compete with plant pathogens for resources, such as space and nutrients, and produce antibiotics or other compounds that impede the proliferation and maturation of the pathogen. Conversely, endophytic fungi reside within plant tissues harmlessly, and they have the ability to activate the defense mechanisms of plants against pathogens. Mycoparasitic fungi are those that attack and kill other fungi by parasitizing them (Clemons et al. 2016).

Fungal biocontrol agents have several advantages over traditional synthetic pesticides. First, they are environmentally friendly, as they are made from natural substances and do not leave harmful residues in the soil or water. Second, they have a low risk of resistance development, as they have complex modes of action that are difficult for pathogens to overcome. Third, they have a diverse array of activity against a wide variety of pathogens, making them a versatile option for disease management in crops. Fourth, they can be easily integrated into other management strategies, such as cultural and physical control methods (Huilgol et al. 2022; Zehnder et al. 2007).

According to a study by Pathak et al. (2022), the prolonged utilization of synthetic pesticides has resulted in the emergence of resistance within pest populations, making them less effective over time, leading to an increased reliance on stronger and more toxic pesticides, which can have even more negative impacts on the

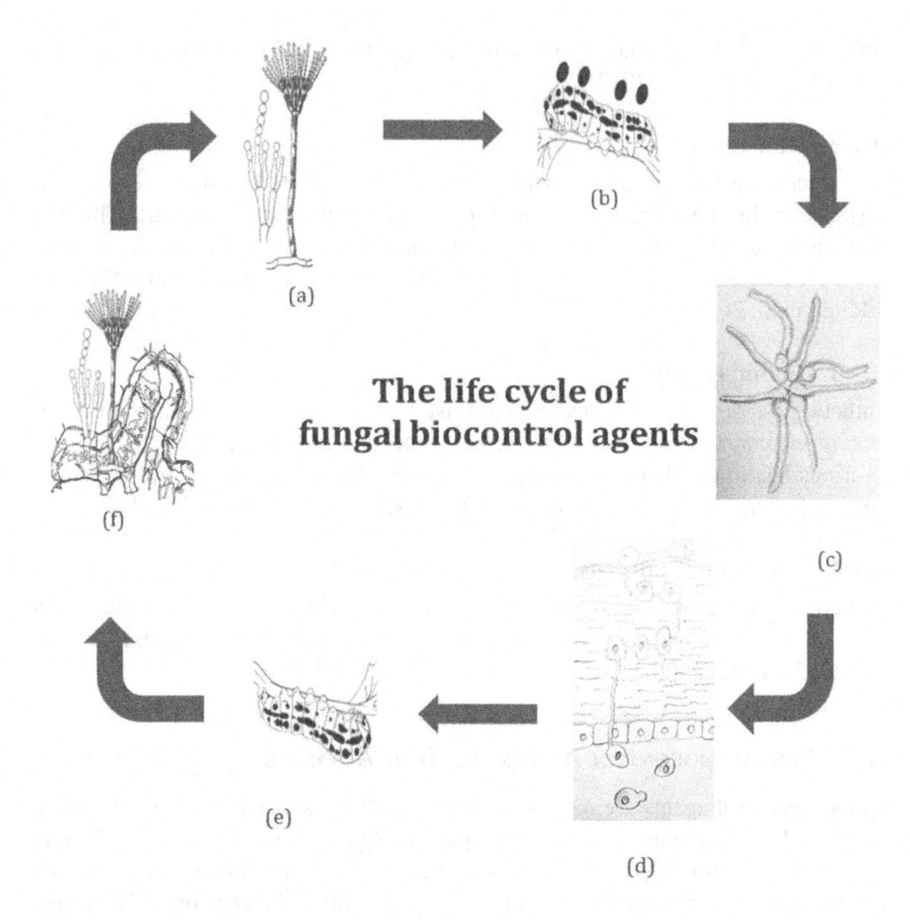

FIGURE 9.1 The life cycle of fungal biocontrol agent. (a) Conidia of fungi, (b) attachment of conidia, (c) germination of conidia, (d) enzyme production and lysis of cuticle, (e) penetration of mycelium and secondary metabolite production, and (f) death of insect and emergence of conidiophore over insect body.

environment and human well-being. As an alternative, fungal biocontrol agents have several advantages over traditional synthetic pesticides, such as being environmentally friendly and having a low risk of resistance development (Palmieri et al. 2022).

Overall, fungal biocontrol agents represent a promising alternative to traditional synthetic pesticides for managing plant diseases in crops. Ongoing research is being conducted to improve the efficacy, practicality, and sustainability of these agents in the field.

9.2 TYPES OF FUNGAL BIOCONTROL AGENTS

9.2.1 Antagonistic Fungi

Antagonistic fungi are biocontrol agents that can control plant pathogens through direct competition for resources and the production of antifungal compounds. The main mode of action of antagonistic fungi is through mycoparasitism, in which the biocontrol agent parasitizes the pathogen, leading to its death. Another mode of action is through the synthesis of secondary metabolites, such as antibiotics and volatile organic compounds (VOCs), which can inhibit the growth and development of the pathogen.

Metarhizium and *Beauveria* species are examples of antagonistic fungi that have been used as biological control agents targeting a diverse spectrum of plant pathogens. According to Pandey et al. (2022), *Metarhizium* has been identified as a potent biocontrol agent against Fusarium wilt in tomato plants. The study also highlighted the function of secondary metabolites generated by *Metarhizium* in restraining the pathogen's growth. *Beauveria bassiana*, on the other hand, is recognized for producing various secondary metabolites, such as beauvericin and bassianolide, which have antifungal efficacy against various plant pathogens (Pedrini 2022; Wang et al. 2021). *B. bassiana* can also produce enzymes like chitinases and proteases, which can break down the cell walls of the pathogen (Bhadani et al. 2021).

In addition to *Metarhizium* and *Beauveria* species, *Trichoderma*, *Gliocladium*, and *Coniothyrium* species are other examples of antagonistic fungi that have been used successfully as biocontrol agents against plant pathogens (Bamisile et al. 2021; Elshahawy & El-Mohamedy 2019; Kredics et al. 2003). *Trichoderma* has demonstrated its effectiveness as a biocontrol agent against *Fusarium oxysporum* in cotton, wheat, and muskmelon (Sivan & Chet 1986). The studies found that *Trichoderma* produces several antifungal compounds, including chitinases, which have played a pivotal role in managing the pathogen (Loc et al. 2019; Mukhopadhyay & Kumar 2020). Table 9.1 shows the list of some antagonistic fungi and their target organisms.

In summary, antagonistic fungi are biocontrol agents that can control plant pathogens through mycoparasitism and the production of secondary metabolites. Ongoing research is being carried out to further comprehend the functional processes of antagonistic fungi and to develop more effective biocontrol strategies using these agents.

9.2.2 Endophytic Fungi

Endophytic fungi encompass a wide array of fungal species that reside within plant tissues without inflicting any damage to the host plant. These fungi have been found in all types of plants, including agricultural crops, and are known to provide a range of benefits to their host, including enhanced plant development and resilience against both biotic and abiotic pressures. The mechanisms of operation of endophytic fungi are varied and include direct antagonism against pathogens that affect plants, induction of the defense mechanisms of plants, production of substances that enhance plant growth, and improvement of plant nutrient uptake (Akram et al. 2023; Fadiji & Babalola 2020; Verma et al. 2022).

Endophytic fungi can synthesize various secondary metabolites, including alkaloids, terpenoids, and phenolic compounds, which have been shown to have antifungal, antibacterial, and antiviral properties (Wen et al. 2022). One example of an endophytic fungus that has been extensively studied for its biocontrol potential is *Trichoderma*. *Trichoderma* species have been shown to inhabit the roots and leaves

TABLE 9.1
List of Some Antagonistic Fungi and Their Target Organisms

Antagonistic Fungi	Target Organism	Reference
Trichoderma spp.	Phytopathogenic fungi and soilborne pathogens	Harman (2006)
B. bassiana	Insect pests, including whiteflies, aphids, and thrips	Jaronski (2010)
M. anisopliae	Insect pests, including beetles, weevils, and grasshoppers	Sharma and Sharma (2021); Clifton et al. (2020); Mantzoukas et al. (2019)
Pochonia chlamydosporia	Root-knot nematodes	Ghahremani et al. (2019)
Penicillium oxalicum	Pearl millet against downy mildew disease	Murali and Amruthesh (2015)
Purpureocillium lilacinum	Root-knot nematodes	Khan and Tanaka (2023)
Lecanicillium lecanii	Whiteflies, mealybugs, and other insect pests	Gopal et al. (2021); Gangireddy Eswara Reddy (2021)
Aspergillus flavus (Atoxinogenic)	Aflatoxin-producing fungi	Lavkor et al. (2023); Patel et al. (2014)
Verticillium lecanii	Plant pathogenic fungi, aphids, and other insect pests	Gangireddy Eswara Reddy (2021)
Coniothyrium minitans	*Sclerotinia sclerotiorum* and other plant pathogenic fungi	Albert et al. (2022)

of various crop plants and provide protection against plant pathogens through competition for resources, mycoparasitism, and production of antifungal compounds (Akram et al. 2023). *Trichoderma* species can also generate plant growth-promoting compounds like indole acetic acid and gibberellins, which have been demonstrated to improve plant growth and yield (Nieto-Jacobo et al. 2017; Tyśkiewicz et al. 2022).

Another example of an endophytic fungus with biocontrol potential is *Fusarium oxysporum*. This fungus has been shown to inhabit the root systems of tomato plants, as well as provide protection against Fusarium wilt, a destructive ailment brought about by the soilborne pathogen *F. oxysporum* f. sp. *lycopersici*. The endophytic fungus works by inducing plant protection mechanisms, like the production of proteins related to pathogenesis and activation of the jasmonic acid signaling pathway (Adeleke et al., 2022; Akram et al. 2023). Table 9.2 shows the list of some endophytic fungi and their target organisms.

In summary, endophytic fungi constitute a diverse array of fungal species that reside within plant tissues and offer a variety of advantages to their host plant, including enhanced plant growth and enhanced resistance to both biotic and abiotic stresses. The methods of operation of endophytic fungi include direct antagonism against agents responsible for causing plant diseases, induction of mechanisms plants use to defend against threats, production of substances that stimulate plant growth, and improvement of plant nutrient uptake. *Trichoderma* and *F. oxysporum*

TABLE 9.2
List of Some Endophytic Fungi and Their Target Organisms

Endophytic Fungi	Target Organism	Reference
Trichoderma harzianum	*Fusarium oxysporum*	Bubici et al. (2019)
B. bassiana	*Helicoverpa armigera*	Qayyum et al. (2015)
M. anisopliae	*Rhipicephalus microplus*	Beys-da-Silva et al. (2020)
Penicillium janthinellum	*Heterodera glycines Ichinohe* (soybean cyst nematode)	Yan et al. (2021)
Aspergillus terreus	*Rhizoctonia solani*	Abdelaziz et al. (2023)
Streptomyces sp.	*Colletotrichum gloeosporioides*	Kim et al. (2014)
Trichoderma viride	*Botrytis cinerea*	Mónaco et al. (2009)
Cladosporium oxysporum	*Fusarium oxysporum* f. sp. *ciceris*	Kumari et al. (2023)
Fusarium solani	*Meloidogyne incognita*	Patil et al. (2021)
Paecilomyces lilacinus	*Radopholus similis*	Mendoza et al. (2004)
Purpureocillium lilacinum	*Meloidogyne incognita*	Khan and Tanaka (2023)
Trichoderma asperellum	*Phomopsis azadirachtae*	Patil et al. (2021)
Aspergillus flavus	*Sclerotium rolfsii*	Safari Motlagh et al. (2022)

are examples of endophytic fungi that have been thoroughly investigated for their potential in biocontrol applications. Ongoing research is being conducted to further understand the mechanisms underlying the beneficial effects of endophytic fungi and to develop more effective biocontrol strategies using these agents.

9.2.3 Mycoparasitic Fungi

Mycoparasitic fungi are a type of fungi that have been extensively studied due to their potential as agents to manage fungal pathogens in plants (Harman 2006; Thambugala et al. 2020). These fungi attack and parasitize other fungi, providing a range of benefits in the control of diseases in plants brought about by fungal pathogens. The modes of action of mycoparasitic fungi are varied and include direct antagonism against fungal pathogens, parasitism of fungal hyphae, and production of extracellular enzymes that degrade the cell walls of fungal pathogens.

One example of a mycoparasitic fungus with biocontrol potential is *Trichoderma harzianum*. This fungus has been shown to parasitize and lyse the hyphae of a wide range of fungal invaders, including *Rhizoctonia solani* and *Botrytis cinerea*. *T. harzianum* produces a variety of extracellular enzymes, including glucanases and chitinases, which break down the cell walls of fungal pathogens, rendering them more vulnerable to the attack of the mycoparasitic fungus (Abdullah et al. 2021; Ting & Chai 2015).

Another example of a mycoparasitic fungus with biocontrol potential is *Coniothyrium minitans*. This fungus has been shown to parasitize and degrade the sclerotia of *Sclerotinia* species, including *Sclerotinia sclerotiorum*, a devastating fungal pathogen of a wide range of crops. *C. minitans* produces extracellular enzymes, such as proteases and chitinases, which degrade the cell walls of the sclerotia of

TABLE 9.3
List of Some Mycoparasitic Fungi and Their Target Organisms

Mycoparasitic Fungi	Target Organism	Reference
Trichoderma spp.	Various fungal pathogens	Harman (2011)
Ampelomyces quisqualis	Powdery mildew fungi	Glawe (2008)
Clonostachys rosea	*Botrytis cinerea* (gray mold)	Sarven et al. (2020)
Coniothyrium minitans	*S. sclerotiorum* (white mold)	Budge and Whipps (2001)
Mycogone perniciosa	*Agaricus bisporus* (button mushroom)	Li et al. (2019)
Trichoderma harzianum	Various fungal pathogens, including *R. solani* and *B. cinerea*	Yao et al. (2023); Mayo et al. (2015)
Coniothyrium minitans	Sclerotinia species, including *S. sclerotiorum*	Li et al. (2006); Abdullah et al. 2008; Budge and Whipps (2001); Whipps et al. (2007)

Sclerotinia species, leading to their eventual destruction (Abdullah et al. 2008; Budge & Whipps 2001; Whipps et al. 2007). Table 9.3 shows the list of some mycoparasitic fungi and their target organisms.

In summary, mycoparasitic fungi are a promising group of biocontrol agents for controlling fungal pathogens in plants. *T. harzianum* and *C. minitans* are examples of mycoparasitic fungi with biocontrol potential, and ongoing research is being conducted to further understand their mechanisms of action and develop more effective biocontrol strategies using these agents.

9.3 MECHANISMS OF ACTION OF FUNGAL BIOCONTROL AGENTS

9.3.1 Direct Inhibition of Pathogen Growth and Development

One of the key approaches through which biocontrol agents apply their positive impacts on plant health is through the direct inhibition of pathogen growth and development. This process involves the production of a range of compounds that are toxic or otherwise detrimental to the growth and survival of fungal and bacterial pathogens.

An example of a biocontrol agent that directly impedes fungal pathogen growth and development is the fungus *Trichoderma* species. *Trichoderma* species produce a variety of antifungal metabolites, such as trichodermin and harzianum A, known for their remarkable capacity to disrupt the cell membranes of fungal pathogens, ultimately causing cell death and thwarting the growth of these harmful organisms (Hermosa et al. 2012; Mukherjee et al. 2012).

Another example of a biocontrol agent that directly inhibits pathogen growth and development is the fungus *T. harzianum*. This fungus produces a range of extracellular enzymes, including chitinases and glucanases, which have been identified for their capacity to break down the cell walls of fungal pathogens. *T. harzianum* also produces antimicrobial compounds, such as harzianic acid, that directly inhibit pathogen growth and development (Vinale et al. 2013).

Direct inhibition of pathogen growth and development is an important mechanism by which biocontrol agents have the potential to decrease the occurrence and intensity of plant diseases. By producing a range of antimicrobial compounds, these agents are able to directly attack and kill fungal and bacterial pathogens, leading to healthier and more productive plants. Ongoing research is being conducted to identify and characterize new biocontrol agents with direct inhibitory effects on pathogen growth and development and to optimize the use of existing agents for more effective disease control in agriculture and horticulture.

9.3.2 Indirect Stimulation of Plant Defense Mechanisms

Indirect stimulation of plant defense mechanisms is another important mechanism by which biocontrol agents benefit plant health. This process involves the activation or enhancement of the plant's natural defense mechanisms, which in turn help to safeguard the plant from attacks by pathogens.

An exemplar of a biocontrol agent that indirectly stimulates plant defense mechanisms is the bacterium *Pseudomonas fluorescens*. This bacterium produces a range of VOCs that have been shown to trigger systemic resistance in plants against an array of pathogens. These VOCs are thought to activate the innate defense mechanisms of the plant, resulting in the synthesis of defense-related proteins and other substances that aid in safeguarding the plant from subsequent pathogen attacks.

Yet another instance of a biocontrol agent that indirectly stimulates plant defense mechanisms is the fungus *Piriformospora indica*. This fungus establishes a mutualistic association with the plant and has been shown to enhance plant growth and maturation, while also enhancing the plant's ability to withstand various biotic and abiotic pressures. *P. indica* has demonstrated the capacity to stimulate systemic resistance in plants against an array of fungal and bacterial pathogens and is thought to do so by activating the plant's natural defense mechanisms.

Indirect stimulation of plant defense mechanisms serves as a significant approach through which biocontrol agents can help to protect plants from pathogen attack. By activating the plant's natural defense mechanisms, these agents can enhance the plant's ability to resist and recover from disease, leading to healthier and more productive plants. Ongoing research is being conducted to identify and characterize new biocontrol agents that can indirectly stimulate plant defense mechanisms and to optimize the use of existing agents for more effective disease control in agriculture and horticulture.

One such study by Brotman et al. (2013) demonstrated the effectiveness of *P. fluorescens* in inducing systemic defense throughout tomato plants against the fungal pathogen *Botrytis cinerea*. The study showed that treatment with *P. fluorescens* led to a significant reduction in disease severity compared to control plants and also triggered the synthesis of compounds related to defense in the treated plants. Similarly, a study by Jacobs et al. (2011) illustrated the efficacy of *P. indica* in promoting the growth and resilience of *Arabidopsis* plants against various biotic and abiotic pressures. The study demonstrated that the application of *P. indica* resulted in an increase in the activation of genes associated with defense and the synthesis of defense-related compounds in the treated plants. Figure 9.2 shows the influence of fungal biocontrol on quality and crop yield.

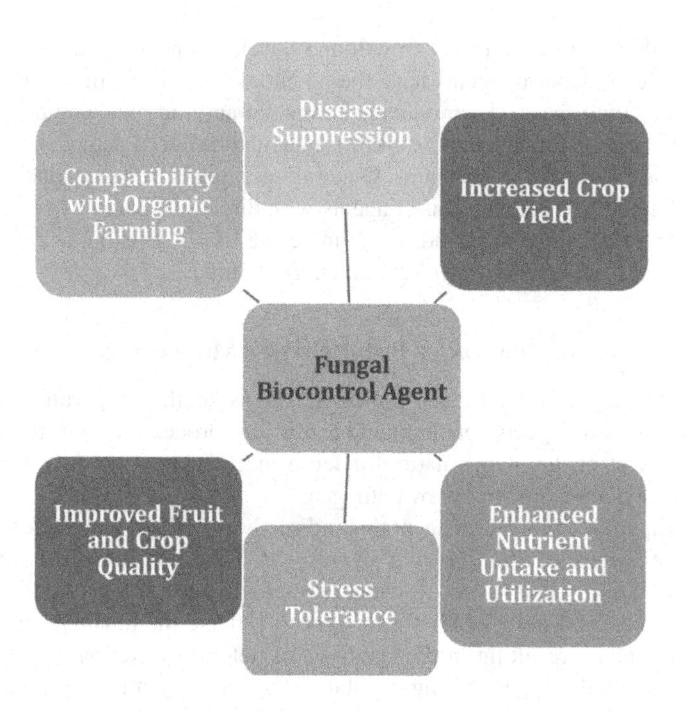

FIGURE 9.2 The influence of fungal biocontrol on quality and crop yield.

Another recent study by Khairullina et al. (2023) investigated the mechanism by which the fungus *Clonostachys rosea*, a biocontrol agent, induced systemic defense in oat plants against the fungal pathogen *Fusarium graminearum* that causes Fusarium head blight. The study showed that treatment with *C. rosea* led to the upregulation of genes associated with plant protection and the synthesis of compounds related to defense mechanisms in the treated plants, indicating that the fungus was able to activate the plant's natural defense mechanisms. The study also identified activated expression of WRKY23-like transcription factor and genes encoding four PR-proteins by *C. rosea*, which played a key function in eliciting systemic resistance within the plants and also detoxifying the trichothecene deoxynivalenol (DON) mycotoxin.

These recent studies highlight the potential that biocontrol agents possess to indirectly stimulate plant defense mechanisms and enhance plant health. Further research is needed to fully understand the mechanisms underlying this process and to identify new biocontrol agents with similar effects.

9.4 FACTORS AFFECTING EFFICACY OF FUNGAL BIOCONTROL AGENTS

Fungal biocontrol agents (BCAs) hold the potential to become effective and environmentally friendly alternatives to chemical pesticides for managing plant diseases. However, their efficacy can be affected by several factors.

9.4.1 Selection of Appropriate Agent for Target Pathogen

The selection of the appropriate fungal BCA for the target pathogen is critical for effective biocontrol. The efficacy of the BCA is determined through its ability to compete with the pathogen for nutrients, generate antifungal substances, and establish a presence on the plant's surface. Therefore, selecting a BCA with a specific mechanism of action against the targeted pathogen is vital (Köhl et al. 2019). Trichoderma species, for instance, are opportunistic plant symbionts that produce antifungal metabolites and induce systemic immunity within plants against phytopathogens (Harman et al. 2004).

9.4.2 Timing and Method of Application

The timing and method of application of the BCA are also important factors that can affect its efficacy. Applying the BCA when applied appropriately and effectively ensures that it establishes itself on the plant surface and has the opportunity to interact with the pathogen. For example, applying the BCA early in the disease cycle can help prevent infection, while applying it later in the disease cycle can help control the spread of the pathogen (He et al., 2021; Lahlali et al. 2022).

9.4.3 Environmental Conditions

Environmental conditions, such as temperature, humidity, and rainfall, can also affect the efficacy of fungal BCAs. Fungal BCAs have optimal growth conditions, and when these conditions are not met, the BCA may not be able to establish itself on the plant surface or produce enough antifungal compounds to control the pathogen. In addition, environmental stresses, such as drought or flooding, can also affect the interactions between the BCA and the plant, which can ultimately impact the effectiveness of the BCA (Mannaa & Kim 2018). Hence, comprehending the environmental conditions necessary for the optimal growth and effectiveness of the BCA is imperative.

Careful consideration of the selection of appropriate BCA, timing and method of application, and environmental conditions is necessary for the successful implementation of biocontrol strategies. Incorporating these factors will lead to highly effective and sustainable alternative methods for managing plant diseases (Pandit et al. 2022; Whipps 2001).

9.5 CURRENT STATUS OF FUNGAL BIOCONTROL AGENTS IN CROP DISEASE MANAGEMENT

Fungal BCAs have gained increased attention in recent years as an environmentally friendly alternative to synthetic pesticides for crop disease management. Fungal BCAs are naturally occurring or artificially developed/modified fungi with the capacity to inhabit plant surfaces and produce antifungal compounds that can suppress plant pathogens.

There are numerous examples of successful use of fungal BCAs in different crops (Table 9.4). One such example is the use of *Trichoderma harzianum* as a BCA

TABLE 9.4
Successful Implementation of Fungal Biocontrol in Different Crop Species

Fungal Biocontrol Agent	Target Organism	Controlled Disease/Pathogen	Crop Species	Reference
T. harzianum	*Fusarium* spp.	Fusarium wilt	Tomato	Harman et al. (2004)
B. bassiana	*Plutella xylostella*	Diamondback moth	Cabbage	Bathina and Bonam (2020)
Pseudomonas fluorescens	*Rhizoctonia solani*	Rice sheath blight	Rice	Kabdwal et al. (2023); Li et al. (2021)
Streptomyces sp.	*Botrytis cinerea*	Gray mold	Strawberry	Daojing et al. (2022)
Gliocladium roseum	*Pythium* spp.	Damping off	Wheat	Abdelzaher (2004)
Ampelomyces quisqualis	*Erysiphe necator*	Powdery mildew	Grapevine	Gadoury et al. (2012)
Purpureocillium lilacinum	*Verticillium dahliae*	Verticillium wilt	Eggplant	Lan et al. (2017)
Talaromyces flavus	*Sclerotinia sclerotiorum*	Sclerotinia wilt	Sunflower	McLaren et al. (1994)
Coniothyrium minitans	*Sclerotinia minor*	Lettuce drop	Lettuce	Chitrampalam et al. (2011)
Trichoderma aggressivum f. *europaeum*	*Phytophthora capsici*	Phytophthora blight	Pepper	Santos et al. (2023)
T. harzianum	*Phytophthora colocasiae*	Taro leaf blight	Taro	Nath et al. (2014)

against Fusarium wilt in tomato crops. Studies have shown that the application of *T. harzianum* can diminish disease occurrence and boost yield (Harman et al. 2004). Another example is the use of *Beauveria bassiana* as a BCA against the coffee berry borer, a pest that causes significant damage to coffee crops. Application of *B. bassiana* has been shown to significantly reduce the population of the coffee berry borer and improve coffee yields (Bayman et al. 2021).

In comparison to synthetic pesticides, fungal BCAs offer several advantages. They are generally safer for human health and the environment, as they do not leave harmful residues on crops or in the soil. Fungal BCAs also have the potential for long-term effectiveness, as they can establish themselves on plant surfaces and persist over time, whereas synthetic pesticides may need to be reapplied frequently. Additionally, the use of fungal BCAs can promote plant growth and enhance soil health, further contributing to sustainable agriculture.

However, there are also limitations and challenges associated with the use of fungal BCAs. One limitation is the specificity of fungal BCAs, as each BCA might exhibit effectiveness against only a restricted array of plant pathogens. In addition, the efficacy of fungal BCAs can be affected by environmental factors, such as temperature and humidity, which can impact the growth and activity of the fungi. The cost of production and application of fungal BCAs can also be a limiting factor for widespread adoption, as they may be more expensive than synthetic pesticides.

While there are limitations and challenges associated with fungal BCAs, their potential benefits make them a promising alternative to synthetic pesticides for crop disease management. Further research and development are needed to address the challenges and optimize the use of fungal BCAs in agriculture.

9.6 FUTURE PROSPECTS AND DIRECTIONS FOR RESEARCH

Fungal BCAs have gained significant attention in recent years as environmentally friendly alternatives to synthetic pesticides for managing plant diseases. The potential for developing new fungal BCAs and integrating them with other control methods can lead to sustainable agriculture practices. Recent studies have identified various fungal species, such as *Trichoderma*, *Beauveria*, and *Metarhizium*, as potential candidates for biocontrol agents because of their capability to generate antifungal substances and establish on the plant surface (Sabbahi et al. 2022).

Incorporating fungal BCAs with other control methods, such as cultural practices, can enhance their efficacy and provide a more holistic approach to disease management. For instance, the amalgamation of biocontrol agents with the utilization of compost and crop rotation has been shown to reduce disease incidence and increase crop yields (Collinge et al. 2022; Larkin & Brewer 2020). Moreover, the use of BCAs in combination with reduced pesticide applications can lead to a significant reduction in environmental contamination and promote sustainable agriculture (Lahlali et al. 2022).

The role of fungal BCAs in sustainable agriculture extends beyond disease management. Recent studies have shown that BCAs can enhance plant growth and increase nutrient uptake (Akram et al. 2023; Pirttilä et al. 2021). Furthermore, the use of BCAs can improve soil health and promote a balanced microbial community, which is essential for maintaining soil fertility and productivity (He et al. 2021).

The potential for developing new fungal BCAs and integrating them with other control methods can lead to sustainable agriculture practices. The use of BCAs not only provides an effective means for managing plant diseases but also promotes plant growth and soil health (Altomare et al., 1999). By implementing these practices, we can ensure a more sustainable and environmentally friendly approach to agriculture.

9.7 CONCLUSION

In conclusion, fungal biocontrol agents have emerged as effective and eco-friendly alternatives to synthetic pesticides for controlling and managing plant diseases in crops. Through the production of antifungal compounds, colonization of plant surfaces, and the ability to enhance nutrient uptake and soil health, fungal biocontrol agents have shown great potential in promoting sustainable agriculture practices. While there are still some limitations and challenges associated with their use, ongoing research and development in this field suggest promising possibilities for the future. Incorporating fungal biocontrol agents alongside other control strategies, like cultural practices, can further enhance their efficacy and provide a more holistic approach to disease management. By promoting utilizing fungal biocontrol agents and sustainable agriculture practices, we can ensure a healthier environment and a more secure food supply for future generations.

REFERENCES

Abdelaziz, A. M., El-Wakil, D. A., Hashem, A. H., Al-Askar, A. A., AbdElgawad, H., & Attia, M. S. (2023). Efficient role of endophytic *Aspergillus terreus* in biocontrol of *Rhizoctonia solani* causing damping-off disease of *Phaseolus vulgaris* and *Vicia faba*. *Microorganisms*, *11*(6), 1487. https://doi.org/10.3390/microorganisms11061487

Abdelzaher, H. M. (2004). Occurrence of damping-off of wheat caused by *Pythium diclinum* tokunaga in El-Minia, Egypt and its possible control by *Gliocladium roseum* and *Trichoderma harzianum*. *Archives of Phytopathology and Plant Protection*, *37*(2), 147–159. https://doi.org/10.1080/03235400420000205893

Abdulkhair, W. M., & Alghuthaymi, M. A. (2016). *Plant Pathogens*. InTech. https://doi.org/10.5772/65325

Abdullah, M. T., Ali, N. Y., & Suleman, P. (2008). Biological control of *Sclerotinia sclerotiorum* (Lib.) de Bary with *Trichoderma harzianum* and *Bacillus amyloliquefaciens*. *Crop Protection*, *27*(10), 1354–1359. https://doi.org/10.1016/j.cropro.2008.05.007

Abdullah, N. S., Doni, F., Mispan, M. S., Saiman, M. Z., Yusuf, Y. M., Oke, M. A., & Suhaimi, N. S. M. (2021). Harnessing Trichoderma in agriculture for productivity and sustainability. *Agronomy*, *11*(12), 2559. https://doi.org/10.3390/agronomy11122559

Adeleke, B. S., Ayilara, M. S., Akinola, S. A., et al. (2022). Biocontrol mechanisms of endophytic fungi. *Egyptian Journal of Biological Pest Control*, *32*, 46. https://doi.org/10.1186/s41938-022-00547-1

Agrios, G. N. (2005). *Plant Pathology* (5th ed.). Elsevier Academic Press, London, UK.

Akram, S., Ahmed, A., He, P., He, P., Liu, Y., Wu, Y., Munir, S., & He, Y. (2023). Uniting the role of endophytic fungi against plant pathogens and their interaction. *Journal of Fungi*, *9*, 72. https://doi.org/10.3390/jof9010072

Albert, D., Dumonceaux, T., Carisse, O., Beaulieu, C., & Filion, M. (2022). Combining desirable traits for a good biocontrol strategy against *Sclerotinia sclerotiorum*. *Microorganisms*, *10*(6), 1189. https://doi.org/10.3390/microorganisms10061189

Altomare, C., Norvell, W. A., Björkman, T., & Harman, G. E. (1999). Solubilization of phosphates and micronutrients by the plant-growth-promoting and biocontrol fungus *Trichoderma harzianum* rifai 1295-22. *Applied and Environmental Microbiology*, *65*(7), 2926–2933. https://doi.org/10.1128/AEM.65.7.2926-2933.1999

Atehnkeng, J., Ojiambo, P. S., Ikotun, T., Sikora, R. A., Cotty, P. J., & Bandyopadhyay, R. (2008). Evaluation of atoxigenic isolates of *Aspergillus flavus* as potential biocontrol agents for aflatoxin in maize. *Food Additives & Contaminants. Part A, Chemistry, Analysis, Control, Exposure & Risk Assessment*, *25*(10), 1264–1271. https://doi.org/10.1080/02652030802112635

Ayilara, M. S., Adeleke, B. S., Akinola, S. A., Fayose, C. A., Adeyemi, U. T., Gbadegesin, L. A., Omole, R. K., Johnson, R. M., Uthman, Q. O., & Babalola, O. O. (2023). Biopesticides as a promising alternative to synthetic pesticides: A case for microbial pesticides, phytopesticides, and nanobiopesticides. *Frontiers in Microbiology*, *14*. https://doi.org/10.3389/fmicb.2023.1040901

Bamisile, B. S., Akutse, K. S., Siddiqui, J. A., & Xu, Y. (2021). Model application of entomopathogenic fungi as alternatives to chemical pesticides: Prospects, challenges, and insights for next-generation sustainable agriculture. *Frontiers in Plant Science*, *12*. https://www.frontiersin.org/articles/10.3389/fpls.2021.741804

Bansal, A., Sharma, M., Pandey, A., & Shankar, J. (2023). Aflatoxins: Occurrence, biosynthesis pathway, management, and impact on health. In Singh, I., Rajpal, V. R., Navi, S. S. (eds), *Fungal Resources for Sustainable Economy*. Springer, Singapore.

Bathina, P., & Bonam, R. (2020). Effect of endophytic isolates of *Beauveria bassiana* (Balsamo) Vuillemin and *Metarhizium anisopliae* (Metchnikoff) Sorokin on *Plutella xylostella* (L.) (Lepidoptera: Plutellidae) in cabbage. *Egyptian Journal of Biological Pest Control, 30*, 142. https://doi.org/10.1186/s41938-020-00342-w

Bayman, P., Mariño, Y. A., García-Rodríguez, N. M., Oduardo-Sierra, O., & Rehner, S. A. (2021). Local isolates of *Beauveria bassiana* for control of the coffee berry borer *Hypothenemus hampei* in Puerto Rico: Virulence, efficacy, and persistence. *Biological Control, 155*, 104533. https://doi.org/10.1016/j.biocontrol.2021.104533

Bebber, D., Ramotowski, M., & Gurr, S. (2013). Crop pests and pathogens move polewards in a warming world. *Nature Climate Change, 3*, 985–988. https://doi.org/10.1038/nclimate1990

Beys-da-Silva, W. O., Rosa, R. L., Berger, M., Coutinho-Rodrigues, C. J. B., Vainstein, M. H., Schrank, A., Bittencourt, V. R. E. P., & Santi, L. (2020). Updating the application of *Metarhizium anisopliae* to control cattle tick *Rhipicephalus microplus* (Acari: Ixodidae). *Experimental Parasitology, 208*, 107812. https://doi.org/10.1016/j.exppara.2019.107812

Bhadani, R. V., Gajera, H. P., Hirpara, D. G., Kachhadiya, H. J., & Dave, R. A. (2021). Metabolomics of extracellular compounds and parasitic enzymes of *Beauveria bassiana* associated with biological control of whiteflies (*Bemisia tabaci*). *Pesticide Biochemistry and Physiology, 176*, 104877. https://doi.org/10.1016/j.pestbp.2021.104877

Brotman, Y., Landau, U., Cuadros-Inostroza, Á., Tohge, T., Fernie, A. R., Chet, I., Viterbo, A., & Willmitzer, L. (2013). Trichoderma-plant root colonization: Escaping early plant defense responses and activation of the antioxidant machinery for saline stress tolerance. *PLoS Pathogens, 9*(3), e1003221. https://doi.org/10.1371/journal.ppat.1003221

Bubici, G., Kaushal, M., Prigigallo, M. I., Cabanás, C. G., & Mercado-Blanco, J. (2019). Biological control agents against fusarium wilt of banana. *Frontiers in Microbiology, 10*. https://doi.org/10.3389/fmicb.2019.00616

Budge, S. P., & Whipps, J. M. (2001). Potential for integrated control of *Sclerotinia sclerotiorum* in glasshouse lettuce using *Coniothyrium minitans* and reduced fungicide application. *Phytopathology, 91*(2), 221–227. https://doi.org/10.1094/PHYTO.2001.91.2.221

Chitrampalam, P., Wu, B. M., Koike, S. T., & Subbarao, K. V. (2011). Interactions between *Coniothyrium minitans* and *Sclerotinia minor* affect biocontrol efficacy of *C. minitans*. *Phytopathology, 101*(3), 358–366. https://doi.org/10.1094/PHYTO-06-10-0170

Clemons, K. V., Shankar, J., & Stevens, D. A. (2016). Mycologic endocrinology. In Lyte, M. (ed.), *Microbial Endocrinology: Interkingdom Signaling in Infectious Disease and Health* (pp. 337–363), Advances in Experimental Medicine and Biology, Vol. 874). Springer. https://doi.org/10.1007/978-3-319-20215-0_16

Clifton, E. H., Jaronski, S. T., & Hajek, A. E. (2020). Virulence of commercialized fungal entomopathogens against Asian Longhorned Beetle (*Coleoptera: Cerambycidae*). *Journal of Insect Science (Online), 20*(2), 1. https://doi.org/10.1093/jisesa/ieaa006

Collinge, D. B., Jensen, D. F., Rabiey, M., Sarrocco, S., Shaw, M. W., & Shaw, R. H. (2022). Biological control of plant diseases – What has been achieved and what is the direction? *Plant Pathology, 71*, 1024–1047. https://doi.org/10.1111/ppa.13555

Damalas, C. A., & Eleftherohorinos, I. G. (2011). Pesticide exposure, safety issues, and risk assessment indicators. *International Journal of Environmental Research and Public Health, 8*(5), 1402–1419. https://doi.org/10.3390/ijerph8051402

Daojing, Y., Yue, L., Kai, G., Yingying, Y., Shuai, Z., Qiong, D., Cailing, R., Aiying, L., Jun, F., Jinfeng, N., Youming, Z., & Ruijuan, L. (2022). Biocontrol of strawberry gray mold caused by *Botrytis cinerea* with the termite-associated *Streptomyces sp. sdu1201* and actinomycin D. *Frontiers in Microbiology, 13*. https://doi.org/10.3389/fmicb.2022.1051730

Elshahawy, I. E., & El-Mohamedy, R. S. (2019). Biological control of Pythium damping-off and root-rot diseases of tomato using *Trichoderma* isolates employed alone or in combination. *Journal of Plant Pathology, 101*(3), 597–608. https://doi.org/10.1007/s42161-019-00248-z

Fadiji, A. E., & Babalola, O. O. (2020). Elucidating mechanisms of endophytes used in plant protection and other bioactivities with multifunctional prospects. *Frontiers in Bioengineering and Biotechnology, 8*. https://doi.org/10.3389/fbioe.2020.00467

Gadoury, D. M., Cadle-Davidson, L., Wilcox, W. F., Dry, I. B., Seem, R. C., & Milgroom, M. G. (2012). Grapevine powdery mildew (*Erysiphe necator*): A fascinating system for the study of the biology, ecology and epidemiology of an obligate biotroph. *Molecular Plant Pathology, 13*(1), 1–16. https://doi.org/10.1111/j.1364-3703.2011.00728.x

Gangireddy Eswara Reddy, S. (2021). Lecanicillium spp. *for the Management of Aphids, Whiteflies, Thrips, Scales and Mealy Bugs: Review*. IntechOpen. https://doi.org/10.5772/intechopen.94020

Garcia, L. (2020). Ecological and economic benefits and risks of using botanical insecticides in Tanzanian farms. *Independent Study Project (ISP) Collection, 3372*. https://digitalcollections.sit.edu/isp_collection/3372

Ghahremani, Z., Escudero, N., Saus, E., Gabaldón, T., & Sorribas, F. J. (2019). *Pochonia chlamydosporia* induces plant-dependent systemic resistance to *Meloidogyne incognita. Frontiers in Plant Science, 10*, 945. https://doi.org/10.3389/fpls.2019.00945

Glawe, D. A. (2008). The powdery mildews: A review of the world's most familiar (yet poorly known) plant pathogens. *Annual Review of Phytopathology, 46*, 27–51. https://doi.org/10.1146/annurev.phyto.46.081407.104740

Gopal, G. S., Venkateshalu, B., Nadaf, A. M., et al. (2021). Management of the grape mealy bug, *Maconellicoccus hirsutus* (Green), using entomopathogenic fungi and botanical oils: A laboratory study. *Egyptian Journal of Biological Pest Control, 31*, 100. https://doi.org/10.1186/s41938-021-00444-z

Harman, G. E. (2006). Overview of mechanisms and uses of *Trichoderma* spp. *Phytopathology, 96*(2), 190–194. https://doi.org/10.1094/PHYTO-96-0190

Harman, G.E. (2011), Multifunctional fungal plant symbionts: New tools to enhance plant growth and productivity. *New Phytologist, 189*, 647–649. https://doi.org/10.1111/j.1469-8137.2010.03614.x

Harman, G. E., Howell, C. R., Viterbo, A., Chet, I., & Lorito, M. (2004). *Trichoderma* species–opportunistic, avirulent plant symbionts. *Nature Reviews. Microbiology, 2*(1), 43–56. https://doi.org/10.1038/nrmicro797

He, D. C., He, M. H., Amalin, D. M., Liu, W., Alvindia, D. G., & Zhan, J. (2021). Biological control of plant diseases: An evolutionary and eco-economic consideration. *Pathogens (Basel, Switzerland), 10*(10), 1311. https://doi.org/10.3390/pathogens10101311

Hermosa, R., Viterbo, A., Chet, I., & Monte, E. (2012). Plant-beneficial effects of *Trichoderma* and of its genes. *Microbiology (Reading, England), 158*(Pt 1), 17–25. https://doi.org/10.1099/mic.0.052274-0

Huilgol, S. N., Nandeesha, K. L., & Banu, H. (2022). Fungal biocontrol agents: An eco-friendly option for the management of plant diseases to attain sustainable agriculture in India. In Rajpal, V.R., Singh, I., Navi, S.S. (eds), *Fungal Diversity, Ecology and Control Management*. Springer, Singapore. https://doi.org/10.1007/978-981-16-8877-5_22

Jacobs, S., Zechmann, B., Molitor, A., Trujillo, M., Petutschnig, E., Lipka, V., Kogel, K. H., & Schäfer, P. (2011). Broad-spectrum suppression of innate immunity is required for colonization of Arabidopsis roots by the fungus *Piriformospora indica. Plant Physiology, 156*(2), 726–740. https://doi.org/10.1104/pp.111.176446

Jaronski, S. T. (2010). Ecological factors in the inundative use of fungal entomopathogens. *BioControl*, *55*, 159–185. https://doi.org/10.1007/s10526-009-9248-3

Kabdwal, B. C., Sharma, R., Kumar, A., et al. (2023). Efficacy of different combinations of microbial biocontrol agents against sheath blight of rice caused by *Rhizoctonia solani*. *Egyptian Journal of Biological Pest Control*, *33*, 29. https://doi.org/10.1186/s41938-023-00671-6

Khairullina, A., Micic, N., Jørgensen, H. J. L., Bjarnholt, N., Bülow, L., Collinge, D. B., & Jensen, B. (2023). Biocontrol effect of *Clonostachys rosea* on *Fusarium graminearum* infection and mycotoxin detoxification in oat (*Avena sativa*). *Plants*, *12*(3), 500. http://doi.org/10.3390/plants12030500

Khan, M., & Tanaka, K. (2023). *Purpureocillium lilacinum* for plant growth promotion and biocontrol against root-knot nematodes infecting eggplant. *PLoS One*, *18*(3), e0283550. https://doi.org/10.1371/journal.pone.0283550

Kim, H. J., Lee, E. J., Park, S. H., Lee, H., & Chung, N. (2014). Biological control of anthracnose (*Colletotrichum gloeosporioides*) in pepper and cherry tomato by *Streptomyces* sp. A1022. *Journal of Agricultural Science*, *6*(2). https://doi.org/10.5539/jas.v6n2p54

Köhl, J., Kolnaar, R., & Ravensberg, W. J. (2019). Mode of action of microbial biological control agents against plant diseases: Relevance beyond efficacy. *Frontiers in Plant Science*, *10*, 845. https://doi.org/10.3389/fpls.2019.00845

Kredics, L., Antal, Z., Dóczi, I., Manczinger, L., Kevei, F., & Nagy, E. (2003). Clinical importance of the genus *Trichoderma*. A review. *Acta Microbiologica et Immunologica Hungarica*, *50*(2–3), 105–117. https://doi.org/10.1556/AMicr.50.2003.2-3.1

Krupke, C. H., Hunt, G. J., Eitzer, B. D., Andino, G., & Given, K. (2012). Multiple routes of pesticide exposure for honey bees living near agricultural fields. *PloS One*, *7*(1), e29268. https://doi.org/10.1371/journal.pone.0029268

Kumari, D., Yadav, N. K., Kumhar, K. C., Kumar, A., Vashisht, P., & Garima. (2023). Antagonistic potential and growth promoting activities of native *Trichoderma* isolates against *Fusarium oxysporum* f. sp. *ciceri*. *The Indian Journal of Agricultural Sciences*, *93*(7), 780–785. https://doi.org/10.56093/ijas.v93i7.137287

Lahlali, R., Ezrari, S., Radouane, N., Kenfaoui, J., Esmaeel, Q., El Hamss, H., Belabess, Z., & Barka, E. A. (2022). Biological control of plant pathogens: A global perspective. *Microorganisms*, *10*(3), 596. https://doi.org/10.3390/microorganisms10030596

Lan, X., Zhang, J., Zong, Z., Ma, Q., & Wang, Y. (2017). Evaluation of the biocontrol potential of *Purpureocillium lilacinum* QLP12 against *Verticillium dahliae* in eggplant. *BioMed Research International*, *2017*, 4101357. https://doi.org/10.1155/2017/4101357

Larkin, R. P., & Brewer, M. T. (2020). Effects of crop rotation and biocontrol amendments on rhizoctonia disease of potato and soil microbial communities. *Agriculture*, *10*(4), 128. http://doi.org/10.3390/agriculture10040128

Lavkor, I., Ay, T., Sobucovali, S., Var, I., Saghrouchni, H., Salamatullah, A. M., & Mekonnen, A. B. (2023). Non-aflatoxigenic *Aspergillus flavus*: A promising biological control agent against aflatoxin contamination of corn. *ACS Omega*, *8*(19), 16779–16788. https://doi.org/10.1021/acsomega.3c00303

Li, G., Huang, H., Miao, H., Erickson, R., Jiang, Daohong, & Xiao, Y. N. (2006). Biological control of sclerotinia diseases of rapeseed by aerial applications of the mycoparasite *Coniothyrium minitans*. *European Journal of Plant Pathology*, *114*, 345–355. https://doi.org/10.1007/s10658-005-2232-6

Li, D., Li, S., Wei, S., & Sun, W. (2021). Strategies to manage rice sheath blight: Lessons from interactions between rice and *Rhizoctonia solani*. *Rice (New York, N.Y.)*, *14*(1), 21. https://doi.org/10.1186/s12284-021-00466-z

Li, D., Sossah, F. L., Yang, Y., et al. (2019). Genetic and pathogenic variability of *Mycogone perniciosa* isolates causing wet bubble disease on *Agaricus bisporus* in China. *Pathogens*, *8*(4), 179. http://doi.org/10.3390/pathogens8040179

Loc, N. H., Huy, N. D., Quang, H. T., Lan, T. T., & Thu Ha, T. T. (2019). Characterisation and antifungal activity of extracellular chitinase from a biocontrol fungus, *Trichoderma asperellum* PQ34. *Mycology, 11*(1), 38–48. https://doi.org/10.1080/21501203.2019.1703839

Mannaa, M., & Kim, K. D. (2018). Effect of temperature and relative humidity on growth of *Aspergillus* and *Penicillium* spp. and biocontrol activity of *Pseudomonas protegens* AS15 against aflatoxigenic *Aspergillus flavus* in stored rice grains. *Mycobiology, 46*(3), 287–295. https://doi.org/10.1080/12298093.2018.1505247

Mantzoukas, S., Lagogiannis, I., Mpekiri, M., Pettas, I., & Eliopoulos, P. A. (2019). Insecticidal action of several isolates of entomopathogenic fungi against the granary weevil *Sitophilus granarius. Agriculture, 9*(10), 222. http://doi.org/10.3390/agriculture9100222

Mauro, A., Battilani, P., & Cotty, P.J. (2015). Atoxigenic *Aspergillus flavus* endemic to Italy for biocontrol of aflatoxins in maize. *BioControl, 60*, 125–134.

Mayo, S., Gutiérrez, S., Malmierca, M. G., De La Varga, A. L., Rodríguez, M. P. C., Hermosa, R., & Casquero, P. A. (2015). Influence of *Rhizoctonia solani* and *Trichoderma* spp. in growth of bean (*Phaseolus vulgaris* L.) and in the induction of plant defense-related genes. *Frontiers in Plant Science, 6*. https://doi.org/10.3389/fpls.2015.00685

McLaren, D. L., Huang, H. C., Kozub, G. C., & Rimmer, S. R. (1994). Biological control of sclerotinia wilt of sunflower with *Talaromyces flavus* and *Coniothyrium minitans. Plant Disease, 78*(3), 231–235. https://doi.org/10.1094/PD-78-0231

Mendoza, A., Sikora, R. A., & Kiewnick, S. (2004). Efficacy of *Paecilomyces lilacinus* (strain 251) for the control of *Radopholus similis* in banana. *Communications in Agricultural and Applied Biological Sciences, 69*(3), 365–372.

Mónaco, C., Dal Bello, G., Rollán, M., Ronco, L., Lampugnani, G., Arteta, N., Abramoff, C., Aprea, A., Larran, S., & Stocco, M. (2009). Biological control of *Botrytis cinerea* on tomato using naturally occurring fungal antagonists. *Archives of Phytopathology and Plant Protection, 42*, 729–737. https://doi.org/10.1080/03235400701390646

Mukherjee, M., Mukherjee, P. K., Horwitz, B. A., Zachow, C., Berg, G., & Zeilinger, S. (2012). Trichoderma-plant-pathogen interactions: Advances in genetics of biological control. *Indian Journal of Microbiology, 52*(4), 522–529. https://doi.org/10.1007/s12088-012-0308-5

Mukhopadhyay, R., & Kumar, D. (2020). Trichoderma: A beneficial antifungal agent and insights into its mechanism of biocontrol potential. *Egyptian Journal of Biological Pest Control, 30*, 133. https://doi.org/10.1186/s41938-020-00333-x

Murali, M., & Amruthesh, K. N. (2015). Plant growth-promoting fungus *penicillium oxalicum* enhances plant growth and induces resistance in pearl millet against downy mildew disease. *Journal of Phytopathology, 163*, 743–754. https://doi.org/10.1111/jph.12371

Nath, V. S., John, N. S., Anjanadevi, I. P., et al. (2014). Characterization of *Trichoderma spp.* antagonistic to *Phytophthora colocasiae* associated with leaf blight of taro. *Annals of Microbiology, 64*, 1513–1522. https://doi.org/10.1007/s13213-013-0794-7

Ngoune Liliane, T., & Shelton Charles, M. (2020). *Factors Affecting Yield of Crops.* IntechOpen. https://doi.org/10.5772/intechopen.90672

Nicolopoulou-Stamati, P., Maipas, S., Kotampasi, C., Stamatis, P., & Hens, L. (2016). Chemical pesticides and human health: The urgent need for a new concept in agriculture. *Frontiers in Public Health, 4*. https://doi.org/10.3389/fpubh.2016.00148

Nieto-Jacobo, M. F., Steyaert, J. M., Salazar-Badillo, F. B., et al. (2017). Environmental growth conditions of *Trichoderma* spp. affects indole acetic acid derivatives, volatile organic compounds, and plant growth promotion. *Frontiers in Plant Science, 8*. https://doi.org/10.3389/fpls.2017.00102

Palmieri, D., Ianiri, G., Del Grosso, C., et al. (2022). Advances and perspectives in the use of biocontrol agents against fungal plant diseases. *Horticulturae*, *8*(7), 577. MDPI AG. Retrieved from http://doi.org/10.3390/horticulturae8070577

Pandey, A. K., Kumar, A., Dinesh, K., Varshney, R., & Dutta, P. (2022). The hunt for beneficial fungi for tomato crop improvement – Advantages and perspectives. *Plant Stress*, *6*, 100110. https://doi.org/10.1016/j.stress.2022.100110

Pandit, M. A., Kumar, J., Gulati, S., et al. (2022). Major biological control strategies for plant pathogens. *Pathogens (Basel, Switzerland)*, *11*(2), 273. https://doi.org/10.3390/pathogens11020273

Patel, T. K., Anand, R., Singh, A. P., et al. (2014). Evaluation of aflatoxin B1 biosynthesis in *A. flavus* isolates from central India and identification of atoxigenic isolates. *Biotechnology and Bioprocess Engineering*, *19*(6), 1105–1113. https://doi.org/10.1007/s12257-014-0464-z

Pathak, V. M., Verma, V. K., Rawat, B. S., Kaur, B., Babu, N., Sharma, A., Dewali, S., Yadav, M., Kumari, R., Singh, S., Mohapatra, A., Pandey, V., Rana, N., & Cunill, J. M. (2022). Current status of pesticide effects on environment, human health and it's eco-friendly management as bioremediation: A comprehensive review. *Frontiers in Microbiology*, *13*, 962619. https://doi.org/10.3389/fmicb.2022.962619

Patil, J. A., Yadav, S., & Kumar, A. (2021). Management of root-knot nematode, *Meloidogyne incognita*, and soil-borne fungus, *Fusarium oxysporum*, in cucumber using three bioagents under polyhouse conditions. *Saudi Journal of Biological Sciences*, *28*(12), 7006–7011. https://doi.org/10.1016/j.sjbs.2021.07.081

Pedrini, N. (2022). The entomopathogenic fungus *Beauveria bassiana* shows its toxic side within insects: Expression of genes encoding secondary metabolites during pathogenesis. *Journal of Fungi (Basel)*, *8*(5), 488. https://doi.org/10.3390/jof8050488

Pirttilä, A. M., Mohammad Parast Tabas, H., Baruah, N., & Koskimäki, J. J. (2021). Biofertilizers and biocontrol agents for agriculture: How to identify and develop new potent microbial strains and traits. *Microorganisms*, *9*(4), 817. https://doi.org/10.3390/microorganisms9040817

Qayyum, M. A., Wakil, W., Arif, M., Sahi, S. T., & Dunlap, C. A. (2015). Infection of *Helicoverpa armigera* by endophytic *Beauveria bassiana* colonizing tomato plants. *Biological Control*, *90*, 200–207. https://doi.org/10.1016/j.biocontrol.2015.04.005

Richard, B., Qi, A., & Fitt, B. D. L. (2022). Control of crop diseases through integrated crop management to deliver climate-smart farming systems for low- and high-input crop production. *Plant Pathology*, *71*, 187–206. https://doi.org/10.1111/ppa.13493

Rizzo, D. M., Lichtveld, M., Mazet, J. A. K., et al. (2021). Plant health and its effects on food safety and security in a one health framework: Four case studies. *One Health Outlook*, *3*, 6. https://doi.org/10.1186/s42522-021-00038-7

Sabbahi, R., Hock, V., Azzaoui, K., Saoiabi, S., & Hammouti, B. (2022). A global perspective of entomopathogens as microbial biocontrol agents of insect pests. *Journal of Agriculture and Food Research*, *10*, 100376. https://doi.org/10.1016/j.jafr.2022.100376

Safari Motlagh, M. R., Farokhzad, M., Kaviani, B., & Kulus, D. (2022). Endophytic fungi as potential biocontrol agents against *Sclerotium rolfsii* Sacc.-The causal agent of Peanut White Stem Rot disease. *Cells*, 11(17), 2643. https://doi.org/10.3390/cells11172643

Santos, M., Diánez, F., Sánchez-Montesinos, B., et al. (2023). Biocontrol of diseases caused by *Phytophthora capsici* and *P. parasitica* in pepper plants. *Journal of Fungi*, *9*(3), 360. http://doi.org/10.3390/jof9030360

Sarven, M. S., Hao, Q., Deng, J., Yang, F., Wang, G., Xiao, Y., & Xiao, X. (2020). Biological control of tomato gray mold caused by *Botrytis cinerea* with the entomopathogenic fungus *Metarhizium anisopliae*. *Pathogens (Basel, Switzerland)*, *9*(3), 213. https://doi.org/10.3390/pathogens9030213

Shankar. J. (2021). Food habit associated mycobiota composition and their impact on human health. *Frontiers in Nutrition*, *8*, 1–8. https://doi.org/10.3389/fnut.2021.773577

Shankar, J., Tiwari, S., Shishodia, S.K., Gangwar, M., Hoda, S., Thakur, R., & Vijayaraghavan, P. (2018). Molecular insights into development and virulence determinants of *Aspergilli*: A proteomic perspective. *Frontiers in Cellular and Infection Microbiology*, *8*, 180.

Sharma, R., & Sharma, P. (2021). Fungal entomopathogens: A systematic review. *Egyptian Journal of Biological Pest Control*, *31*, 57. https://doi.org/10.1186/s41938-021-00404-7

Siddiqui, J. A., Fan, R., Naz, H., Bamisile, B. S., Hafeez, M., Ghani, M. I., Wei, Y., Xu, Y., & Chen, X. (2023). Insights into insecticide-resistance mechanisms in invasive species: Challenges and control strategies. *Frontiers in Physiology*, *13*, 1112278. https://doi.org/10.3389/fphys.2022.1112278

Sivan, A., & Chet, I. (1986). Biological control of *Fusarium spp.* in cotton, wheat and muskmelon by *Trichoderma harzianum*. *Journal of Phytopathology*, *116*, 39–47. https://doi.org/10.1111/j.1439-0434.1986.tb00892.x

Thambugala, K. M., Daranagama, D. A., Phillips, A. J. L., Kannangara, S. D., & Promputtha, I. (2020). Fungi vs. fungi in biocontrol: An overview of fungal antagonists applied against fungal plant pathogens. *Frontiers in Cellular and Infection Microbiology*, *10*. https://doi.org/10.3389/fcimb.2020.604923

Ting, A. S. Y., & Chai, J. Y. (2015). Chitinase and β-1,3-glucanase activities of *Trichoderma harzianum* in response toward pathogenic and non-pathogenic isolates: Early indications of compatibility in consortium. *Biocatalysis and Agricultural Biotechnology*, *4*(1), 109–113. https://doi.org/10.1016/j.bcab.2014.10.003

Tiwari, S., & Shankar, J. (2018) Integrated proteome and HPLC analysis revealed quercetin-mediated inhibition of aflatoxin B1 biosynthesis in *Aspergillus flavus*. *3 Biotech*, *8*, 47.

Tudi, M., Daniel Ruan, H., Wang, L., Lyu, J., Sadler, R., Connell, D., Chu, C., & Phung, D. T. (2021). Agriculture development, pesticide application and its impact on the environment. *International Journal of Environmental Research and Public Health*, *18*(3), 1112. https://doi.org/10.3390/ijerph18031112

Tyśkiewicz, R., Nowak, A., Ozimek, E., & Jaroszuk-Ściseł, J. (2022). Trichoderma: The current status of its application in agriculture for the biocontrol of fungal phytopathogens and stimulation of plant growth. *International Journal of Molecular Sciences*, *23*(4), 2329. https://doi.org/10.3390/ijms23042329

Verma, A., Shameem, N., Jatav, H. S., Sathyanarayana, E., Parray, J. A., Poczai, P., & Sayyed, R. Z. (2022). Fungal endophytes to combat biotic and abiotic stresses for climate-smart and sustainable agriculture. *Frontiers in Plant Science*, *13*. https://doi.org/10.3389/fpls.2022.953836

Vinale, F., Nigro, M., Sivasithamparam, K., Flematti, G., Ghisalberti, E. L., Ruocco, M., Varlese, R., Marra, R., Lanzuise, S., Eid, A., Woo, S. L., & Lorito, M. (2013). Harzianic acid: A novel siderophore from *Trichoderma harzianum*. *FEMS Microbiology Letters*, *347*(2), 123–129. https://doi.org/10.1111/1574-6968.12231

Wang, H., Peng, H., Li, W., Cheng, P., & Gong, M. (2021). The toxins of *Beauveria bassiana* and the strategies to improve their virulence to insects. *Frontiers in Microbiology*, *12*. https://doi.org/10.3389/fmicb.2021.705343

Wen, J., Okyere, S. K., Wang, S., Wang, J., Xie, L., Ran, Y., & Hu, Y. (2022). Endophytic fungi: An effective alternative source of plant-derived bioactive compounds for pharmacological studies. *Journal of Fungi (Basel, Switzerland)*, *8*(2), 205. https://doi.org/10.3390/jof8020205

Whipps, J. M. (2001). Microbial interactions and biocontrol in the rhizosphere. *Journal of Experimental Botany*, *52*(Special issue), 487–511. https://doi.org/10.1093/jexbot/52.suppl_1.487

Whipps, J. M., Sreenivasaprasad, S., Muthumeenakshi, S., Rogers, C. W., & Challen, M. P. (2007). Use of *Coniothyrium minitans* as a biocontrol agent and some molecular aspects of sclerotial mycoparasitism. In Collinge, D. B., Munk, L., Cooke, B. M. (eds),

Sustainable Disease Management in a European Context. Springer, Dordrecht. https://doi.org/10.1007/978-1-4020-8780-6_11

Yan, J., Xing, Z., Lei, P., et al. (2021). Evaluation of Scopoletin from *Penicillium janthinellum* Snef1650 for the Control of *Heterodera glycines* in Soybean. *Life, 11*(11), 1143. http://doi.org/10.3390/life11111143

Yao, X., Guo, H., Zhang, K., Zhao, M., Ruan, J., & Chen, J. (2023). Trichoderma and its role in biological control of plant fungal and nematode disease. *Frontiers in Microbiology, 14.* https://doi.org/10.3389/fmicb.2023.1160551

Zehnder, G., Gurr, G. M., Kühne, S., Wade, M. R., Wratten, S. D., & Wyss, E. (2007). Arthropod pest management in organic crops. *Annual Review of Entomology, 52,* 57–80. https://doi.org/10.1146/annurev.ento.52.110405.091337

10 Microbes as Biocontrol Agents for Sustainable Development

Sandeep Patra, Nidhi Verma*, Sneh Priya, Aishani Gupta, and Vandana Gupta*

10.1 INTRODUCTION

The organisms or microbes that suppress the plant pathogen are known as biocontrol agents or biological control agents (BCAs) (Heimpel and Mills, 2017). They are essential in controlling plant pest populations including insects, nematodes, bacteria, and fungi and protecting the crop from numerous diseases caused by these pests such as crown gall, root rot, leaf wilt, and curling disease. They destroy the pest population and prevent them from spreading (Koul et al., 2022). The use of biocontrol agents is widespread since they kill a number of insects naturally while not harming the primary crop. The public's concern over the use of dangerous chemical pesticides has contributed to an increase in fascination with the biological management of plant pathogens, but it has also risen because it may be able to control diseases better than other control methods. Numerous researches on the biological control of plant diseases by antagonistic bacteria have been conducted so far, and many of them concentrate on the suppressive impacts of single strains that are frequently introduced into soil or on plant matter at quite high densities. Contrary to this strategy, crop rotations and organic fertilisers have received a smaller amount of focus for management and manipulation of naturally occurring antagonistic microbial communities, despite the fact that these tactics have produced extremely effective biological control (Hoitink and Boehm, 1999). Many microbes, including bacteria, fungus, virus, and protozoa, are used as microbial biocontrol agents (MBAs) or microbial biological control agents (MBCAs). Some of the target pathogenic species include *Alternaria citri*; *Botryosphaeria* sp.; *Botrytis cinerea*; *Colletotrichum gloesporioides*; *Erwinia amylovora*; *Fusicoccum aromaticum*; *Fusarium oxysporum* f. sp.; *Gaeumannomyces tritici*; *Lasidiplodia theobromae*; *Macrophomina phaseolina*; *Monilia fructicola*; *M. laxa*; *Penicillium crustosum*; *Phomopsis perse*; *Phytophthora cactorum*; *Podosphaera fusca*; *Pseudomonas syringae*; *P. syringae* pv *kiwi*; *Rosellinia necatrix*; *Rhizoctonia solani*; *Verticillium dahliae*; *Xanthomonas arboricola*; and *X. fragariae* (Bonaterra et al., 2022). Additionally, MBCAs can

*Both the authors have contributed equally to the manuscript and both are the first authors.

DOI: 10.1201/9781003407683-10

interact with the pathogen directly through antibiosis or as cells of bacterial pathogens are invaded and killed by hyperparasites (Ghorbanpour et al., 2018). Another direct route of action is the production of secondary metabolites having antibacterial properties (Raaijmakers and Mazzola, 2012). It is possible to choose isolates of helpful microorganisms active against pathogens and grow them on artificial media. "Augmentative biological control" refers to the use of such well-chosen and mass-produced MBCAs at high concentrations once or multiple times during a growth season (Eilenberg et al., 2001; Heimpel and Mills, 2017; van Lenteren et al., 2018). This study on potential microbes as biocontrol agents will highlight major aspects of sustainable biocontrol mechanisms and their future prospects.

10.2 POTENTIAL BIOCONTROL MICROORGANISMS AND THE MECHANISMS OF BIOCONTROL

Microbes including several genera of bacteria and fungi are among the most extensively studied and reported biocontrol agents, which can utilise multiple mechanisms to help limit the development of plant diseases. Several microorganisms and microbe-based products are already approved and commercialised as bioinsecticides and biopesticides (Figure 10.1). Some of the popular commercial biocontrol products used in the plant disease management and their mechanisms of action are listed in Table 10.1.

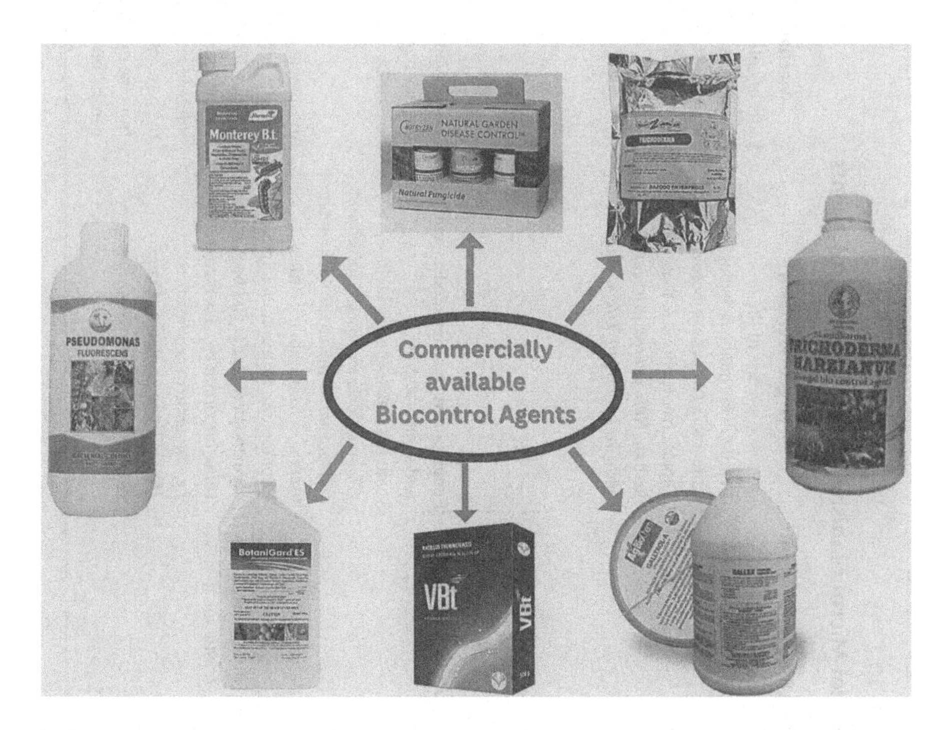

FIGURE 10.1 Commercially available microbial biocontrol agent preparations.

TABLE 10.1
Potential Microorganisms Used as Biocontrol Agents

S. No.	Biocontrol Agent	Target Organism	Mechanism of Action	Current Status	References
1.	*Bacillus thuringiensis*	Arthropod species of the order Lepidoptera (moths & butterflies), Coleoptera (weevils & beetles), Diptera (mosquitoes & flies); species include *Helicoverpa armigera*, *Spilosoma obliqua*, *Olepa ricini*	Lysis of columnar epithelial cells in the midgut of the insects by Bt toxin, thereby forming pores in the gut epithelial tissue	Available in various formulations for direct application in the fields. Transgenic Bt-cotton expressing Cry1Ac has been approved for cultivation in India; Bt rice, chickpea, brinjal, and tomato are at various stages of trials in India; cloning strategies using Cry genes are being employed to overcome resistance development in Lepidopterans	Kumar et al., (2021); Sujayanand et al. (2021); Jain et al. (2016)
2.	*Pseudomonas* spp.	Fungi, oomycetes, nematodes, aphids, and some bacteria Some important species include, *Globodera rostochiensis*, *Fusarium oxysporum*, *Clavibacter michiganensis*	Competition for nutrient and space, antibiosis, induced systemic resistance	Available in various formulations for direct application in the fields –	Lahlali et al. (2022); Bakker et al. (2007); Lanteigne et al. (2012)
3.	*Beauveria bassiana* *Beauveria brongniartii*	Insect species within the order Lepidoptera, Coleoptera, Diptera, Hymenoptera; fungi and bacteria; some important species include *Leptinotarsa decemlineata*, *Helicoverpa armigera*, *Cydia pomonella*, *Ostrinia Mubilalis*	Attachment of the spore to the insect cuticle; germination of the spore on the cuticle; penetration through the cuticle; annihilating the host immune response; expansion within the host; saprophytic outgrowth from the deceased host; and generation of new conidia	Being used as a mycoinsecticide and biopesticide, an alternative to synthetic pesticides, and is being sold under the names BotaniGard, Mycotrol, and Naturalis	Sandhu et al. (2012); Singh et al. (2015); Keswani et al. (2013)

(*Continued*)

TABLE 10.1 (*Continued*)
Potential Microorganisms Used as Biocontrol Agents

S. No.	Biocontrol Agent	Target Organism	Mechanism of Action	Current Status	References
4.	*Metarhizium brunneun*, *Metarhizium anisopliae*, *Metarhizium acridum*	Termites (*Isoptera*), mosquitoes (*Anopheles* spp. & *Aedes* spp.), ticks (*Rhipicephalus* spp. & *Haemaphysalis* spp.), grasshoppers and locusts (*Spodoptera litura*)	Same as *Beauveria*	Being used as an active ingredient in mycoinsecticides and mycoacaricides; BIO 1020, a mycoinsecticide against Coleoptera species; Green Guard®, a mycoacaricides against Orthoptera species	Aw and Hue (2017); De Faria and Wraight (2007)
5.	*Trichoderma* spp.	Fungi, bacteria, arthropods, & nematodes; Some important species are, *Rhizoctonia solani*, *Verticillium dahliae*, *Fusarium* spp., *Botrytis cinerea*, *Ralstonia solanacearum*, *Meloidogyne javanica*, *Leucinodes orbonalis*	Competition for nutrients and space; ability to produce/resist metabolites that may hinder fungal spore germination (fungistatic), cell lysis (antibiosis), or modify the rhizosphere (soil acidification); it may also result from direct physical contact with the target as in mycoparasitism	Used as a biofungicide, biofertiliser, and antiparasitic (insecticidal and nematicidal) agent in various agricultural practices; sold under various trade names including Binab-T WP, SoilGard, Trichodex, Trichopel, Harzian 20, Bio-Fungus, Root shield, etc.	Tyśkiewicz et al. (2022); Benítez et al. (2004); Gajera et al. (2013); Ghazanfar et al. (2018)
6.	*Coniothyrium minitans*	Fungal pathogens like *Sclerotinia sclerotiorum*, *Sclerotinia minor*	Biocontrol mechanism is by the production of various antifungal metabolites and lytic enzymes including β-1,3 glucanases, and chitinases and an antibiotic macrosphelide A	Used commercially as biofungicide by the trade name Contans WG and KONI	de Vrije et al. (2001); Tomprefa et al. (2009); Albert et al. (2022)

(*Continued*)

TABLE 10.1 (*Continued*)
Potential Microorganisms Used as Biocontrol Agents

S. No.	Biocontrol Agent	Target Organism	Mechanism of Action	Current Status	References
7.	*Pantoea agglomerans*	Fungal and bacterial pathogens like *Neofusicoccum parvum*, *Ralstonia solanacearum*, *Erwinia amylovora*, *Alternaria solani*	Primary biocontrol mechanism is by competitive exclusion and by the production of various antibacterials including pantocins, herbicolins, and phenazines	Being used in biocontrol of fire blight, bacterial wilt, grapevine trunk disease, tomato early blight, etc.; commercially registered strains for biocontrol are BlossomBless P10c, BlightBan C9-1, Bloomtime E325	Abo-Elyousr and Hassan, (2021); Haidar et al. (2021); Rezzonico et al. (2009)
8.	*Bacillus amyloliquefaciens*	Fungal and bacterial pathogens like *Ralstonia solanacearum*, *Fusarium oxysporum*, *Alternaria alternata*, *Phytophthora sojae*	Induction of systemic resistance by bacterial metabolites including surfactin and VOCs; antifungal activity due to bacillomycin D and fengycin; antibacterial activity due to bacilysin, bacteriocins, and polyketides; competition for root colonisation and nutrients	Currently being used as biofungicide and biofertiliser; commercially available by various trade names, RhizoVital 42, Taegro, AmyloX, & Double Nickel 55	Luo et al. (2022); Chowdhury et al. (2015); Ngalimat et al. (2021)
9.	*Lysobacter capsici*	*Pythium ultimum*, *Rhizoctonia solani*, *Sclerotinia minor*, *Phytophthora infestans*	Biocontrol mechanisms include the production of lytic agents active against phytopathogens like bacteria, fungi, and Oomycetes	–	Lazazzara et al. (2017); Vlassi et al. (2020); Afoshin et al. (2020)
10.	*Lactiplantibacillus plantarum*	*Botrytis cinerea*, *Fusarium oxysporum*, *Pseudomonas syringae* pv. *actinidiae*, *Xanthomonas arboricola* pv. *pruni*, *Xanthomonas fragariae*, and *Ascosphaera apis*	Competition for key nutrients and production of bioactive molecules like organic acids, ethanol, hydrogen peroxide, acetoin, diacetyl, bacteriocins, etc. through their antimicrobial activity is deployed	Currently being used as a biocontrol agent in the agricultural sector and as microbial starters and probiotics in the food industry	Daranas et al. (2019); Arena et al. (2016); De Simone et al. (2021); Iorizzo et al. (2021)

(*Continued*)

TABLE 10.1 (*Continued*)
Potential Microorganisms Used as Biocontrol Agents

S. No.	Biocontrol Agent	Target Organism	Mechanism of Action	Current Status	References
11.	*Coelomomyces*	Mosquito larvae and other aquatic dipterans genera include *Anopheles, Aedes, Culex*, & *Opifex*	Biocontrol mechanism is exerted by its generalised life cycle in the mosquito larvae; it infects hemocoel of larvae to form hyphae which develop into resting sporangia, followed by the release of meiospores which infect Copepods.	Used as a mycoinsecticide in the biological control of mosquitoes	Scholte et al. (2004); Rueda-Páramo et al. (2017)
12.	Baculovirus	Arthropods like *Cydia pomonella, Spodoptera exigua, Homona magnanima, Helicoverpa armigera, Spodoptera litura*	Infects and kills target organisms in a two-stage process: an initial infection and amplification stage in the insect midgut, subsequently followed by a systemic infection culminating with the formation of crystalline occlusion bodies (OBs)	Used as bioinsecticides and biopesticides in agriculture to control insect pathogens; commercially available with trade names Spod-X, Gemstar LC, Elcar, Spodopterin, etc.	Sabbahi et al. (2022); Grzywacz (2017); Popham et al. (2016); Kalha et al. (2014)
13.	Endosymbiotic *Wolbachia*	Primarily affects mosquitoes including *Aedes aegypti, Aedes albopictus, Anopheles* spp., *Culex* spp., *Culiseta morsitans*	Manipulate host reproduction by feminisation, male killing, parthenogenesis, cytoplasmic incompatibility	Currently being used in the control of arboviral disease; biological control of mosquito vectors	Jeffries and Walker (2016); Manoj et al. (2021)
	Nematophagous fungi, e.g., *Purpureocillium lilacinum* and *Arthrobotrys dactyloides*	Plant parasitic nematodes such as *Meloidogyne incognita* and *Meloidogyne javanica*	Formation of adhesive knobs, adhesive nets, and constrictive rings to immobilise nematodes	Various phytonematodes	Pedro et al. (2022)

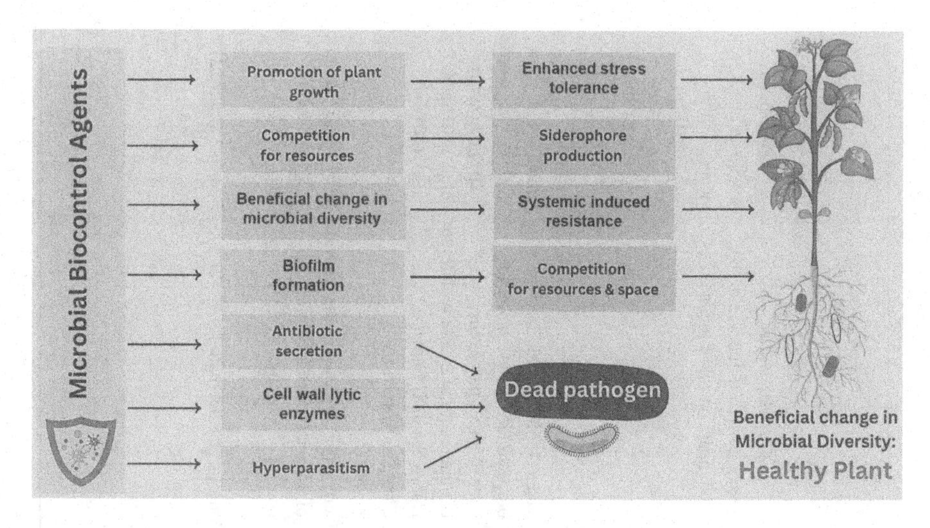

FIGURE 10.2 Mechanisms of action of biocontrol agents.

The mechanisms of biocontrol can be direct or indirect. The direct methods include antibiosis, production of cell lytic enzymes, and hyperparasitism. Nematophagous fungi, e.g., *Purpureocillium lilacinum* and *Arthrobotrys dactyloides* destroy plant parasitic nematodes such as *Meloidogyne incognita* and *Meloidogyne javanica*. The fungi entrap nematodes through the formation of adhesive knobs, adhesive nets, and constrictive rings that kill the nematodes. Various microorganisms are capable of producing a wide range of volatile organic compounds (VOCs) which are effective in biocontrol due to their antimicrobial effects. VOCs are organic chemicals with low molecular weight, low solubility in water, and high vapour pressure. Indirect methods include induction of systemic resistance in the plant through the release of metabolites by MBCA, competition for the nutrients (e.g., scavenging iron through the production of high affinity siderophores by the MBCA) and space (biofilm formation on the root surface thus occupying all the space on the root surface by the MBCA), and promoting plant growth (Figure 10.2). Some of the antibiotics and other compounds responsible for direct biocontrol produced by bacterial BCAs are iturin, fengycin, surfactin, cyanide, chitinases, proteases, pyoluteorin, phenazines, pyrroinitrin, bacilysin, amphisin, voscosinamide, mycosubtilin, beta-glucanases, tensin, subtilosin, etc. Fungal BCAs are reported to be producing ketones, aldehydes, hydrogen cyanide, alcohols, gliotoxin, gliovirin, harzianolide, harzianopyridone, ethylene, heptelidic acid, valinotrocin, viridiol, viridian, 1-hydroxy-3 methylanthraquinone, T39 butanolide, etc. (reviewed in Tariq et al., 2020).

10.3 IMPORTANT FORMULATIONS AND APPLICATIONS METHODS OF MBCAs: STEPPING OUT OF LAB

Many formulations of BCAs prepared in a wide range of carriers and adjuvants are available, with powdered, granulated, or liquid forms being most common in commercial preparations. The substance utilised to formulate the biocontrol agent for field application has a significant impact on both its capacity to survive and its ability to effectively control infections and pests. It has been studied how to formulate microbial biocontrol agents using biopolymers such as alginate, starch, chitosan, etc. (Saberi-Riseh et al., 2021). Bacterial encapsulation strategies result in the creation of a physical barrier between the internal material and its surroundings in order to shield them from hostile conditions such as moisture fluctuations, pH changes, and oxidation. This approach provides benefits, such as regulated material discharge and protection of encapsulated compounds against environmental alterations. Some of the commonly used formulations of BCAs are powder and granules; cell immobilisation systems and liquid formulations.

10.3.1 Powder and Granules

These are prepared using organic/inorganic carriers and/or soil. The particle size of powder inoculants varies from a few μm to about a hundred μm, and the particle sizes of granules typically range from 0.1 to 2.5 mm. Granules are subdivided into microgranules, fine granules, and large granules with particle sizes of 100–600 μm, 0.3–2.5 mm, up to 6 mm, respectively (Bejarano and Puopolo, 2020).

Powder formulations are prepared by mixing BCAs with a finely ground carrier and adjuvants in a blender. Lyophilisation and spray drying can also be used to produce powder formulations. Lyophilisation preserves the viability of BCAs like *Pseudomonas. fluorescens* and other species. Spray-drying is suitable only for sturdy spore-forming BCAs, such as *B. subtilis* (Bejarano and Puopolo, 2020). Peat is the carrier of choice for both bacterial and fungal BCAs. Peat can also be modified using other carriers such as coir dust, charcoal, compost, sand, sawdust, and sugarcane bagasse. Inorganic carriers such as talc and kaolin exhibited suitability for usage in BCA formulations.

10.3.2 Cell Immobilisation Systems

Immobilisation of BCA cells in a shell or capsule of polymeric substance has been reported to be an effective method for formulation of MBCAs for field applications. Alginate polymer has been rated as the best polymer for such formulations for environmental and agricultural applications. Other natural polysaccharides such as agarose/agar, gums (obtained from gellan, guar bean, gum arabic); α- and k-carrageenan, starch and starch-based materials, chitosan, xanthan, pectin, lignin, cellulose and its derivatives such as carboxymethylcellulose (CMC) and ethylcellulose); complex lipids such as waxes; proteinaceous substances such as gelatine or whey have been used to increase the viability and shelf life of the MBCAs (reviewed in Bejarano and Puopolo, 2020).

10.3.3 Liquid Formulations

Liquid formulations primarily consist of microbial suspensions in water, polymers, mineral or organic oils, or combinations. Suspension concentrates can be easily prepared by adding the microbial inoculants to an aqueous solution. The preparation is usually stabilised by adding 1–3% thickeners and 1–5% dispersants. They also prevent sedimentation of the active microbial cells. Wetting agents at a concentration of 3–8% inhibit clustering of the particles in suspension. Oil dispersion presents a better version of the suspension medium. Here the solid active ingredients (microbial cells or granules) are dispersed in oil such as vegetable oil, paraffinic to aromatic solvent type oils, methylated seed oils, etc. Ideally, biodegradable oils are more suited for the application. Emulsions (oil in water or water in oil) are prepared using an emulsifier. Suspo-emulsions prepared using a mixture of a suspension concentrate and an emulsion exhibit superior properties (reviewed in Bejarano and Puopolo, 2020).

10.3.4 Additives

Adjuvants, usually surfactants and oils, that enable mixing, handling, etc. and enhance the effectiveness of the MBCA in the fields are also added to the formulations. Activator adjuvants improve the spread of MBCA in the field and it is better retained with enhanced penetration and decreased evaporation rate. Non-ionic surfactants such as polyoxyethylene and block copolymer surfactants are most frequently used in agriculture. Utility adjuvants such as dispersants (celluloses and PVP); co-solvents (glycol and alcohol), buffering agents (potassium phosphate and citric acid); thickeners and stickers (gums and celluloses); coupling agents (organosilanes); stabilising agents (ammonium sulphate, EDTA, triethanolamine); agents for foaming and defoaming (dimethylsiloxane); humectants (PEG, glycerol, ethylene glycol); and UV adsorbents (zinc oxide and titanium dioxide) improve a formulation's physical properties and are commonly used in MBCA formulations (reviewed in Bejarano and Puopolo, 2020).

Various methods are used for the application of MBCA formulations as discussed below (reviewed in Prasad et al., 2023):

1. **Soil Application:** For 1 acre of land 2–2.5 kg of powder or 500–1000 mL of liquid formulation of MBCA is properly mixed with 25–50 kg organic manure. It is kept for 2–3 weeks in shade and covered with straw for multiplication of MBCA with mixing every 3–4 days along with maintenance of proper moisture. It is applied 2 weeks prior to sowing in the field.
2. **Cutting/Seedling's Root Dip Application:** 10 g of powder or 10 mL of the liquid formulation is mixed in 1 L of water. Before transplantation cuttings and roots of seedlings are dipped in it for about 30 minutes.
3. **Nursery Bed Treatment:** 10–15 kg well decomposed organic manure prepared as in point 1 is applied in a one-acre area in the evening time along with proper moisture conditions.
4. **Foliar Application:** 10 g of powder formulation/10 mL of liquid formulation in 1 L of water can be sprayed 2–3 times uniformly in cereals, pulses, and oilseeds on diseased plants during morning or evening hours.
5. **Seed Bio-priming:** Seeds are treated with BCA prepared with gum arabica, jaggery, or other thickening agent. Seeds are mixed properly in the above BCA preparation. Treated seeds are filled in polythene bags, heaped,

covered with moist jute sacks, and allowed to germinate. The seed coat is covered by a protective layer during this period. This technique leads to speedy and even seedling development.

6. **Seed Treatment:** The required quantity of seeds is mixed with BCA formulation. Evenly coated seeds are dried. Seeds are kept in the shade for 20–30 min before sowing. This method protects the seedlings from seed and soilborne pathogens (reviewed in Prasad et al., 2023).

10.4 RECENT ADVANCES AND CASE STUDIES: SUCCESSFUL APPLICATIONS OF MBCAs

Plant pathogens pose an imminent threat to the forestry and agricultural sectors as they transmit diseases that have a substantial impact on the economy and environment (Bonaterra et al., 2022). Their impact has currently escalated as a result of market globalisation and global climate change, which permit the manifestation and rapid spread of emerging diseases. Consequently, there has been an upsurge in research over the past few decades into the selection, assessment, characterisation, and usage of MBCAs. The efficacy of MBCAs as a green approach to control plant diseases, promote plant development and performance, and enhance yield are well proven (El-Saadony et al., 2022). Plant growth-promoting rhizobacteria/fungi (PGPR/PGPF) not only promote plant development but also prevent diseases of plants by synthesising inhibitory compounds and stimulating plant immune responses to phytopathogens (El-Saadony et al., 2022). The effectiveness of PGPR as BCAs in controlled conditions emphasises their utility in greenhouse cultivation systems and their potential in commercial horticulture. It has been shown that *Bacillus subtilis*, *Bacillus amyloliquefaciens*, and *Pseudomonas stutzeri* are among those that are successful in colonising cucumber roots and preventing the disease by *Phytophthora capsici* in cucumbers (Islam et al., 2016). During the post-harvest stage, *B. subtilis* can protect the tomatoes from infection by *Rhizopus stolonifer* and *Penicillium* sp. (Zamir et al., 2016).

As antagonists, nematophagous and endoparasitic fungi have been successfully used to minimise root-knot nematode disease. It was found that the talc-based formulation of *Paecilomyces lilacinus* was more effective at minimising the overall population of *Meloidogyne incognita* in soils where tomato plants were cultivated (Priya and Kumar, 2006). *Trichoderma harzianum*, *T. viride*, *P. lilacinus*, *Pochonia chlamydosporia*, and *Pseudomonas fluorescens* at 20 g/kg Okra seed significantly lowered the nematode number in soil and enhanced plant growth development (Kumar et al., 2012).

Recently, the use of biomimics of the VOCs to manage fungal pathogens has been introduced with better efficacy and regulation while revoking the time and cost of culturing the organisms in large amounts (Gabriel et al., 2018). *Rhodococcus rhodochrous* strain DAP 96253 is capable of inhibiting several fungal pathogens including *Pseudogymnoascus destructans* and *Ophidiomyces ophiodiicola* that are the causative agents of white-nose syndrome (WNS) of bats and snake fungal disease (SFD) of reptiles, respectively (Cornelison et al., 2014, 2016). In a study, VOC formulations from the fungus *Muscodor crispans* strain B-23 used to treat WNS were tested and the synthetic formulation, commercially known as Flavorzon 185B (Jeneil Biotech, Inc.), successfully inhibited the growth of diverse fungal pathogens (Mitchell et al., 2010). The constituents of Flavorzon include acetaldehyde, 2-butanone, propanoic acid, 2-methyl-, methyl ester, etc., and it could serve as an example to develop and commercialise antimicrobial VOCs for biocontrol (Gabriel et al., 2018; Mitchell et al., 2010).

In a study, several strains of *Bacillus* were tested against the bacterial citrus canker (CBC) causing agent *Xanthomonas citri* subsp. *citri* (*Xcc*) and strains closely related to *Bacillus velezensis* produced high antibacterial activity. *B. velezensis* Bv-21 strain displayed a reduction of CBC by 26.30% and pathogen density by 81.68% compared to control due to its effective antagonistic property indicating its use as an effective biocontrol agent (Rabbee et al., 2022).

Another study exhibited the use of *Bacillus ginsengihumi* S38 strain for the control of necrotrophic fungus *Botrytis cinerea* that leads to Botrytis bunch rot (BBR) disease in wine and table grapes. The mechanism of action was antibiosis and the severity of the disease was reduced by an average of 35–60% compared to untreated control. The strain displayed high consistency and potential as results were maintained throughout the seasons and were as good as other commercial products (Calvo-Garrido et al., 2019).

Fungal pathogen (oomycete) *Phytophthora infestans* is the cause of the late blight disease of potatoes and tomatoes worldwide. It is noted that during the pre-infection phase outside the host, it is most susceptible to attacks by biocontrol agents. In a case study, the researchers discovered that *Trichoderma* spp., *Pythium oligandrum, Bacillus, Pseudomonas,* and *Streptomyces* were able to successfully control the potato plant pathogen, *P. infestans* through various mechanisms including antibiosis, mycoparasitism, secretion of lytic enzymes, secondary metabolites, and VOCs (Hashemi et al., 2022).

The bacterium *Curtobacterium flaccumfaciens* pv. *flaccumfaciens* (Cff) responsible for bacterial wilt of *Phaseolus vulgaris* (common bean) was targeted by rhizobacterial strains with potential antagonistic biocontrol activity and the experiment revealed that foliar administration accompanied by treatment of seeds using strains of *Pseudomonas fluorescens*, *Bacillus cereus*, and *Paenibacillus polymyxa* produced a significant reduction in the frequency and severity of disease (Munene et al., 2023).

Use of fungal biocontrol agents against fungal pathogen *Plasmodiophora brassicae*, the causal agent of clubroot disease of cruciferous plants was studied. *Trichoderma* strains were found to be predominant among the root isolates and two strains Hz36 and Hk3 determined to be *T. guizhouense* and *T. koningiopsis*, respectively, significantly inhibited the resting spore germination, growth, and enzyme activity of the fungal pathogen while promoting the development of rapeseed. Strain Hz36 displayed biocontrol efficacy of clubroot in rapeseed and *Arabidopsis thaliana* to be 44.29% and 52.18%, respectively, while Hk37 displayed effectiveness of 57.30% on and 68.01% towards clubroot disease in rapeseed and *A. thaliana*, respectively, hence suggesting its significant potential as a biocontrol agent of clubroot (Zhao et al., 2022).

Besides the use of biocontrol during pre-harvest agricultural practices, the use of microbial biocontrol in managing post-harvest product quality is also gaining attention as a sustainable practice (Sellitto et al., 2021). Research on post-harvest conditions in citrus fruits as a result of *Penicillium digitatum* and *P. italicum* causing green and blue moulds assessed over 180 yeasts and bacterial isolates from citrus fruit peels. Five isolates (two yeasts and three bacteria), *Candida oleophila*, *Debaryomyces hansenii*, *Bacillus amyloliquefaciens*, *B. pumilus*, and *B. subtilis*, were found to prevent infection by moulds and showed high antagonistic efficiency causing a reduction in the spoilage of citrus fruits while maintaining the quality and longevity of fruits. These strains displayed various protective mechanisms like biofilm formation and production of antifungal lipopeptides, lytic enzymes, and VOCs and have the potential to serve as successful biocontrol agents for post-harvest protection of citrus fruits (Hammami et al., 2022).

10.5 FUTURE PROSPECTS IN MICROBIAL BIOCONTROL

The use of MBCAs in agriculture is expected to increase further over the next few years. Microbial antagonist-based biocontrol products are a safer substitute for chemical products. As a result, research on the selection, characterisation, and commercialisation of MBCAs has grown substantially over the past few years (Palmieri et al., 2022). Significant progress appears to have been achieved towards biological and integrated management of post-harvest diseases on fruits. Research and development in biological control, including both augmentative and traditional biocontrol, is accelerating in developing nations like Brazil. Brazil has seen extensive and effective usage of biological control in recent years (Barratt et al., 2018). Novel MBCA needs to be identified in the future, and methods of screening that can assess many candidates MBCAs rapidly and thoroughly must be developed. Furthermore, an extensive study of model MBCA employing comparative genome, transcriptome, and proteome analysis would provide an effective framework enabling a thorough examination of the biological processes of BCA and to develop strategies boosting its advantageous effect (Bonaterra et al., 2022). Some of the recently discovered MBCAs with commercial potential are discussed further.

Endophytic *Bacillus* strains isolated from cotton roots have shown an inhibitory effect against *Verticillium dahliae* strain VD-080 in a dual culture bioassay (Hasan et al., 2020). Two *Bacillus* strains (HNH7 and HNH9) were treated on cotton plants, and when compared to control treatments, they significantly reduced the severity of verticillium wilt (Hasan et al., 2020). Both *Bacillus megaterium* (SB-9) and *Bacillus subtilis* (BCB-19) significantly reduced larval growth and mortality in *Helicoverpa armigera* and *Spodoptera litura* (Gopalakrishnan et al., 2011). *B. amyloliquefaciens* SF14 and *B. amyloliquefaciens* SP10 inhibit *Monilinia fructigena* and *Monilinia laxa*, which induce brown rot disease in fruits, respectively (Lahlali et al., 2022). *Rhizoctonia solani*, which causes lettuce bottom rot, has been substantially reduced by the commercially accessible *B. amyloliquefaciens* FZB42 product (Lahlali et al., 2022). These *Bacillus* species have shown promising results and can be employed in the future for sustainable biocontrol applications.

Pseudomonas chlororaphis strains are employed as biopesticides in agriculture due to their ability to protect plants from a variety of microbiological infections, insects, and nematodes (Anderson and Kim, 2018). These isolates actively inhibit insects, nematodes, and microbial pathogens by producing several kinds of metabolites. *Pseudomonas aeruginosa* isolated from the rhizosphere of a banana field produced antifungal chemicals such as bacteriocin, HCN, and siderophore which suppressed invasion by phytopathogens including *Fusarium oxysporum*, *Aspergillus niger*, *Aspergillus flavus*, and *Alternaria alternata* (Lahlali et al., 2022). According to Lanteigne et al., DAPG and HCN are the compounds that *Pseudomonas* sp. LBUM300 produces to control the growth and proliferation of *Clavibacter michiganensis* subsp. *michiganensis* in vitro and tomato bacterial canker under natural soil conditions (Lanteigne et al., 2012).

Both plant endophytic fungi and insect pest entomopathogens have been related to several fungal species (Poveda, 2021). A common fungus genus called *Trichoderma* can be found as saprotrophs, mycoparasites, soil residents, and plant symbionts and is by far the most extensively researched fungal biocontrol agent, and certain species have been employed as biopesticides and biofertilisers in the past (Alfiky and Weisskopf, 2021; Lahlali et al., 2022). *Trichoderma* is a possible future alternative for the advancement of sustainable agriculture as it is effective not just against plant infections but also against insect pests (Poveda, 2021). It has been shown that *Trichoderma* species effectively diminish *Sclerospora graminicola*, the cause of pearl millet downy mildew disease (Lahlali et al., 2022). It has been clearly shown that they play a significant role in indirectly controlling the pathogen in the rhizosphere and developing systemic resistance (Nandini et al., 2021). In order to find a potential yeast that had potent biocontrol abilities against post-harvest and wilt diseases, Fernandez-San Millan et al. examined the in vivo and in vitro impacts of 69 strains of yeast isolated from Spanish vineyards against *Alternaria alternata*, *Penicillium expansum*, and *B. cinerea*, as well as soilborne diseases *V. dahliae* and *Fusarium oxysporum* (Fernandez-San Millan et al., 2021). Consequently, *Wickerhamomyces anomalus* Wa-32 decreased the extent of *V. dahliae* infection by up to 40% and *F. oxysporum* infection by up to 50% (Fernandez-San Millan et al., 2021). The same study found *Candida lusitaniae* Cl-28, *Candida oleophila* Co-13, *Debaryomyces hansenii* Dh-67, and *Hypopichia pseudoburtonii* Hp-54 to be the most effective against *P. expansum* (Fernandez-San Millan et al., 2021).

10.6 ETHICAL, SAFETY, AND ENVIRONMENTAL CONSIDERATIONS IN MICROBIAL BIOCONTROL

Compared with the conventional chemical methods, biocontrol is considered better as in comparison to the chemical methods it is a much safer alternative since it is less likely to harm any non-target pathogens and produce harmful substances that are detrimental to the environment (Collatz et al., 2021). However excessive use of microbes could have repercussions on the environment and human and animal health. MBCAs can grow, reproduce, and spread; therefore, controlling them after the release is complicated. This makes it essential to have policies for risk management to avoid unmanageable situations. Extensive data regarding environmental factors, survival, and spread of MBCA needs to be inspected before its release in the market (Köhl et al., 2019). However, these regulations also create difficulty in the implementation of microbial biocontrol in practice. The regulatory institutions also overlong the process of evaluation due to inexpertise and inexperience with MBCAs (Sundh and Eilenberg, 2021). There are certain prerequisites that must be fulfilled in order for a commercial MBCA to be publicly recognised or registered. Therefore, experts ought to enhance the capability of BCA for managing diseases (Bashan et al., 2014). This can be accomplished by developing a BCA with numerous beneficial characteristics and modes of action as conceivable (El-Saadony et al., 2022). These features may include, but are not limited to, the ability of BCA to grow rapidly in laboratory conditions, produce a variety of bioactive metabolites, exhibit excellent rhizosphere competence abilities, improve plant growth performance, be ecologically safe, be compatible with other rhizobacteria/fungi, and be resilient to abiotic stresses (Lyu et al., 2020). When identifying efficient BCA isolates, survival and colonisation are essential factors. Disease management is effective and persistent with plant growth promoting rhizobacteria-based biocontrol. The procedure, formulations, shipping, and storage conditions all have an impact on PGPR stability. One should advance formulation technology, prolong the MBCA product's shelf life, maximise the production of specific microbial types, and accomplish low-cost, mass production to obtain high levels of BCA. High production costs and regulatory expenses of shipping new BCAs for commercial use are one of the most serious challenges faced (El-Saadony et al., 2022). A number of ethical and legal concerns that could jeopardise the local biodiversity are raised if MCBAs are to be used to control diseases (Hajek et al., 2016). To enable the evaluation and marketing of novel BCAs and their products, BCA registration necessitates close collaboration among government institutes, universities, and industrial domains (El-Saadony et al., 2022). Farmers may see little to no economic benefit as opposed to chemical pesticides, which tend to be more dependable and predictable. Awareness of the use of BCA in particular areas of agriculture may expand due to such programmes, which include local workshops, training sessions, and free seminars (El-Saadony et al., 2022).

Commercially available microbes as biofertilisers and BCAs are at times seen to be less efficient in fields compared to laboratory or greenhouse experiments (Gosal et al., 2020). Adaptability of BCAs to a foreign environment is often low and efficacy against multiple targets is also unsatisfactory resulting in their limited usage (Jaiswal et al., 2022). The introduction of non-native microbes could also lead to ecological and environmental problems if the organism becomes invasive and causes outbreaks or grows excessively, leading to ecological imbalance. Agricultural management approaches including the use of biocontrol agents and biofertilisers cause changes in the microbiome which can be identified by large-scale DNA sequencing based metabarcoding studies, but most species detected by these metabarcoding analyses have never been functionally defined, thus making it infeasible to recognise the implications of the changes observed in the microbiome (Collatz et al., 2021). The safety and effectiveness of MBCAs require consideration of the multifaceted interactions that occur between the microbes, plants, and the environment. The cellular, molecular, and genetic level comprehension of the biocontrol agent could help in resolving the safety issues associated with it (Zehra et al., 2021). Prior to introducing a BCA into the environment, its potential risk to the environment, direct or indirect non-target effects, and its impact on biogeochemical cycles must be thoroughly assessed along with their safety issues. The magnitude of direct non-target attack (NTA) brought on by BCA fluctuates greatly, with population-level impacts ranging from negligible to detrimental (Hinz et al., 2020). The criterion for releasing the MBCA into the ecosystem is that, in addition to successful disease suppression, its impacts on non-target species are manageable, if not minimal (Hinz et al., 2020; Winding et al., 2004).

The introduction of BCAs into the environment poses significant risks, encompassing changes in the quantity or distribution of native species, disruption of native natural enemies' effectiveness against pest species due to competitive displacement, transmission of pathogenic organism to native species, loss of biodiversity, dilution of native species' genes through hybridisation with closely related non-native species, and potential disturbance of ecological functions such as pollination resulting from significant alterations in the balance of native insect species (De Clercq et al., 2011).

Organisms intended for use as biological control of an arthropod, nematode, or weed may also cause unintended pathogenic effects on beneficial and synergistic organisms or may cause opportunistic human diseases (Cook et al., 1996). Airborne BCAs and their spores may cause hypersensitive reactions in factory workers who are routinely exposed to fungal spores of *Beauveria* or *Metarhizium* spp. (Sandhu et al., 2012).

Furthermore, certain microbes produce a broad spectrum of potentially toxic substances that might influence the growth, development, or behaviour of non-target species that naturally inhabit that environment. Various microbial secondary metabolites might not only have acute toxicity levels that are substantially higher than those of antibiotics and fungicides but also have potent carcinogenic effects (Deising et al., 2017). The Integrated Pest Management (IPM) principle also recognises the safety affiliated with the use of biological control and considers chemical control to be the last resort where the use of biological control is not practicable (Collatz et al., 2021). To prevent the harmful impacts of using pesticides and chemicals, endophytes could be employed as a BCA if they could be successfully transferred from laboratory to land (Tewari et al., 2019). The exclusively biological formulations obtained from endophytic microorganisms would be non-toxic, biodegradable, non-polluting, release no carbon traces, and thus safe for the ecology of soil, environment, and human health (Tewari et al., 2019). The repeated commercial usage of the same BCAs could result in a significant selection of infections and the emergence of novel strains of pathogens that can escape or mitigate the adverse effects of the BCAs (Barratt et al., 2018).

If environmental conditions change significantly, the BCAs could become ineffective (Lahlali et al., 2022). Thus, the degree of environmental patchiness, biocontrol activity, and survival of biocontrol agent associated with the external settings along with the mechanism of action are required to be understood and predict the result of biocontrol (Xu and Jeger, 2020). Studies from the interaction between biocontrol strains and pathogens also reveal that the biocontrol activity should be considered as a mode of behaviour based on prevalent conditions instead of a characteristic of the bacterial strain (Cray et al., 2016).

It is necessary that the ecological niche of the MBCA should be similar to that of the target pathogen for its survival and effectiveness especially when the mode of action employed is competition. Effective colonisation in the host's environment is necessary or else repeated application of the BCA would be required to prevent infection from the pathogen (Xu and Jeger, 2020). Unsuccessful inhibition of pathogens and insufficient colonisation of the host due to the biotic and abiotic components of the rhizosphere environment prevent the disease controlling functions of the MBCAs (Niu et al., 2020). Therefore. a successful MBCA must adapt to both the environmental factors(abiotic) and host-specific factors (biotic) to improve plant health (Legein et al., 2020).

Genetic manipulation of the microbes increases their survival and efficacy in soil and makes them potential biocontrol agents but the ethical implications make their acceptance a major issue. They are considered dangerous for the environment as a horizontal transfer of genes such as the antibiotic resistance genes to the local microorganisms could lead to mutations and thus further research on effective and environmentally safe recombinants is needed (Gosal et al., 2020).

10.7 REGULATION OF MICROBIAL BIOCONTROL

Biofungicide and biopesticide commercialisation involves multiple steps and is subject to a number of limitations (Lahlali et al., 2022). Prior to receiving authorisation for commercialisation, MBCAs are subject to risk analyses just like synthetic pesticides. Plant protection product marketing regulations are laid out by European Regulation (EC) No. 1107/2009 based on a risk assessment (Lahlali et al., 2022). Regulation (EC) No 540/2011 sets a list of approved microorganisms for biocontrol application throughout Europe (Villaverde et al., 2014).

In June 2011, Regulation EC1107/2009 became effective to stop the approval of chemicals that pose unreasonable hazards to the environment and the health of humans and animals If a compound is a persistent organic pollutant (POP), persistent, bioaccumulative, and toxic (PBT), or very persistent, very bioaccumulative (vPvB), it cannot be registered (Villaverde et al., 2014).

According to Article 4(3) of Regulation EC 1107/2009 (2011), a plant protection product must comply with the following prerequisites: (i) it must be adequately effective; (ii) it must not adversely affect human or animal health either immediately or later; (iii) it must not have any inadmissible negative effects on plants or plant products; (iv) it must not cause unnecessary pain and discomfort to vertebrates; and (v) it must not adversely influence the environment and ecosystem (Villaverde et al., 2014). Commercial approvals for biopesticides are then granted via a lengthy process that involves a series of examinations, including toxicological and environmental tests as well as efficiency (Villaverde et al., 2014). However, the current trend towards reducing the use of chemical pesticides and simplifying the regulatory procedure for low-risk substances may enable the production and use of MBCAs on a global scale.

10.8 CONCLUSION AND FUTURE PROSPECTS

Biological control or biocontrol is one of the efficient and sustainable ways of controlling fungal and bacterial diseases in plants. Among the most extensively studied biocontrol agents, bacteria and fungi are in the limelight, which can utilise multiple mechanisms to help limit the development of plant diseases. Several sustainable and environment-friendly MBCAs are produced and commercialised. Moreover, numerous studies on their mechanisms of action and potential to limit pathogens have been conducted. In the approaching years, there will be a high demand for such microbes and their products to meet the increasing food demand in an ecologically balanced way. The inconsistent performance of MBCAs in disease causing pathogen control limits their widespread use in open fields. Several factors, including abiotic and biotic, influence the mode of action of biocontrol agents and multitrophic interactions between plant pathogens and MBCAs. This chapter showcases some relevant cases of extensively studied and reported bacterial biocontrol agents. Furthermore, recent advances and challenges in the development of bacterial biocontrol agents are emphasised. In addition, some future directions in the development of MBCAs are discussed.

Over the last decade, there have been imperative propels about the information of BCAs for improvement of commercial preparations for fungal and bacterial disease elimination in plants. However, the large-scale production and commercialisation are limited by the translational research on potent MBCAs. Future prospects in this direction warrant the identification and screening of novel MBCAs, requiring reliable and robust methods capable of high-throughput screening of potential candidates. In addition, detailed studies of model MBCAs using comparative omics studies and detailed analysis of mechanisms of actions of BCAs and the development of strategies to propagate their beneficial effects might provide a valuable framework for a sustainable future. This deep and detailed study will allow to critically analyse the potential genera as BCAs, their ecological impact to further ensure its biological as well as environmental safety, and to understand how the microscopic organisms can be modulated to improve the efficacy of controlling pathogenic microbiome.

REFERENCES

Abo-Elyousr, K. A. M., & Hassan, S. A. (2021). Biological control of *Ralstonia solanacearum* (Smith), the causal pathogen of bacterial wilt disease by using *Pantoea* spp. *Egyptian Journal of Biological Pest Control, 31*, 113. https://doi.org/10.1186/s41938-021-00460-z

Afoshin, A. S., Kudryakova, I. V., Borovikova, A. O., et al. (2020). Lytic potential of *Lysobacter capsici* VKM B-2533T: Bacteriolytic enzymes and outer membrane vesicles. *Scientific Reports, 10*, 9944. https://doi.org/10.1038/s41598-020-67122-2

Albert, D., Dumonceaux, T., Carisse, et al. (2022). Combining desirable traits for a good biocontrol strategy against *Sclerotinia sclerotiorum. Microorganisms, 10*(6), 1189. https://doi.org/10.3390/microorganisms10061189

Alfiky, A., & Weisskopf, L. (2021). Deciphering Trichoderma–plant–pathogen interactions for better development of biocontrol applications. *Journal of Fungi, 7*(1), 61. https://doi.org/10.3390/jof7010061

Anderson, A. J., & Kim, Y. C. (2018). Biopesticides produced by plant-probiotic *Pseudomonas chlororaphis* isolates. *Crop Protection, 105*, 62–69. https://doi.org/10.1016/j.cropro.2017.11.009

Arena, M. P., Silvain, A., Normanno, G., et al. (2016). Use of *Lactobacillus plantarum* strains as a bio-control strategy against food-borne pathogenic microorganisms. *Frontiers in Microbiology, 7*, 464.

Aw, K. M. S., & Hue, S. M. (2017). Mode of infection of *Metarhizium* spp. fungus and their potential as biological control agents. *Journal of Fungi (Basel, Switzerland), 3*(2), 30. https://doi.org/10.3390/jof3020030

Bakker, P., Pieterse, C. M. J., & Loon, L. (2007). Induced systemic resistance by fluorescent *Pseudomonas* spp. *Phytopathology, 97,* 239–243. https://doi.org/10.1094/PHYTO-97-2-0239

Barratt, B. I. P., Moran, V. C., Bigler, F., et al. (2018) The status of biological control and recommendations for improving uptake for the future. *BioControl, 63,* 155–167. https://doi.org/10.1007/s10526-017-9831-y

Bashan, Y., de-Bashan, L. E., Prabhu, S. R., et al. (2014) Advances in plant growth-promoting bacterial inoculant technology: Formulations and practical perspectives (1998–2013). *Plant Soil, 378,* 1–33. https://doi.org/10.1007/s11104-013-1956-x

Bejarano, A., & Puopolo, G. (2020). Bioformulation of microbial biocontrol agents for a sustainable agriculture. In De Cal, A., Melgarejo, P., Magan, N. (eds), *How Research Can Stimulate the Development of Commercial Biological Control Against Plant Diseases. Progress in Biological Control,* Volume 21. Springer, Cham. https://doi.org/10.1007/978-3-030-53238-3_16

Benítez, T., Rincón, A. M., Limón, M. C., et al. (2004). Biocontrol mechanism of *Trichoderma strains. International Microbiology: the Official Journal of the Spanish Society for Microbiology, 7,* 249–260.

Bonaterra, A., Badosa, E., Daranas, N., et al. (2022). Bacteria as biological control agents of plant diseases. *Microorganisms, 10*(9), 1759. https://doi.org/10.3390/microorganisms10091759

Calvo-Garrido, C., Roudet, J., Aveline, et al. (2019). Microbial antagonism toward Botrytis bunch rot of grapes in multiple field tests using one *Bacillus ginsengihumi* strain and formulated biological control products. *Frontiers in Plant Science, 10,* 105.

Chowdhury, S. P., Hartmann, A., Gao, X., & Borriss, R. (2015). Biocontrol mechanism by root-associated *Bacillus amyloliquefaciens* FZB42 – a review. *Frontiers in Microbiology, 6,* 780. https://doi.org/10.3389/fmicb.2015.00780

Collatz, J., Hinz, H., Kaser, J. M., et al. (2021). Benefits and risks of biological control. In *Biological Control: Global Impacts, Challenges and Future Directions of Pest Management* (pp. 142–165). CSIRO Publishing: Clayton South, Australia.

Cook, R. J., William, L., Bruckart, J. R., et al. (1996). Safety of microorganisms intended for pest and plant disease control: A framework for scientific evaluation. *Biological Control, 7*(3), 333–351. https://doi.org/10.1006/bcon.1996.0102

Cornelison, C. T., Cherney, B., Gabriel, K. T., et al. (2016). Contact-independent antagonism of *Ophidiomyces ophiodiicola,* the causative agent of snake fungal disease by *Rhodococcus rhodochrous* DAP 96253 and select volatile organic compounds. *Journal of Veterinary Science and Technology, 7,* 1–6.

Cornelison, C. T., Keel, M. K., Gabriel, K. T., et al. (2014). A preliminary report on the contact-independent antagonism of *Pseudogymnoascus destructans* by *Rhodococcus rhodochrous* strain DAP96253. *BMC Microbiology, 14,* 246. https://doi.org/10.1186/s12866-014-0246-y

Cray, J. A., Connor, M. C., Stevenson, A., et al. (2016). Biocontrol agents promote growth of potato pathogens, depending on environmental conditions. *Microbial Biotechnology, 9*(3), 330–354. https://doi.org/10.1111/1751-7915.12349

Daranas, N., Roselló, G., Cabrefiga, J., et al. (2019). Biological control of bacterial plant diseases with *Lactobacillus plantarum* strains selected for their broad-spectrum activity. *Annals of Applied Biology, 174*(1), 92–105. https://doi.org/10.1111/aab.12476.

De Clercq, P., Mason, P. G., & Babendreier, D. (2011). Benefits and risks of exotic biological control agents. *BioControl, 56,* 681–698. https://doi.org/10.1007/s10526-011-9372-8

de Faria, M. R., & Wraight, S. P. (2007). Mycoinsecticides and Mycoacaricides: A comprehensive list with worldwide coverage and international classification of formulation types. *Biological Control, 43*(3), 237–256. https://doi.org/10.1016/j.biocontrol.2007.08.001

De Simone, N., Capozzi, V., de Chiara, M. L. V., et al. (2021). Screening of lactic acid bacteria for the bio-control of *Botrytis cinerea* and the potential of *Lactiplantibacillus plantarum* for eco-friendly preservation of fresh-cut kiwifruit. *Microorganisms*, *9*(4), 773. https://doi.org/10.3390/microorganisms9040773

de Vrije, T., Antoine, N., Buitelaar, R. M., et al. (2001). The fungal biocontrol agent *Coniothyrium minitans*: Production by solid-state fermentation, application and marketing. *Applied Microbiology and Biotechnology*, *56*(1–2), 58–68. https://doi.org/10.1007/s002530100678

Deising, H. B., Gase, I., & Kubo, Y. (2017). The unpredictable risk imposed by microbial secondary metabolites: How safe is biological control of plant diseases? *Journal of Plant Diseases and Protection*, *124*, 413–419. https://doi.org/10.1007/s41348-017-0109-5

Eilenberg, J., Hajek, A., & Lomer, C. (2001). Suggestions for unifying the terminology in biological control. *BioControl*, *46*, 387–400.

El-Saadony, M. T., Saad, A. M., Soliman, S. M., et al. (2022). Plant growth-promoting microorganisms as biocontrol agents of plant diseases: Mechanisms, challenges and future perspectives. *Frontiers in Plant Science*, *13*, 923880. https://doi.org/10.3389/fpls.2022.923880

Fernandez-San Millan, A., Larraya, L., Farran, I., et al. (2021). Successful biocontrol of major postharvest and soil-borne plant pathogenic fungi by antagonistic yeasts. *Biological Control*, *160*, 104683. https://doi.org/10.1016/j.biocontrol.2021.104683

Gabriel, K. T., Joseph Sexton, D., & Cornelison, C. T. (2018). Biomimicry of volatile-based microbial control for managing emerging fungal pathogens. *Journal of Applied Microbiology*, *124*(5), 1024–1031. https://doi.org/10.1111/jam.13667

Gajera, H., Domadiya, R., Patel, S., et al. (2013). Molecular mechanism of *Trichoderma* as bio-control agents against phytopathogen system – A review. *Current Research in Microbiology and Biotechnology*, *1*(4), 133–142.

Ghazanfar, M. U., Raza, M., Raza, W., et al. (2018). Trichoderma as potential biocontrol agent, its exploitation in agriculture: A review. *Plant Protection*, *2*(3), 109–135.

Ghorbanpour, M., Omidvari, M., Abbaszadeh-Dahaji, P., et al. (2018). Mechanisms underlying the protective effects of beneficial fungi against plant diseases. *Biological Control*, *117*, 147–157.

Gopalakrishnan, S., Rao, G. V. R., Humayun, P., et al. (2011) Efficacy of botanical extracts and entomopathogens on control of *Helicoverpa armigera* and *Spodoptera litura*. *African Journal of Biotechnology*, *10*, 16667–16673.

Gosal, S. K., Kaur, J., & Kaur, J. (2020). Microbial biotechnology: A key to sustainable agriculture. In Kumar, M., Kumar, V., Prasad, R. (eds), *Phyto-Microbiome in Stress Regulation. Environmental and Microbial Biotechnology* (pp. 219–243). Springer, Singapore https://doi.org/10.1007/978-981-15-2576-6_11

Grzywacz, D. (2017). Basic and applied research: Baculovirus. In *Microbial Control of Insect and mite Pests* (pp. 27–46). Academic Press, Amsterdam. doi.org/10.1016/B978-0-12-803527-6.00003-2

Haidar, R., Amira, Y., Roudet, J., et al. (2021). Application methods and modes of action of *Pantoea agglomerans* and *Paenibacillus* sp. to control the grapevine trunk disease-pathogen, *Neofusicoccum parvum*. *OENO One*, *55*(3), 1–16. https://doi.org/10.20870/oeno-one.2021.55.2.4530

Hajek, A. E., Hurley, B. P., Kenis, M., et al. (2016) Exotic biological control agents: A solution or contribution to arthropod invasions? *Biological Invasions*, *18*, 953–969. https://doi.org/10.1007/s10530-016-1075-8

Hammami, R., Oueslati, M., Smiri, M., et al. (2022). Epiphytic yeasts and bacteria as candidate biocontrol agents of green and blue molds of citrus fruits. *Journal of Fungi*, *8*(8), 818.

Hasan, N., Farzand, A., Heng, Z., et al. (2020). Antagonistic potential of novel endophytic Bacillus strains and mediation of plant defense against verticillium wilt in upland cotton. *Plants*, *9*(11), 1438. https://doi.org/10.3390/plants9111438

Hashemi, M., Tabet, D., Sandroni, M., et al. (2022). The hunt for sustainable biocontrol of oomycete plant pathogens, a case study of *Phytophthora infestans*. *Fungal Biology Reviews*, *40*, 53–69.

Heimpel, G. E., & Mills, N. J. (2017). *Biological Control: Ecology and applications*. Cambridge University Press, England. https://doi.org/10.1017/9781139029117

Hinz, H. L., Winston, R. L., & Schwarzländer, M. (2020). A global review of target impact and direct nontarget effects of classical weed biological control. *Current Opinion in Insect Science*, *38*, 48–54. https://doi.org/10.1016/j.cois.2019.11.006

Hoitink, H. A. J., & Boehm, M. J. (1999). Biocontrol within the context of soil microbial communities: A substrate-dependent phenomenon. *Annual Review of Phytopathology*, *37*, 427–446.

Iorizzo, M., Testa, B., Ganassi, S., et al. (2021). Probiotic properties and potentiality of *Lactiplantibacillus plantarum* strains for the biological control of chalkbrood disease. *Journal of Fungi (Basel, Switzerland)*, *7*(5), 379. https://doi.org/10.3390/jof7050379

Islam, S., Akanda, A. M., Prova, A., et al. (2016). Isolation and identification of plant growth promoting rhizobacteria from cucumber rhizosphere and their effect on plant growth promotion and disease suppression. *Frontiers in Microbiology*, *6*, 1360. https://doi.org/10.3389/fmicb.2015.01360

Jain, D., Saharan, V., & Pareek, S. (2016). Current status of *Bacillus thuringiensis*: Insecticidal crystal proteins and transgenic crops. In Al-Khayri, J., Jain, S., Johnson, D. (eds), *Advances in Plant Breeding Strategies: Agronomic, Abiotic and Biotic Stress Traits*. Springer, Cham. https://doi.org/10.1007/978-3-319-22518-0_18

Jaiswal, D. K., Gawande, S. J., Soumia, P. S., et al. (2022). Biocontrol strategies: An eco-smart tool for integrated pest and diseases management. *BMC Microbiology*, *22*(1), 1–5.

Jeffries, C. L., & Walker, T. (2016). Wolbachia biocontrol strategies for arboviral diseases and the potential influence of resident *Wolbachia* strains in mosquitoes. *Current Tropical Medicine Reports*, *3*, 20–25. https://doi.org/10.1007/s40475-016-0066-2

Kalha, C. S., Singh, P. P., Kang, S. S., et al. (2014). Entomopathogenic Viruses and Bacteria for Insect-Pest Control. https://doi.org/10.1016/B978-0-12-398529-3.00013-0, https://api.semanticscholar.org/CorpusID:80824153.

Keswani, C., Singh, S., & Singh, H. (2013). *Beauveria bassiana*: Status, mode of action, applications and safety issues. *Biotech Today*, *3*, 16. https://doi.org/10.5958/j.2322-0996.3.1.002

Köhl, J., Booij, K., Kolnaar, R., et al. (2019). Ecological arguments to reconsider data requirements regarding the environmental fate of microbial biocontrol agents in the registration procedure in the European Union. *BioControl*, *64*, 469–487.

Koul, B., Chopra, M., & Lamba, S. (2022). Microorganisms as biocontrol agents for sustainable agriculture. In *Relationship Between Microbes and the Environment for Sustainable Ecosystem Services*, Volume 1 (pp. 45–68). Elsevier, Amsterdam. https://doi.org/10.1016/B978-0-323-89938-3.00003-7

Kumar, P., Kamle, M., Borah, R., et al. (2021). *Bacillus thuringiensis* as microbial biopesticide: Uses and application for sustainable agriculture. *Egyptian Journal of Biological Pest Control*, *31*, 95. https://doi.org/10.1186/s41938-021-00440-3

Kumar, V., Singh, A. U., & Jain, R. K. (2012). Comparative efficacy of bioagents as seed treatment for management of *Meloidogyne* incognita infecting okra. *Nematologia Mediterranea*, *40*, 209–211.

Lahlali, R., Ezrari, S., Radouane, N., et al. (2022). Biological control of plant pathogens: A global perspective. *Microorganisms, 10*(3), 596. http://dx.doi.org/10.3390/microorganisms10030596

Lanteigne, C., Gadkar, V. J., Wallon, T., et al. (2012) Production of DAPG and HCN by *Pseudomonas* sp. LBUM300 contributes to the biological control of bacterial canker of tomato. *Phytopathology, 102*(10), 967–973. https://doi.org/10.1094/PHYTO-11-11-0312

Lazazzara, V., Perazzolli, M., Pertot, I., et al. (2017). Growth media affect the volatilome and antimicrobial activity against *Phytophthora infestans* in four Lysobacter type strains. *Microbiological Research, 201*, 52–62. https://doi.org/10.1016/j.micres.2017.04.015

Legein, M., Smets, W., Vandenheuvel, D., et al. (2020). Modes of action of microbial biocontrol in the phyllosphere. *Frontiers in Microbiology, 11*, 1619. https://doi.org/10.3389/fmicb.2020.01619

Luo, L., Zhao, C., Wang, E., et al. (2022). *Bacillus amyloliquefaciens* as an excellent agent for biofertilizer and biocontrol in agriculture: An overview for its mechanisms. *Microbiological Research, 259*, 127016. https://doi.org/10.1016/j.micres.2022.127016

Lyu, D., Backer, R., Subramanian, S., et al. (2020). Phytomicrobiome coordination signals hold potential for climate change-resilient agriculture. *Frontiers in Plant Science, 11*, 634. https://doi.org/10.3389/fpls.2020.00634

Manoj, R. R. S., Latrofa, M. S., Epis, S., et al. (2021). Wolbachia: Endosymbiont of onchocercid nematodes and their vectors. *Parasites & Vectors, 14*, 245. https://doi.org/10.1186/s13071-021-04742-1

Mitchell, A. M., Strobel, G. A., Moore, E., et al. (2010). Volatile antimicrobials from *Muscodor crispans*, a novel endophytic fungus. *Microbiology (Reading, England), 156*(Pt 1), 270–277. https://doi.org/10.1099/mic.0.032540-0

Munene, L., Mugweru, J., & Mwirichia, R. (2023). Management of bacterial wilt caused by *Curtobacterium flaccumfaciens* pv. flaccumfaciens in common bean (*Phaseolus vulgaris*) using rhizobacterial biocontrol agents. *Letters in Applied Microbiology, 76*(1), ovac011. https://doi.org/10.1093/lambio/ovac011

Nandini, B., Puttaswamy, H., Saini, R. K., et al. (2021). Trichovariability in rhizosphere soil samples and their biocontrol potential against downy mildew pathogen in pearl millet. *Scientific Reports, 11*, 9517. https://doi.org/10.1038/s41598-021-89061-2

Ngalimat, M. S., Yahaya, R. S. R., Baharudin, M. M. A. A., et al. (2021). A review on the biotechnological applications of the operational group *Bacillus amyloliquefaciens*. *Microorganisms, 9*(3), 614. https://doi.org/10.3390/microorganisms9030614

Niu, B., Wang, W., Yuan, Z., et al. (2020). Microbial interactions within multiple-strain biological control agents impact soil-borne plant disease. *Frontiers in Microbiology, 11*, 585404. https://doi.org/10.3389/fmicb.2020.585404

Palmieri, D., Ianiri, G., Del Grosso, C., et al. (2022). Advances and perspectives in the use of biocontrol agents against fungal plant diseases. *Horticulturae, 8*(7), 577. https://doi.org/10.3390/horticulturae8070577

Pedro M. de G., Fabio, R. B., & Jackson, V. de A. (2022). Nematophagous fungi, an extraordinary tool for controlling ruminant parasitic nematodes and other biotechnological applications. *Biocontrol Science and Technology, 32*(7), 777–793. https://doi.org/10.1080/09583157.2022.2028725

Popham, H. J., Nusawardani, T., & Bonning, B. C. (2016). Introduction to the use of baculoviruses as biological insecticides. *Methods in Molecular Biology (Clifton, N.J.), 1350*, 383–392. https://doi.org/10.1007/978-1-4939-3043-2_19

Poveda, J. (2021). Trichoderma as biocontrol agent against pests: New uses for a mycoparasite. *Biological Control, 159*, 104634. https://doi.org/10.1016/j.biocontrol.2021.104634

Prasad, D., Singh, R. P., & Singh, R. P. (2023). Trichoderma: Mode of action and application methods for crop disease management. *Biotica Research Today, 5*(2), 166–171.

Priya, M. S., & Kumar, S. (2006). Dose optimization of *Paecilomyces lilacinus* for the control of Meloidogyne incognita on tomato. *Indian Journal of Nematology*, *36*, 27–31.

Raaijmakers, J. M., & Mazzola, M. (2012). Diversity and natural functions of antibiotics produced by beneficial and plant pathogenic bacteria. *Annual Review of Phytopathology*, *50*, 403–424.

Rabbee, M. F., Islam, N., & Baek, K. H. (2022). Biocontrol of citrus bacterial canker caused by *Xanthomonas citri* subsp. *citri* by *Bacillus velezensis*. *Saudi Journal of Biological Sciences*, *29*(4), 2363–2371. https://doi.org/10.1016/j.sjbs.2021.12.005

Rezzonico, F., Smits, T. H., Montesinos, E., et al. (2009). Genotypic comparison of *Pantoea agglomerans* plant and clinical strains. *BMC Microbiology*, *9*, 204. https://doi.org/10.1186/1471-2180-9-204

Rueda-Páramo, M. E., Montalva, C., Arruda, W., et al. (2017). First report of *Coelomomyces santabrancae* sp. nov. (Blastocladiomycetes: Blastocladiales) infecting mosquito larvae (Diptera: Culicidae) in central Brazil. *Journal of Invertebrate Pathology*, *149*, 114–118. https://doi.org/10.1016/j.jip.2017.08.010

Sabbahi, R., Hock, V., Azzaoui, K., et al. (2022). A global perspective of entomopathogens as microbial biocontrol agents of insect pests. *Journal of Agriculture and Food Research*, *10*, 100376.

Saberi-Riseh, R., Moradi-Pour, M., Mohammadinejad, R., et al. (2021). Biopolymers for biological control of plant pathogens: Advances in microencapsulation of beneficial microorganisms. *Polymers*, *13*(12), 1938. https://doi.org/10.3390/polym13121938

Sandhu, S., Sharma, A., Beniwal, V., et al. Myco-biocontrol of insect pests: Factors involved, mechanism, and regulation. *Journal of Pathogens*, *2012*, 126819. https://doi.org/10.1155/2012/126819

Scholte, E. J., Knols, B. G., Samson, R. A., et al. (2004). Entomopathogenic fungi for mosquito control: A review. *Journal of Insect Science*, *4*(1), 19. https://doi.org/10.1673/031.004.1901

Sellitto, V. M., Zara, S., Fracchetti, F., et al. (2021). Microbial biocontrol as an alternative to synthetic fungicides: Boundaries between pre-and postharvest applications on vegetables and fruits. *Fermentation*, *7*(2), 60.

Singh, H. B, Keswani, C, Ray, S, et al. (2015). *Beauveria bassiana*: Biocontrol beyond lepidopteran pests. *Biocontrol of Lepidopteran Pests*, *43*, 219–235. https://doi.org/10.1007/978-3-319-14499-3_10

Sujayanand, G. K., Akram, M., Konda, A., et al. (2021). Distribution and toxicity of *Bacillus thuringiensis* (Berliner) strains from different crop rhizosphere in Indo-Gangetic plains against polyphagous lepidopteran pests. *International Journal of Tropical Insect Science*, *41*, 2713–2731. https://doi.org/10.1007/s42690-021-00451-5

Sundh, I., & Eilenberg, J. (2021). Why has the authorization of microbial biological control agents been slower in the EU than in comparable jurisdictions? *Pest Management Science*, *77*(5), 2170–2178.

Tariq, M., Khan, A., Asif, M., et al. (2020). Biological control: A sustainable and practical approach for plant disease management. *Acta Agriculturae Scandinavica, Section B: Soil & Plant Science*, *70*(6), 507–524. https://doi.org/10.1080/09064710.2020.1784262

Tewari, S., Shrivas, V. L., Hariprasad, P., et al. (2019). Harnessing endophytes as biocontrol agents. In Ansari, R., Mahmood, I. (eds), *Plant Health Under Biotic Stress: Volume 2: Microbial Interactions* (pp. 189–218). Springer, Singapore. https://doi.org/10.1007/978-981-13-6040-4_10

Tomprefa, N., McQuilken, M. P., Hill, R. A., et al. (2009). Antimicrobial activity of *Coniothyrium minitans* and its macrolide antibiotic macrosphelide A. *Journal of Applied Microbiology*, *106*(6), 2048–2056. https://doi.org/10.1111/j.1365-2672.2009.04174.x

Tyśkiewicz, R., Nowak, A., Ozimek, E., et al. (2022). *Trichoderma*: The current status of its application in agriculture for the biocontrol of fungal phytopathogens and stimulation of plant growth. *International Journal of Molecular Sciences*, *23*(4), 2329. https://doi.org/10.3390/ijms23042329

van Lenteren, J. C., Bolckmans, K., Köhl, J., et al. (2018). Biological control using invertebrates and microorganisms: Plenty of new opportunities. *BioControl*, *63*, 39–59.

Villaverde, J. J., Sevilla-M, B., Sandín-E, P., et al. (2014). Biopesticides in the framework of the European pesticide regulation (EC) no. 1107/2009. *Pest Management Science*, *70*(1), 2–5. https://doi.org/10.1002/ps.3663

Vlassi, A., Nesler, A., Perazzolli, M., et al. (2020). Volatile organic compounds from *Lysobacter capsici* AZ78 as potential candidates for biological control of soil-borne plant pathogens. *Frontiers in Microbiology*, *11*, 1748. https://doi.org/10.3389/fmicb.2020.01748

Winding, A. Binnerup, S. J., & Pritchard, H. (2004). Non-target effects of bacterial biological control agents suppressing root pathogenic fungi. *FEMS Microbiology Ecology*, *47*(2), 129–141. https://doi.org/10.1016/S0168-6496(03)00261-7

Xu, X., & Jeger, M. (2020). More ecological research needed for effective biocontrol of plant pathogens. In De Cal, A., Melgarejo, P., Magan, N. (eds), *How Research Can Stimulate the Development of Commercial Biological Control Against Plant Diseases, Progress in Biological Control*, Volume 21 (pp. 15–30). Springer, Cham. https://doi.org/10.1007/978-3-030-53238-3_2.

Zamir, K., Punja, G. R., & Ananchanok, T. (2016) Effects of *Bacillus subtilis* strain QST 713 and storage temperatures on post-harvest disease development on greenhouse tomatoes. *Crop Protection*, *84*, 98–104. https://doi.org/10.1016/j.cropro.2016.02.011

Zehra, A., Raytekar, N. A., Meena, M., et al. (2021). Efficiency of microbial bio-agents as elicitors in plant defense mechanism under biotic stress: A review. *Current Research in Microbial Sciences*, *2*, 100054.

Zhao, Y., Chen, X., Cheng, J., et al. (2022). Application of *Trichoderma* Hz36 and Hk37 as biocontrol agents against clubroot caused by *Plasmodiophora brassicae*. *Journal of Fungi (Basel, Switzerland)*, *8*(8), 777. https://doi.org/10.3390/jof8080777

11 Potential of Extremophilic and Native Microbial Consortia in the Bioleaching of Heavy Metals on Mining Process

Edwin Hualpa-Cutipa, Andi Solórzano Acosta, Jorge Johnny Huayllacayan Mallqui, Heidy Mishey Aguirre Catalan, Andrea León Chacón, and Lucero Katherine Castro Tena

11.1 INTRODUCTION

Minerals and metals are important for the economic and social development of many countries, as well as for modern life (Carvalho, 2017). Therefore, the importance of the mining sector has been highlighted from an economic perspective for many years; however, in current circumstances, environmental aspects are of great relevance (Mancini and Sala, 2018). Perhaps the most significant environmental problem in mining is water pollution, which is considered a universal issue (Orandi, 2017). Meanwhile, among the waste materials posing the greatest risks to human health and ecosystem balance are heavy metals, due to their toxicity at low concentrations as they are both bioaccumulative and non-biodegradable (Rehman et al., 2021). Metallic substances have a significant atomic weight and can be toxic even at low concentrations (Al Osman et al., 2019). Metals are not able to break, so they may remain in the environment for long periods, causing serious environmental and health issues (Jan et al., 2015).

In mining, metals can be extracted in an aqueous phase from solid matrices through leaching, which allows for the availability of the metal for recovery (Mwewa et al., 2022). In recent years, the increase in reserves of minerals with refractory characteristics has required the use of unconventional processes such as bioleaching for the recovery of precious metals. Currently, facing environmental pollution, the development of alternative technologies that aim to mitigate pollution and increase metal recovery due to competitive operational costs is a challenge (Larrabure et al., 2021). The process of bioleaching encompasses the utilization of living creatures, predominantly microbes,

DOI: 10.1201/9781003407683-11

for the purpose of extracting or mobilizing metals ex situ. These metals are often found in low concentrations or are considered challenging to manage or process in commercially feasible rocks, soils, or solid wastes (Cornu et al., 2017).

In this particular case, the bioleaching of heavy metals is an interesting topic to address because it implies not only the traditional concept of recovering precious metals such as gold without resorting to elements like mercury but also involves the bioremediation of resulting from the extraction of these minerals. Consequently, they can be disposed of with lower concentrations of these toxic elements, resulting in a reduced likelihood of contaminating events (Yang et al., 2020). Currently, the utilization of bacterial consortia in bioleaching is widely acknowledged as an environmentally sustainable alternative to conventional mining methods. This is primarily attributed to its economic viability, practicality, and capacity for long-term use (Quach et al., 2022). The resistance of microorganisms to metal toxicity, coupled with their significant contributions to the resource cycle and energy dynamics within ecosystems, further underscores their importance in this context (Sharma et al., 2021). However, the bioleaching of metals, in general, becomes less efficient under extreme conditions and requires organisms not only to adapt to adverse weather conditions but also capable of withstanding the presence of heavy metals (Peng et al., 2019). Extremophiles, a class of microorganisms, have emerged as powerful tools in the field of biotechnology (Chiacchiarini et al., 2016).

Extremophilic and native microorganisms are attractive for this purpose because they possess mechanisms of multidrug resistance that make them versatile. Particularly, there is interesting evidence regarding acidophilic microorganisms and the fact that microbial diversity in the presence of heavy metals is not affected in native consortia that exhibit broad ranges of adaptation and versatility (Massello and Donati, 2021). The acidophilic communities indicated above exhibit robust growth in low pH conditions, often ranging from pH 2.0 to 4.0. These populations play a crucial role in facilitating the extraction of metals from solid waste phases, promoting their transition into the liquid phase. Bioleaching bacteria, such as *Acidithiobacillus ferrooxidans*, *A. thiooxidans*, *Leptospirillum ferrooxidans*, and *Sulfolobus* sp., have been recognized by bioleaching consortia. Additionally, fungal species such as *Penicillium* sp. and *Aspergillus niger* contribute to the mineral release process (Mishra and Rhee, 2010).

From an environmental perspective and knowing that microorganisms have a great capacity for the biological removal of environmental pollutants, bioleaching of heavy metals becomes an acceptable strategy for detoxifying mining residues (Begum et al., 2022). Heavy metal remediation with microbes has proven as a viable and cost-effective strategy; therefore, the objective of this chapter is to review the potential use of extremophilic microbial consortia in the bioleaching of heavy metals.

11.2 OLD AND NEW PHYSICOCHEMICAL RECOVERY STRATEGIES OF HEAVY METALS IN MINING PROCESS

11.2.1 Hydrometallurgy

Metals can be recovered by chemical reactions conducted in aqueous or organic solutions within this procedure. The procedure typically involves three main steps:

leaching, concentration/purification, and recovery (Nakhjiri et al., 2022). The basic hydrometallurgical process includes the application of various reagents in the acid leaching procedure. These reagents include sulfuric acid (H_2SO_4), nitric acid (HNO_3), hydrochloric acid (HCl), hydrogen peroxide (H_2O_2), thiourea (CH_4N_2S), ferric chloride ($FeCl_2$), aqua regia (HNO_3 + 3 HCl), potassium isocyanate (KOCN), potassium iodide (KI), iodine (I), a mixture of nitrite iodide, thiosulfate (S_2O_2), and cyanides (CN^-). The concentration and purification steps of the process include the use of many extraction techniques, including cementation, solvent extraction, filtering, precipitation, ion exchange, and distillation, in order to recover and concentrate the desired metals. The last stage in a hydrometallurgical cycle involves the retrieval of metals, which may be accomplished by several methods such as electrolysis, gas reduction, and precipitation (Krishnan et al., 2021).

11.2.2 Electrometallurgy

According Evans (2014), electrometallurgy is a method employed in the mining sector to recover metals by means of electrochemical procedures. Metals are recovered from minerals by subjecting an aqueous solution containing metal ions to an electric current. Also, urban mines have the capability to do direct electrochemical recovery in a single, uncomplicated process (Krishnan et al., 2021).

11.2.3 Leaching

Leaching is a physicochemical technique used in the mining industry for the recovery of heavy metals from ores and concentrates. In this process, metals are extracted from the ore by dissolving them in an aqueous solution. Chemical reagents, such as acids, bases, or complex salts, are used to dissolve the metals. The choice of reagent depends on the nature of the ore and the metal to be recovered. The dissolved metals can be recovered through precipitation processes, electrolysis, or solvent extraction (Kiprono et al., 2023).

11.3 BIOLOGICAL STRATEGIES FOR RECOVERY OF HEAVY METALS FROM MINING WASTEWATER

There are biological methods that employ techniques for the removal of heavy metals in the environment, known as absorption (Kumar et al., 2021). This phenomenon refers to a physicochemical process wherein living tissues have the ability to chemically interact with pollutants, hence facilitating their subsequent removal. According to Razzak et al. (2022), biosorption is the process by which various substances, including microbes, plant-derived compounds, waste products, and biopolymers, bind and accumulate ions through reversible interactions. It occurs through the binding of ions to functional groups located on the exterior surface of the biosorbent. By connecting ions to functional groups on the biosorbent's outside, contaminants are taken out of a solution.

11.4 BIOMINERALIZATION EXPERIENCES USING EXTREMOPHILIC MICROORGANISMS

The first microbes employed in biomineralization were mainly mesophiles or moderate thermophiles, but after sequencing using the 16S rRNA gene, thermophilic archaea were identified. These extremophilic bacteria belonged to the group of acidophiles with bioleaching capacity, belonging to the *Acidithiobacillus* genus, previously classified as *Thiobacillus*. These species were the first isolates fulfilling the role of sulfur or iron oxidation. Among them, mesophilic *A. thiooxidans* and *A. ferrooxidans* were identified, along with a moderate thermophile, *A. caldus* (Chiacchiarini et al., 2016). The utilization of acidophilic microorganisms with the capacity for iron oxidation is advantageous in the context of generating and recycling reagents employed as lixiviants. This is primarily attributed to their capability to effectively mitigate the surplus presence of iron, sulfate, and other impurities within hydrometallurgical solvents (Yaashikaa et al., 2022).

According to a study conducted by Gumulya et al. in 2018, *A. ferrooxidans* ATCC 23270 demonstrates certain responses when subjected to acid stress. These responses include the expulsion of Na^+ ions, retention of K^+ ions, and the regulation of gene expression associated with cytoplasmic pH buffering. Specifically, the genes involved in carbonic anhydrase and polyamine anabolism are found to be regulated under acid stress conditions. Also, according to Peng et al. (2017), genomic investigations of *A. ferrivorans* have revealed the presence of certain genes that have evolved to function well in environments with low and mesophilic temperatures (6 and 28°C) and the pH kept decreasing. These genes include rusA, hdrA, cyoC1, doxDA, and cycA1. The study found that rusA is crucial in the iron-oxidation pathway, with iron oxidation being more prominent in the mid-log phase. However, expression of hdrA and cyoC1 increased in the stationary phase, indicating cells mainly use sulfur as energy. The gene doxDA, responsible for thiosulfate quinone oxidoreductase, was low due to chalcopyrite oxidation via the polysulfide pathway.

The most commonly employed chemolithoautotrophic archaea in biomineralization include *Acidianus brierleyi*, *Metallosphaera sedula*, and *Sulfolobus metallicus*. These species are capable of withstanding high concentrations of metals during the processing of mineral sulfides (Martínez-Bussenius et al., 2017). Research has proposed these extremophiles as alternative chassis to produce chemicals and fuels based on renewable resources (Kernan et al., 2016).

11.5 BIOLEACHING OF HEAVY METALS BY NATIVE EXTREMOPHILE MICROBIAL CONSORTIA

11.5.1 Potential of Bioleaching

The bioleaching processes method was first used in metal recuperation, where microorganisms were used for oxidizing reduced sulfur and iron components (Sajjad et al., 2019). In recent years, bioleaching has arisen as a bioremediation technique utilized for the purpose of reducing heavy metal pollution present in hazardous industrial wastes, such as sludge derived from the treatment of mining wastewater (Mani and Kumar, 2014). The recovered minerals could serve as secondary raw materials, thereby reducing the demand for primary metal mining and helping to improve environmental quality (Igogo et al., 2021).

Heavy metals are mostly soluble in water, which makes their separation challenging through physical separation methods (Gunatilake, 2015). Various techniques are employed for the removal of heavy metals, including chemical precipitation, oxidation or reduction, filtration, ion exchange, reverse osmosis, membrane filtering technology, evaporation, and electrochemical approaches (Pabón et al., 2020). Nevertheless, the efficacy of these approaches may diminish when the amounts of heavy metals fall below the threshold of 100 mg/L (Yenial and Bulut, 2017). Bioleaching has emerged as a viable and effective alternative to conventional techniques such as chemical extraction or thermal stabilization for addressing heavy metal-contaminated settings (Okoh et al., 2018). The utilization of physicochemical techniques at low concentrations of heavy metals is characterized by high costs and inefficiency (Arora and Khosla, 2021).

The main benefit of effectively using bioleaching with the most recent technology available is the minimal environmental damage it causes (Krishnan et al., 2021). As a result, it has been shown to minimize the release of greenhouse gases in the environment. The microbes employed may be critical for organic material remineralization, reusing of living biomass, contributing to the overall redox state of our planet's surface, and permanent biological engineering of the natural system (Wang et al., 2021).

In general, compared to mineral smelting, the bioleaching mechanism releases fewer toxic substances into the environment and damages the land less. Furthermore, it will provide a cost-effective tool for recovering high-value metals in mineral extraction (Gopikrishnan et al., 2020). As a result, the industry has a need to take advantage of the chances given by bioleaching, as the method incorporates multiple fields including geology, microbiology, molecular biology, and biochemistry, which could be used in the future.

11.5.2 Mining Environments as a Source of Extreme Microbial Consortia Isolation

The extraction conditions are distinguished by a diversity of microbes designed for remaining under conditions of high metal concentrations, extreme temperatures, and elevated levels of acidic conditions (Dopson and Holmes, 2014). In these locations, initial research succeeded in isolating and identifying pure cultures such as *T. ferrooxidans*, the main bacteria involved in bioleaching processes (Auerbach et al., 2019). However, it is not the only microorganism present in bioleaching ecosystems, and numerous studies have reported environments and residues associated with mining activities as habitats created by humans for metal recovery operations or as deposits of mineral waste where these species are found, such as yeasts, amoebas, and other types of bacteria in mine drainage waters, in addition to *Thiobacillus* bacteria (Hedrich and Schippers, 2021).

Other species such as *Leptospirillum sp.* along with different moderate thermophilic and extreme thermophilic microorganisms, such as bacteria of the *Sulfolobus* genus, have shown evidence of mineral interaction with a mixed population of microorganisms. In addition to autotrophic microorganisms, other heterotrophic microorganisms, as well as algae, fungi, and protozoa with broad ranges of tolerance to heavy metals, temperatures, and acidic conditions, are involved (Johnson and Aguilera, 2019).

11.5.3 Metabolic Capacity of Extremophile Consortia

Heavy metals, as well as other contaminants, exert significant selection pressure on microbial communities, and their presence in intensely exposed environments can shape highly specialized communities. For example, extremophilic species have the ability to oxidize iron and reduce inorganic sulfur compounds, which are used as sources of energy in different kinds of metabolism. These microorganisms play a catalytic role in the bioleaching process, which is essential for metals recovery (Gao et al., 2021).

Currently, there have been documented a minimum of four mechanisms of resistance to heavy metals, namely, metal ion active transport (efflux), metal ion enzymatic reduction, extracellular barrier, and intracellular sequestration. Certain bacterial species have the potential to form complexes or chelates with extracellular polymers, resulting in a reduction in the permeability of metals (Gallo et al., 2021).

11.5.4 Heavy Metals and Bioleaching

In fact, several definitions of "heavy metal" refer to atomic weight, and a heavy metal is defined as an element of chemistry with an atomic weight ranging from 63.55 (Cu) to 200.59 (Hg). Another classification is based on atomic number and refers to metals with densities ranging from 4 to 7 g/cm^3 (Al-Attar et al., 2022).

Not all high-density metals are particularly toxic at normal concentrations; in fact, some of them are necessary for human beings (Sonone et al., 2020). Nevertheless, there exists a multitude of heavy metals that are recognized for their capacity to cause significant environmental challenges, such as mercury (Hg), lead (Pb), cadmium (Cd), thallium (Tll), as well as copper (Cu), zinc (Zn), and chromium (Cr). Sometimes, when discussing heavy metal pollution, other toxic light elements such as beryllium (Be) or aluminum (Al), or semimetals like arsenic (As), are included (Vardhan et al., 2019).

In the case of mercury, this metal continues to receive attention from various disciplines due to its high mobility and toxicity to human health and the environment. Solid waste sources containing mercury can come from non-ferrous metallurgy, mercury extraction (Xie et al., 2020), and particularly gold mining. The use of mercury as a method for gold recovery in artisanal mining involves three significant drawbacks: (1) low gold recovery when it is covered with pyrite; (2) the environmental impact it generates; and (3) its harmful effects on human beings (Fashola et al.,

2016). To date, the feasibility of mercury bioleaching from cinnabar has been examined based on temperature, initial pH, and dilution rate. Wang et al. (2013) were able to extract 1.38 g/L of mercury from 5.0 g of cinnabar in the presence of 100 mL of culture solution. Apparently, mercury bioleaching is not as effective due to the slow leaching kinetics (Xie et al., 2020).

It is important to remember that heavy metals are combined in the mineral matrix, which is why consortia are important because metabolic diversity is required. Ghassa et al. (2014), for example, achieved 98.5% zinc recovery from high-grade Zn-Pb ore after 25 days of treatment (pulp density 50 g/L, initial pH 1, and $FeSO_4 \cdot 7H_2O$ concentration of 75 g/L) employing a combination culture of thermophilic iron and sulfur oxidizing microbial species. However, due to the lower solubility of Pb, the lead dissolution was only 0.027% under the same conditions, and they also recovered cadmium (98%).

Bioleaching efficiency has also been tested with fungi and has been optimized using fractionation techniques employing *Aspergillus niger.* Zeng et al. (2015) achieved the highest extraction efficiency of Pb (65.4%) in sediments through two-step fractionation, as well as for Cd (99.5%). Interestingly, the remaining metals in the sediment were mainly found in stable fractions, and the amounts were reduced well below the levels in two Chinese standards, with lower toxicity observed in wheat and earthworms.

According Lee et al. (2015), to date, multiple studies have been reported on the bioleaching of heavy metals in mine tailings, with *Acidithiobacillus* spp. being the most commonly used microorganism. The use of pure cultures or consortia is also addressed. For example, *Acidithiobacillus* spp. and *Leptospirillum* spp. have been employed for chalcopyrite and *A. thiooxidans* and *A. ferrooxidans* for arsenic (Lee et al., 2015).

In general, bioleaching studies so far have focused on the treatment of sediments, mine tailings, slag, and even more broadly on the treatment and recovery of metals in electronic waste (Wheaton et al., 2015; Hu et al., 2022). Regarding the microorganisms employed, acidophilic sulfur-oxidizing bacteria, such as *A. ferrooxidans* and *A. thiooxidans*, stand out as model organisms due to their ability to transform toxic metal sulfides into less toxic sulfates. Oligoelements undergo decomposition by sulfur-eliminating fungi, such as *Aspergillus* sp., which possess exceptional proficiency in bioleaching and the extraction of Fe, Sn, and Au. The extraction of copper, gold, iron, manganese, and lead from contaminated settings commonly involves the use of sulfur-oxidizing microorganisms to break down and reduce the metals, as a preparatory step in the recovery process of dangerous metal ions (Chaudhary and Goyal, 2019).

11.6 BIOLIXIVIATION MECHANISMS

The mechanisms of bacteria and fungi on minerals are based on their ability to metabolize metals using three principles: acidolysis, complexolysis, and redoxolysis (Okoh et al., 2018) (Figure 11.1).

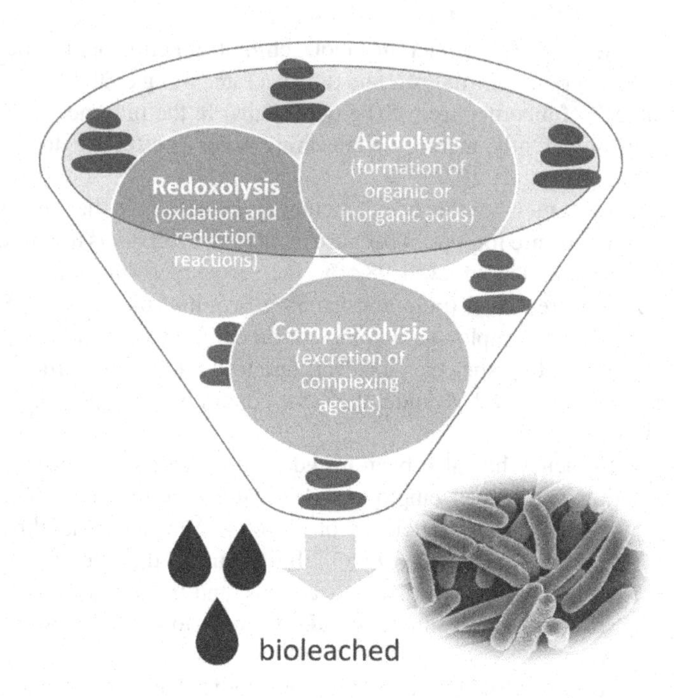

FIGURE 11.1 Some mechanics of bioleaching.

Figure 11.1 illustrates the primary mechanisms and processes associated with the bioleaching process facilitated by microorganisms.

11.6.1 Use of Consortia

Regarding the use of selected microorganism species, the use of consortia has proven to be more efficient, and all investigations so far have concentrated on the application of consortium leaching (Latorre et al., 2016). When single-species bacteria and consortia cultures were compared, the results revealed that consortium bacteria leached more copper and arsenic than single bacteria and that each bacterium and mixed bacteria mediated an order of heavy metal appearance. Due to the synergistic effect of consortia, higher extraction rates can be achieved in biological leaching (Sajjad et al., 2020).

11.6.2 Factors Influencing Bioleaching

As in any biotechnological process involving microorganisms, factors such as dissolved oxygen and carbon dioxide, as well as pH and particle size, are important for process efficiency. Other variables affecting heavy metal biological leaching, especially from tailings, have been reported; thus, operation-related conditions such as elevated pressure and salinity deserve to be considered to achieve a greater level of heavy metal bioleaching (Gao et al., 2021) (Figure 11.2).

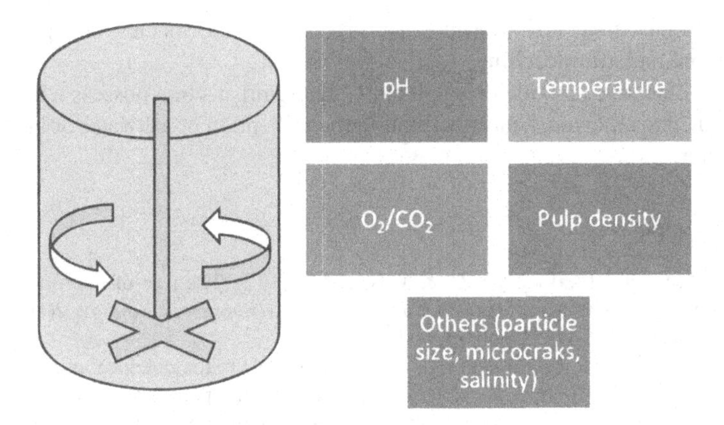

FIGURE 11.2 Factors affecting bioleaching.

11.7 FUTURE OUTLOOKS

11.7.1 Elucidation of Mechanisms for the Biolixiviation of Heavy Metals

Because of its ability to recover valuable metals, biolixiviation has been widely used, but its cleaning mechanisms require further investigation (Okoh et al., 2018). As the demand for less expensive and more environmentally friendly metal solubilization methods grows, biolixiviation will become even more important (Roy et al., 2021).

11.7.2 Genetic Improvement of Biolixiviator Microorganisms

There is a need for enhancement in the technological and biological components pertaining to the capabilities of microorganisms (Abatenh et al., 2017). The utilization of nanofiber technology has demonstrated noteworthy enhancements in the efficiency of heavy metal oxygen reduction reaction (ORR) and oxygen evolution reaction (OER) processes (Song et al., 2022). The significance of conducting research in this field is paramount due to its ability to further the scientific understanding of biolixiviation, therefore enabling its use in mining operations. This would entail transferring the risk associated with mining activities to microorganisms, thereby creating a safer working environment for human laborers (Okoh et al., 2018). Conducting research aimed at identifying microorganisms that are better adapted to high temperature environments is vital, as it has the potential to significantly enhance the efficacy of biolixiviation processes (Mahajan et al., 2017).

11.7.3 Criteria for the Selection of Biolixiviator Microorganisms

In coming years, the choice of microorganisms for biolixiviation could potentially be improved in several ways (Gao et al., 2021):

a. Conduct a search for microorganisms that have the ability to undergo oxidation processes inside alkaline conditions.

b. Examine microbial morphology, processes, and species changes both before and after leaching.
c. The objective is to identify and cultivate organisms that possess a high tolerance to chlorine, hence facilitating the utilization of saltwater or brackish water.

REFERENCES

Abatenh, E., Gizaw, B., Tsegaye, Z., & Wassie, M. (2017). The role of microorganisms in bioremediation – A review. *Open Journal of Environmental Biology*, *2*(1), 038–046. https://doi.org/10.17352/ojeb.000007

Al-Attar, A., Zeid, I. A., & Felemban, L. (2022). An update to toxicological profiles of heavy metals, especially lead as hazardous environmental pollutant: A review. *Current Science International*, *11*(01), Article 01. https://www.curresweb.com/index.php/CSI1/article/view/36

Al Osman, M., Yang, F., & Massey, I. Y. (2019). Exposure routes and health effects of heavy metals on children. *Biometals*, *32*, 563–573.

Arora, V., & Khosla, B. (2021). Conventional and contemporary techniques for removal of heavy metals from soil. In *Biodegradation Technology of Organic and Inorganic Pollutants*. IntechOpen.

Auerbach, R., Bokelmann, K., Stauber, R., Gutfleisch, O., Schnell, S., & Ratering, S. (2019). Critical raw materials–Advanced recycling technologies and processes: Recycling of rare earth metals out of end of life magnets by bioleaching with various bacteria as an example of an intelligent recycling strategy. *Minerals Engineering*, *134*, 104–117.

Begum, S., Rath, S. K., & Rath, C. C. (2022). Applications of microbial communities for the remediation of industrial and mining toxic metal waste: A review. *Geomicrobiology Journal*, *39*(3–5), 282–293. https://doi.org/10.1080/01490451.2021.1991054

Carvalho, F. P. (2017). Mining industry and sustainable development: Time for change. *Food and Energy Security*, *6*(2), 61–77.

Chaudhary, S., & Goyal, S. (2019). Sulphur oxidizing fungus: A review. *Journal of Pharmacognosy and Phytochemistry*, *8*(6), 40–43. https://www.phytojournal.com/archives/2019.v8.i6.9986/sulphur-oxidizing-fungus-a-review

Chiacchiarini, P., Lavalle, L., Urbieta, M. S., Ulloa, R., Donati, E., & Giaveno, A. (2016). Springer International Publishing, Cham, pp. 185–204. https://doi.org/10.1007/978-3-319-42801-7_12. Extremophilic Patagonian microorganisms working in biomining. In *Biology and Biotechnology of Patagonian Microorganisms* (pp. 185–204).

Cornu, J. Y., Huguenot, D., Jézéquel, K., Lollier, M., & Lebeau, T. (2017). Bioremediation of copper-contaminated soils by bacteria. *World Journal of Microbiology and Biotechnology*, *33*, 1–9.

Dopson, M., & Holmes, D. S. (2014). Metal resistance in acidophilic microorganisms and its significance for biotechnologies. *Applied Microbiology and Biotechnology*, *98*(19), 8133–8144. https://doi.org/10.1007/s00253-014-5982-2

Evans, J. (2014). Introduction and the significance of electrometallurgy. In *Reference Module in Materials Science and Materials Engineering*. https://doi.org/10.1016/b978-0-12-803581-8.03594-3

Fashola, M. O., Ngole-Jeme, V. M., & Babalola, O. O. (2016). Heavy metal pollution from gold mines: Environmental effects and bacterial strategies for resistance. *International Journal of Environmental Research and Public Health*, *13*(11), 1047.

Gallo, G., Puopolo, R., Carbonaro, M., Maresca, E., & Fiorentino, G. (2021). Extremophiles, a nifty tool to face environmental pollution: From exploitation of metabolism to genome engineering. *International Journal of Environmental Research and Public Health*, *18*(10), 5228.

Gao, X., Jiang, L., Mao, Y., Yao, B., & Jiang, P. (2021). Progress, challenges, and perspectives of bioleaching for recovering heavy metals from mine tailings. *Adsorption Science & Technology, 2021*, 1–13.

Ghassa, S., Boruomand, Z., Abdollahi, H., Moradian, M., & Akcil, A. (2014). Bioleaching of high grade Zn–Pb bearing ore by mixed moderate thermophilic microorganisms. *Separation and Purification Technology, 136*, 241–249.

Gopikrishnan, V., Vignesh, A., Radhakrishnan, M., Joseph, J., Shanmugasundaram, T., Doble, M., & Balagurunathan, R. (2020). Microbial leaching of heavy metals from e-waste: Opportunities and challenges. In Krishnaraj Rathinam, N. & Sani, R. K. (eds), *Biovalorisation of Wastes to Renewable Chemicals and Biofuels* (pp. 189–216). Elsevier. https://doi.org/10.1016/B978-0-12-817951-2.00010-9

Gumulya, Y., Boxall, N. J., Khaleque, H. N., Santala, V., Carlson, R. P., & Kaksonen, A. H. (2018). In a quest for engineering acidophiles for biomining applications: Challenges and opportunities. *Genes, 9*(2), Article 2. https://doi.org/10.3390/genes9020116

Gunatilake, S. K. (2015). Methods of removing heavy metals from industrial wastewater. *Methods, 1*(1), 14.

Hedrich, S., & Schippers, A. (2021). Distribution of acidophilic microorganisms in natural and man-made acidic environments. *Current Issues in Molecular Biology, 40*(1), 25–48.

Hu, X., Wu, C., Shi, H., Xu, W., Hu, B., & Lou, L. (2022). Potential threat of antibiotics resistance genes in bioleaching of heavy metals from sediment. *Science of The Total Environment, 814*, 152750.

Igogo, T., Awuah-Offei, K., Newman, A., Lowder, T., & Engel-Cox, J. (2021). Integrating renewable energy into mining operations: Opportunities, challenges, and enabling approaches. *Applied Energy, 300*, 117375.

Jan, A. T., Azam, M., Siddiqui, K., Ali, A., Choi, I., & Haq, Q. M. R. (2015). Heavy metals and human health: Mechanistic insight into toxicity and counter defense system of antioxidants. *International Journal of Molecular Sciences, 16*(12), 29592–29630.

Johnson, D. B., & Aguilera, A. (2019). Extremophiles and acidic environments. In Schmidt, T. M. (ed), *Encyclopedia of Microbiology* (Fourth Edition) (pp. 206–227). Academic Press. https://doi.org/10.1016/B978-0-12-809633-8.90687-3

Kernan, T., Majumdar, S., Li, X., Guan, J., West, A.C., & Banta, S. (2016). Engineering the iron-oxidizing chemolithoautotroph *Acidithiobacillus ferrooxidans* for biochemical production. *Biotechnology and Bioengineering, 113*(1), 189–197. https://doi.org/10.1002/bit.25703

Kiprono, N. R., Smolinski, T., Rogowski, M., & Chmielewski, A.G. (2023). The state of critical and strategic metals recovery and the role of nuclear techniques in the separation technologies development: Review. *Separations, 10*(2), 112. https://doi.org/10.3390/separations10020112

Krishnan, S., Zulkapli, N. S., Kamyab, H., et al. (2021). Current technologies for recovery of metals from industrial wastes: An overview. *Environmental Technology & Innovation, 22*, 101525.

Krishnan, S., Zulkapli, N. S., Din, M. F. B. M., Majid, Z. A., Nasrullah, M., & Sairan, F. M. (2023). Photocatalytic degradation of methylene blue dye and fungi *Fusarium equiseti* using titanium dioxide recovered from drinking water treatment sludge. *Biomass Conversion and Biorefinery, 13*(12), 10853–10863. https://doi.org/10.1007/s13399-021-01990-0

Kumar, M., Seth, A., Singh, A. K., et al. (2021). Remediation strategies for heavy metals contaminated ecosystem: A review. *Environmental and Sustainability Indicators, 12*(100155), 100155. https://doi.org/10.1016/j.indic.2021.100155

Larrabure, G., Chero-Osorio, S., Silva-Quiñones, D., et al. (2021). Surface processes at a polymetallic (Mn-Fe-Pb) sulfide subject to cyanide leaching under sonication conditions and with an alkaline pretreatment: Understanding differences in silver extraction with X-ray photoelectron spectroscopy (XPS). *Hydrometallurgy, 200*, 105544. https://doi.org/10.1016/j.hydromet.2020.105544

Latorre, M., Cortés, M. P., Travisany, D., et al. (2016). The bioleaching potential of a bacterial consortium. *Bioresource Technology*, *218*, 659–666.

Lee, M. H., Park, H. J., & Lee, J. U. (2015). Biolixiviación de arsénico y metales pesados de relaves mineros mediante cultivos puros y mixtos de *Acidithiobacillus* spp. *Revista de Química Industrial y de Ingeniería*, *21*, 451–458.

Mahajan, S., Gupta, A., & Sharma, R. (2017). Bioleaching and biomining. In Singh, R. (eds), *Principles and Applications of Environmental Biotechnology for a Sustainable Future. Applied Environmental Science and Engineering for a Sustainable Future*. Springer, Singapore. https://doi.org/10.1007/978-981-10-1866-4_13

Mancini, L., & Sala, S. (2018). Social impact assessment in the mining sector: Review and comparison of indicators frameworks. *Resources Policy*, *57*, 98–111.

Mani, D., & Kumar, C. (2014). Biotechnological advances in bioremediation of heavy metals contaminated ecosystems: An overview with special reference to phytoremediation. *International Journal of Environmental Science and Technology*, *11*, 843–872.

Martínez-Bussenius, C., Navarro, C. A., & Jerez, C. A. (2017). Microbial copper resistance: Importance in biohydrometallurgy. *Microbial Biotechnology*, *10*(2), 279–295. https://doi.org/10.1111/1751-7915.12450

Massello, F. L., & Donati, E. (2021). Effect of heavy metal-induced stress on two extremophilic microbial communities from Caviahue-Copahue, Argentina. *Environmental Pollution*, *268*, 115709.

Mishra, D., & Rhee, Y. H. (2010). Current research trends of microbiological leaching for metal recovery from industrial wastes. *Current Research, Technology and Education Topics in Applied Microbiology and Microbial Biotechnology*, *2*, 1289–1292.

Mwewa, B., Tadie, M., Ndlovu, S., Simate, G. S., & Matinde, E. (2022). Recovery of rare earth elements from acid mine drainage: A review of the extraction methods. *Journal of Environmental Chemical Engineering*, *10*(3), 107704.

Nakhjiri, A. T., Sanaeepur, H., Omidkhah, M., & Mahdi, M. (2022). Recovery of precious metals from industrial wastewater towards resource recovery and environmental sustainability: A critical review. *Desalination*, *527*, 115510. https://doi.org/10.1016/j.desal.2021.115510

Okoh, M. P., Olobayetan, I. W., & Machunga-Mambula, S. S. (2018). Bioleaching, a technology for metal extraction and remediation: Mitigating health consequences for metal exposure. *International Journal of Development and Sustainability*, *7*, 2103–2118.

Orandi, S. (2017). Biosorption of metals by microorganisms in the bioremediation of toxic metals. In *Handbook of Metal-Microbe Interactions and Bioremediation* (pp. 281–297). CRC Press.

Pabón, S. E., Benítez, R., Sarria, R. A., & Gallo, J. A. (2020). Contaminación del agua por metales pesados, métodos de análisis y tecnologías de remoción. Una revisión. *Entre Ciencia e Ingeniería*, *14*(27), 9–18.

Peng, T., Chen, L., Wang, J., et al. (2019). Dissolution and passivation of chalcopyrite during bioleaching by Acidithiobacillus ferrivorans at low temperature. *Minerals*, *9*(6), 332.

Peng, T., Ma, L., Feng, X., Tao, J., Nan, M., Liu, Y., Jiaokun, L., Shen, L., Wu, X., Runlan, Y., Liu, X., Qiu, G., & Zeng, W. (2017). Genomic and transcriptomic analyses reveal adaptation mechanisms of an *Acidithiobacillus ferrivorans* strain YL15 to alpine acid mine drainage. *PLoS One*, *12*(5), e0178008. https://doi.org/10.1371/journal.pone.0178008.

Quach, N. T., Pham-Ngoc, C., Bui, T. L., Tran, T. H., Tran, T. A., Chu, H. H., & Phi, Q. T. (2022). Bioleaching potential of indigenous bacterial consortia from gold-bearing sulfide ore of Ta Nang Mine in Vietnam. *Polish Journal of Environmental Studies*, *31*(1), 803–813.

Razzak, S. A., Faruque, M. O., Alsheikh, Z., et al. (2022). A comprehensive review on conventional and biological-driven heavy metals removal from industrial wastewater. *Environmental Advances*, *7*, 100168. https://doi.org/10.1016/j.envadv.2022.100168

Rehman, A. U., Nazir, S., Irshad, R., Tahir, K., ur Rehman, K., Islam, R. U., & Wahab, Z. (2021). Toxicity of heavy metals in plants and animals and their uptake by magnetic iron oxide nanoparticles. *Journal of Molecular Liquids*, *321*, 114455.

Roy, J. J., Cao, B., & Madhavi, S. (2021). A review on the recycling of spent lithium-ion batteries (LIBs) by the bioleaching approach. *Chemosphere*, *282*, 130944.

Sajjad, W., Zheng, G., Din, G., Ma, X., Rafiq, M., & Xu, W. (2019). Metals extraction from sulfide ores with microorganisms: The bioleaching technology and recent developments. *Transactions of the Indian Institute of Metals*, *72*, 559–579.

Sajjad, W., Zheng, G., Ma, X., Wang, X., Ali, B., Rafiq, M., Zada, S., Irfan, M., & Zeman, J. (2020). Dissolution of Cu and Zn-bearing ore by indigenous iron-oxidizing bacterial consortia supplemented with dried bamboo sawdust and variations in bacterial structural dynamics: A new concept in bioleaching. *Science of the Total Environment*, *709*, 136136. https://doi.org/10.1016/j.scitotenv.2019.136136

Sharma, P., Dutta, D., Udayan, A., & Kumar, S. (2021). Industrial wastewater purification through metal pollution reduction employing microbes and magnetic nanocomposites. *Journal of Environmental Chemical Engineering*, *9*(6), 106673.

Song, P., Xu, D., Yue, J., Ma, Y., Dong, S., & Feng, J. (2022). Recent advances in soil remediation technology for heavy metal contaminated sites: A critical review. *Science of the Total Environment*, *838*, 156417. https://doi.org/10.1016/j.scitotenv.2022.156417

Sonone, S. S., Jadhav, S., Sankhla, M. S., & Kumar, R. (2020). Water contamination by heavy metals and their toxic effect on aquaculture and human health through food chain. *Letters in Applied NanoBioScience*, *10*(2), 2148–2166.

Vardhan, K. H., Kumar, P. S., & Panda, R. C. (2019). A review on heavy metal pollution, toxicity and remedial measures: Current trends and future perspectives. *Journal of Molecular Liquids*, *290*, 111197.

Wang, Y. J., Yang, Y. J., Li, D. P., Hu, H. F., Li, H. Y., & He, X. H. (2013). Bioxidative dissolution of cinnabar by iron-oxidizing bacteria. *Biochemical Engineering Journal*, *74*, 102–106.

Wang, Z., Liu, T., Duan, H., et al. (2021). Post-treatment options for anaerobically digested sludge: Current status and future prospect. *Water Research*, *205*, 117665.

Wheaton, G., Counts, J., Mukherjee, A., Kruh, J., & Kelly, R. (2015). The confluence of heavy metal biooxidation and heavy metal resistance: Implications for bioleaching by extreme thermoacidophiles. *Minerals*, *5*(3), 397–451.

Xie, F., Dong, K., Wang, W., & Asselin, E. (2020). Leaching of mercury from contaminated solid waste: A mini-review. *Mineral Processing and Extractive Metallurgy Review*, *41*(3), 187–197.

Yaashikaa, P. R., Priyanka, B., Senthil Kumar, P., Karishma, S., Jeevanantham, S., & Indraganti, S. (2022). A review on recent advancements in recovery of valuable and toxic metals from e-waste using bioleaching approach. *Chemosphere*, *287*, 132230. https://doi.org/10.1016/j.chemosphere.2021.132230

Yang, W., Song, W., Li, J., & Zhang, X. (2020). Bioleaching of heavy metals from wastewater sludge with the aim of land application. *Chemosphere*, *249*, 126134.

Yenial, Ü., & Bulut, G. (2017). Examination of flotation behavior of metal ions for process water remediation. *Journal of Molecular Liquids*, *241*, 130–135.

Zeng, X., Wei, S., Sun, L., et al. (2015). Bioleaching of heavy metals from contaminated sediments by the *Aspergillus niger* strain SY1. *Journal of Soils and Sediments*, *15*, 1029–1038.

12 Bioprospecting and Exploration of Extremophilic Enzymes in Bioremediation of Wastewater Polluted

Edwin Hualpa-Cutipa, Andi Solórzano Acosta, Milagros Estefani Alfaro Cancino, Fiorella Maité Arquíñego-Zárate, Nikol Gianella Julca Santur, María José Mayhua, and Lucero Katherine Castro Tena

12.1 INTRODUCTION

Water is a fundamental resource for the development of life on the planet, enabling the growth of various living organisms in different terrestrial and aquatic ecosystems. Currently, the health of these aquatic ecosystems has been affected due to the increase in residual toxic agents produced by industries dedicated to various activities such as coal conversion, oil refining, resin and plastic production, textile industry, oil extraction, tanning, mining, and pulp and paper manufacturing. All of these activities have a negative impact on aquatic environments through the discharge of polluting effluents (Bashir et al., 2020). These wastewater streams often have a high degree of alkalinity and a considerable chemical and biological oxygen demand (Yadav et al., 2022). To counteract these impacts, various physicochemical techniques have been used, but they have not proven to be as effective, resulting in further losses due to their high cost and inadequate removal of contaminants.

Over the years, waste mitigation methodologies that are environmentally sustainable (bioremediation) have been developed, facilitating the use and application of living organisms (microorganisms and plants) with the capacity to biodegrade and/or accumulate harmful contaminants (Jain et al., 2022). This biological strategy has demonstrated greater effectiveness in waste elimination compared to conventional techniques, with the additional advantage of not generating harmful waste that can accumulate in the aquatic ecosystem. Furthermore, due to its low operational and maintenance costs, it has generated great interest among researchers in the field (Giovanella et al., 2020). The fundamental principle of this process is based

DOI: 10.1201/9781003407683-12

on the hydrolysis of contaminants composed of large molecules, which are biologically reduced through aerobic and anaerobic metabolic pathways until they become soluble (Aragaw, 2020). The biodegradation and generation of intermediate products show variability that is related to the type of microorganism used (anaerobic or aerobic) (Yin et al., 2019). Additionally, optimal environmental conditions play a fundamental role in the efficiency of the biological process (Saxena, Kishor and Bharagava, 2020). The bioremediation process is successful when microbial components (enzymes) efficiently convert contaminants into harmless products for the aquatic ecosystem.

The microorganisms involved in bioremediation processes must possess tolerance and adaptation characteristics to cope with the presence of compounds of different nature, which inhibit their growth and activity. The literature frequently reports various bacterial species evaluated and applied in bioremediation processes (*Pseudomonas*, *Bacillus*, *Escherichia coli*, among others). However, their tolerance and resistance are almost negligible when facing high-stress conditions, resulting in very limited biodegradation outcomes (Ren, Lee and Na, 2020). In this regard, microbes that can tolerate and adapt to severe environmental conditions are required. An interesting group is extremophilic microorganisms, which are capable of residing and adapting to extreme environmental conditions, often found in hostile natural environments (Satyanarayana et al., 2005; Torsvik and Øvreås, 2008). This group of microbes presents a clear metabolic advantage when applied to the treatment of toxic contaminants because they not only eliminate harmful waste through versatile mechanisms but also possess the ability to withstand the extreme conditions required for these biological processes (Bakermans, 2015).

The different categories of extremophilic microorganisms (thermophiles, acidophiles, hyperthermophiles, among others) encompass species such as *Geobacillus stearothermophilus*, *Sulfolobus sulfataricus*, and *Thermus thermophilus*, which produce a variety of enzymes that play a crucial role in the bioremediation process (Moussa and Khalil, 2022). On the other hand, the catalytic biomolecules of these extremophilic microorganisms, known as "extremozymes," a term provided for the enzymes of extremophiles, have been extensively studied and applied in various biological treatments, primarily due to their high degree of specificity, great stability at high temperatures, pH tolerance, survival in high concentrations of metals and salts, etc., and their rapid reaction rate, which reduces processing costs. Therefore, this chapter aims to provide updated information on sustainable wastewater bioremediation processes using extremophilic microorganisms and the enzymes they contain and also consider their properties, characteristics, stability, and biodegradation capacity of their extremophilic enzymes. In addition, the potential and efficacy of extremophiles are summarized through genetic modifications and editions that allow obtaining biomolecules with greater versatility and efficacy.

12.2 CURRENT STATUS OF WASTEWATER POLLUTION

Currently, one of the most concerning environmental issues is the availability and scarcity of water on the planet, along with the sanitary conditions in which it is found in different ecosystems. This is a globally reaching problem because its impact is at a global scale (Zhang et al., 2016).

There are several factors involved in the pollution of aquatic ecosystems; however, the main factor is associated with the increase in urban populations. As the population rate increases, pollution will also increase in general, due to the consumerism generated by human beings (Afolalu et al., 2022). Among the negative effects of pollution is the weakening of the ozone layer (Ahmed et al., 2018), which significantly impacts ecosystems and leads to climate disorder with accelerated global warming.

The causes of water pollution and wastewater generation are associated with the release of different effluents from industrial activities and/or processes, which transport contaminants to aquatic ecosystems (rivers, lakes, seas) (Afolalu et al., 2022). While it is true that the components of an aquatic ecosystem (bacteria, fungi, algae, protozoa, etc.) have the ability to naturally remove and/or degrade pollutants (heavy metals, pesticides, pharmaceuticals, etc.), this capacity has a limit that is exceeded by the high pollutant load of the discharged effluents. This problem worsens over time, leading to a crisis of contaminants in aquatic systems and causing disruption to aquatic organisms (Jan et al., 2023).

One of the issues of water pollution is the high presence and accumulation of organic matter that continuously and permanently enters, causing eutrophication of aquatic systems (Sönmez, Akarsu and Sivri, 2023). This eutrophic process, originated by the influx of domestic wastewater, triggers the proliferation of anaerobic microorganisms, which in turn critically reduces the levels of dissolved oxygen in water, aiding microbial action in the anaerobic decomposition of organic matter (Yang et al., 2008). In addition to this ecological problem, the impact of eutrophication on aquatic flora and fauna is ongoing. The proliferation of algae with accelerated metabolism leads to the production of toxins that, when ingested by fish in these environments, cause their mortality (Bashir, 2020).

In order to alleviate this issue, awareness programs targeted at the population have been developed, which, through their governing bodies, have allowed the implementation of strategies to mitigate the impacts generated by water pollution. Additionally, necessary measures have been implemented for environmental protection (Zhao et al., 2023). An example of the application of these strategies is the nation of China, a country with high economic development and considered one of the global powers. Thanks to its governmental measures and environmental policies to counteract pollution in wastewater, they have achieved a significant improvement in a relatively short period. This demonstrates the efficiency of various treatment plants for wastewater, establishing China as the possessor of the world's second largest wastewater treatment capacity, after the United States (Zhang et al., 2016).

Wastewater treatment plants (WWTPs) are currently recognized as a potential prominent contributor to microplastic (MP) contamination in aquatic ecosystems (Sadia et al., 2022). The presence and subsequent contamination of minute residues in organisms can be attributed to the aggregation of these MPs, a phenomenon that is expected to escalate with the ongoing utilization and disposal of these artificial polymers within the environment (Chuah et al., 2022). A significant proportion of synthetic polymers originate from commonly used consumer items, including facial cleansers, exfoliators, sunscreen, nail polish, hair colors, eyeshadows, shower gels, and toothpaste (Suaria et al., 2016, 2020).

Many MPs are expelled and discharged into the aquatic environment through treated effluents in WWTPs. Several studies have determined the presence and accumulation of MPs in aquatic matrices, causing toxicity issues to flora and fauna (Prata, 2018; Sun et al., 2019). Aquatic species are at an increased risk of entanglement, asphyxiation, and physical harm due to the diminutive dimensions of microplastic particles (Wang et al., 2019; Saeed et al., 2020). Within WWTPs, MPs have the ability to adsorb various chemicals or diseases, so serving as carriers for hydrophobic persistent organic pollutants (POPs). These POPs are known to have ecotoxicological impacts on the aquatic environment, as highlighted by McCormick et al. (2014).

12.3 EXTREMOPHILIC ENZYMES (EXTREMOZYMES) AND THEIR POTENTIAL IN BIOREMEDIATION

Extremophilic microorganisms exhibit adaptation processes that enable them to flourish in diverse and harsh environmental circumstances, including severe temperatures (both low and high), high pressure, salt, alkalinity, acidity, and other challenging factors. The metabolic flexibility of these bacteria has been shown to have a significant role in bioremediation (Kochhar et al., 2022). Therefore, microorganisms have the ability to break down a diverse array of toxins and contaminants, rendering them very important instruments for the remediation of damaged environments.

Halophilic microorganisms, which have adapted to thrive in high-salt environments, play a crucial role in the treatment of industrial waste with extreme salinity. These microorganisms possess unique physiological and biochemical characteristics that enable them to tolerate and even utilize high salt concentrations. Their ability to maintain high salt concentrations inside their cells helps to control osmotic pressure and prevent water from leaving the cells. This adaptation makes them particularly well-suited for treating saline waste, as they can thrive in conditions that would be inhibitory to other microorganisms. By harnessing the metabolic capabilities of halophilic microorganisms, it is possible to develop innovative treatment strategies that are tailored to the specific challenges posed by industrial waste with extreme salinity (Borthakur et al., 2022).

These particular microbes exhibit a remarkable ability to alter their outer membrane and increase the hydrophobic nature of their cell surface, so facilitating the uptake of hydrocarbons and utilizing them as a source of carbon. In addition, it has been observed that halophilic bacteria possess the capability to synthesize a significant quantity of biosurfactants, which have been found to augment the process of biodegradation of petroleum hydrocarbons (Shukla and Singh, 2020). These extremophiles are employed for bioremediation and often represent the solution to various environmental pollution problems. Therefore, there is a high interest in their study through isolation and molecular identification for applications in bioremediation (Figure 12.1).

Figure 12.1 shows us the sources for obtaining extremophilic microorganisms, as well as the techniques and approaches used for their identification. Additionally, it depicts the products and/or biomolecules that can be used in various biological processes. Below, we provide some reports on the identification and application of extremophilic microorganisms in the treatment of contaminated environments.

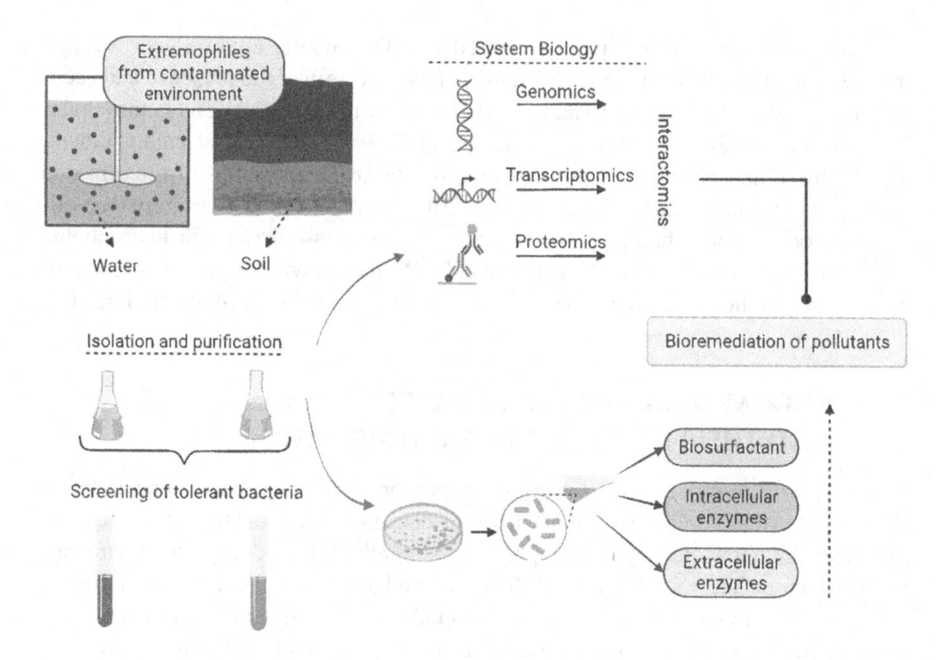

FIGURE 12.1 Steps for the isolation of extremophiles and their application in bioremediation.

In order to verify the potential of *Bacillus* enzymes for biotechnological applications, a study was conducted on 56 bacterial isolates obtained from five thermal sources in South Africa. The study aimed to search for and identify enzymes with high biodegradation capacity for contaminants in wastewater (Jardine et al., 2018). These enzymes have shown great potential for breaking down lignocellulosic materials, such as plant biomass, which is important for various industrial processes. In this study, the researchers aimed to find enzymes that can efficiently break down contaminants in wastewater. "Kochhar et al. (2022) conducted a study on wastewater treatment processes and discovered a new species of bacteria called *Anoxibacillus* sp. This species has the ability to produce enzymes that can break down harmful compounds in wastewater. This finding suggests that *Anoxibacillus* sp. could be a valuable asset in improving the efficiency and quality of wastewater treatment processes."

Acid mine drainage (AMD) is a process that involves the oxidation of sulfide minerals to sulfuric acid when exposed to water and in the presence of oxygen. These harmful compounds are then mobilized toward the surface and groundwater. This is one of the main causes of water pollution and the dispersion of heavy metals in the aquatic and terrestrial environment. Research focused on mitigating this contaminant phenomenon is based on the search for acidophilic microorganisms capable of oxidizing and reducing the concentration of iron and sulfur. Additionally, it has been identified that these acid-tolerant microorganisms are resistant to the presence of toxic metals such as cadmium, chromium, nickel, and arsenic. Furthermore, bioremediation strategies applied to AMD do not employ individual enzymes; on the contrary, whole cells (iron-oxidizing acidophiles) are used and multiplied through the use of bioreactors (Mesbah, 2022).

Lignocellulosic pretreatment, according to Zhu et al. (2022), is a crucial step in the production of biofuels and other value-added products from plant biomass. It involves the breakdown of lignin and hemicellulose, which are complex polymers present in plant cell walls, to make the cellulose more accessible for enzymatic hydrolysis. However, during this process, the generation of phenolic compounds can pose a challenge due to their toxicity and potential environmental impact. To mitigate the negative effects of phenolic compounds, researchers have been exploring various strategies. One promising approach is the integration of advanced oxidation processes (AOPs) during lignocellulosic pretreatment. AOPs utilize powerful oxidants to break down the phenolic compounds into less toxic substances. Furthermore, the integration of AOPs not only addresses environmental concerns but also enhances the efficiency of lignocellulosic pretreatment by facilitating the removal of other undesirable components. Overall, the progress made in AOPs in lignocellulosic pretreatment is very encouraging for long-lasting and effective ways to treat wastewater in biofuel production.

Acidophiles are particularly valuable in the cleanup of contaminated soil and water due to their ability to reduce substances, absorb them biologically, and attach metals to jarosites. These processes allow acidophiles to remove harmful contaminants from the environment, effectively reducing pollution levels and improving the overall quality of soil and water resources (Orellana et al., 2018). Further research on acidophiles has revealed their capacity to biodegrade phenol at low pH levels. This finding is significant because it suggests that acidophiles can play a crucial role in the treatment of wastewater contaminated with oil under acidic conditions (Arulazhagan et al., 2017). This indicates the potential for utilizing acidophiles in environmental remediation efforts.

Figure 12.2 shows us the mechanisms used by extremophilic microorganisms when they encounter an environment contaminated with heavy metals and/or other toxic pollutants. The highlighted microbial mechanisms are associated with biotransformation, bio-adsorption, bioleaching, biomineralization, bioaccumulation, and biodegradation of heavy metals.

12.4 EXTREMOPHILIC MICROORGANISMS FOR WASTEWATER TREATMENT: DESIGN AND GENETIC MODIFICATIONS

Studies on extremophiles conducted by Shahi et al. (2021) identified *Cohnella* sp. a thermophilic bacterium present in wastewater from shrimp farming wastewater located in Choebdeh-Abadan (southwestern of Iran). Bioinformatics analysis revealed a high similarity and identity of the amino acid sequences belonging to the superoxide dismutase isoenzymes related to the CaSOD genes of *Cohnella* sp. These sequencing results revealed amino acid conservation in all sequences, as well as key patterns of amino acid residues that could help predict the protein's binding capacity with metal ions.

Regarding pollution in aquatic environments, another factor that generates wastewater is the discharge of microplastics derived from the petrochemical industry. To address this problem, a possible solution would be to use microbial polyhydroxyalkanoates (PHA), degradable polymers that have an intracellular storage function in

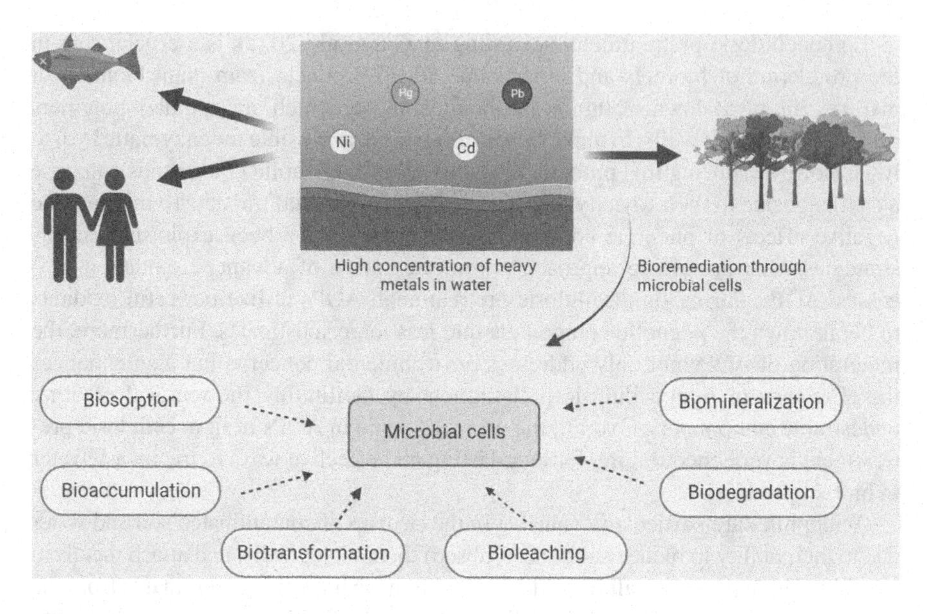

FIGURE 12.2 Main microbial mechanisms in the bioremediation of heavy metals when are present in a contaminated environment.

some extremophiles (Kourmentza et al., 2017). These bioplastic processors act as a replacement for single use petro-plastics (packaging materials, straws, plates, disposable cutlery, and some cosmetic particles). Renewable plastics require a high-cost and energy-intensive level of purity, which is obtained through proper downstream processing (DSP) for product recovery and refinement.

According to Koller and Rittmann (2022), the use of halophilic extremophilic strains, such as haloarchaea, in industrial production would allow for a high degree of sterility due to their ability to tolerate high and/or low temperatures, salinity, and pH values. An additional advantage of these extremophiles is that they do not produce lipopolysaccharides (LPS), which are endotoxins found in Gram-negative cell walls and contaminate the obtained products, causing inflammatory reactions and increasing costs when seeking proper purity. In relation to this, genetic engineering approaches have been developed for *Haloferax mediterranei* cultures, a type of haloarchaeon, with the aim of identifying and characterizing genes involved in the biosynthesis of Class III PHA synthase enzymes capable of withstanding high temperatures and salinity to produce degrading PHA granules. These bioproducts were obtained through differential centrifugation, and positive results were subsequently detected using PCR. For a better genetic design of haloarchaea, the goal is to eliminate the group of genes responsible for the formation of by-products, such as high-value extracellular polysaccharide (EPS) pigments, in order to increase PHA synthesis yield and facilitate its application in high purity, for example, in biomedical applications.

Extremophilic microorganisms, including bacteria, fungi, and microalgae, have proven to be of great interest in various research studies (Chai et al., 2021). For instance, a study conducted by Zheng et al. (2021) in China examined the fungal strain *Monascus pilosus* YX-1125 and its potential use in treating water contaminated with high concentrations of polysaccharides, lactic acid, acetic acid, ethanol, and other organic compounds. These fungal strains require a high chemical demand for oxygen, which can be addressed through an aerobic fermentation process that converts these compounds into biodegradable metabolites.

In addition, the possibility of using metabolic engineering for a more detailed metabolic analysis has been raised, focusing on the study of enzymes such as reverse beta and their energy capacity in oxidation cycles (Soni, 2022). A particularly relevant aspect is the study of the DNA sequence of these strains, which demonstrates tolerance to high concentrations of ethanol, with the aim of optimizing their performance (Chan and Vogel, 2010). In summary, the study of extremophilic microorganisms and their application in various fields, such as the treatment of contaminated water, has opened up new possibilities for the development of sustainable and efficient solutions.

Extremophilic microalgae play a fundamental role in the biological treatment of wastewater contaminated with organic and inorganic impurities, such as industrial waste, pathogens, heavy metals, and pesticides (Rathod, 2015). These microalgae employ a waste biotransformation mechanism that involves the absorption of nitrogen and phosphorus to reduce or eliminate the pollutant load. This process requires several steps, in addition to CO_2 fixation. Some notable species include *Chlorogleopsis* sp., a thermophilic cyanobacteria that has demonstrated high tolerance for successful growth under both high and low light intensity conditions. On the other hand, *Galdieria sulfuraria* is capable of thriving in extremely acidic environments with a pH as low as 1.8, as well as at thermal conditions of 56°C.

Finally, green algae have demonstrated bioremediation capabilities in a thermophilic environment. In a study conducted with the acidophilic species *Chlamydomonas acidophila*, their tolerance to acidity and their ability to remove heavy metals such as cadmium were evident (Puente-Sánchez et al., 2018). By applying genetic engineering, the differential gene expression was used in combination with other extremophilic species. Furthermore, analysis was carried out to detect genes exclusive to green algae, and codon usage studies were performed to identify proteins that provide adaptation mechanisms to acidic environments. In summary, the use of genetic approaches allowed for obtaining and collecting more comprehensive data on orthologous genes present in green algae, which could contribute to the identification of individual genes of interest related to the purification or detoxification of wastewater.

12.5 POTENTIAL APPLICATION OF SYNTHETIC BIOLOGY TO EXTREMOPHILIC MICROORGANISMS FOR WASTEWATER TREATMENT

Technological advancements in recent decades have had a significant impact in various areas, and the field of research is no exception. New biological strategies such as synthetic biology are being implemented with the help of information technologies (Lorenzo et al., 2018). According to Rylott and Bruce (2020), the synthetic biology is focused on designing and modifying the metabolic processes of different microorganisms with the aim of enhancing their biodegradation abilities or production of industrially relevant compounds. Therefore, applying this tool in the field of extremophilic microorganisms would provide a very interesting benefit in bioremediation processes.

Extremophiles, according to Zhu et al. (2020), are organisms that thrive in extreme environments. Halophiles are a type of extremophile that can survive in highly saline environments. Compatible solutes, or osmolytes, are substances that halophiles accumulate inside their cells to help them survive in high salt concentrations. Thermosomes are proteins that thermophiles produce to maintain the structure and functionality of other proteins in extremely hot or cold conditions.

Also, Zhu et al. (2020) reported that halophiles with the purpose of perpetuating themselves in a high salt concentration environment have developed various strategies to maintain their osmotic balance and retain water. Furthermore, through genetic tools, genetic modification technology, and synthetic biology, the performance of these microorganisms can be enhanced to accelerate their growth and enzyme production.

Research conducted by di Cicco (2021), reported to the genus *Galdieria* which is an extremophile that inhabits highly toxic environments with high acidity and warm climates, also, are considered thermoacidic environments. They are usually found near volcanic areas, sulfuric environments, and acidic thermal waters. These microorganisms have the ability to develop and modify a thermotolerant cell wall, complemented by high-affinity metal ion transport systems, allowing them to survive in the presence of heavy metals through biosorption, which enables them to remove heavy metals and contaminated waste.

A specific type of this genre is *G. sulfaria*, which, in consort with heterotrophic bacteria, can biodegrade ammonia and phosphates from wastewater (Rylott and Bruce, 2020) and genome studies of this species have determined that 5% of its genetic code encodes membrane transport proteins, some of which are specific transporters for divalent metal cations, allowing them to selectively capture essential metals.

12.6 METHODS FOR PROCESSING AND MODIFICATION OF ENZYMES FOR WASTEWATER TREATMENT

The development of methods for enzyme modification is a key factor in enhancing their stability and resistance to optimize productivity. There are several methods that enable a proper process: free enzymatic method, immobilized enzymatic method, and enzymatic mediator methods.

12.6.1 Free Enzymatic Method

This procedure presents different treatments and outcomes according to the type of enzyme used (crude or purified enzyme). Regardless of the enzyme employed, its efficiency in the results is positive, but in varying proportions (Feng et al., 2021). Crude enzymes exhibit higher removal activity compared to purified enzymes due to the presence of mediators. However, unused nutrients and mediators can lead to greater contamination in the treatment of wastewater. As an example, in the elimination of naproxen, it was found that crude laccase produced by *Trametes versicolor* achieved complete suppression, while commercial laccase obtained from *Myceliophthora thermophila* showed a 60% suppression rate (Lloret et al., 2010; Tran, Urase and Kusakabe, 2010). Another consideration to consider is that purified enzymes have a larger size, which requires additional processing such as size exclusion chromatography and membrane separation (Feng et al., 2021).

12.6.2 Immobilized Enzymatic Method

This methodology presents the physical coupling between matrices and enzymes to achieve proper system manipulation, as enzymes often exhibit inhibition behavior in wastewater. Since the catalytic potential of the enzyme is preserved, once the biomolecule is immobilized, it can be used in different cycles within the procedure, generating a positive impact on the environment and reducing treatment costs (Feng et al., 2021). Other effects resulting from immobilization are associated with the extension of enzymatic lifespan and a decrease in enzymatic activity loss (Fan et al., 2018; Wang et al., 2019).

12.6.3 Support Materials

The immobilization supports must be affordable biomaterials compatible with the processes developed in biological treatment. Some characteristics that must be taken into account include mechanical and thermal consistency, high specific surface area, suitable pore size, active groups, and potential for recovery and reuse (Harguindeguy et al., 2020). Usually, organic materials, inorganic materials, or a combination of both in varying amounts are used for the supports. Similarly, materials are classified as synthetic or natural. Among the most used synthetic inorganic supports are minerals such as kaolinite, stevensite, and halloysite, and metal oxides such as Fe_3O_4, SiO_2, and TiO_2. Among the natural organic supports, the use of alginate, gelatin, corn cob, rice straw, and chitosan is reported. Synthetic organic polymers such as ion exchange polystyrene resins and polyacrylonitrile are also used (Mohammadi et al., 2022).

The combination of support materials is much more efficient compared to using a single matrix. In catalysis processes, these combinations optimize electron transfer, favor separation, and increase the number of active groups and the surface area. Magnetization is often performed on support materials, which influences the increase in surface area by enhancing the electron transfer rate and enabling the dissociation of biocatalysts from a solution (Dai et al., 2016; Darwesh, Matter and Eida, 2019).

Lately, support systems based on nanomaterials have been developed, which incorporate nanoscale polymer beads, metallic nanoparticles, and carbon nanotubes. These supports cover a large surface area and allow enzymes to be loaded at high concentrations, which must be available for pollutants. Additionally, these supports have qualities that enable them to avoid enzymatic leaching, achieve optimal catalytic activity, and, in the case of magnetic iron oxide nanoparticles, easily separate from the solution and be reusable (Darwesh et al., 2019; Naguib and Badawy, 2020).

During the enzymatic immobilization process, it is crucial to achieve an appropriate bond between the support and the enzyme. The presence of certain functional groups can have an impact on the effectiveness of the process. To attain effective immobilization, it is desirable to promote the development of electron-donating chemical moieties, such as alkoxy, alkyl, amines, carbonyl, carboxyl, diol, epoxy, and hydroxyl groups. In contrast, the inclusion of electron-accepting functional groups, such as amide, halogen, and nitro, has been observed to have an adverse impact on immobilization, as reported by Simón-Herrero et al. (2019).

12.6.4 Enzyme-Mediated Systems

Enzyme mediators are small, stable molecules that aid in the transport of electrons between substrates and enzymes. These mediators enhance the variety of enzyme substrates by creating extremely reactive radicals (Naghdi et al., 2018). Furthermore, due to their small size, these particles promote the development of the enzyme used in the process, such as lacasse, which catalyzes the formation of lignin, increasing the efficiency with which the selected contaminants are eliminated (Munk, Andersen and Meyer, 2018).

These mediators are divided into natural and synthetic ones. Among the natural enzymatic mediators are 4-hydroxybenzyl alcohol, acetovanillone, syringaldehyde, and methyl vanillate. On the other hand, synthetic mediators include violuric acid, 1-hydroxybenzotriazole, and 2,2′-azinobis-(3-ethylbenzothiazoline)-6-sulfonic acid, among others. Although these mediators have the potential to enhance enzymatic processes, their mixture can result in low productivity and modify the treatment mechanism, making control difficult (Varga et al., 2019). Therefore, it is important to identify cases where the intervention of enzymatic mediators would be essential, such as in the treatment of TCS and oxybenzone. However, it's important to note that these mediators are toxic, which can lead to increased contamination, in addition to being costly.

12.7 METHODS FOR PROCESSING AND MODIFICATION OF ENZYMES FOR WASTEWATER TREATMENT

Extremophilic bacteria possess a high degree of adaptability, rendering them a very desirable reservoir of biocatalysts for the purpose of bioremediation. This technology holds significant importance in the realm of environmental conservation as it facilitates the elimination of harmful substances from the environment. There has been a growing focus on the separation, identification, and characterization of biocatalysts derived from extremophilic microbes. Most of these biocatalysts are enzymes referred to as extremozymes, owing to their efficacy in highly challenging environments. The enzymes possess notable characteristics that are anticipated to facilitate the integration of chemical and biological industrial procedures. Therefore, to fully harness the capacity of enzymes in wastewater treatment, it is necessary to acquire knowledge about treatment methods and enzyme modification.

Reports from an investigation were able to identify and express a new azoreductase (AzoRed2) from *Streptomyces* sp. S27 for the decolorization of dyes in industrial wastewater. This advancement is of great importance because wastewater often contains a mixture of dyes, organic solvents, ions, and organic matter, which can affect the activity of biocatalysts. AzoRed2 demonstrated excellent stability against pH changes, organic solvents, and methyl red, suggesting that it could maintain its stability under adverse conditions. Furthermore, a study was conducted to construct an efficient whole-cell biocatalyst that contained both AzoRed2 and BSGDH. This system achieved the goal of completely degrading the azo dye, reaching a degradation percentage of 99% without the need for externally adding NAD+ (Dong et al., 2019).

A study was conducted to estimate the immobilization of marine halophilic crops using a strain of *Bacillus subtilis* AAK through trapping and adsorption techniques. Additionally, partial purification and immobilization of the laccase enzyme (benzenediol:oxygen oxidoreductase; EC 1.10.3.2) were performed. The findings indicate that the immobilization of the biocatalyst and its laccase enzyme holds significant potential as a strategy to enhance the degradation of several phenolic compounds, including 2,4-DCP, among other hazardous phenolic and aromatic chemicals. The cost-effective implementation of this bioremediation approach was facilitated by the utilization of the biocatalyst and its laccase enzyme, as demonstrated in the study conducted by Farag, El-Naggar and Ghanem (2022).

Enzymes from extremophilic microorganisms are applied in the treatment of wastewater for remediation purposes. Among them, cold-adapted lipases stand out, which have numerous biotechnological applications in this field. Lipases from psychrophilic bacteria, such as those found in *Pseudomonas* sp., are capable of decomposing fats and have developed specific characteristics that give them thermal flexibility and high specific activity at low temperatures. The presence of lipase genes in the samples can be determined using a real-time quantitative PCR technique. To increase the production of lipase in psychrophilic strains, it is necessary to investigate and optimize different parameters, both physical and nutritional, until achieving optimal production (Salwoom et al., 2019).

To examine and determine the optimal conditions for lipase production by bacteria, certain steps need to be followed. In the case of the thermophilic-halophilic extremophilic bacterium PLS80, isolated from an underwater thermal source, lipase detection was carried out in agar media supplemented with olive oil and rhodamine B. This strain exhibited lipase activity, generating a red-orange, fluorescent area around the colonies under ultraviolet light. In all the assays conducted, it was determined that the strain had a higher lipase activity of approximately 24.3 U/mL (Febriani, Saidi and Iqbalsyah, 2019).

12.8 LIMITATIONS AND FUTURE PERSPECTIVES

The treatment of wastewater using enzymes offers various benefits, such as the biological transformation of contaminants present in wastewater and the reduction of their toxicity. Extremophilic enzymes are particularly relevant in this context as they originate from living organisms and contribute to environmental sustainability by being biodegradable. These enzymes have the potential to replace or reduce the use of hazardous chemicals in different industries. However, their application poses challenges due to their limited availability and behavior under stressful conditions, which still require further research. Nevertheless, extremophiles represent a special group of microorganisms with genetic and metabolic opportunities that can be harnessed for contaminant removal.

When used alongside traditional approaches, contemporary technologies continue to necessitate substantial exertion to address constraints related to their own metabolic processes, the intricate interplay among microorganisms, and external factors affecting the organisms. These external factors include the need for bacteria-specific organic acid production for calibration purposes, the inclusion of organic solvents in water to counterbalance osmotic pressure, the presence of sugars in polluted sites that serve as constituents of bioplastics, and the concentrations of toxic contaminants (Pham et al., 2022).

The use of extremophilic microorganisms in biotechnological processes represents a viable and sustainable alternative for development. However, a joint effort is needed from the scientific community to search for new sources of extremophiles and develop techniques that involve genetic modification and structural and functional changes in the biomolecules derived from these organisms. These efforts will translate into benefits for both present and future generations (Oliart-Ros, Manresa-Presas and Sánchez-Otero, 2016).

The growing global population has increased the demand for clean water, which presents challenges in the provision of clean water services. In this context, the use of enzymes in wastewater treatment stands out as a primary option for the cost-effective and efficient removal of most dyes. Several studies have been conducted to evaluate the capability of immobilized enzymes in nanoparticles in treating wastewater with specific dyes, yielding positive results. However, the feasibility of these systems is still unclear since all studies have been conducted at the laboratory scale (Wong et al., 2019).

Finding the optimal conditions for environmental remediation is a significant challenge due to the lack of evaluation of its effectiveness in treating wastewater under different pollution conditions. Laboratory results may differ from real-world outcomes achieved through various strategies, such as nanotechnology, which enables the removal of dyes and colorants from wastewater. Therefore, it is crucial to conduct laboratory-scale experiments under actual industrial conditions to overcome this challenge. Furthermore, the sensitivity of enzymes to environmental conditions requires an appropriate immobilization strategy, which still faces obstacles (insufficient understanding of enzymes involved in nanotechnology, recovery of immobilized enzymes, and stability of the nanomaterials used). These additional challenges turn this sustainable measure into a major undertaking.

REFERENCES

Afolalu, S.A., Ikumapayi, O.M., Ogedengbe, T.S., Kazeem, R.A. and Ogundipe, A. (2022). Waste pollution, wastewater and effluent treatment methods – An overview. Materials Today: Proceedings, 62, pp. 3282–3288. https://doi.org/10.1016/j.matpr.2022.04.231

Ahmed, N., Khan, T.I. and Augustine, A. (2018). Climate changed and environmental degradation: A serious threat to global security. European Journal of Social Sciences Studies, 3. https://doi.org/10.5281/zenodo.1307227

Aragaw, T.A. (2020). Functions of various bacteria for specific pollutants degradation and their application in wastewater treatment: A review. International Journal of Environmental Science and Technology, 18(7), pp. 2063–2076. https://doi.org/10.1007/s13762-020-03022-2

Arulazhagan, P., Al-Shekri, K., Huda, Q., et al. (2017). Biodegradation of polycyclic aromatic hydrocarbons by an acidophilic *Stenotrophomonas maltophilia* strain AJH1 isolated from a mineral mining site in Saudi Arabia. Extremophiles, 21(1), pp. 163–174. https://doi.org/10.1007/s00792-016-0892-0

Bakermans, C. (Ed.). (2015). Microbial Evolution under Extreme Conditions. De Gruyter. https://doi.org/10.1515/9783110340716

Bashir, I., Lone, F.A., Bhat, R.A., Mir, S.A., Dar, Z.A. and Dar, S. A. (2020). Concerns and threats of contamination on aquatic ecosystems. In K. R. Hakeem, R. A. Bhat, & H. Qadri (Eds.), Bioremediation and Biotechnology: Sustainable Approaches to Pollution Degradation (pp. 1–26). Springer International Publishing. https://doi.org/10.1007/978-3-030-35691-0_1

Borthakur, D., Rani, M., Das, K., et al. (2022). Bioremediation: An alternative approach for detoxification of polymers from the contaminated environment. Letters in Applied Microbiology, 75(4), pp. 744–758. https://doi.org/10.1111/LAM.13616

Chai, W.S., Tan, W.G., Munawaroh, H.S.H., et al. (2021). Multifaceted roles of microalgae in the application of wastewater biotreatment: A review. Environmental Pollution (Barking, Essex: 1987), 269, p. 116236. https://doi.org/10.1016/j.envpol.2020.116236

Chan, D.I. and Vogel, H.J. (2010). Current understanding of fatty acid biosynthesis and the acyl carrier protein. The Biochemical Journal, 430(1), pp. 1–19. https://doi.org/10.1042/BJ20100462

Chuah, L.F., Mokhtar, K., Bakar, A.A., Othman, M.R., Osman, N.H., Bokhari, A., Mubashir, M., Abdullah, M.A. and Hasan, M. (2022). Marine environment and maritime safety assessment using Port State Control database. Chemosphere, 304, p. 135245. https://doi.org/10.1016/j.chemosphere.2022.135245

Dai, Y., Yao, J., Song, Y., Liu, X., Wang, S. and Yuan, Y. (2016). Enhanced performance of immobilized laccase in electrospun fibrous membranes by carbon nanotubes modification and its application for bisphenol A removal from water. Journal of Hazardous Materials, 317, pp. 485–493. https://doi.org/10.1016/j.jhazmat.2016.06.017

Darwesh, O.M., Matter, I.A. and Eida, M.F. (2019). Development of peroxidase enzyme immobilized magnetic nanoparticles for bioremediation of textile wastewater dye. Journal of Environmental Chemical Engineering, 7(1), p. 102805. https://doi.org/10.1016/j.jece.2018.11.049

di Cicco, M.R., Iovinella, M., Palmieri, M., et al. (2021). Extremophilic microalgae *Galdieria* gen. For urban wastewater treatment: Current state, the case of 'POWER' system, and future prospects. Plants, 10(11), p. 2343. https://doi.org/10.3390/plants10112343

Dong, H., Guo, T., Zhang, W., Ying, H., Wang, P., Wang, Y. and Chen, Y. (2019). Biochemical characterization of a novel azoreductase from *Streptomyces* sp.: Application in eco-friendly decolorization of azo dye wastewater. International Journal of Biological Macromolecules, 140, pp. 1037–1046. Available at: https://doi.org/10.1016/j.ijbiomac.2019.08.196 [Accessed 26 Mar. 2023].

Fan, X., Hu, M., Li, S., Zhai, Q., Wang, F. and Jiang, Y. (2018). Charge controlled immobilization of chloroperoxidase on both inner/outer wall of NHT: Improved stability and catalytic performance in the degradation of pesticide. Applied Clay Science, 163, pp. 92–99. https://doi.org/10.1016/j.clay.2018.07.016

Farag, A.M., El-Naggar, M.Y. and Ghanem, K.M. (2022). 2,4-Dichlorophenol biotransformation using immobilized marine halophilic *Bacillus* subtilis culture and laccase enzyme: Application in wastewater treatment. Journal of Genetic Engineering and Biotechnology, 20(1). Available at: https://link.springer.com/article/10.1186/s43141-022-00417-1 [Accessed 26 Mar. 2023].

Febriani, Ulwiyyah, N.H., Saidi, N. and Iqbalsyah, T.M. (2019). Screening and Production of Lipase from a Thermo-halophilic Bacterial Isolate of Pria Laot Sabang 80 Isolated from Under Water Hot Spring. KnE Engineering, 4(2). https://doi.org/10.18502/keg.v1i2.4436

Feng, S., Hao Ngo, H., Guo, W., Woong Chang, S., Duc Nguyen, D., Cheng, D., Varjani, S., Lei, Z. and Liu, Y. (2021). Roles and applications of enzymes for resistant pollutants removal in wastewater treatment. Bioresource Technology, 335, p. 125278. https://doi.org/10.1016/j.biortech.2021.125278

Giovanella, P., Vieira, G.A.L., Ramos Otero, I.V., Pais Pellizzer, E., de Jesus Fontes, B. and Sette, L.D. (2020). Metal and organic pollutants bioremediation by extremophile microorganisms. Journal of Hazardous Materials, 382, p. 121024. https://doi.org/10.1016/j.jhazmat.2019.121024

Gunjal Aparna, B., Waghmode Meghmala, S. and Patil Neha, N. (2021). Role of extremozymes in bioremediation. Research Journal of Biotechnology, 16, p. 3. [Accessed 31 Mar. 2023].

Harguindeguy, M., Antonelli, C., Belleville, M., Sanchez-Marcano, J. and Pochat-Bohatier, C. (2020). Gelatin supports with immobilized laccase as sustainable biocatalysts for water treatment. Journal of Applied Polymer Science, 138(2), p. 49669. https://doi.org/10.1002/app.49669

Jan, S., Mishra, A.K., Bhat, M.A., Bhat, M.A. and Jan, A.T. (2023). Pollutants in aquatic system: A frontier perspective of emerging threat and strategies to solve the crisis for safe drinking water. Environmental Science and Pollution Research, 30(53), pp. 113242–113279. https://doi.org/10.1007/s11356-023-30302-4

Jain, M., Khan, S.A., Sharma, K., Jadhao, P.R., Pant, K.K., Ziora, Z.M. and Blaskovich, M.A.T. (2022). Current perspective of innovative strategies for bioremediation of organic pollutants from wastewater. Bioresource Technology, 344, p. 126305. https://doi.org/10.1016/j.biortech.2021.126305

Jardine, J.L., Stoychev, S., Mavumengwana, V., et al. (2018). Screening of potential bioremediation enzymes from hot spring bacteria using conventional plate assays and liquid chromatography – Tandem mass spectrometry (Lc-Ms/Ms). Journal of Environmental Management, 223, pp. 787–796. https://doi.org/10.1016/J.JENVMAN.2018.06.089

Kochhar, N., Kavya, I.K., Shrivastava, S., et al. (2022). Perspectives on the microorganism of extreme environments and their applications. Current Research in Microbial Sciences, 3, p. 100134. https://doi.org/10.1016/J.CRMICR.2022.100134

Koller, M. and Rittmann, S.K.-M.R. (2022). Haloarchaea as emerging big players in future polyhydroxyalkanoate bioproduction: Review of trends and perspectives. Current Research in Biotechnology, 4, pp. 377–391. https://doi.org/10.1016/j.crbiot.2022.09.002

Kourmentza, C., Plácido, J., Venetsaneas, N., Burniol-Figols, A., Varrone, C., Gavala, H.N. and Reis, M.A.M. (2017). Recent advances and challenges towards sustainable Polyhydroxyalkanoate (PHA) production. Bioengineering, 4(2). https://doi.org/10.3390/bioengineering4020055

Lloret, L., Eibes, G., Lú-Chau, T.A., Moreira, M.T., Feijoo, G. and Lema, J.M. (2010). Laccase-catalyzed degradation of anti-inflammatories and estrogens. Biochemical Engineering Journal, 51(3), pp. 124–131. https://doi.org/10.1016/j.bej.2010.06.005

Lopez-Lopez, O., Cerdan, E.M. and Siso, I.G.M. (2014). New extremophilic lipases and esterases from metagenomics. Current Protein & Peptide Science, 15(5), pp. 445–455. https://doi.org/10.2174/1389203715666140228153801

de Lorenzo, V., Prather, K.L.J., Chen, G.-Q., et al. (2018). The power of synthetic biology for bioproduction, remediation and pollution control: The UN's Sustainable Development Goals will inevitably require the application of molecular biology and biotechnology on a global scale. EMBO Reports, 19(4), p. e45658. https://doi.org/10.15252/embr.201745658

McCormick, A.R., Hoellein, T.J., Mason, S.A., Schluep, J. and Kelly, J.J. (2014). Microplastic is an abundant and distinct microbial habitat in an urban river. ScholarSphere (Penn State Libraries), 48(20), pp. 11863–11871. https://doi.org/10.1021/es503610r

Mesbah, N.M. (2022). Industrial biotechnology based on enzymes from extreme environments. Frontiers in Bioengineering and Biotechnology, 10, p. 542. https://doi.org/10.3389/fbioe.2022.870083

Mohammadi, S.A., Najafi, H., Zolgharnian, S., Sharifian, S. and Asasian-Kolur, N. (2022). Biological oxidation methods for the removal of organic and inorganic contaminants from wastewater: A comprehensive review. Science of The Total Environment, 843, p. 157026. https://doi.org/10.1016/j.scitotenv.2022.157026

Moussa, T.A.A. and Khalil, N.M. (2022). Chapter 10—Extremozymes from extremophilic microorganisms as sources of bioremediation. In M. Kuddus (Ed.), Microbial Extremozymes (pp. 135–146). Academic Press. https://doi.org/10.1016/B978-0-12-822945-3.00005-1

Munk, L., Andersen, M.L. and Meyer, A.S. (2018). Influence of mediators on laccase catalyzed radical formation in lignin. Enzyme and Microbial Technology, 116, pp. 48–56. https://doi.org/10.1016/j.enzmictec.2018.05.009

Naghdi, M., Taheran, M., Brar, S.K., Kermanshahi-pour, A., Verma, M. and Surampalli, R.Y. (2018). Removal of pharmaceutical compounds in water and wastewater using fungal oxidoreductase enzymes. Environmental Pollution, 234, pp. 190–213. https://doi.org/10.1016/j.envpol.2017.11.060

Naguib, D.M. and Badawy, N.M. (2020). Phenol removal from wastewater using waste products. Journal of Environmental Chemical Engineering, 8(1), p. 103592. https://doi.org/10.1016/j.jece.2019.103592

Oliart-Ros, R.M., Manresa-Presas, Á and Sánchez-Otero, M.G. (2016). Utilización de microorganismos de ambientes extremos y sus productos en el desarrollo biotecnológico. CienciaUAT, 11(1), pp. 79–90. Available at: https://www.scielo.org.mx/scielo.php?script=sci_arttext&pid=S2007-78582016000200079 [Accessed 1 Apr. 2023].

Orellana, R., Macaya, C., Bravo, G., et al. (2018). Living at the frontiers of life: Extremophiles in Chile and their potential for bioremediation. Frontiers in Microbiology, 9, p. 2309. https://doi.org/10.3389/fmicb.2018.02309

Pal, S. and Debanshi, S. (2022). Exploring the effect of wastewater pollution susceptibility towards wetland provisioning services. Ecohydrology and Hydrobiology, 23(1), pp. 162–176. https://doi.org/10.1016/j.ecohyd.2022.12.003

Pham, V.H.T., Kim, J., Chang, S. and Chung, W. (2022). Bacterial biosorbents, an efficient heavy metals green clean-up strategy: Prospects, challenges, and opportunities. Microorganisms, 10(3), p. 610. https://doi.org/10.3390/microorganisms10030610

Prata, J.C. (2018). Microplastics in wastewater: State of the knowledge on sources, fate and solutions. Marine Pollution Bulletin, 129(1), pp. 262–265. https://doi.org/10.1016/j.marpolbul.2018.02.046

Puente-Sánchez, F., Díaz, S., Penacho, V., et al. (2018). Basis of genetic adaptation to heavy metal stress in the acidophilic green alga *Chlamydomonas acidophila*. Aquatic Toxicology (Amsterdam, Netherlands), 200, pp. 62–72. https://doi.org/10.1016/j.aquatox.2018.04.020

Rathod, H. (2015, August). Algae based wastewater treatment. In A Seminar Report of Master of Technology in Civil Engineering. Roorkee, Uttarakhand, India.

Ren, J., Lee, J. and Na, D. (2020). Recent advances in genetic engineering tools based on synthetic biology. Journal of Microbiology, 58, pp. 1–10. https://doi.org/10.1007/s12275-020-9334-x

Rylott, E.L. and Bruce, N.C. (2020). How synthetic biology can help bioremediation. Current Opinion in Chemical Biology, 58, pp. 86–95. https://doi.org/10.1016/j.cbpa.2020.07.004

Sadia, M., Mahmood, A., Ibrahim, M., Irshad, M.K., Ali, H., Bokhari, A., Mubashir, M., Chuah, L.F. and Show, P.L. (2022). Microplastics pollution from wastewater treatment plants: A critical review on challenges, detection, sustainable removal techniques and circular economy. Environmental Technology and Innovation, 28, pp. 102946. https://doi.org/10.1016/j.eti.2022.102946

Saeed, T., Al-Jandal, N., Al-Mutairi, A. and Taqi, H. (2020). Microplastics in Kuwait marine environment: Results of first survey. Marine Pollution Bulletin, 152, p. 110880. https://doi.org/10.1016/j.marpolbul.2019.110880

Salwoom, L., Raja Abd Rahman, R.N.Z., Salleh, A.B., Mohd. Shariff, F., Convey, P., Pearce, D. and Mohamad Ali, M.S. (2019). Isolation, characterisation, and lipase production of a cold-adapted bacterial strain *Pseudomonas* sp. LSK25 Isolated from Signy Island, Antarctica. Molecules, 24(4). https://doi.org/10.3390/molecules24040715

Satyanarayana, T., Raghukumar, C. and Shivaji, S. (2005). Extremophilic microbes: Diversity and perspectives. Current Science, 89(1), pp. 78–90. http://www.jstor.org/stable/24110434

Saxena, G., Kishor, R. and Bharagava, R.N. (2020). Application of microbial enzymes in degradation and detoxification of organic and inorganic pollutants. In G. Saxena & R. Bharagava (Eds.), Bioremediation of Industrial Waste for Environmental Safety. Springer, Singapore. https://doi.org/10.1007/978-981-13-1891-7_3

Shahi, Z.K.M., Takalloo, Z., Mohamadzadeh, J., et al. (2021). Thermophilic iron containing type superoxide dismutase from *Cohnella* sp. A01. International Journal of Biological Macromolecules, 187, pp. 373–385. https://doi.org/10.1016/j.ijbiomac.2021.07.150

Shukla, A.K. and Singh, A.K. (2020). Exploitation of potential extremophiles for bioremediation of xenobiotics compounds: A biotechnological approach. Current Genomics, 21(3), p. 161. https://doi.org/10.2174/1389202921999200422122253

Simón-Herrero, C., Naghdi, M., Taheran, M., Kaur Brar, S., Romero, A., Valverde, J.L., Avalos Ramirez, A. and Sánchez-Silva, L. (2019). Immobilized laccase on polyimide aerogels for removal of carbamazepine. Journal of Hazardous Materials, 376, pp. 83–90. https://doi.org/10.1016/j.jhazmat.2019.05.032

Soni, S. (2022). Trends in lipase engineering for enhanced biocatalysis. Biotechnology and Applied Biochemistry, 69(1), pp. 265–272. https://doi.org/10.1002/bab.2105

Sönmez, V.Z., Akarsu, C. and Sivri, N. (2023). Impact of coastal wastewater treatment plants on microplastic pollution in surface seawater and ecological risk assessment. Environmental Pollution, 318, pp. 120922. https://doi.org/10.1016/j.envpol.2022.120922

Suaria, G., Avio, C.G., Mineo, A., Lattin, G.L., Magaldi, M.G., Belmonte, G., Moore, C.J., Regoli, F. and Aliani, S. (2016). The Mediterranean plastic soup: Synthetic polymers in Mediterranean surface waters. Scientific Reports, 6(1). https://doi.org/10.1038/srep37551

Suaria, G., Achtypi, A., Perold, V., Lee, J.R., Pierucci, A., Bornman, T.G., Aliani, S. and Ryan, P.G. (2020). Microfibers in oceanic surface waters: A global characterization. Science Advances, 6(23), eaay8493. https://doi.org/10.1126/sciadv.aay8493

Sun, J., Dai, X., Wang, Q., van Loosdrecht, M.C.M. and Ni, B.-J. (2019). Microplastics in wastewater treatment plants: Detection, occurrence and removal. Water Research, 152, pp. 21–37. https://doi.org/10.1016/j.watres.2018.12.050

Tkavc, R., Matrosova, V.Y., Grichenko, O.E., Gostinčar, C., Volpe, R.P., Klimenkova, P., Gaidamakova, E.K., Zhou, C.E., Stewart, B.J., Lyman, M.G., Malfatti, S.A., Rubinfeld, B., Courtot, M., Singh, J., Dalgard, C.L., Hamilton, T., Frey, K.G., Gunde-Cimerman, N., Dugan, L. and Daly, M.J. (2018). Prospects for fungal bioremediation of acidic radioactive waste sites: Characterization and genome sequence of *Rhodotorula taiwanensis* MD1149. Frontiers in Microbiology, 8. https://doi.org/10.3389/fmicb.2017.02528

Torsvik, V. and Øvreås, L. (2008). Microbial diversity, life strategies, and adaptation to life in extreme soils. In P. Dion & C. S. Nautiyal (Eds.), Microbiology of Extreme Soils (pp. 15–43). Springer. https://doi.org/10.1007/978-3-540-74231-9_2

Tran, N.H., Urase, T. and Kusakabe, O. (2010). Biodegradation characteristics of pharmaceutical substances by whole fungal culture Trametes versicolor and its laccase. Journal of Water and Environment Technology, 8(2), pp. 125–140. https://doi.org/10.2965/jwet.2010.125

Varga, B., Somogyi, V., Meiczinger, M., Kováts, N. and Domokos, E. (2019). Enzymatic treatment and subsequent toxicity of organic micropollutants using oxidoreductases – A review. Journal of Cleaner Production, 221, pp. 306–322. https://doi.org/10.1016/j.jclepro.2019.02.135

Venkateswar Reddy, M., Kumar, G., Mohanakrishna, G., et al. (2020). Review on the production of medium and small chain fatty acids through waste valorization and CO_2 fixation. Bioresource Technology, 309, p. 123400. https://doi.org/10.1016/j.biortech.2020.123400

Wang, J., Yu, S., Feng, F. and Lu, L. (2019). Simultaneous purification and immobilization of laccase on magnetic zeolitic imidazolate frameworks: Recyclable biocatalysts with enhanced stability for dye decolorization. Biochemical Engineering Journal, 150, p. 107285. https://doi.org/10.1016/j.bej.2019.107285

Wong, J.K.H., Tan, H.K., Lau, S.Y., Yap, P.-S. and Danquah, M.K. (2019). Potential and challenges of enzyme incorporated nanotechnology in dye wastewater treatment: A review. Journal of Environmental Chemical Engineering, 7(4), p.103261. doi:https://doi.org/10.1016/j.jece.2019.103261

Yadav, A.N., Suyal, D.C., Kour, D., Rajput, V.D., Rastegari, A.A. and Singh, J. (2022). Bioremediation and waste management for environmental sustainability. Journal of Applied Biology & Biotechnology, pp. 1–5. https://doi.org/10.7324/jabb.2022.10s201

Yang, X., Wu, X., Hao, H. and He, Z. (2008). Mechanisms and assessment of water eutrophication. Journal of Zhejiang University SCIENCE B, 9(3), pp. 197–209. https://doi.org/10.1631/jzus.B0710626

Yin, X., Sun, X., Yang, Y. and Ding, H. (2019). In-situ bioremediation of soil pollution with electric heating temperature regulation bio-ventilation. IOP Conference Series: Earth and Environmental Science, 242, p. 042011. https://doi.org/10.1088/1755-1315/242/4/042011

Zhang, Q., Yang, W., Ngo, H.H., Guo, W., Jin, P., Dzakpasu, M., Yang, S., Wang, Q., Wang, X.F. and Ao, D. (2016). Current status of urban wastewater treatment plants in China. Environment International, 92–93, pp. 11–22. https://doi.org/10.1016/j.envint.2016.03.024

Zhao, M.M., Zheng, G.G., Kang, X., Zhang, X., Guo, J., Zhang, M., Zhang, J., Chen, Y. and Xue, L. (2023). Arsenic pollution remediation mechanism and preliminary application of arsenic-oxidizing bacteria isolated from industrial wastewater. Environmental Pollution, 324, pp. 121384–121384. https://doi.org/10.1016/j.envpol.2023.121384

Zheng, Y., Zhang, T., Lu, Y., et al. (2021). Monascus pilosus YX-1125: An efficient digester for directly treating ultra-high-strength liquor wastewater and producing short-chain fatty acids under multiple-stress conditions. Bioresource Technology, 331, p. 125050. https://doi.org/10.1016/j.biortech.2021.125050.

Zhu, D., Adebisi, W.A., Ahmad, F., et al. (2020). Recent development of extremophilic bacteria and their application in biorefinery. Frontiers in Bioengineering and Biotechnology, 8, p. 483. https://doi.org/10.3389/fbioe.2020.00483.

Zhu, D., Qaria, M.A., Zhu, B., et al. (2022). Extremophiles and extremozymes in lignin bioprocessing. Renewable and Sustainable Energy Reviews, 157, p. 112069. https://doi.org/10.1016/J.RSER.2021.112069

13 Exploration of Microbial Enzymes in Bioremediation

Divya Bajaj, Varsha Yadav, Anamika Dhyani, and Neetu Kukreja Wadhwa

13.1 INTRODUCTION

The global population has seen steady growth over the past few decades. This rise in population is driving economic demand toward higher resource extraction for the supply of food and energy. Industrial and agricultural expansion and other anthropogenic activities have expedited the generation of potentially toxic products/wastes. Inappropriate disposal of such wastes has led to the accumulation of compounds such as plastics, heavy metals, synthetic dyes, insecticides, chemical fertilizers, pesticides, etc. in nature (Markandeya and Shukla, 2022). This coupled with other wastes generated by construction and demolition, mining activities, municipal waste, etc. has compounded the problem of pollution of natural resources. Contamination of our resources with xenobiotics/micropollutants is posing hazards to all life, affecting ecosystems adversely (Liu et al., 2020). Thus, there is a need to develop systems for better processing of wastes, to eliminate toxic components. Further, the current waste disposal methods have limited processing capacity and are unable to completely turn potentially toxic compounds. These harmful pollutants persist in the environment, posing a serious threat to all living organisms. The prolific biomineralization and transformation of such effluents remain an arduous challenge for environmentalists.

Several inorganic and organic compounds have been reported as soil and water contaminants. Inorganic contaminants include trace elements like arsenic (As), cadmium (Cd), copper (Cu), chromium (Cr), lead (Pb), manganese (Mn), mercury (Hg), and zinc (Zn). BTEX compounds (benzene, ethylbenzene, toluene, and xylene), pesticides like dichlorodiphenyltrichloroethane (DDT), dieldrin, and hexachlorobenzene (HCB), hexa-chlorohexane (HCH), polychlorinated biphenyls (PCBs) and polycyclic aromatic hydrocarbons (PAHs) are some of the organic contaminants (Kim et al., 2011; Kozak et al., 2017). Several pollutants have been implicated in severe health illnesses. Nondegradable pollutants can enter and biomagnify in the food chain. These pollutants can be consumed or absorbed by wildlife/aquatic life influencing their survival and reproduction. Further, some of these in turn may be consumed by humans. Crops/plants grown on contaminated soil can absorb pollutants and then pass them on to their consumers. The toxins in the soil can percolate

 DOI: 10.1201/9781003407683-13

slowly and steadily into the water table leading to contamination of the underground water. Consumption of meat/plant food/water loaded with such contaminants can cause ill effects on our health (https://www.conserve-energy-future.com/causes-and-effects-of-soil-pollution.php). Pollution of air, water, and soil accounts for at least 9 million deaths every year (Munzel et al., 2023). Thus, there is a need for contriving approaches to degrade the hazardous pollutants abating the environment. Various methods, including physical, chemical, and biological, are used to restore natural resources. Conventional methods of remediation such as chemical and physical methods (oxidizing agents, chemical precipitation, membrane filtration, electrochemical treatments, adsorption of pollutants, electrokinesis, photocatalysis, photo-oxidation, and ion exchange, etc.) are very expensive and non-sustainable (Muharrem and Ince, 2017; Bisht et al., 2017; Dixit et al., 2015; Lee et al., 2007; Venkata et al., 2015; Yadav et al., 2017). Remediation using biological agents can help in the cost-effective elimination of these contaminants. Further, such methods are effective in the degradation/detoxification, eradication, or immobilization of hazardous materials. The use of microbes and their enzymes for the removal of pollutants thus comes across as environmentally benign, cost-effective, and eco-friendly (Paul et al., 2005; Raghunandan et al., 2018, Agrawal et al., 2020a). The current chapter explores the role of such microbial enzymes in bioremediation.

13.2 TYPES OF BIOREMEDIATION

Bioremediation can be employed for treating contaminated soil, groundwater, and other resources at the site with minimal disturbance (*in situ* methods) and also to the soil and water removed from the site by excavation or pumping (*ex situ* methods). Several species of bacteria, fungi, and algae with degradative capacities have been employed for degrading and converting complex pollutants to less toxic forms (Table 13.1) These microbes can, however, be propagated under conditions suitable for their growth. Physical characteristics like pH, temperature, oxygen, moisture, availability of nutrients, and presence of other toxic compounds tend to influence the activity of the organisms. Maintaining optimal conditions in open settings to achieve microbial growth thus presents a challenge to successful bioremediation. Further, many microbes effective as bioremediation agents under laboratory conditions may not be so productive under natural conditions (Dave and Das, 2021; Dua et al., 2002; Vidali, 2001).

13.3 MICROBIAL ENZYMES: THE WORKHORSES OF BIOREMEDIATION

Microbes are equipped with enzymes with diverse substrate specificities and thus are the most suitable natural agents to remediate the environment of toxic contaminants. Bacteria, as well as molds (fungi), use a range of intracellular and extracellular enzymes to remediate stubborn compounds (Karigar and Rao, 2011; Vidali, 2001).

Cultivation of whole microbes for bioremediation involves the provision of continuous aeration, nutrition, and environmental conditions. Individual enzymes offer

TABLE 13.1
Microbes with Biodegradative Properties

Microbe	Example	Compound Metabolized/ Degraded by Microbe	Reference
Bacteria	*Alcaligenes*, *Achromobacter*, *Acinetobacter*, *Alcanivorax*, *Alteromonas*, *Arthrobacter*, *Burkholderia*, *Bacillus*, *Cycloclasticus*, *Enterobacter*, *Flavobacterium*, *Pseudomonas*, *Acinetobacter*, *Sphingomonas*, *Nocardia*, *Flavobacterium*, *Rhodococcus*, and *Mycobacterium*	Alkanes, hydrocarbons, pesticides, and polyaromatic compounds	Dell Anno et al., 2021
Fungi	*Aspergillus*, *Curvularia*, *Drechslera*, *Fusarium*, *Lasiodiplodia*, *Mucor*, *Penicillium*, *Rhizopus*, *Trichoderma*	Organochlorinated pesticide, polycyclic aromatic hydrocarbons (PAHs)	Balaji et al., 2014; Llado et al., 2013; Chang et al., 2016
Algae	*Chlorella*, *Selenastrum*, *Scenedemus*	Polycyclic aromatic hydrocarbons (naphthalene, phenanthrene, and pyrene)	Lei et al., 2007; Takacova et al., 2014; De Llasera et al., 2016; Ghosal et al., 2016

greater activity per unit, specificity, efficient handling and storage, and types of reactions including oxidation, reduction, and hydrolytic cleavage of the contaminant. However, the low yield of such enzymes is a limiting factor affecting their use. Enzyme effectiveness, activity, stability, substrate selectivity, and shelf life can be improved by using sustainable approaches. Different enzymes have been categorized into six different groups by the International Union of Biochemistry. These classes include oxidoreductases, transferases, hydrolases, lyases, isomerases, and ligases. Bioremediating enzymes derived from different sources belonging to different classes are listed in Table 13.2 (Mousavi et al., 2021). Several enzymes from bacteria, molds, and plants have been exploited for bioremediation of organic pollutants owing to their ability to metabolize diverse compounds. Microbial enzymes can catalyze several different types of reactions including oxidation, reduction, and hydrolytic cleavage of the contaminant.

13.3.1 Oxidoreductases (EC 1)

Oxidoreductase enzymes catalyze the transfer of electrons from a molecule (electron donor/reductant) to another molecule (electron acceptor/oxidant). Several strategies employ different subclasses of oxidoreductases (dehydrogenases, oxidases, oxygenases, oxidative deaminases, hydroxylase, and peroxidase) to remediate the pollutants.

TABLE 13.2
Class Wise Distribution of Enzymes Employed in Bioremediation

Class of Enzyme	Subclass		Examples	Use in Bioremediation	References
OXIDOREDUCTASE (EC 1)	Dehydrogenases		Aldehyde Dehydrogenases	Oxidative inactivation and catabolism of chloramphenicol	Zhang et al., 2023
	Oxidases		Laccases	Oxidation of aromatic amines, diamines, phenolic compounds, lignin-related compounds, PAHs, dyes, pesticides benzenediol	Arregui et al., 2019; Mousavi et al., 2021; Roohi et al., 2016; Agrawal et al., 2020b
	Oxygenases	Monooxygenases	P-450 cytochrome	Xenobiotic detoxification, catabolism of PAHs like naphthalene, oxidative, peroxidative, metabolism of steroids, bile acids, fatty acids, vitamins, prostaglandins, & leukotrienes	England et al., 1998; Syed et al., 2010
		Dioxygenases	Naphthalene Dioxygenase, benzoate dioxygenase, toluene dioxygenase, Catechol dioxygenase	Chlorophenol oxidation, diuron degradation	Jouanneau et al., 2011; Karigar and Rao, 2011; Rao, 2010
	Oxidative deaminases		Monoamine oxidase	Degradation of monocyclic aromatic amines like aminophenols, chloramphenicol, nitroaniline	Arora, 2015
	Hydroxylases		Alkane hydroxylases	Degradation of oil, chlorinated hydrocarbons, fuel additives, etc.	Nie et al., 2014; Ji et al., 2013; Elumalai et al., 2017
	Peroxidases		Lignin peroxidase and manganese peroxidase	Degradation of dioxins, polychlorinated biphenyls pesticides, petroleum hydrocarbons, trinitrotoluene, industrial dye effluents, herbicides	Bansal and Kanwar, 2013; Sellami et al., 2022

(*Continued*)

TABLE 13.2 (*Continued*)
Class Wise Distribution of Enzymes Employed in Bioremediation

Class of Enzyme	Subclass	Examples	Use in Bioremediation	References
TRANSFERASES (EC 2)	Kinases	Polyphosphate kinase	Bioremediation of phosphate and heavy metal contamination in municipal waste	Keasling et al., 1998; Ruiz et al., 2011
	Acetyl transferases	Arylamine N-acetyltransferase 2	Detoxification of aromatic amine pesticide residue 3,4-dichloroaniline	Silar et al., 2011; Martins et al., 2009
	Others	Rhodanese	Cyanide detoxification	Itakorode et al., 2022
HYDROLASES (EC 3)	Esterases	Phosphotriesterase, carboxylesterase, polyurethanase, esterase	Biodegradation of oil spill, food waste, plastic waste, organophosphate pesticides, insecticides, diethyl glycol adipate, polyurethanes, and aromatic and aliphatic polyesters	Mousavi et al., 2021; Howard et al., 1999
	Lipases		Oils, fats, and protein of greasy effluents petroleum contaminants	Karigar and Rao, 2011
	Peptidases	Leucyl aminopeptidase, arginyl aminopeptidase, prolyl aminopeptidase	Degradation of several organic compounds and polymers like PHB (polyhydroxybutyrate), oil spill control, crustacean waste deproteinization	Roohi et al., 2017; Schmidt et al., 2021; Rathore et al., 2022
	Phosphatases, nitrilase, dehalogenase	Cyanide hydratase (nitrilase)	Degrade cyanide to formamide used for bioremediation of wastewater from coal coking	Martinkova et al., 2015

(*Continued*)

TABLE 13.2 (*Continued*)
Class Wise Distribution of Enzymes Employed in Bioremediation

Class of Enzyme	Subclass	Examples	Use in Bioremediation	References
LYASE (EC 4)	Aldolases	Hydratase aldolase	Biodegradation of hydrocarbons like phenanthrene, pyrene, and naphthalene	Radhakrishnan et al., 2023
	Decarboxylases	Manganese decarboxylase	Metabolism of lignocellulosic compounds	Bala et al., 2022
	Dehydratases	Threonine dehydratase	Biomineralization of cadmium sulfide nanocrystals	Ma et al., 2021
ISOMERASE (EC 5)	Isomerases, epimerases, racemases, intramolecular transferases	α-Methylacyl-CoA racemase	Catabolism of natural rubber degradation and methyl-branched hydrocarbons	Sarkar and Mandal, 2020
LIGASE (EC 6)	Carboxylases, ligases, etc.	Isophthalyl-CoA ligase	Isophthalate activation and degradation	Junghare et al., 2022

13.3.1.1 Dehydrogenases

Aldehyde dehydrogenases have recently been reported to be involved in oxidative inactivation and catabolism of chloramphenicol (Zhang et al., 2023).

13.3.1.2 Oxidases

Oxidases facilitate oxidation-reduction reactions by transferring hydrogen from a substrate to oxygen. This results in the formation of either water or hydrogen peroxide. Polyphenol oxidases such as laccases (multicopper oxidases) and tyrosinases catalyze the oxidation of various phenolic and non-phenolic aromatic compounds into less toxic quinones. Laccases are widely distributed in higher plants, bacteria, fungi, and insects and can oxidize phenolic as well as non-phenolic molecules including aromatic amines, diamines, and lignin-related molecules. These are used for pulp delignification, waste detoxification, degradation of pesticides or insecticides, remediation of dyes, and treatment of other environmental xenobiotics and biosensors (Alsukaibi, 2022; Jain & Ramteke, 2016). Laccase from *Bacillus licheniformis* can decolorize azo, indigo, and anthraquinone dyes by 80% in an hour using acetosyringone as a mediator (Bu et al., 2020). Further, these enzymes have been utilized using nano biotechnological approaches.

13.3.1.3 Oxygenases

Oxygenases play a crucial role in the metabolism of xenobiotic compounds. Oxidative degradation of aromatic compounds involves monooxygenases and dioxygenases.

13.3.1.3.1 Monooxygenases

Several aerobic biodegrading microbes utilize monooxygenases, like methane monooxygenase, toluene monooxygenase, ammonia monooxygenase, and cytochrome P450s (Hazen et al., 2009). Studies indicate the role of methane monooxygenase enzymes of methanotrophs in bioremediation of trichloroethylene (TCE). Ammonia monooxygenase has been implicated in the oxidation of ammonia to nitrite by *Nitrosomonas europaea* (https://clu-in.org/download/techfocus/biochlor/hazen_cometabolic_bio_2009.pdf).

Cytochrome P450s (CYPs) play a crucial role in the metabolism of various endogenous and exogenous compounds through oxidation, peroxidation, and reduction. Compounds such as steroids, bile acids, fatty acids, prostaglandins, vitamins, leukotrienes, and xenobiotics and organic pollutants such as dioxins, PCBs (polychlorinated biphenyls), PCDDs (polychlorinated dibenzo-p-dioxins), PCDFs (polychlorinated dibenzofurans), PAHs, aliphatic hydrocarbons, etc. can be metabolized using cytochromes (Eibes et al., 2015; Kumar, 2010; Sakaki et al., 2013; Thatoi et al., 2014). Native and engineered P450s have been identified for metabolizing PAHs such as chrysene, naphthalene, phenanthrene, and pyrene (England et al., 1998; Sideri et al., 2013; Syed et al., 2010; Syed et al., 2013). Bacterial and eukaryotic CYPs have been reported to oxidize and degrade aliphatic hydrocarbons (Pinto et al., 2020). Class I CYPs are well conserved and are active in the metabolism of procarcinogens, xenobiotics, and drugs, e.g., CYP1A1, CYP1A2, and CYP3A4. Class II CYPs are highly polymorphic and active in the metabolism of drugs, but not of pre-carcinogens, e.g., CYP2B6 and CYP2C9 (Lamb et al., 2000).

13.3.1.3.2 Dioxygenases

Dioxygenases incorporate molecular oxygen into diverse substrates. Depending on the site of cleavage of the aromatic ring, these are classified as estradiol and intradiol dioxygenases. Extradiol dioxygenases channelize the substrates to a meta-pathway whereas intradiol dioxygenases channelize their substrates into an ortho-pathway. Catechol 1,2-dioxygenases derived from *Acinetobacter calcoaceticus*, *Pseudomonas putida,* and several other microbes are implicated in the cleavage of compounds with aromatic carbon rings (Guzik et al., 2011). Hydroxyquinol metabolizing dioxygenase is critical for breaking down aromatic xenobiotics, such as polychlorinated phenols, as well as amino- and nitrophenols.

13.3.1.4 Oxidative Deaminases

These are amine oxidases that catalyze the oxidative deamination of amino acids. Monoamine oxidases have been reported to play a role in the transformation of aromatic amines into ammonia and the corresponding aromatic acids.

13.3.1.5 Hydroxylase

Hydroxylases catalyze oxidation reactions by adding hydroxyl groups to their substrate. Alkane hydroxylases can catalyze the degradation of oils, fuel additives, chlorinated hydrocarbons, and many other compounds (Elumalai et al., 2017; Ji et al., 2013; Nie et al., 2014).

13.3.1.6 Peroxidase

Peroxidase enzymes catalyze the reduction of peroxides, such as hydrogen peroxide (H_2O_2), and the oxidation of several organic and inorganic compounds (Hamid, 2009; Chanwun et al., 2013). Peroxidases produced by white rot fungi possess the unique ability to decompose a variety of environmental pollutants such as dioxins, polychlorinated biphenyls, petroleum hydrocarbons, munitions wastes (such as trinitrotoluene), industrial dye effluents, herbicides, and pesticides (Marco-Urrea and Reddy, 2012). Peroxidases derived from bacteria (*B. sphaericus*, *B. subtilis*, *Citrobacter* sp., *Pseudomonas* sp.), cyanobacteria (*Anabaena* sp.), molds (*Candida krusei*, *Coprinopsis cinerea*, *Phanerochaete chrysosporium*), actinomycetes (*Streptomyces* sp., *Thermobifidafusca*), and yeast are utilized in the paper-pulp industry for lignin degradation, for dye decolorization, for sewage treatment, and also as biosensors. Ligninolytic peroxidase enzymes are used for the degradation and detoxification of lignocellulosic waste in the environment (Hong et al., 2011). Manganese peroxidase, dye-decolorizing peroxidase, and versatile peroxidases are also employed extensively for bioremediation (Bansal and Kanwar, 2013; Sellami et al., 2022).

13.3.2 Transferases (EC 2)

Transferases play a crucial role in biodegradation by transferring specific groups from a donor molecule to a recipient. Phosphotransferases (polyphosphate kinase) have been employed for bioremediation of phosphate contamination in municipal wastewater and in heavy metal tolerance (Keasling et al., 1998). Arylamine N-acetyltransferase catalyzes the chemical modification of the amine group with an acetyl group. This enzyme is implicated in the detoxification and metabolic activation of numerous pesticide-derived aromatic amines and other xenobiotic compounds including drugs and environmental carcinogens (Zhou et al., 2013). A thiosulfate sulfurtransferase (rhodanese) of *Klebsiella oxytoca* has been reported to have the ability to detoxify cyanides from polluted environments (Itakorode et al., 2022).

13.3.3 Hydrolases (EC 3)

These enzymes can hydrolyze upon esters, peptide bonds, carbon-halide bonds, ureas, and thioesters bonds by adding –H or –OH groups of H_2O. Various types of hydrolases are utilized in the bioremediation of chemicals, including herbicides, nitrile compounds, polymers, and organophosphorus compounds such as pesticides (Chaudhry et al., 1988). Esterases are employed in bioremediation of organophosphorus herbicides and pesticides. Carbofuran hydrolase produced by *Pseudomonas* sp. organophosphate hydrolase enzymes is used for metabolizing organophosphate pesticides (Chaudhry et al., 1988). Esterases and lipases have been effectively used in biodegrading food waste, plastic waste, oil spills, and organophosphates. Nitrilases degrade herbicides, plastics, and polymers by hydrolyzing their nitrile group to carboxylic acid and ammonia. Cyanide hydratase (a nitrilase) of *Fusarium lateritium*, *Neurospora crassa*, and *Gloeocercospora sorghi* has been reported to degrade cyanide to formamide. These enzymes are utilized for the bioremediation of wastewater from coal-coking and metal-plating processes (Martinkova et al., 2015).

Dehalogenases can catalyze the biodegradation of organohalide compounds (chlorinated, brominated, and some iodinated pollutants) by catalyzing cofactor-independent cleavage of carbon–halogen bonds of toxic environmental pollutants (Klages et al., 1983). Proteases owing to their hydrolytic power are used widely for the bioremediation of food industry waste and domestic wastewater (Bhandari et al., 2021). Keratinases are serine hydrolases that can have the ability to break down various types of keratin proteins which are commonly found in domestic wastewater and poultry waste. These are also used for hydrolyzing keratin from chicken feathers generated by poultry-processing industries. Serine proteases of *Bacillus licheniformis* can process waste feather to protein hydrolysate which can be used as animal feed or organic fertilizer (Bhandari et al., 2021). Lipases hydrolyze ester bonds of triglycerides into fatty acid and glycerol and thus play a role in the degradation of lipids derived from microbes, animal plants, oil residues, petroleum contaminants, effluents, and soil recovery. Lipases isolated from microbial and fungal species such as *Candida cylindracea*, *Pseudomonas aeruginosa*, *Pseudomonas fluorescens*, *Penicillium simplicissimum*, and *Rhizopus delemar* have been reported for their ability to degrade micro-nano plastics that are potentially hazardous to living organisms in all ecosystems (Ali et al., 2023).

13.3.4 Lyases (EC 4)

Lyases can cleave upon carbon–carbon, carbon–oxygen, carbon–nitrogen, and other bonds by elimination, leaving double bonds. Lyases play a key role in degradation pathways of PAHs (LeVieux et al., 2018). An isomerase organomercurial lyase has been implicated in breaking carbon mercury bonds in organo-mercuric compounds (Krout et al., 2022).

13.3.5 Isomerases (EC 5)

Isomerases have been implicated in the conversion of stable isomeric forms of xenobiotics to their alternative forms. α-Methylacyl coenzyme, a racemase of *Gordonia polyisoprenivorans*, has been implicated in the catabolism of natural rubber and methyl-branched hydrocarbons (Arenskotter et al., 2008).

13.3.6 Ligases (EC 6)

Ligases catalyze the formation of C–O, C–N, and C–S bonds. Several ligases have been utilized in bioremediation. Isophthalate (coenzyme A ligase) has been used in the degradation of isophthalate compounds that are used widely in plastics, coatings, packaging, and pharmaceuticals. The isophthalate coenzyme A ligase of *Syntrophorhabdus aromaticivorans* activates the degradation of isophthalate by catalyzing the formation of isophthalyl-CoA (Junghare et al., 2022). Similarly, aryl carboxylate CoA ligase activates intermediates for anaerobic degradation of monocyclic aromatic compounds such as benzenes, benzoates, cresols, phenols, and aromatic amino acids.

13.4 RECENT TRENDS IN THE USE OF ENZYMES FOR BIOREMEDIATION

Purified enzymes offer the advantage of high efficiency, with the potential to degrade an array of molecules. However, a high cost of production, poor reusability, and low stability associated with such enzymes limit their usage. Here are some key trends in novel remediation approaches: nanoparticle-enzyme hybrids can be created by conjugating or immobilizing enzymes onto nanoparticles, producing nanozymes that catalyze reactions (Figure 13.1).

13.4.1 Nanoparticle-Enzyme Hybrids

Enzymes from various sources, like microorganism extracts, can be conjugated or immobilized onto nanoparticles, creating nanozymes which are nanoparticles with the enzymatic ability. Such nanoparticles offer the advantage of improved catalysis and greater enzyme stability, higher surface area for enhanced contact with pollutants, and provide controlled release of enzymes in the environment. They can also be engineered to target specific contaminants or be used in combination with other remediation techniques, such as nano-remediation.

Bioremediation processes that use biosurfactants for enhancing the solubility of hydrophobic contaminants at contaminated marine sites have been improvised with nanoparticles (Rando et al., 2022). For instance, iron-based nanoparticles are reported to stimulate the production of biosurfactants by bacteria like the marine actinobacterium *Nocardiopsis* MSA13A (Kiran et al., 2014). Studies indicate that

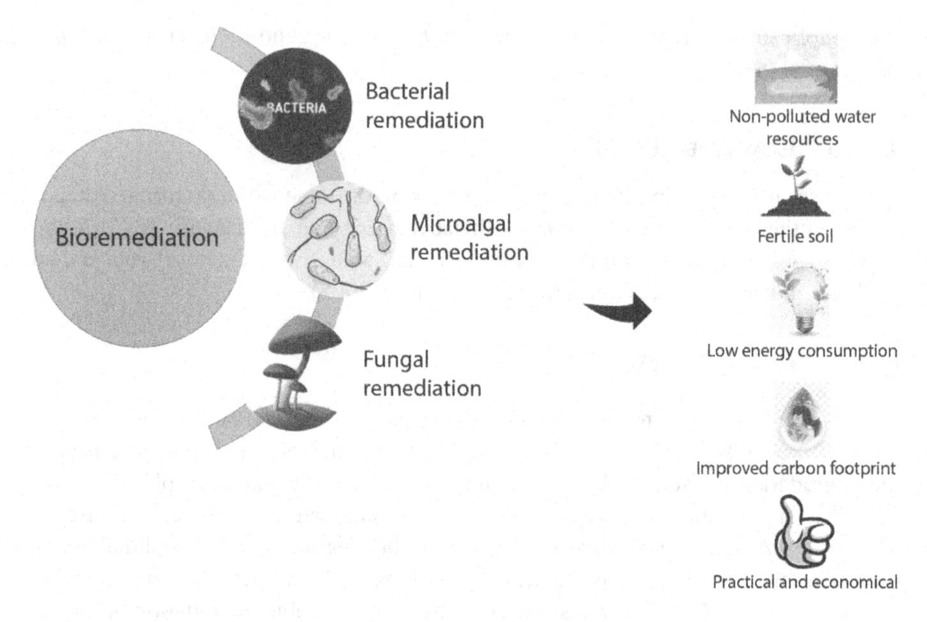

FIGURE 13.1 Bioremediation: key players and advantages.

heavy metals in contaminated soil can be absorbed and transformed with nanoparticles (Rizwan and Ahmed, 2018). According to Wang et al. (2017), gold nanozymes can catalyze the generation of hydroxyl radicals (OH) from the decomposition of H_2O_2 under acidic conditions.

Nanozymes exhibit exceptional advantages such as robustness, high stability, and low-cost production along with ease of scale-up (Maduraiveeran and Jin, 2017). The use of nanozymes provides several advantages: they are low-cost, robust, highly stable, have a longer shelf life, and are easily mass-produced, as compared to traditional enzymes (Wei and Wang, 2013).

13.4.2 Enzyme-Mediated Biofilm Reactors

Biofilms are organized communities of microorganisms which are aggregations of single or mixed microbial cells that attach to biotic and abiotic surfaces in aqueous environments. Owing to their flexibility, genetic diversity, and simple cellular structure, microbial communities show promise to repair damaged environments. Biofilm-forming bacteria can outcompete other bacteria for nutrients and tolerate contaminants, creating a protective environment for cells. For the elimination of persistent pollutants, enzymes are also added to the biofilm matrix or secreted by biofilm-forming microorganisms to enhance the degradation rates and stability of the biofilm system. It is considered an innovative method for effectively biodegrading environmental contaminants.

Biofilm microorganisms sorb and metabolize pollutants and heavy metals via quorum sensing-regulated genes. Biofilms are also engineered using biotechnological techniques. Advancements in synthetic biology have enabled the reprogramming of biofilms, enhancing their function and biodegradation capabilities (Mohsin et al., 2021). Genetic engineering enables the biofilms to be used as biocatalysts in biotransformation applications. Huang et al. developed a *B. subtilis* biofilm that was able to degrade highly toxic mono(2-hydroxyethyl) terephthalic acid to less toxic terephthalic acid (Huang et al., 2019). This technique represents a new frontier in environmental sustainability and bioremediation (Mitra and Mukhopadhyay, 2016). Bacterial biofilms are a reliable indicator of heavy metal pollution in marine and terrestrial ecosystems (Maurya and Raj, 2020; Mosharaf et al., 2018). These also find application as biological control agents against phytopathogens and biofertilizers for the betterment of the quality and productivity of crops (Timmusk et al., 2017). Biofilm-mediated remediation is an affordable, eco-friendly, effective, and organized option for the removal/refinement of pollutants. However, the rate of bioremediation using this approach is markedly low when compared to the chemical treatment (Mitra and Mukhopadhyay, 2016).

13.4.3 Membrane Biofilm Reactors

Membrane biofilm reactors (MBRs) involve the breakdown of biological waste components with physical separation and biofilms, aided by a membrane for secondary settlement. MBRs are effective in treating urban and industrial wastewater by successfully eradicating both inorganic and organic matter (Di Fabio et al., 2013).

In contrast to the traditional filtration membranes, MBRs do not mediate the separation of liquids from solids, but the transmission of a gaseous substrate such as oxygen or hydrogen. It also demonstrates flexibility with regard to disinfection efficiency, tremendous volumetric loading, and influent fluctuation (Lin et al., 2012; Martin and Nerenberg, 2012). Novel MBRs utilize inexpensive and abundant gases like methane and carbon dioxide, in addition to hydrogen and oxygen. Methane-based MBRs allow the growth of methanotrophic bacteria that are known to co-metabolize organic molecules and donate electrons by generating decay products (Modin et al., 2008a, 2088b).

13.4.4 Genetically Engineered Enzymes for Bioremediation

Genetically engineering desired enzymes is an emerging approach to metabolize xenobiotics for energy and biomass (Bhandari et al., 2021). GE enzymes are designed or modified to exhibit some of the following specific properties that make them highly effective in bioremediation processes.

1. Enhanced catalytic activity: Improved catalytic activity enables enzymes to break down pollutants more efficiently (Mousavi et al., 2021). For example, certain enzymes can be modified to have a broader substrate specificity, allowing them to degrade a wider range of contaminants.
2. Increased stability and tolerance: Resistance to harsh environmental conditions and factors such as temperature, pH, salinity, and the presence of inhibitors can be incorporated in variants of natural enzymes. This improves their performance and longevity in contaminated environments.

3. Targeted pollutant degradation: GE enzymes can be tailored to target specific pollutants. This specificity allows for more precise and effective remediation of polluted sites.
4. Adaptation to new environments: Bioremediation often involves the introduction of organisms to contaminated sites. GE enzymes can be modified to help these organisms adapt to new environments.
5. Metabolic pathway optimization: GE enzymes can be utilized to optimize metabolic pathways in microorganisms involved in bioremediation (Aquino et al., 2011). By introducing or modifying enzymes involved in pollutant degradation pathways, the efficiency of the overall process can be improved.

13.4.5 Enzyme-Assisted Phytoremediation

Phytoremediation uses plant-associated microbial communities to remove or transform toxic contaminants from various environments. This can be enhanced by the administration of enzymes directly to the plants or the rhizosphere (root zone) for degrading contaminants or increasing the bioavailability of pollutants for plant uptake. The addition of enzymes is seen to improve the efficiency of phytoremediation and increase the range of contaminants for remediation (Truu et al., 2015).

A variety of phytoremediation strategies are employed for the bioremediation of heavy metal-contaminated soils, like (i) phytostabilization, which refers to the process of using plants to decrease the level of heavy metal bioavailability in the soil, (ii) phytoextraction, which involves the use of plants to extract and eliminate heavy metals from the soil, (iii) phytovolatilization, that utilizes plants to absorb heavy metal from soil and discharge it into the atmosphere as volatile compounds, and (iv) phytofiltration, which uses hydroponically cultured plants to absorb or adsorb heavy metal ions from sources like groundwater and aqueous waste (Salt et al., 1995; Ernst, 2005; Marques et al., 2009).

13.4.6 Enzyme-Assisted Electro-Bioremediation

Electrokinetic (EK) remediation involves applying a direct electric current across electrodes kept in the contaminated soil. Electrokinetic processes such as electrophoresis, electromigration, and electroosmosis allow interaction between microorganisms, nutrients, and contaminants present in the soil (Paillat et al., 2000; Rodrigo et al., 2014). The EK approach is particularly recommended for polluted soils with low permeability where the traditional methods do not allow groundwater to transport the pollutants along the soil (Reddy and Cameselle 2009). However, this method also presents certain limitations like heating up of the soil by the Joule effect, development of extreme pH zones around electrodes, or reduced mobility of non-polar pollutants in the soil. Hence, to overcome these challenges, a coupled technology known as electro-bioremediation is adopted (Yeung and Gu, 2011).

Electro-bioremediation involves the use of electrical currents to drive the movement of contaminants and enhance microbial activity. Enzymes can be introduced to the electrochemical system to facilitate the degradation of pollutants. The electric field can aid in the transport of enzymes to targeted locations, improving their effectiveness in pollutant degradation (Barba et al. 2020, Wick et al., 2007). The electric field is used to supply nutrients, oxygen, or other substances known to improve the microbial activity in the soils. The electrodes supply the soil with electron acceptors and donors for metabolic reactions. The electric field could also allow the mobilization and transport of pollutants and microflora, enhancing the bioavailability of the pollutants and hence the rate of degradation. The success of electro-bioremediation lies with the appropriate application of the electric field to obtain the required environmental conditions enhancing the metabolic activity of the soil microflora. Any significant changes in the physicochemical properties of soil that could compromise the survival of the microorganisms need to be avoided.

13.4.7 Enzymatic Biosensors for Monitoring

Biosensors have promising applications in detecting pollutants and transducing signals in the environment. Biosensors typically comprise two key components – a biosensing element and a transducer. The biological sensory component can be oligonucleotide probes, antibodies enzymes, and cell receptors that bind and react with the analyte. The transducer could be an optical, physicochemical, or piezo-electric compound that can convert biological signals to optical and electrical signals (Nguyen et al., 2019). IUPAC nomenclature describes biosensors as integrated receptor–transducer devices, capable of producing selective quantitative or semiquantitative analytical information utilizing a biological recognition element (Tangahu et al., 2011; Thevenot et al., 2001). It is recognized for several biological applications including the detection of several contaminants such as pollutants, microbial load, metabolites, etc. (Neethirajan et al., 2018).

Enzymes can be utilized in biosensors to detect and monitor the levels of contaminants in real time during bioremediation processes. These biosensors employ enzymes that react specifically with target pollutants, generating measurable signals that indicate the presence and concentration of contaminants. This allows for efficient monitoring and adjustment of bioremediation strategies. Further, the association of nanoparticles with biosensors has helped to improve various parameters like reliability, validity, lower detection limit, residence time, stability, sensitivity, etc. (Malekzad et al., 2017).

13.4.8 Enzyme-Mediated Microbial Fuel Cells (MFCs)

This technique was first identified by M.C. Potter, who proved that microbes can be used for degrading pollutants and obtaining energy from bacteria (Rahimnejad et al., 2015). It was later reported that bacteria harbor electrochemically active redox proteins capable of transferring electrons to the anode (Drendel et al., 2018). MFCs are here hence considered as an eco-friendly mechanism for the generation of electricity along with keeping a check on pollutants from wastewater. The working of MFCs involves the transfer of electrons from the exoelectrogenic bacteria to the anode and then to the cathode utilizing an external circuit. There is direct movement of protons from anode to cathode in an oxidative environment (Kim et al., 2015). The electrotroph microbes present in the MFCs accept electrons from electrodes and in turn convert the toxic pollutants into less toxic ones (Palanisamy et al., 2019). Enzyme-mediated MFCs are bio-electrochemical systems capable of generating electricity using the metabolic activity of microorganisms. Enzymes are associated with the MFC systems to enhance the elimination of contaminants while successfully producing electrical energy.

MFCs effectively employ the catalytic ability of microorganisms in harnessing the energy from chemical bonds of organic compounds (pollutants) to generate electrical energy. Electroactive bacteria, also known as EAB, are a group of bacterial and archaeal species that are typically found in various environments such as water bodies, soil, and sediments. These microorganisms are able to interact electrically with their extracellular environments as well as with each other, even in extreme conditions. EABs are reported to generate an electrical current in MFCs. Biodegradable organic matter is utilized by EABs for the generation of energy. MFCs also find application in remediating wastewater and soil and sediments from organic and inorganic pollutants (Abbas and Rafatullah, 2021; Goto et al., 2015).

13.4.9 Immobilization of Enzymes

Immobilization of enzymes involves anchoring or entrapping biocatalysts on an inert carrier or matrix, making them durable and reusable. This allows the exchange of media containing the enzyme's substrate and/or the activator or inhibitor (Zhu, 2007). These methods have gained attention for their potential to enhance the shelf life and reusability of enzymes in bioremediation.

Immobilized enzymes can be attached to solid supports or encapsulated within materials, allowing for better control and retention of their catalytic activity. Cross-linking agents are used to create a stable network of polymers or hydrogels around the enzymes, Techniques like glutaraldehyde cross-linking, enzyme encapsulation in calcium alginate beads, and covalent binding to polymer matrices have been employed to improve the efficacy of this approach.

The enzyme immobilization techniques can be broadly categorized as adsorption, binding to a surface, cross-linking or copolymerization, entrapment, and encapsulation (Bayat et al., 2015; Kourkoutas et al., 2004). Studies with cyanide detoxifying enzymes (rhodanese) of *K. oxytoca* reported increased thermal stability and its reusability on immobilization of enzyme (Caterina et al., 2013, Itakorode et al., 2022).

13.5 BIOINFORMATICS APPROACHES

Biological data from various repositories, such as databases of chemical structure and composition, RNA/protein expression, organic compounds, catalytic enzymes, microbial degradation pathways, and comparative genomics are being explored for interpretation of the degradative metabolic pathways of different organisms (Yergeau et al., 2012). Further, genomics, transcriptomics, metabolomics, and proteomics approaches have facilitated the profiling of the bioremediation potential of microorganisms. Such knowledge can be utilized in characterizing microbial proteins/enzymes with the potential to degrade contaminants (Figure 13.2).

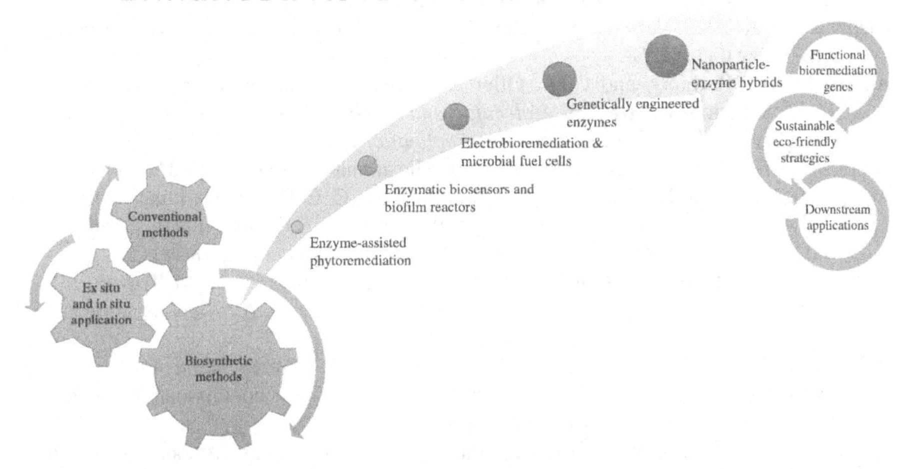

FIGURE 13.2 New approaches in bioremediation.

13.6 CONCLUSION

Rapid urbanization and massive synthetic waste generation have led to the accumulation of toxic/hazardous components in various resources. The cleanest approach for decontamination is offered by bioremediation. Several different microorganisms and their enzymes have been investigated for their biodegradative potential and several different enzymes have been proposed for such usage. New cost-effective strategies involving nanoparticle-based enzymes, recombinant enzymes, biosensors, etc. have been attracting great interest. Proper utilization of the ideal candidate enzyme or key battery of enzymes using the technologies described is required for efficient bioremediation. Further technological advancements can be utilized to enhance enzyme stability and efficiency for special conditions or particular substrates (Festa et al., 2008; Theerachat et al., 2012). While omics technologies have played a significant role in these developments, interventions such as enzyme-mediated MFCs, immobilized enzymes, and nano enzymes hold promise for effective remediation of the environment (Ufarte et al., 2015).

REFERENCES

Abbas, S. Z., and M. Rafatullah. 2021. Recent advances in soil microbial fuel cells for soil contaminants remediation. *Chemosphere* 272:129691.

Agrawal, K., J. Shankar, R. Kumar, and P. Verma. 2020a. Insight into multicopper oxidase laccase from *Myrothecium verrucaria* ITCC-8447: A case study using in silico and experimental analysis. *Journal of Environmental Science and Health, Part B* 55(12): 1048–1060.

Agrawal, K., J. Shankar, and P. Verma. 2020b. Multicopper oxidase (MCO) laccase from *Stropharia* sp. ITCC-8422: An apparent authentication using integrated experimental and in silico analysis. *3 Biotech* 10:413.

Ali, S., S. A. Khan, M. Hamayun, and I.-J. Lee. 2023. The recent advances in the utility of microbial lipases: A review. *Microorganisms* 11(2):510.

Alsukaibi, A. K. 2022. Various approaches for the detoxification of toxic dyes in wastewater. *Processes* 10(10):1968.

Aquino, E., C. Barbieri, and C. A. Oller Nascimento. 2011. Engineering Bacteria for Bioremediation. *Progress in Molecular and Environmental Bioengineering – From Analysis and Modeling to Technology Applications*. InTech.

Arora, P. K. 2015. Bacterial degradation of monocyclic aromatic amines. *Front Microbiol* 6:820.

Arregui, L., M. Ayala, X. Gomez-Gil, G. Gutierrez-Soto, C. E. Hernandez-Luna, M. Herrera de Los Santos, and N. A. Valdez-Cruz. 2019. Laccases: Structure, function, and potential application in water bioremediation. *Microbial Cell Factories* 18(1):1–33.

Bala, S., D. Garg, B. Thirumalesh, M. Sharma, K. Sridhar, B. Inbaraj, and M. Tripathi. 2022. Recent strategies for bioremediation of emerging pollutants: A review for a green and sustainable environment. *Toxics* 10(8):1–24.

Balaji, V., P. Arulazhagan, and P. Ebenezer. 2014. Enzymatic bioremediation of polyaromatic hydrocarbons by fungal consortia enriched from petroleum contaminated soil and oil seeds. *Journal of Environmental Biology* 35:521–529.

Bansal, N., and S. S. Kanwar. 2013. Peroxidase(s) in environment protection. *The Scientific World Journal* 714639. Volume 2013, Article ID 714639, 9 pages.

Barba, S., J. Villasenor, M. A. Rodrigo, and P. Canizares. 2020. Towards the optimization of electro-bioremediation of soil polluted with 2,4-dichlorophenoxyacetic acid. *Environmental Technology & Innovation* 20:101156.

Bayat, Z., M. Hassanshahian, and S. Cappello. 2015. Immobilization of microbes for bioremediation of crude oil polluted environments: A mini review. *The Open Microbiology Journal* 9:48–54.

Bhandari, S., D. K. Poudel, R. Marahatha, S. Dawadi, K. Khadayat, S. Phuyal, S. Shrestha, S. Gaire, K. Basnet, U. Khadka, and N. Parajuli. 2021. Microbial enzymes used in bioremediation. *Journal of Chemistry* 2021:1–17.

Bisht, R., M. Agarwal, and K. Singh. 2017. Methodologies for removal of heavy metal ions from wastewater: An overview. *Interdisciplinary Environmental Review* 18(2):124–142.

Bu, T., R. Yang, Y. Zhang, Y. Cai, Z. Tang, C. Li, C. Li, Q. Wu, and H. Chen. 2020. Improving decolorization of dyes by laccase from *Bacillus licheniformis* by random and site-directed mutagenesis. *PeerJ* 8:e10267.

Caterina, G. C. M. N., H. E. Toma, and L. H. Andrade. 2013. Superparamagnetic nanoparticles as versatile carriers and supporting materials for enzymes. *Journal of Molecular Catalysis B: Enzymatic* 85–86:71–92.

Chang, Y. T., J. F. Lee, K. H. Liu, Y. F. Liao, and V. Yang. 2016. Immobilization of fungal laccase onto a nonionic surfactant-modified clay material: Application to PAH degradation. *Environmental Science and Pollution Research* 23:4024–4035.

Chanwun, T., N. Muhamad, N. Chirapongsatonkul, and N. Churngchow. 2013. *Hevea brasiliensis* cell suspension peroxidase: Purification, characterization and application for dye decolorization. *AMB Express* 3:14.

Chaudhry, G. R., A. N. Ali, and W. B. Wheeler. 1988. Isolation of a methyl parathion-degrading *Pseudomonas* sp. that possesses DNA homologous to the opd gene from a *Flavobacterium* sp. *Applied and Environmental Microbiology* 54(2):288–293.

Dave, S., and J. Das. 2021. Role of Microbial Enzymes for Biodegradation and Bioremediation of Environmental Pollutants: Challenges and Future Prospects. *Bioremediation for Environmental Sustainability* (pp. 325–346). Published by Elsevier. https://doi.org/10.1016/B978-0-12-820524-2.00013-4

De Llasera, M. P. G., J. J. Olmos-Espejel, G. Diaz-Flores, and A. Montano-Montiel. 2016. Biodegradation of benzo(a)pyrene by two freshwater microalgae *Selenastrum capricornutum* and *Scenedesmus acutus*: A comparative study useful for bioremediation. *Environmental Science and Pollution Research* 23:3365–3375.

Dell Anno, F., E. Rastelli, C. Sansone, C. Brunet, A. Ianora, and A. Dell Anno. 2021. Bacteria, fungi and microalgae for the bioremediation of marine sediments contaminated by petroleum hydrocarbons in the omics era. *Microorganisms* 9(8):1695.

Di Fabio, S., S. Lampis, L. Zanetti, F. Cecchi, and F. Fatone. 2013. Role and characteristics of problematic biofilms within the removal and mobility of trace metals in a pilot-scale membrane bioreactor. *Process Biochemistry* 48:1757–1766.

Dixit, R., X. Wasiullah, D. Malaviya, K. Pandiyan, U. B. Singh, A. Sahu, and D. Paul. 2015. Bioremediation of heavy metals from soil and aquatic environment: An overview of principles And criteria of fundamental processes. *Sustainability* 7(2):2189–2212.

Drendel, G., E. R. Mathews, L. Semenec, and A. E. Franks. 2018. Microbial fuel cells, related technologies, and their applications. *Applied Sciences* 8:2384.

Dua, M., A. Singh, N. Sethunathan, and A. K. Johri. 2002. Biotechnology and bioremediation: Successes and limitations. *Applied Microbiology and Biotechnology* 59(2–3):143–152.

Eibes, G., A. Arca-Ramos, G. Feijoo, J. M. Lema, and M. T. Moreira. 2015. Enzymatic technologies for remediation of hydrophobic organic pollutants in soil. *Applied Microbiology and Biotechnology* 99(21):8815–8829.

Elumalai, P., P. Parthipan, O. P. Karthikeyan, and A. Rajasekar. 2017. Enzyme-mediated biodegradation of long-chain n-alkanes (C_{32} and C_{40}) by thermophilic bacteria. *3 Biotech* 7:116.

England, P. A., C. F. Harford-Cross, J. A. Stevenson, D. A. Rouch, and L. L. Wong. 1998. The oxidation of naphthalene and pyrene by cytochrome P450 (cam). *FEBS Letters* 424:271–274.

Ernst, W. H. 2005. Phytoextraction of mine wastes–options and impossibilities. *Chem Erde Geochem* 65:29–42.

Festa, G., F. Autore, F. Fraternali, P. Giardina, and G. Sannia. 2008. Development of new laccases by directed evolution: Functional and computational analyses. *Proteins: Structure, Function, and Bioinformatics* 72(1):25–34.

Ghosal, D., S. Ghosh, T. K. Dutta, and Y. Ahn. 2016. Current state of knowledge in microbial degradation of polycyclic aromatic hydrocarbons (PAHs): A review. *Frontiers in Microbiology* 7:1369.

Goto, Y., N. Yoshida, Y. Umeyama, T. Yamada, R. Tero, and A. Hiraishi. 2015. Enhancement of electricity production by graphene oxide in soil microbial fuel cells and plant microbial fuel cells. *Frontiers in Bioengineering and Biotechnology* 3(1-8).

Guzik, U., I. Gren, K. Hupert-Kocurek, and D. Wojcieszynska. 2011. Catechol 1,2-dioxygenase from the new aromatic compounds-degrading *Pseudomonas putida* strain N6. *International Biodeterioration & Biodegradation* 65(3):504–512.

Hamid, M. 2009. Potential applications of peroxidases. *Food Chemistry* 115(4):1177–1186.

Hazen, T. C., R. Chakraborty, J. M. Fleming, I. R. Gregory, J. P. Bowman, L. Jimenez, D. Zhang, S. M. Pfiffner, F. J. Brockman, and G. S. Sayler. 2009. Use of gene probes to assess the impact and effectiveness of aerobic in situ bioremediation of TCE. *Archives in Microbiology* 191:221–232.

Hong, Y., M. Dashtban, S. Chen, R. Song, and W. Qin. 2011. Enzyme production and lignin degradation by four basidiomycetous fungi in submerged fermentation of peat containing medium. *International Journal of Biology* 4(1):172.

Howard, G. T., C. Ruiz, and N. P. Hilliard. 1999. Growth of *Pseudomonas chlororaphis* on a polyester–polyurethane and the purification and characterization of a polyurethanase–esterase enzyme. *International Biodeterioration and Biodegradation* 43:7–12.

Huang, J., S. Liu, C. Zhang, X. Wang, J. Pu, F. Ba, S. Xue, H. Ye, T. Zhao, K. Li, and Y. Wang. 2019. Programmable and printable bacillus subtilis biofilms as engineered living materials. *Nature Chemical Biology* 15(1):34–41.

Itakorode, B. O., R. E. Okonji, and N. Torimiro. 2022. Cyanide bioremediation potential of *Klebsiella oxytoca* JCM 1665 rhodanese immobilized on alginate-glutaraldehyde beads. *Biocatalysis and Biotransformation* 41:1–10.

Jain, N., and P. W. Ramteke. 2016. Microbial laccase and its applications in bioremediation. *Current Biochemical Engineering* 3(2):110–121.

Ji, Y., G. Mao, Y. Wang, and M. Bartlam. 2013. Structural insights into diversity and n-alkane biodegradation mechanisms of alkane hydroxylases. *Frontiers in Microbiology* 4:1–13.

Jouanneau, Y., F. Martin, S. Krivobok, and J. C. Willison. 2011. Ring-Hydroxylating Dioxygenases Involved in PAH Biodegradation: Structure, Function, Biodiversity. *Microbial Bioremediation of Non-Metals: Current Research* (pp. 149–175). Caister Academic Press.

Junghare, M., J. Frey, K. M. Naji, D. Spiteller, G. Vaaje-Kolstad, and B. Schink. 2022. Isophthalate: Coenzyme a ligase initiates anaerobic degradation of xenobiotic isophthalate. *BMC Microbiology* 22(1):227.

Karigar, C. S., and S. S. Rao. 2011. Role of microbial enzymes in the bioremediation of pollutants: A review. *Enzyme Research* 7:805187.

Keasling, J. D., S. J. Van Dien, and J. Pramanik. 1998. Engineering polyphosphate metabolism in *Escherichia coli*: Implications for bioremediation of inorganic contaminants. *Biotechnology and Bioengineering* 58(2–3):231–239.

Kim, E. J., S. D. Choi, and Y. S. Chang. 2011. Levels and patterns of polycyclic aromatic hydrocarbons (PAHs) in soils after forest fires in South Korea. *Environmental Science and Pollution Research* 18:1508–1517.

Kim, K. Y., W. Yang, and B. E. Logan. 2015. Impact of electrode configurations on retention time and domestic wastewater treatment efficiency using microbial fuel cells. *Water Research* 80:41–46.

Kiran, G. S., L. A. Nishanth, S. Priyadharshini, K. Anitha, and J. Selvin. 2014. Effect of fe nanoparticle on growth and glycolipid biosurfactant production under solid state culture by marine *Nocardiopsis* sp. MSA13A. *BMC Biotechnology* 14(1):1–10.

Klages, U., S. Krauss, and F. Lingens. 1983. 2-Haloacid dehalogenase from a 4-chlorobenzoate-degrading *Pseudomonas* spec. CBS 3. *Hoppe-Seyler's Zeitschrift für physiologische Chemie* 364(5):529–535.

Kourkoutas, Y., A. Bekatorou, I. M. Banat, R. Marchant, and A. A. Koutinas. 2004. Immobilization technologies and support materials suitable in alcohol beverages production: A review. *Food Microbiology* 21(4):377–397.

Kozak, K., M. Ruman, K. Kosek, G. Karasinski, L. Stachnik, and Z. Polkowska. 2017. Impact of volcanic eruptions on the occurrence of PAHs compounds in the aquatic ecosystem of the southern part of West Spitsbergen (Hornsund Fjord, Svalbard). *Water* 9(1):42.

Krout, I. N., T. Scrimale, D. Vorojeikina, E. S. Boyd, and M. D. Rand. 2022. Organomercurial lyase (merB)-mediated demethylation decreases bacterial methylmercury resistance in the absence of mercuric reductase (merA). *Applied and Environmental Microbiology* 88(6):10–22.

Kumar, S. 2010. Engineering cytochrome P450 biocatalysts for biotechnology, medicine and bioremediation. *Expert Opinion on Drug Metabolism & Toxicology* 6(2):115–131.

Lamb, D. C., D. E. Kelly, S. Masaphy, G. L. Jones, and S. L. Kelly. 2000. Engineering of heterologous cytochrome P450 in Acinetobacter sp.: Application for pollutant degradation. *Biochemical and Biophysical Research Communications* 276(2):797–802.

Lee, I.-H., Y. C. Kuan, and J. M. Chern. 2007. Equilibrium and kinetics of heavy metal ion exchange. *Journal of the Chinese Institute of Chemical Engineers* 38(1):71–84.

Lei, A. P., Z. L. Hu, Y. S. Wong, and N. F. Y. Tam. 2007. Removal of fluoranthene and pyrene by different microalgal species. *Bioresource Technology* 98:273–280.

LeVieux, J. A., B. Medellin, W. H. Johnson Jr, K. Erwin, W. Li, I. A. Johnson, and C. P. Whitman. 2018. Structural characterization of the hydratase-aldolases, NahE and PhdJ: Implications for the specificity, catalysis, and N-acetylneuraminate lyase subgroup of the aldolase superfamily. *Biochemistry* 57(25):3524–3536.

Lin, H., W. Gao, F. Meng, B. Q. Liao, K. T. Leung, L. Zhao, and H. Hong. 2012. Membrane bioreactors for industrial wastewater treatment: A critical review. *Critical Reviews in Environmental Science and Technology* 42(7):677–740.
Liu, J., Y. Song, M. Tang, Q. Lu, and G. Zhong. 2020. Enhanced dissipation of xenobiotic agrochemicals harnessing soil microbiome in the tillage-reduced rice-dominated agroecosystem. *Journal of Hazardous Materials* 398:122954.
Llado, S., S. Covino, A. M. Solanas, M. Vinas, M. Petruccioli, and A. D'annibale. 2013. Comparative assessment of bioremediation approaches to highly recalcitrant PAH degradation in a real industrial polluted soil. *Journal of Hazardous Materials* 248–249: 407–414.
Ma, N., R. Cai, and C. Sun. 2021. Threonine dehydratase enhances bacterial cadmium resistance via driving cysteine desulfuration and biomineralization of cadmium sulfide nanocrystals. *Journal of Hazardous Materials* 417.
Maduraiveeran, G., and W. Jin. 2017. Nanomaterials based electrochemical sensor and biosensor platforms for environmental applications. *Trends in Environmental Analytical Chemistry* 13:10–23.
Malekzad, H., P. SahandiZangabad, H. Mirshekari, M. Karimi, and M. R. Hamblin. 2017. Noble metal nanoparticles in biosensors: Recent studies and applications. *Nanotechnology* Reviews 6(3):301–329.
Marco-Urrea, E., and C. A. Reddy. 2012. Degradation of chloro-organic pollutants by white rot fungi microbial degradation of xenobiotics. *Environmental Science and Engineering* 2:31–66.
Markandeya, D. M., and S. P. Shukla. 2022. Hazardous consequences of textile mill effluents on soil and their remediation approaches. *Cleaner Engineering and Technology* 7:100434.
Marques, A. P., A. O. Rangel, and P. M. Castro. 2009. Remediation of heavy metal contaminated soils: Phytoremediation as a potentially promising clean-up technology. *Critical Reviews in Environmental Science and Technology* 39:622–654.
Martin, K. J., and R. Nerenberg. 2012. The membrane biofilm reactor (MBfR) for water and wastewater treatment: Principles, applications, and recent developments. *Bioresource Technology* 122:83–94.
Martinkova, L., A. B. Vesela, A. Rinagelova, and M. Chmatal. 2015. Cyanide hydratases and cyanide dihydratases: Emerging tools in the biodegradation and biodetection of cyanide. *Applied Microbiology and Biotechnology* 99:8875–8882.
Martins, M., F. Rodrigues-Lima, J. Dairou, A. Lamouri, F. Malagnac, P. Silar, and J. M. Dupret. 2009. An acetyltransferase conferring tolerance to toxic aromatic amine chemicals: Molecular and functional studies. *Journal of Biological Chemistry* 284(28): 18726–18733.
Maurya, A., and A. Raj. 2020. Recent Advances in the Application of Biofilm in Bioremediation of Industrial Wastewater and Organic Pollutants. *Microorganisms for Sustainable Environment and Health* (pp. 81–118). Elsevier.
Mitra, A., and S. Mukhopadhyay. 2016. Biofilm mediated decontamination of pollutants from the environment. *AIMS Bioengineering* 3(1):44–59.
Modin, O., K. Fukushi, F. Nakajima, and K. Yamamoto. 2008a. A membrane biofilm reactor achieves aerobic methane oxidation coupled to denitrification (AME-D) with high efficiency. *Water Science and Technology 58*(1):83–87.
Modin, O., K. Fukushi, and K. Yamamoto. 2008b. Simultaneous removal of nitrate and pesticides from groundwater using a methane-fed membrane biofilm reactor. *Water Science and Technology 58*(6):1273–1279.
Mohsin, M. Z., R. Omer, J. Huang, A. Mohsin, M. Guo, J. Qian, and Y. Zhuang. 2021. Advances in engineered bacillus subtilis biofilms and spores, and their applications in

bioremediation, biocatalysis, and biomaterials. *Synthetic and Systems Biotechnology* 6(3):180–191.

Mosharaf, M. K., M. Z. H. Tanvir, M. M. Haque, M. A. Haque, M. A. A. Khan, A. H. Molla, M. Z. Alam, M. S. Islam, and M. R. Talukder. 2018. Metal-adapted bacteria isolated from wastewaters produce biofilms by expressing proteinaceous curli fimbriae and cellulose nanofibers. *Frontiers in Microbiology* 9:1334.

Mousavi, S. M., S. A. Hashemi, S. M. Iman Moezzi, N. Ravan, A. Gholami, C. W. Lai, W. H. Chiang, N. Omidifar, K. Yousefi, and G. Behbudi. 2021. Recent advances in enzymes for the bioremediation of pollutants. *Biochemistry Research International* 2021:12. https://doi.org/10.1155/2021/5599204

Muharrem, I. N. C. E., and O. K. Ince. 2017. An overview of adsorption technique for heavy metal removal from water/wastewater: A critical review. *International Journal of Pure and Applied Sciences* 3(2):10–19.

Munzel, T., O. Hahad, A. Daiber, and P. J. Landrigan. 2023. Soil and water pollution and human health: What should cardiologists worry about? *Cardiovascular Research* 119(2): 440–449.

Neethirajan, S., V. Ragavan, X. Weng, and R. Chand. 2018. Biosensors for sustainable food engineering: Challenges and perspectives. *Biosensors* 8(1):23.

Nguyen, H. H., S. H. Lee, U. J. Lee, C. D. Fermin, and M. Kim. 2019. Immobilized enzymes in biosensor applications. *Materials* 12(1):1–34.

Nie, Y., C. Q. Chi, H. Fang, J. L. Liang, S. L. Lu, G. L. Lai, Y. Q. Tang, and X. L. Wu. 2014. Diverse alkane hydroxylase genes in microorganisms and environments. *Scientific Reports 4*(1): 4968.

Paillat, T., E. Moreau, P. O. Grimaud, and G. Touchard. 2000. Electrokinetic phenomena in porous media applied to soil decontamination. *IEEE Transactions on Dielectrics and Electrical Insulation* 7(5):693–704.

Palanisamy, G., H. Y. Jung, T. Sadhasivam, M. D. Kurkuri, S. C. Kim, and S. H. A. Roh. 2019. Comprehensive review on microbial fuel cell technologies: Processes, utilization, and advanced developments in electrodes and membranes. *Journal of Cleaner Production* 221:598–621.

Paul, D., G. Pandey, J. Pandey, and R. K. Jain. 2005. Accessing microbial diversity for bioremediation and environmental restoration. *TRENDS in Biotechnology 23*(3):135–142.

Pinto, E. S. M., M. Dorn, and B. C. Feltes. 2020. The tale of a versatile enzyme: Alpha-amylase evolution, structure, and potential biotechnological applications for the bioremediation of n-alkanes. *Chemosphere* 250(3). https://doi.org/10.1016/j.chemosphere.2020.126202

Radhakrishnan, A., P. Balaganesh, M. Vasudevan, N. Natarajan, A. Chauhan, J. Arora, A. Ranjan, V. D. Rajput, S. Sushkova, T. Minkina, R. Basniwal, R. Kapardar, and R. Srivastav. 2023. Bioremediation of hydrocarbon pollutants: Recent promising sustainable approaches, scope, and challenges. *Sustainability* 15(7):5847.

Raghunandan, K., A. Kumar, S. Kumar, K. Permaul, and S. Singh. 2018. Production of gellan gum, an exopolysaccharide, from biodiesel-derived waste glycerol by *Sphingomonas* spp. *3 Biotech* 8:1–13.

Rahimnejad, M., A. Adhami, S. Darvari, A. Zirepour, and S. E. Oh. 2015. Microbial fuel cell as new technology for bioelectricity generation: A review. *Alexandria Engineering Journal* 54:745–756.

Rando, G., S. Sfameni, M. Galletta, D. Drommi, S. Cappello, and M. R. Plutino. 2022. Functional nanohybrids and nanocomposites development for the removal of environmental pollutants and bioremediation. *Molecules* 27(15):4856.

Rao, M. A. 2010. Role of enzymes in the remediation of polluted environments. *Journal of Soil Science and Plant Nutrition* 10(3):333–353.

Rathore, S., A. Varshney, S. Mohan, and P. Dahiya. 2022. An innovative approach of bioremediation in enzymatic degradation of xenobiotics. *Biotechnology and Genetic Engineering Reviews* 38:1–32.

Reddy, K. R., and C. Cameselle. 2009. *Electrochemical Remediation Technologies for Polluted Soils, Sediments and Groundwater* (p. 732). John Wiley & Sons. https://doi.org/10.1002/9780470523650

Rizwan, M., and M. U. Ahmed. 2019. Nanobioremediation: Ecofriendly Application of Nanomaterials. In: Martínez, L., Kharissova, O., Kharisov, B. (eds) *Handbook of Ecomaterials*. Springer. https://doi.org/10.1007/978-3-319-68255-6_97

Rodrigo, M. A., N. Oturan, and M. A. Oturan. 2014. Electrochemically assisted remediation of pesticides in soils and water: A review. *Chemical Reviews 114*(17):8720–8745.

Roohi, K. Bano, M. Kuddus, M. R. Zaheer, Q. Zia, M. F. Khan, G. M. Ashraf, A. Gupta, and G. Aliev. 2017. Microbial enzymatic degradation of biodegradable plastics. *Current Pharmaceutical Biotechnology* 18(5):429–440. https://doi.org/10.2174/1389201018666170523165742

Roohi, N. Jain, and R. Pramod. 2016. Microbial laccase and its applications in bioremediation. *Current Biochemical Engineering* 3:110–121.

Ruiz, O. N., D. Alvarez, G. G. Ruiz, and C. Torres. 2011. Characterization of mercury bioremediation by transgenic bacteria expressing metallothionein and polyphosphate kinase. *BMC Biotech* 11(1):82.

Sakaki, T., K. Yamamoto, and S. Ikushiro. 2013. Possibility of application of cytochrome P450 to bioremediation of dioxins. *Biotechnology and Applied Biochemistry* 60(1): 65–70.

Salt, D. E., M. Blaylock, N. P. B. A. Kumar, V. Dushenkov, B. D. Ensley, and I. Chet. 1995. Phytoremediation: A novel strategy for the removal of toxic metals from the environment using plants. *Nature Biotechnology* 13:468–474.

Sarkar, B., and S. Mandal. 2020. Microbial Degradation of Natural and Synthetic Rubbers. *Microbial Bioremediation & Biodegradation*. Springer. 527–550.

Schmidt, J. M., T. M. Royalty, K. G. Lloyd, and A. D. Steen. 2021. Potential activities and long lifetimes of organic carbon-degrading extracellular enzymes in deep subsurface sediments of the Baltic Sea. *Frontiers in Microbiology* 12. https://doi.org/10.3389/fmicb.2021.702015

Sellami, K., A. Couvert, N. Nasrallah, R. Maachi, M. Abouseoud, and A. Amrane. 2022. Peroxidase enzymes as green catalysts for bioremediation and biotechnological applications: A review. *Science of the Total Environment* 806(1):150500.

Sideri, A., A. Goyal, G. Di Nardo, G. E. Tsotsou, and G. Gilardi. 2013. Hydroxylation of non-substituted polycyclic aromatic hydrocarbons by cytochrome P450 BM3 engineered by directed evolution. *Journal of Inorganic Biochemistry* 120:1–7.

Silar, P., J. Dairou, A. Cocaign, F. Busi, F. R. Lima, and J. M. Dupret. 2011. Fungi as a promising tool for bioremediation of soils contaminated with aromatic amines, a major class of pollutants. *Nature Reviews Microbiology* 9(6):477.

Syed, K., H. Doddapaneni, V. Subramanian, Y. W. Lam, and J. S. Yadav. 2010. Genome-to-function characterization of novel fungal P450 monooxygenases oxidizing polycyclic aromatic hydrocarbons (PAHs). *Biochemical and Biophysical Research Communications* 399:492–497.

Syed, K., A. Porollo, D. Miller, and J. S. Yadav. 2013. Rational engineering of the fungal P450 monooxygenase CYP5136A3 to improve its oxidizing activity toward polycyclic aromatic hydrocarbons. *Protein Engineering, Design and Selection* 26:553–557.

Takacova, A., M. Smolinska, J. Ryba, T. Mackulak, J. Jokrllova, P. Hronec, and G. Cik. 2014. Biodegradation of benzo[*a*]pyrene through the use of algae. *Central European Journal of Chemistry* 12:1133–1143.

Tangahu, B. V., S. R. Sheikh Abdullah, H. Basri, M. Idris, N. Anua, and M. Mukhlisin. 2011. A review on heavy metals (As, Pb, and Hg) uptake by plants through phytoremediation. *International Journal of Chemical Engineerning* 2011:31. https://doi.org/10.1155/2011/939161

Thatoi, H., S. Das, J. Mishra, B. P. Rath, and N. Das. 2014. Bacterial chromate reductase, A potential enzyme for bioremediation of hexavalent chromium: A review. *Journal of Environmental Management* 146:383–399.

Theerachat, M., S. Emond, and E. Cambon. 2012. Engineering and production of laccase from *Trametes versicolor* in the yeast *Yarrowia lipolytica. Bioresource Technology* 125:267–274.

Thevenot, D. R., K. Toth, R. A. Durst, and G. S. Wilson. 2001. Electrochemical biosensors: Recommended definitions and classification. *Biosensors Bioelectronics* 16(1–2): 121–131. https://doi.org/10.1016/s0956-5663(01)00115-4

Timmusk, S., L. Behers, J. Muthoni, A. Muraya, and A. C. Aronsson. 2017. Perspectives and challenges of microbial application for crop improvement. *Frontiers in Plant Science* 8:49.

Truu, J., M. Truu, M. Espenberg, H. Nolvak, and J. Juhanson. 2015. Phytoremediation and plant-assisted bioremediation in soil and treatment wetlands: A review. *The Open Biotechnology Journal* 9(1):85–92.

Ufarte, L., E. Laville, S. Duquesne, and G. Potocki-Veronese. 2015. Metagenomics for the discovery of pollutant degrading enzymes. *Biotechnology Advances* 33(8):1845–1854.

Venkata, S. P., S. N. Prakash, and B. S. Nagahanumantharao. 2015. Electrokinetic removal of heavy metals from soil. *Journal of Electrochemical Science and Engineering* 5(1):47–65.

Vidali, M. 2001. Bioremediation. An overview. *Pure and Applied Chemistry* 73(7):1163–1172.

Wang, C., D. Chen, Q. Wang, and R. Tan. 2017. Kanamycin detection based on the catalytic ability enhancement of gold nanoparticles. *Biosensors and Bioelectronics* 91:262–267.

Wei, H., and E. Wang. 2013. Nanomaterials with enzyme-like characteristics (nanozymes): Next-generation artificial enzymes. *Chemical Society Reviews* 14:6060–6093.

Wick, L. Y., L. Shi, and H. Harms. 2007. Electro-bioremediation of hydrophobic organic soil-contaminants: A review of fundamental interactions. *Electrochimica Acta* 52(10): 3441–3448.

Yadav, K., N. Gupta, V. Kumar, and J. Singh. 2017. Bioremediation of heavy metals from contaminated sites using potential species: A review. *Indian Journal of Environmental Protection* 37:65–84.

Yergeau, E., S. Sanschagrin, D. Beaumier, and C. W. Greer. 2012. Metagenomic analysis of the bioremediation of diesel-contaminated Canadian high arctic soils. PloS one 7(1).

Yeung, A. T., and Y. Y. Gu. 2011. A review on techniques to enhance electrochemical remediation of contaminated soils. *Journal of Hazardous Materials* 195:11–29.

Zhang, L., M. Toplak, R. Saleem-Batcha, L. Höing, R. Jakob, N. Jehmlich, M. von Bergen, T. Maier, and R. Teufel. 2023. Bacterial dehydrogenases facilitate oxidative inactivation and bioremediation of chloramphenicol. *ChemBioChem* 24(2):e202200632. https://doi.org/10.1002/cbic.202200632

Zhou, X., Z. Ma, D. Dong, and B. Wu. 2013. Arylamine n-acetyltransferases: A structural perspective. *British Journal of Pharmacology* 169(4):748–760.

Zhu, Y. 2007. Immobilized Cell Fermentation for Production of Chemicals and Fuels. *Bioprocessing for Value-Added Products from Renewable Resources*, 373–396. Published by Elsevier.

14 Exploration of Microbial Hydrolases for Bioremediation

Utsha Ghosh, Kanika Jangir, Parikshana Mathur, Payal Chaturvedi, and Charu Sharma

14.1 INTRODUCTION

The tremendous increase in urbanisation and industrialisation has led to increased land, water, and air pollution. Industries have used heavy metals extensively, which has resulted in their direct or indirect entry into water resources. These are indestructible and can easily accumulate in the surrounding environment. Various industries like tannery, cosmetics, and textiles release harmful environmental pollutants (Srivastava 2021). Therefore, its management and remediation are essential for a sustainable future (Pandey and Singh 2019).

14.2 PHYSICAL AND CHEMICAL METHODS

Several physical and chemical methods are in practice for waste management. The practice of burying waste in a hole and covering it up was a standard method in the past. However, this approach was not viable as it required digging a new hole each time, rendering it highly inefficient. Advancements like high-temperature incineration and chemical decomposition have modified traditional technologies. Although these processes are highly successful at removing a variety of contaminants, they also have several drawbacks (Karigar and Rao 2011). Chlorination, a chemical remediation method, affects the properties of organic matter and generates several by-products (Song et al. 2021). Ozonation is a highly energy-consuming chemical treatment effective against various rebellious pollutants (Srivastava 2021). The drawbacks of these methods have led to efforts to produce the bioremediation process as a better alternative (Karigar and Rao 2011). Moreover, bioremediation technologies are constantly improving, utilising naturally occurring or genetically modified microorganisms to remove toxic substances from contaminated areas (Soccol et al. 2003).

14.3 BIOREMEDIATION

Bioremediation involves methods and actions to biotransform a contaminated space to its original form (Thassitou and Arvanitoyannis 2001). It is a naturally occurring process by which living organisms, primarily microbes, either immobilise or

DOI: 10.1201/9781003407683-14

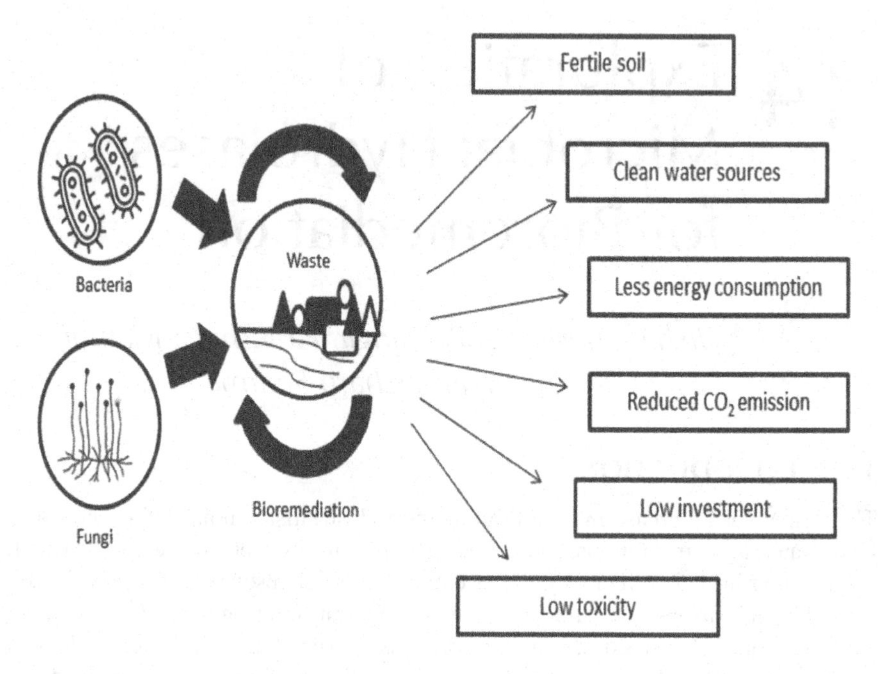

FIGURE 14.1 Microbes for reduction of waste.

transform environmental contaminants into harmless end products. This process has an advantage over all other processes as it removes contaminants permanently from the environment with the help of mineralisation or biochemical transformation. This process is less expensive and prevents the usage of harsh chemical and physical treatments (Soccol et al. 2003) (Figure 14.1).

The microbial bioremediation technique accelerates the growth of indigenous microorganisms by improving the level of available nutrients and maintaining optimum pH conditions. Microbes with biotransforming abilities can also be inoculated on the polluted sites. Moreover, immobilised microbial enzymes can also remove or transform toxic substances (Bollag and Bollag 1995). Microbes play an important role in bioremediation. The community structure of the microbes and conditions of the environment, such as temperature and redox in situ, are crucial factors that control the fate of bioremediation. A low population of functional microorganisms and limited nutrient conditions can also hinder the bioremediation process (Zhang and Yoshikawa 2020).

There are various ways of bioremediation. It is broadly categorised into *ex situ* bioremediation and *in situ* bioremediation as discussed in Figure 14.2. The *ex situ* method is a more efficient method; however, it is less cost-effective. *In situ* remediation is a less effective method but more cost-effective (Williams 2006).

14.3.1 Bioremediation Mediated by Bacterial Enzymes

Microorganisms produce a variety of enzymes that aid in the detoxification and metabolism of hazardous compounds. These enzymes can be oxygenases, hydrolases,

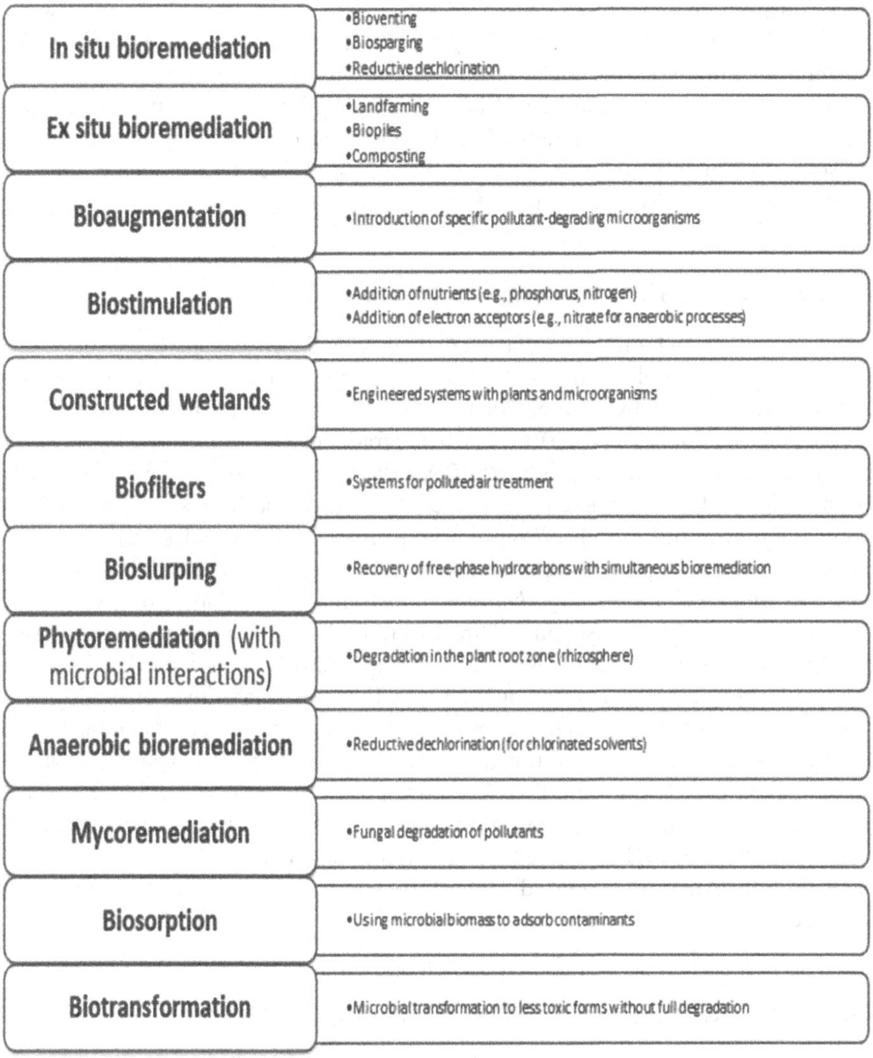

In situ bioremediation	•Bioventing •Biosparging •Reductive dechlorination
Ex situ bioremediation	•Landfarming •Biopiles •Composting
Bioaugmentation	•Introduction of specific pollutant-degrading microorganisms
Biostimulation	•Addition of nutrients (e.g., phosphorus, nitrogen) •Addition of electron acceptors (e.g., nitrate for anaerobic processes)
Constructed wetlands	•Engineered systems with plants and microorganisms
Biofilters	•Systems for polluted air treatment
Bioslurping	•Recovery of free-phase hydrocarbons with simultaneous bioremediation
Phytoremediation (with microbial interactions)	•Degradation in the plant root zone (rhizosphere)
Anaerobic bioremediation	•Reductive dechlorination (for chlorinated solvents)
Mycoremediation	•Fungal degradation of pollutants
Biosorption	•Using microbial biomass to adsorb contaminants
Biotransformation	•Microbial transformation to less toxic forms without full degradation

FIGURE 14.2 Different types of bioremediations.

oxidases, dehydrogenases, or dehalogenase groups (Phale et al. 2019). Bacteria are widespread in the environment because of their catabolic ability. They can be easily cultivated in different environments to generate a variety of enzymes. Aerobic bacteria such as *Sphingomonas*, *Rhodococcus*, *Mycobacterium*, *Alcaligenes*, and *Pseudomonas* produce enzymes that degrade hydrocarbon and pesticides. Similarly, anaerobic bacteria produce enzymes that perform bioremediation of chloroform, trichloroethylene, and polychlorinated biphenyls (Bhandari et al. 2021).

Laccase obtained from thermophilic *Anoxybacillus gonensis* P39 has been used in the removal of textile dyes such as Fuchsine, Reactive Black 5, Acid Red 37, and Allura Red by 5.96%, 22.1%, 1.15%, and 1.34%, respectively, from wastewater (Yanmis et al. 2016). Saranya and colleagues conducted a study in 2019 to explore the potential of a thermostable acidic lipase from *Bacillus pumilus* for treating palm oil wastewater. The enzyme was very effective in biodegrading oil-containing wastewater, even under extreme conditions. *Pseudomonas aeruginosa* SL-72 obtained from mustard-oil-contaminated soil produced lipase that efficiently degraded 82.83% of crude oil in 7 days (Verma, S. et al. 2012).

14.3.2 Bioremediation Mediated by Fungal Enzymes

Fungi are pivotal in bioremediation due to their typical physical characteristics and a broad range of catabolic abilities. Toxic organic compounds from the contaminated sites can be eliminated via bioremediation mediated by fungi, which is a sustainable green route (Deshmukh et al. 2016). The biodegradation capability of *Phanerochaete chrysosporium*, *Bjerkandera adjusta*, *Pleurotus* sp., and *Trametes versicolor* has been explored. Ligninolytic enzymes like peroxidases and laccases obtained from these fungi have been used for the biotransformation of pesticides from contaminated wastewater by promoting microbial activity (Deshmukh et al. 2016). Another group of fungi, i.e., *Pleurotus flabellatus*, *P. ostreatus*, *P. florida*, and *P. sajorcaju*, were used in the decolourisation of Direct Blue 14 dye using manganese peroxidase and laccase (Singh et al. 2013). Verma, A.K., et al. (2012) isolated partially purified laccase from marine fungus and applied over Reactive Blue 4 dye for biodegradation, which recorded colour removal of up to 61% and, in 12 h, a twofold reduction in chemical oxygen demand. Additional analysis confirmed aromatic character change in the parent dye. Mathur et al. (2022) studied the extraction of cellulase enzyme from *Aspergillus niger* in the biodegradation of paper waste and saccharification of bagasse, and significant bioconversion of complex sugars by 83.94% in bagasse and 64.93% in the paper was reported. Table 14.1 shows a few more examples of fungal enzymes used in bioremediation.

14.4 HYDROLASES FOR BIOREMEDIATION

Hydrolases are microbial enzymes that use water to break chemical bonds and convert large compounds into smaller ones, making pollutants less toxic (Sharma et al. 2019). They support the water's ability to break down carbon-carbon, carbon-oxygen, carbon-phosphorus, carbon-nitrogen, sulphur-sulphur, sulphur-phosphorus, sulphur-nitrogen, and other bonds and also catalyse a number of related processes like condensation and alcohol reactions. According to Karigar and Rao (2011), the key benefits of this class of enzymes include their ready accessibility, affordability, environmental friendliness, and lack of cofactor selectivity. Hydrolytic enzymes are widely used in various fields like feed additives, chemical industries, and biomedical sciences (Kumar and Sharma 2019). There are several hydrolytic enzymes that have been studied for their utilisation in bioremediation for various pollutants like textile dye, pesticides, organic wastes, oil spills, and other chemicals (Figure 14.3).

TABLE 14.1
List of Microbial Enzymes Mediating Bioremediation

Name of Bacteria	Enzyme Produced	Name of Pollutant	References
Pseudomonas *Micrococcus* *Rhodococcus*	Esterase	Phthalates Phthalate esters	Phale et al. 2019
Streptomyces cyaneus *Tinea versicolor*	Laccase	Bisphenol A Diclofenac Mefenamic acid	Margot et al. 2013
Geobacillus thermocatenulatus (MS5)	Laccase	Indigo caramine Brilliant green Congo red RBBR Bromophenol blue	Verma and Shirkot 2014
Sphingobacterium sp. *S2*	Lipase	Poly(lactic acid) (PLA)	Satti et al. 2019
Candida antarctica *Humicola insolens*	Lipase Cutinase	Poly(ethylene terephthalate)	Carniel et al. 2017
Tinctoporellus sp. CBMAI 1061 *Marasmiellus* sp. CBMAI 1062 *Peniophora* sp. CBMAI 1063	Lignin peroxidase and Manganese peroxidase	Remazol Brilliant Blue R dye	Bonugli-Santos et al. 2012

14.4.1 Laccase

The economical treatment of waste from land and water provides enormous promise for laccase. Laccase breaks down lignin-related compounds and resistant pollutants in various industries like paper, pulp, textile, and bioremediation (Viswanath et al. 2014). Lu et al. (2007) studied the purification of laccase from the white rot fungus *Pycnoporuss anguineus* in their study. The purified enzyme was observed to decolorise Remazol Brilliant Blue R at critical pH (3) and temperature (65°C) conditions. The ability to break down dye makes it a powerful tool for dye decolourisation from dye-polluted sites. Similar studies have been reported in the past few years to remediate the pollutants using laccase (Agarwal et al. 2022a, 2022b). Niku-Paavola and Viikari (2000) have discussed the extraction of laccase from *Trametes hirsuta* and its use in oxidising alkenes. Researchers (Lu et al. 2013) examined the activity of recombinant laccase made from a gene obtained from *Bacillus licheniformis* and expressed in the bacteria *Pichia pastoris*. The enzyme exhibited notable thermostability at 50–70°C and high tolerance to NaCl. The purified laccase was observed to rapidly decolorise indigo carmine, reactive black 5, and reactive blue 19. Some more such examples are mentioned in Table 14.1.

14.4.2 Peroxidase

Peroxidase enzymes transform aromatic hydrocarbons, phenols, pesticides, polychlorinated biphenyls, and various other xenobiotics by free radical mechanism and

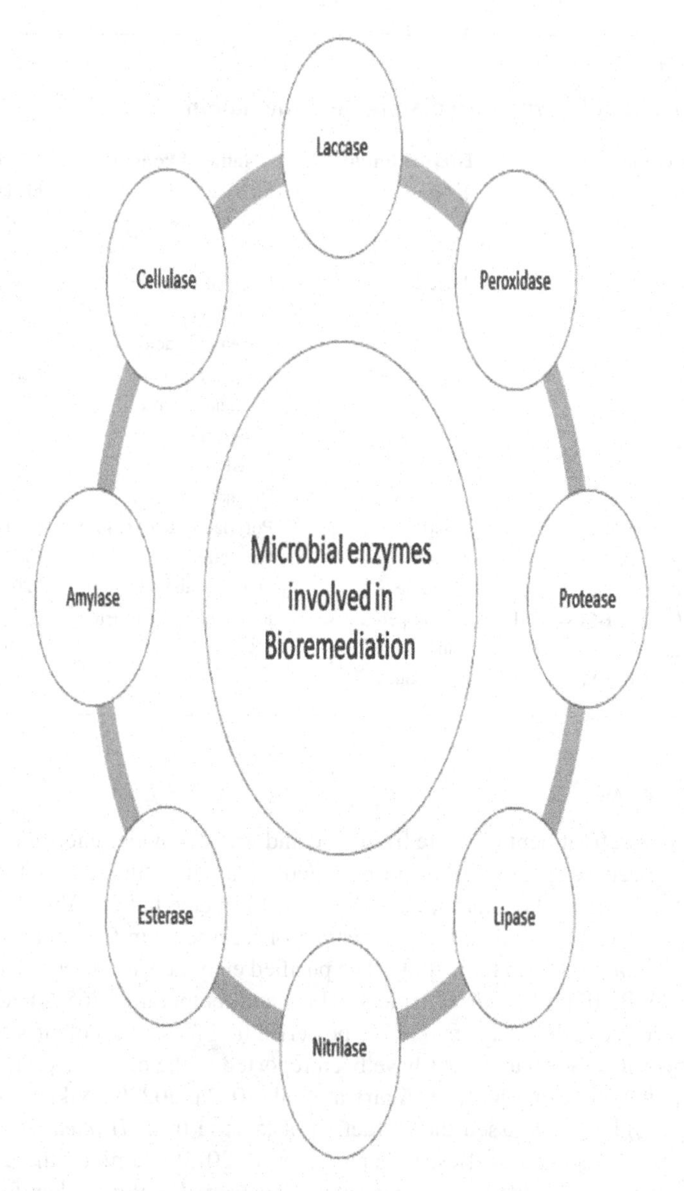

FIGURE 14.3 Important microbial enzymes for bioremediation.

give oxidised or polymerised end products (Shanmugapriya et al. 2019). Chander and Kaur (2015) demonstrated the dye degradation potential of *P. chrysosporium*, *Phlebia floridensis*, and *Phlebia brevispora*. Up to 62% decolourisation of coracryl brilliant blue using cell-free enzyme extract of peroxidase enzyme was observed. In a similar study, 93% and 85% decolourisation of Congo red were observed using lignin peroxidase and manganese peroxidase extracted from *P. chrysosporium* (Bosco et al. 2017). Lignin peroxidase of *P. chrysosporium* has also been studied for its ability to degrade the herbicide diuron. It was seen that it can competently metabolise diuron without accumulation of any toxic products (Coelho-Moreira et al. 2013).

14.4.3 Protease

Proteases are another type of microbial enzymes of the hydrolases family that catalyse protein peptide bonds. Due to their low price, high manufacturing, and effective activity, these enzymes are essential. Owing to the ability of protease to degrade α-ester bonds these can be used to degrade polymers like PLA (Haider et al. 2019). The protease enzyme keratinase can degrade keratin proteins and is used in the bioremediation of poultry waste by breaking down keratin waste and recycling it into useful by-products. Chicken feathers can be biodegraded using the protease enzyme keratinase generated by *Stenotrophomonas maltophilia* KB13 (Bhange et al. 2016). The keratinase enzyme FPF-1, synthesised by *Bacillus* sp., can effectively degrade chicken feathers with a degradation rate of 82.0 ± 1.41%, which confirms its potential to clear the refractory keratin waste biomass (Nnolim et al. 2020). Keratinolysis was effectively carried out by the cooperation of the two enzymes, a *Stenotrophomonas* serine protease and a disulphide reductase, resulting in a 50-fold increase in keratinolytic activity compared to the pure protease. The enzyme disulphide reductase breaks down the disulphide bonds in keratin proteins. The bioremediation of marine crustacean wastes uses protease enzymes to deproteinise and then remove the chitin. The process of deproteinisation involves eliminating proteins from a substance. *B. licheniformis* MP1's alkaline protease was found to be efficient at removing 75% of the proteins from shrimp wastes (Jellouli et al. 2011). Similarly, *Serratia marcescens* FS-3, a bacterium, could remove 84% of the proteins in crab shells during a 7-day procedure (Jo et al. 2008). *P. aeruginosa* K-187 successfully removed 45% of acid-treated SCSP, 78% of natural shrimp shells, and 72% from crab and shrimp shell powder (Oh et al. 2000). Additionally, it was reported that the environmentally acceptable method for the breakdown of crab shell wastes using chitinase BsChi, synthesised by *Bacillus subtilis*, was successful in degrading chitin into N-acetyl-D-glucosamine (Wang et al. 2018). An alkaline protease obtained from alkaliphilic *Bacillus altitudinis* GVC11 was reported by Vijay Kumar and co-workers (2011). According to them, the enzyme demonstrated exceptional dehairing capabilities on goat hide within 18 h without compromising collagen or hair integrity. In a different investigation, Sundararajan and his co-workers (2011) reported another such protease from *Bacillus cereus* VITSN04. According to them, the raw enzyme demonstrated the exceptional dehairing capacity of goat skins.

14.4.4 Lipase

Lipids are the microbial enzymes that break down lipids of various microbial, animal, and plant origin. The lipase activity can result in a significant decrease in the hydrocarbon content of contaminated soil. It is anticipated that research in this field will advance our understanding of oil spill bioremediation (Margesin et al. 1999; Riffaldi et al. 2006). The hydrolysis of triacylglycerols into glycerol and free fatty acids is catalysed by lipases, which are widely distributed enzymes. Lipases derived from the genera *Acinetobacter*, *Mycobacterium*, and *Rhodococcus* are used to treat oil spills (Casas-Godoy et al. 2012). Alternatively, *Pseudomonas* lipases have been used in the bioremediation of industrial oil contaminated soil and the degradation of castor oil (Amara and Salem 2009). Furthermore, *P. aeruginosa* has also been validated in the bioremediation process of crude oil-contaminated wastewater with an 80% reduction in toxicity within a week (Verma, S. et al. 2012). Other bacterial populations have been identified in other studies as potential inoculants of *P. aeruginosa*, *Bacillus*, *Halomonas*, *Citricoccus alkalitolerans*, and *Acinetobacter chaloacetis* for treating oil and gas effluents (Azhdarpoor et al. 2014; Chandra and Singh 2014; Chandra et al. 2016). Using fungal cultures to harness their lipolytic activity has proven to be an effective method for bioremediation. The basidiomycete yeast *Mrakia brollopis* SK-4 has been shown to be efficient in decomposing milk fat at low temperatures (Tsuji et al. 2019). As evidenced by its relationship with glycerol-assimilating *Caulerpa cylindracea* and lipase-secreting *Burkholderia arboris*, yeast symbiosis benefits the biodegradation of oil and gas (Quyen et al. 2012).

14.4.5 Nitrilase

Nitrilases hydrolysed herbicides, polymers, and plastics with carbon and nitrogen triple bonds (the "nitrile group") to carboxylic acid and ammonia. *A. niger*, *Fusarium solani*, *Streptomyces* sp., *Bacillus pallidus*, *Rhodococcus rhodochrous*, and *Pseudomonas fluorescens* are only a few of the species that can express these enzymes (Mousavi et al., 2021). The ability of cyanide-degrading nitrilases to clean up cyanide-contaminated waste, particularly that produced by the gold mining, pharmaceutical, and electroplating sectors, is of interest (Park et al. 2017). Cyanide hydratase is a particular class of enzyme that converts cyanide to formamide, are known to be produced by different filamentous fungal species (Martínková et al. 2015). Cyanide hydratases obtained from *Gibberella zeae*, *Neurospora crassa*, *Gloeocercospora sorghi*, and *Aspergillus nidulans* were made into recombinant forms. The pH activity profiles, relative specific activity, thermal stability, and capacity to clean up cyanide-polluted wastewater from copper and silver electroplating baths were compared for these organisms. The *N. crassa* cyanide hydratase had the best temperature stability, the most comprehensive pH, and more than 50% activity. In addition, it also showed the fastest rate of cyanide breakdown (Basile et al. 2008). Novo-Nordisk unveiled a CynD (cyanide dihydratase) enzyme preparation with commercial potential. Whole *Achromobacter denitrificans* cells were used as the catalyst (Cyanidase®), which had been cross-linked with glutaraldehyde. It was able to lower large cyanide concentrations to trace levels at pH 7–7.6 because of its high affinity for cyanide (Ingvorsen et al. 1992). Campos et al. (2006) analysed and characterised the capacity of *Fusarium oxysporum* CCMI 876 and *Methylobacterium* sp. RXM CCMI 908 to convert formamide and cyanide. *F. oxysporum* CCMI 876 decomposed the cyanide at 0.059 mM/h rate, resulting in 96% conversion of cyanide. *Methylobacterium* sp. RXM CCMI 908, consumed 84% formamide, and came into action with this effluent decreasing the chemical to 0.62 mM, indicating a successful bioremediation of cyanide.

14.4.6 Esterase

Esterases are found in plants, animals, and microbes and are the enzymes that catalyse the hydrolysis of ester linkages. According to Akatin et al. (2011), genuine esterases hydrolyse the esters of short-chain carboxylic acids. The polyester poly(butylene adipate-co-butylene terephthalate) (PBAT) was reported to be hydrolysed by two new esterases from *Clostridium botulinum* ATCC 3502 (CbotuEst_A and CbotuEst_B) (Perz et al. 2016). Phthalate esters, commonly used in industries, are difficult to break down in nature, leading to environmental contamination. *Acinetobacter* LMB-5 strain was obtained from the soil of a greenhouse; by cloning the esterase enzyme from this strain and expressing the genome in *Escherichia coli*, the degradation potential of the strain was examined. It demonstrated the ability to dissolve DBP's (di-n-butyl phthalate) ester linkages. The enzyme showed maximal activity at 40°C and pH 7.0 (Yue et al. 2017). A bacterial strain that uses cypermethrin and has a MIC of 450 ppm has been found in the soil of a pesticide affected agricultural area. The isolated strain was 98% identical to *B. subtilis* strain 1D. After 15 days of aerobic biodegradation under optimal growth conditions, it showed 95% degradation of 1-decanol, 3-(2,2-dichloroethenyl)-2,2-dimethyl cyclopropane carboxylate, cyclododecylamine, phenol, acetic acid, chloroacetic acid, decanoic acid, and cyclopentane palmitoleic acid. With the breakdown of cypermethrin, PCR identification revealed esterase and laccase amplification, which strongly implies the activity of an enzyme in the biodegradation process (Gangola et al. 2018). Another study showed a strong relation between esterase-coding mRNA and biodegradation of cypermethrin (using *B. cereus* SG2, *Bacillus thuringiensis* SG4) and other two compounds sulfosulfuron (using *Bacillus* sp. SA2, *Bacillus* sp. Sulfo3) and fipronil (using *Bacillus endoradicis* strain FA3, *Bacillus* sp. FA4). This indicates that the esterase enzyme helps in the process of decomposition of these compounds by the isolates of bacteria (Bhatt et al. 2019).

14.4.7 Amylases and Cellulase

Amylases are widely divided into three subtypes, of which the first two have proven to be the most beneficial. An enzyme with a quicker action is α-amylase. According to Gopinath et al. (2017), amylases are also referred to as glycoside hydrolases since they break down 1-4 glycosidic bonds. Karimi and Biria (2016) looked into how adding soluble starch affected *B. subtilis* TB1 degraded n-alkanes (C10–C14). In order to analyse the remaining hydrocarbons in the system, gas chromatography was used. Starch was found to increase biodegradation efficiency, and the resulting hydrocarbons were found 53% lower as compared to starch-free samples. The study of the bacterial enzymes revealed that after 72 h of application, the bacteria's amylase may break down hydrocarbons, and the same result was obtained using a commercial α-amylase sample. Some fungi and bacteria produce extracellular cellulases, hemicellulases, and pectinases (Rixon et al. 1992; Adriano-Anaya et al. 2005). The main component of plant biomass is cellulose. As a result, there is a significant amount of cellulose that is either underutilised or wasted in the waste products produced by farms, forests, and other agro-related sectors (Milala et al. 2005). Cellulase enzymes can degrade crystalline cellulose into glucose. The fungi *Humicola* and *Trichoderma* produce acidic and neutral cellulases, while *Bacillus* strains produce alkaline cellulases. Cellulases are utilised in the paper and pulp industries to remove ink from recycled paper (Karigar and Rao 2011). Krishnaswamy et al. (2022) reported two bacterial isolates, *Bacillus pacificus* and *Pseudomonas mucidolens*, from vermicompost and extracted the cellulase enzyme from the bacterial strains. After dialysis, the highest enzyme activity was found to be between 0.12 and 0.17 µm/L. The peak modifications (formation and shifts) were caused by the LDPE degradation, which was also reported to last up to 30 days using FTIR analysis. Not much has been explored on these enzymes, but studies suggest that these enzymes have potential in bioremediation and the clearance of waste from the environment.

14.5 CONCLUSION

The group of microbial hydrolase enzymes is highly versatile and can be utilised for biological remediation to break down stubborn pollutants and reduce environmental toxicity. As a result of their various applications, there has been an upsurge in enzyme production and more cost-effective processes for producing them at an industrial scale. Further research on the mechanisms of action and enzyme activity, as well as the discovery of new hydrolases, might prove to be a more promising method in treating pollutants and creating a healthier environment.

REFERENCES

Adriano-Anaya, M., Salvador-Figueroa, M., Ocampo, J.A., García-Romera, I. 2005. Plant cell-wall degrading hydrolytic enzymes of *Gluconacetobacter diazotrophicus*. *Symbiosis* 40 (3):151–156.

Agrawal, K., J. Shankar, R. Kumar, P. Verma. 2020a. Insight into multicopper oxidase laccase from *Myrothecium verrucaria* ITCC-8447: A case study using in silico and experimental analysis. *Journal of Environmental Science and Health, Part B* 55(12): 1048–1060.

Agrawal, K., J. Shankar, P. Verma. 2020b. Multicopper oxidase (MCO) laccase from *Stropharia* sp. ITCC-8422: An apparent authentication using integrated experimental and in silico analysis. *3 Biotech* 10:413.

Akatin, M.Y., Colak, A., Ertunga, N.S. 2011. Characterization of an esterase activity in *Lycoperdon pyriforms*, an edible mushroom. *Journal of Food Chemistry* 37(2):1–8.

Amara, A.A., Salem, S.R. 2009. Degradation of castor oil and lipase production by *Pseudomonas aeruginosa*. *American-Eurasian Journal of Agricultural and Environmental Sciences* 5 (4):556–563.

Azhdarpoor, A., Mortazavi, B., Moussavi, G. 2014. Oily wastewaters treatment using *Pseudomonas* sp. isolated from the compost fertilizer. *Journal of Environmental Health Science & Engineering* 12 (1):77.

Basile, L.J., Willson, R.C., Sewell, B.T., Benedik, M.J. 2008. Genome mining of cyanide-degrading nitrilases from filamentous fungi. *Applied Microbiology and Biotechnology* 80:427–435.

Bhandari, S., Poudel, D.K., Marahatha, R., Dawadi, S., Khadayat, K., Phuyal, S., Shrestha, S., Gaire, S., Basnet, K., Khadka, U., Parajuli, N. 2021. Microbial enzymes used in bioremediation. *Journal of Chemistry* 2021 (4):1–17.

Bhange, K., Chaturvedi, V., Bhatt, R. 2016. Feather degradation potential of *Stenotrophomonas maltophilia* KB13 and feather protein hydrolysate (FPH) mediated reduction of hexavalent chromium. *3 Biotech* 6 (1):42.

Bhatt, P., Gangola, S., Chaudhary, P., Khati, P., Kumar, G., Sharma, A., Srivastava, A. 2019. Pesticide induced up-regulation of esterase and aldehyde dehydrogenase in indigenous *Bacillus* spp. *Bioremediation Journal* 23 (1):42–52.

Bollag, J.-M., Bollag, W.B. 1995. Soil contamination and the feasibility of biological remediation. In: Skipper, H.D., Turco R.F. (eds), Bioremediation: Science and *Applications* (Vol. 43, pp. 1–12). SSSA Special Publications.

Bonugli-Santos, R.C., Durrant, L.R., Sette, L.D. 2012. The production of ligninolytic enzymes by marine-derived basidiomycetes and their biotechnological potential in the biodegradation of recalcitrant pollutants and the treatment of textile effluents. *Water Air and Soil Pollution* 223:2333–2345.

Bosco, F., Mollea, C., Ruggeri, B. 2017. Decolorization of Congo Red by *Phanerochaete chrysosporium*: The role of biosorption and biodegradation. *Environmental Technology* 38 (20):2581–2588.

Campos, M.G., Pereira, P., Roseiro, J.C. 2006. Packed-bed reactor for the integrated biodegradation of cyanide and formamide by immobilised *Fusarium oxysporum* CCMI 876 and *Methylobacterium* sp. RXM CCMI 908. *Enzyme and Microbial Technology* 38 (6):848–854.

Carniel, A., Valoni, E., Nicomedes, J., Gomes, A.C., de Castro, A.M. 2017. Lipase from *Candida antarctica* (CALB) and cutinase from *Humicola insolens* act synergistically for PET hydrolysis to terephthalic acid. *Process Biochemistry* 59(A):84–90.

Casas-Godoy, L., Duquesne, S., Bordes, F., Sandoval, G., Marty, A. 2012. Lipases: An overview. *Methods in Molecular Biology* 861:3–30.

Chander, M., Kaur I. 2015. An industrial dye decolourisation by *Phlebia* sp. *International Journal of Current Microbiology & Applied Sciences* 4 (5):217–226.

Chandra, P., Rawat, A.P., Singh, D.P. 2016. Isolation of alkaliphilic bacterium *Citricoccus alkalitolerans* CSB1: An efficient biosorbent for bioremediation of tannery waste water. *Cellular and Molecular Biology* 62 (3):135.

Chandra, P., Singh, D.P. 2014. Removal of Cr(VI) by a halotolerant bacterium *Halomonas* sp. CSB 5 isolated from Sambhar Salt Lake Rajasthan (India). *Cellular and Molecular Biology* 60 (5):64–72.

Coelho-Moreira, J.S., Maciel, G.M., Castoldi, R., Mariano, S., Dorneles, F., Bracht, A., Peralta, R. 2013. Involvement of lignin-modifying enzymes in the degradation of herbicides. *Herbicides-Advances in Research* 2013 (1):165–187.

Deshmukh, R., Khardenavis, A.A., Purohit, H.J. 2016. Diverse metabolic capacities of fungi for bioremediation. *Indian Journal of Microbiology* 56 (3):247–264.

Gangola, S., Sharma, A., Bhatt, P., Khati, P., Chaudhary, P. 2018. Presence of esterase and laccase in *Bacillus subtilis* facilitates biodegradation and detoxification of cypermethrin. *Scientific Reports* 8:12755.

Gopinath, S.C.B., Anbu, P., Arshad, M.K.M., Lakshmipriya, T., Voon, C.H., Hashim, U., Chinni, S.V. 2017. Biotechnological processes in microbial amylase production. *BioMed Research International* 2017:1–9.

Haider, T.P., Völker, C., Kramm, J., Landfester, K., Wurm, F.R. 2019. Plastics of the future? The impact of biodegradable polymers on the environment and on society. *Angewandte Chemie International Edition* 58 (1):50–62.

Ingvorsen, K., Godtfredsen, S.E., Hojer-Pedersen, B. 1992. Microbial cyanide converting enzymes, their production and use. European Patent Office Publ. of Application without search report EP19880302186. 11 Mar 1988.

Jellouli, K., Ghorbel-Bellaaj, O., Ayed, H.B., Manni, L., Agrebi, R., Nasri, M. 2011. Alkaline-protease from *Bacillus licheniformis* MP1: Purification, characterization and potential application as a detergent additive and for shrimp waste deproteinization. *Process Biochemistry* 46 (6):1248–1256.

Jo, G.H., Jung, W.J., Kuk, J.H., Oh, K.T., Kim, Y.J., Park, R.D. 2008. Screening of protease-producing *Serratia marcescens* FS-3 and its application to deproteinization of crab shell waste for chitin extraction. *Carbohydrate Polymers* 74 (3):504–508.

Karigar, C.S., Rao, S.S. 2011. Role of microbial enzymes in the bioremediation of pollutants: A review. *Enzyme Research* 2011:805187.

Karimi, M., Biria, D. 2016. The synergetic effect of starch and alpha amylase on the biodegradation of n-alkanes. *Chemosphere* 152 (2016):166–172.

Krishnaswamy, V.G., Sridharan, R., Kumar, P.S., Fathima, M.J. 2022. Cellulase enzyme catalyst producing bacterial strains from vermicompost and its application in low-density polyethylene degradation. *Chemosphere* 288 (2):132552.

Kumar, A., Sharma, S. 2019. *Microbes and Enzymes in Soil Health and Bioremediation, in Microorganisms for Sustainability* (1–17). Springer.

Lu, L., Wang, T.N., Xu, T.F., Wang, J.Y., Wang, C.L., Zhao, M. 2013. Cloning and expression of thermo-alkali-stable laccase of *Bacillus licheniformis* in *Pichia pastoris* and its characterization. *Bioresource Technology* 134:81–86.

Lu, L., Zhao, M., Zhang, B.B., Yu, S.Y., Bian, X.J., Wang, W., Wang, Y. 2007 Purification and characterization of laccase from *Pycnoporus sanguineus* and decolorization of an anthraquinone dye by the enzyme. *Applied Microbiology and Biotechnology* 74 (6): 1232–1239.

Margesin, R., Zimmerbauer, A., Schinner, F. 1999. Soil lipase activity—A useful indicator of oil biodegradation. *Biotechnology Techniques* 13 (12):859–863.

Margot, J., Bennati-Granier, C., Maillard, J., Blánquez, P., Barry, D.A., Holliger, C. 2013. Bacterial versus fungal laccase: Potential for micropollutant degradation. *AMB Express* 3 (1):63.

Martínková, L., Veselá, A.B., Rinágelová, A., Chmátal, M. 2015. Cyanide hydratases and cyanide dihydratases: Emerging tools in the biodegradation and biodetection of cyanide. *Applied Microbiology and Biotechnology* 99:8875–8882.

Mathur, M.B., Peacock, J.R., Robinson, T.N., Gardner, C.D. 2022. Effectiveness of a theory-informed documentary to reduce consumption of meat and animal products: Three randomized controlled experiments. *Nutrients* 14 (13):2672.

Milala, M.A., Shugaba, A., Gidado, A., Ene, A.C., Wafar, J.A. 2005. Studies on the use of agricultural wastes for cellulase enzyme production by *Aspergillus niger*. *Research Journal of Agriculture and Biological Science* 1:325–328.

Mousavi, S.M., Hashemi, S.A., Iman Moezzi, S.M., Ravan, N., Gholami, A., Lai, C.W., Chiang, W.H., Omidifar, N., Yousefi, K., Behbudi, G. 2021. Recent advances in enzymes for the bioremediation of pollutants. *Biochemistry Research International* 2021:5599204.

Niku-Paavola, M.-L., Viikari, L. 2000. Enzymatic oxidation of alkenes. *Journal of Molecular Catalysis B: Enzymatic* 10 (4):435–444.

Nnolim, N.E., Okoh, A.I., Nwodo, U.U. 2020. *Bacillus* sp. FPF-1 produced keratinase with high potential for chicken feather degradation. *Molecules* 25 (7):1505.

Oh, Y.S., Shih, I.L., Tzeng, Y.M., Wang, S.L. 2000. Protease produced by *Pseudomonas aeruginosa* K-187 and its application in the deproteinization of shrimp and crab shell wastes. *Enzyme and Microbial Technology* 27 (1–2):3–10.

Pandey, V.C., Singh, V. 2019. Exploring the potential and opportunities of current tools for removal of hazardous materials from environments. In: Pandey V.C., Bauddh K. (eds), *Phytomanagement of Polluted Sites* (pp. 501–516). Elsevier.

Park, J.M., Sewell, B.T., Benedik, M.J. 2017. Cyanide bioremediation: The potential of engineered nitrilases. *Applied Microbiology and Biotechnology* 101 (8):2309.

Perz, V., Baumschalger, A., Bleymaier, A., Zitzenbacher, S., Hromic, A., Steinkellner, G., Pairitsch, A., Łyskowski, A., Gruber, K., Sinkel, C., Kuper, U., Ribitsch, D., Guebitz, G.M. 2016. Hydrolysis of synthetic polyesters by *Clostridium botulinum* esterases. *Biotechnology and Bioengineering* 113 (5):1024–1034.

Phale, P.S., Sharma, A., Gautam, K. 2019. Microbial degradation of xenobiotics like aromatic pollutants from the terrestrial environments. In: Prasad M.N.V., Vithanage M., Kapley A. (eds), *Pharmaceuticals and Personal Care Products: Waste Management and Treatment Technology* (pp. 259–278). Butterworth-Heinemann.

Quyen, T., Vu, C., Le, G.T. 2012. Enhancing functional production of a chaperone-dependent lipase in *Escherichia coli* using the dual expression cassette plasmid. *Microbial Cell Factories* 11 (1):29.

Riffaldi, R., Levi-Minzi, R., Cardelli, R., Palumbo, S., Saviozzi, A. 2006. Soil biological activities in monitoring the bioremediation of diesel oil-contaminated soil. *Water, Air, and Soil Pollution* 170 (1–4):3–15.

Rixon, J.E., Ferreira, L.M.A., Durrant, A.J., Laurie, J.I., Hazlewood, G.P., Gilbert, H.J. 1992. Characterization of the gene celD and its encoded product 1,4-β-D-glucan glucohydrolase D from *Pseudomonas fluorescens* subsp. cellulosa. *Biochemical Journal* 285 (3):947–955.

Saranya, P., Selvi, P.K., Sekaran, G. 2019. Integrated thermophilic enzyme-immobilized reactor and high-rate biological reactors for treatment of palm oil-containing wastewater without sludge production. *Bioprocess and Biosystems Engineering* 42 (6):1053–1064.

Satti, S.M, Abbasi, A.M., Salahuddin, Rana, Q.A., Marsh, T.L., Auras, R., Hasan, F., Badshah, M., Farman, M., Shah, A.A. 2019. Statistical optimization of lipase production from *Sphingobacterium* sp. strain S2 and evaluation of enzymatic depolymerization of Poly(lactic acid) at mesophilic temperature. *Polymer Degradation and Stability* 160:1–13.

Shanmugapriya, S., Manivannan, G., Selvakumar, G., Sivakumar, N. 2019. Extracellular fungal peroxidases and laccases for waste treatment: Recent improvement. In: Yadav, A., Singh, S., Mishra, S., Gupta, A. (eds), *Recent Advancement in White Biotechnology through Fungi* (pp. 153–187). Springer.

Sharma, A., Sharma, T., Sharma, T., Sharma, S., Kanwar, S.S. 2019. Role of microbial hydrolases in bioremediation. In: Kumar, A., Sharma, S. (eds), *Microbes and Enzymes in Soil Health and Bioremediation*. Microorganisms for Sustainability vol. 16 (pp. 149–164). Springer.

Singh, B. 2014. Review on microbial carboxylesterase: General properties and role in organophosphate pesticides degradation. *Biochemistry & Molecular Biology* 2 (1):1–6.

Singh, M.P., Vishwakarma, S.K., Srivastava, A.K. 2013. Bioremediation of Direct Blue 14 and extracellular ligninolytic enzyme production by white rot fungi: *Pleurotus* spp. *BioMed Research International* 2013:1–4.

Soccol, C.R., Vandenberghe, L.P., Woiciechowski, A.L., Thomaz-Soccol, V., Correia, C.T., Pandey, A. 2003. Bioremediation: An important alternative for soil and industrial wastes clean-up. *Indian Journal of Experimental Biology* 41 (9):1030–1045.

Song, Z.-M., Yang, L.-L., Lu, Y., Wang, C., Liang, J.-K., Du, Y., Li, X.-Z., Hu, Q., Guan, Y.-T., Wu, Q.-Y. 2021. Characterization of the transformation of natural organic matter and disinfection byproducts after chlorination, ultraviolet irradiation and ultraviolet irradiation/chlorination treatment. *Chemical Engineering Journal* 426:131916.

Srivastava, V. 2021. Grand challenges in chemical treatment of hazardous pollutants. *Frontiers in Environmental Chemistry* 2:1–4.

Sundararajan, S., Kannan, C.N., Chittibabu, S. 2011. Alkaline protease from *Bacillus cereus* VITSN04: Potential application as a dehairing agent. *Journal of Bioscience and Bioengineering* 111 (2):128–133.

Thassitou, P.K., Arvanitoyannis, I.S. 2001. Bioremediation: A novel approach to food waste management. *Trends in Food Science & Technology* 12 (5–6):185–196.

Tsuji, M., Kudoh, S., Tanabe, Y., Hoshino. 2019. Basidiomycetous yeast of the genus Mrakia. In: Tiquia-Arashiro, S., Grube, M. (eds), *Fungi in Extreme Environments: Ecological Role and Biotechnological Significance* (pp. 145–156). Springer.

Verma, A., Shirkot, P. 2014. Purification and characterization of thermostable laccase from thermophilic *Geobacillus thermocatenulatus* MS5 and its applications in removal of textile dyes. *Scholars Academic Journal of Biosciences* 2(8):479–485.

Verma, A.K., Chandralata, R.K., Parvatkar, R., Naik, *C.* 2012. A rapid two-step bioremediation of the anthraquinone dye, reactive Blue 4 by a marine-derived fungus. *Water, Air, & Soil Pollution* 223 (6):3499–3509.

Verma, S., Saxena, J., Prasanna, R., Sharma, V., Nain, L. 2012. Medium optimization for a novel crude-oil degrading lipase from *Pseudomonas aeruginosa* SL-72 using statistical approaches for bioremediation of crude-oil. *Biocatalysis and Agricultural Biotechnology* 1 (4):321–329.

Vijay Kumar, E., Srijana, M., Kiran Kumar, K., Harikrishna N., Reddy, G. 2011. A novel serine alkaline protease from *Bacillus altitudinis* GVC11 and its application as a dehairing agent. *Bioprocess and Biosystems Engineering* 34:403–409.

Viswanath, B., Rajesh, B., Janardhan, A., Kumar, A.P., Narasimha, G. 2014. Fungal laccases and their application in bioremediation. *Enzyme Research* 2014:1–21.

Wang, D., Li, A., Han, H., Liu, T., Yang, Q. 2018. A potent chitinase from *bacillus subtilis* for the efficient bioconversion of chitin-containing wastes. *International Journal of Biological Macromolecules* 116:863–868.

Williams, J., 2006. Bioremediation of contaminated soils: a comparison of in situ and ex situ techniques. *Recuperado de/paper/Bioremediation-of-Contaminated-Soils-% 3A-A-Comparison-Williams/4c6afc722040e0d4807a744b7f89a5e7b9dac97f.*

Yanmis, D., Demir, N., Nadaroglu, H., Adiguzel, A., Gulluce, M. 2016. Purification and characterization of laccase from thermophillic *Anoxybacillus gonensis* (p39) and its application of removal textile dyes. *Romanian Biotechnological Letters* 21 (3):11485–11496.

Yue, F., Lishuang, Z., Jing, W., Ying, Z., Bangce, Y.E. 2017. Biodegradation of phthalate esters by a newly isolated *Acinetobacter* sp. strain LMB-5 and characteristics of its esterase. *Pedosphere* 27 (3):606–615.

Zhang, M., Yoshikawa, M. 2020. Bioremediation: Recent advancements and limitations. *Lecture Notes in Civil Engineering* 2020:21–29.

15 Secondary Metabolites in Green Synthesis of Nanoparticles

Rashi Nagar, Parikshana Mathur, Payal Chaturvedi, Charu Sharma, and Pradeep Bhatnagar

15.1 INTRODUCTION

Nanotechnology, the science that deals with matter at the nanoscale, revolves around manipulating matter at the atomic or molecular level to synthesise particles for industrial and scientific purposes. The particles typically in the 1 to 100 nm range are called nanoparticles (Vert et al. 2012) and exist naturally in the environment or can be synthesised. The physical and chemical properties of nanoparticles vary with the change in shape and size, drawing the attention of researchers to them (Khan et al. 2019). Over the years, these nanoparticles have gained tremendous scientific interest due to their potential applications in engineering, biology, and medicine (Zhao et al. 2016).

There are three different methods of synthesis of nanoparticles, namely, physical, chemical, and biological (that includes green synthesis) (Wang et al. 2007; Horwat et al. 2011). The physical nanoparticle synthesis methods include evaporation, sputtering, and laser ablation, whereas the chemical methods include hydrothermal synthesis, microemulsion, precipitation, chemical reduction, and sol-gel (Ijaz et al. 2020).

The physical methods are not eco-friendly and require much time to reach thermal stability while simultaneously consuming a lot of space (in the case of tube furnaces) and energy (Kawasaki and Nishimura 2006; Pal et al. 2019). The chemical methods have limitations as they leave behind toxic ions in the environment (Medici et al. 2021). Green or biological synthesis overcomes the problems associated with physical and chemical approaches for synthesis. Researchers primarily focus on the biological synthesis of nanoparticles using bacteria, yeast, algae, fungi, and plants because of the minimal side effects of the method, sustainability, and safety it provides compared to other approaches (Pal et al. 2019). A typical example of biological synthesis is myco-nanotechnology which implies the usage of fungal biomass and fungus-derived metabolites to synthesise nanoparticles (Mathur et al. 2021). Utilising plant extracts is another option for synthesising nanoparticles (Pal et al. 2019). This green synthesis method is eco-friendly, efficient, economical, and simple to implement. So far, it has been used for the synthesis of metal nanoparticles such as gold (Elia et al. 2014), zinc oxide (Fakhari et al. 2019), and silver (Khalil

DOI: 10.1201/9781003407683-15

et al. 2014) using phytocompounds or secondary metabolites like terpenoids and polyphenols (Kharissova et al. 2013). The nanoparticles so developed are beneficial commercially as well as for research purposes because of their increased catalytic (Edison and Sethuraman 2012), antioxidant (Abdel-Aziz et al. 2014), and antimicrobial (Savithramma et al. 2011) properties. With such tremendous capabilities, there is a broad range of nanoparticle use, with even more new applications being explored continuously by researchers to benefit humankind and the environment.

15.2 NANOPARTICLES

In the literature, nanoparticles have been defined in different ways depending on their properties. In a broader sense, nanoparticles are particles having lengths in two or three dimensions ranging between 1 and 100 nm (Vert et al. 2012). They are also called zero-dimensional particles to distinguish them from one- or two-dimensional particles heavier than them.

They differ from other bulkier particles in size, energy absorption, mobility, chemical reactivity, etc. (Murthy 2007). Nanoparticles exhibit exceptional physiochemical properties due to their size, including high energy, quantum internment, and large specific surface area (Nel et al. 2006; Iravani 2011; Ahmed et al. 2016). They can also form different morphological shapes such as nanochains (Kralj and Makovec 2015), nanospheres (Agam and Guo 2007), nanoreefs (Choy et al. 2004), or nanoboxes (Murphy 2002). Nanoparticles in sub-microscopic particle size have a vast scope in semiconductors, batteries, industrial catalysis, pharmaceuticals, etc. (Benelli and Lukehart 2017).

15.3 CLASSIFICATION OF NANOPARTICLES

Over the years, different types of nanoparticles have been synthesised and discovered by researchers across the globe. These nanoparticles are classified into distinct categories based on criteria such as physical and chemical properties, size, and morphology (Figure 15.1).

15.3.1 Metallic Nanoparticles

Metal nanoparticles comprise only a single element (that can exist as an individual atom or group of atoms). The most common metal nanoparticles are Au, Ag, Zn, Fe, Ni, Ti, Pt, and Cu. They are studied widely due to their unique physical and chemical properties, such as enhanced catalytic activity (Bordiwala 2023).

15.3.2 Semiconductor Nanoparticles

Semiconductor nanoparticles are also known as quantum dots and are intermediate between metal and non-metal nanoparticles. They are fluorescent and have a silica coating to decrease photobleaching. They also possess size-dependent optoelectronic properties (Sharma and Rabinal 2013). Their application ranges from solar energy conversion in solar cells, LEDs, and cellular or biological imaging to ultrasensitive detection (Kargozar et al. 2020).

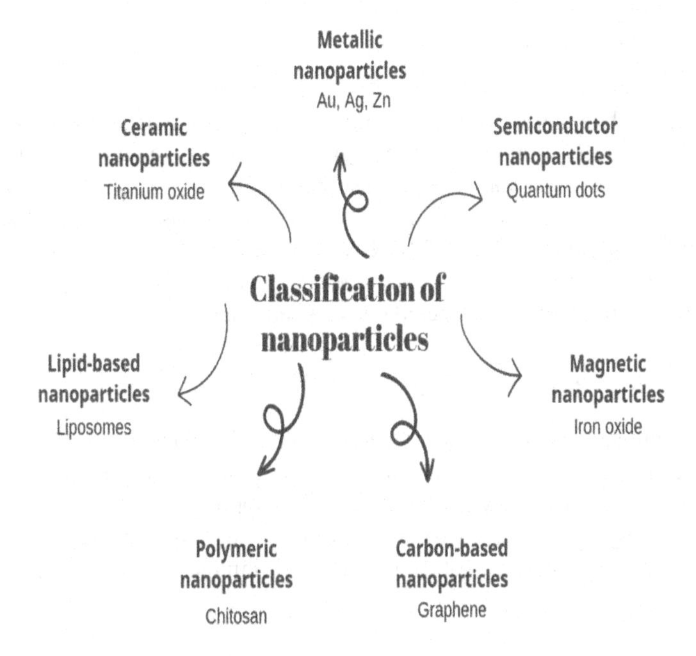

FIGURE 15.1 Classification of nanoparticles.

15.3.3 Magnetic Nanoparticles

Magnetic nanoparticles are a class of nanoparticles that exhibit magnetic properties typically manipulated using magnetic fields. They are made up of two different materials: a chemical component that provides functionality and a magnetic material such as cobalt, iron, or nickel. They are widely used in magnetic resonance imaging (Huang et al. 2019), drug delivery (Kianfar 2021), and environmental remediation (Dan and Chattree 2023).

15.3.4 Carbon-Based Nanoparticles

Carbon-based nanoparticles are known for their exceptional mechanical strength, thermal conductivity, and electrical properties. These properties render the utilisation of these nanoparticles in electronics, energy storage, and nanomedicine. Classic examples of this class of nanoparticles include carbon nanotubes and graphene (Gavali et al. 2023).

15.3.5 Polymeric Nanoparticles

Polymeric nanoparticles have been widely popular for their application in the delivery of bioactive compounds since the 1980s (Pauluk et al. 2019). These nanoscale particles made up of synthetic or natural polymers find applications in various fields such as diagnostics, drug delivery with controlled release, and imaging. These nanoparticles are typically spherical and with sizes varying from 1 to 1000 nm (Ekambaram and Sathali 2011).

15.3.6 Lipid-Based Nanoparticles

Lipid-based nanoparticles (LNPs) are delivery systems made up of lipids routinely used to encapsulate and deliver therapeutic or biological agents like drugs, nucleic acids, or proteins (Xu et al. 2022). They have a lipid bilayer enveloping a hydrophobic core, making them suitable for encapsulating both hydrophobic and hydrophilic compounds. They were also used in COVID-19 mRNA-based vaccines like Pfizer-BioNTech (Suzuki and Ishihara 2021). LNPs are widely studied because they show promising results for siRNA-based therapeutics, drug delivery, and gene therapies due to their biocompatibility, targeted delivery, and stability (Mirahadi et al. 2018).

15.3.7 Ceramic Nanoparticles

Ceramic nanoparticles are made of ceramic materials. They are widely used in research related to materials science as they have enhanced mechanical, electrical, and thermal properties. They also have potential applications in drug delivery systems, bioimaging, and tissue engineering (Kaushik 2021) in the field of biomedicine as well as in energy storage, catalysis, and environmental remediation (Dai et al. 2011).

15.4 APPROACHES FOR THE SYNTHESIS OF NANOPARTICLES

Nanoparticle synthesis involves three primary methods: chemical, physical, and biological (Figure 15.2).

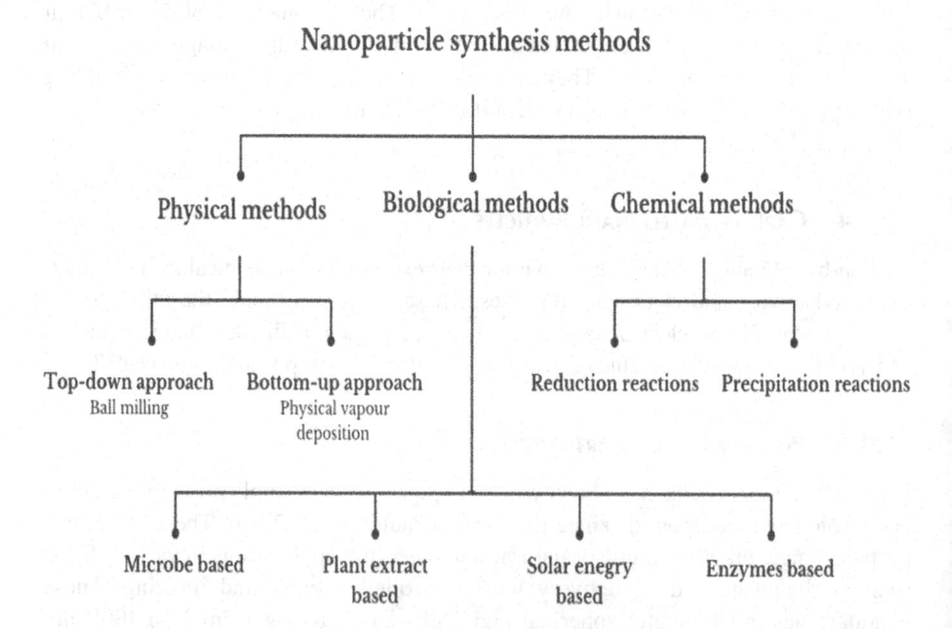

FIGURE 15.2 Various synthesis methods of nanoparticles.

15.4.1 Physical Methods

Physical methods for nanoparticle synthesis refer to manipulating materials through physical or mechanical means to obtain nanoparticles with desired properties and size. The two commonly employed approaches are the top-down and bottom-up methods.

In the top-down approach, the desired bulk material is broken down into fine particles using suitable lithographic techniques such as milling, grinding, and sputtering (Rani et al. 2023). The material is ground using balls in a milling chamber in the high-energy ball milling technique used to synthesise titanium dioxide (TiO_2) nanoparticles (Sun et al. 2023). Another top-down approach is laser ablation, which utilises laser energy to vaporise a target material, thus producing nanoparticles of desired characteristics and size (Marcano et al. 2010).

The bottom-up approach involves the assembly of atoms or molecules to form nanoparticles. It includes methods like physical vapour deposition or condensation and synthesises tungsten nanoparticles (Bai et al. 2023). Nanoparticles such as lead sulphide and fullerene have also been successfully synthesised using evaporation-condensation methods (Abou El-Nour et al. 2010).

15.4.2 Chemical Methods

The chemical methods imply the use of chemical reactions to synthesise nanoparticles. These methods employ techniques like reduction reaction for the synthesis of metal nanoparticles such as silver (Ag) from silver nitrate ($AgNO_3$) using a plant extract as a reducing agent (Mavani and Shah 2013). Another chemical method is precipitation, in which the reactants are mixed to form nanoparticles through nucleation and subsequent growth. Osman and Mustafa (2015) synthesised zinc oxide nanoparticles through precipitation using zinc acetate and sodium hydroxide as precursors.

15.4.3 Biological Methods

Biological methods exploit bacteria, fungi, or plant and plant extracts like secondary metabolites for the biosynthesis of nanoparticles, such as silver nanoparticles with the fungus *Verticillium* (Mukherjee et al. 2001) or within plant tissues (Song and Kim 2009).

15.5 WHY SHIFT TOWARDS GREEN SYNTHESIS?

The recent trends in nanoparticle synthesis have moved towards the green approach which has proved more economical and sustainable. The chemical and physical methods for synthesising nanoparticles have certain disadvantages, which are discussed below:

i. **Harsh reaction conditions:** Chemical methods for synthesising nanoparticles often require harsh conditions, such as high temperatures, extreme pH levels, and pressure. However, these conditions can lead to limitations in the types of nanoparticles that can be produced and can also result in the generation of unwanted by-products (Bhattacharya and Gupta 2005; Iravani 2011).

ii. **Lack of eco-friendliness and sustainability:** Organic solvents and surfactants are often used in the chemical approach of synthesis, which can incorporate impurities or residual chemicals into the synthesised nanoparticles. These impurities are also environmentally detrimental and can affect the properties of nanoparticles (Iravani 2011; Parveen et al. 2016).

iii. **More energy consumption:** The physical methods typically consume more energy and lead to increased cost of production as well as a negative impact on the environment (Bhattacharya and Gupta 2005; Parveen et al. 2016).

iv. **Post-synthesis purification requirements:** In most cases, physical and chemical methods used in nanoparticle synthesis require extra purification steps to remove unwanted by-products, unreacted precursors, or surfactants. These extra steps make the process more complicated, time-consuming, and expensive. As a result, the yield and purity of the nanoparticles are affected (Bhattacharya and Gupta 2005; Miu and Dinischiotu 2022).

v. **Potential toxicity risks:** Sometimes, the use of harmful chemicals and toxic by-products production during physical and chemical synthesis methods can lead to concerns about the toxicity of the resulting nanoparticles (Bhattacharya and Gupta 2005; Parveen et al. 2016).

15.6 WHAT IS THE ROLE OF SECONDARY METABOLITES IN GREEN SYNTHESIS OF NANOPARTICLES?

Secondary metabolites are the bioactive compounds produced by any life form, such as microorganisms, plants, and animals, and tend to play a crucial role in the green synthesis of nanoparticles. Owing to their inherent reducing and stabilising properties that facilitate the formation and stabilisation of nanoparticles, they serve as the essential components of green synthesis approaches (Williams et al. 1989; Zlatić and Stanković 2020) (Table 15.1).

Utilising secondary metabolites for green synthesis offers numerous benefits, including the use of renewable and natural resources, decreased dependency on harmful chemicals, and energy-intensive methods (Ahmed and Mustafa 2020). Secondly, these metabolites are often non-toxic, biocompatible, and biodegradable, making them fit for biomedical and environmental applications (Kaur and Chopra 2018).

Following are different secondary metabolites derived from plant extracts or microbial sources that have been employed in the green synthesis of nanoparticles, leading to significant progress in sustainable nanomaterial production.

15.6.1 FLAVONOIDS

Flavonoids are a diverse group of secondary metabolites present in algae, fungi, mosses, citrus fruits, tea, and more. They have been widely studied due to their antioxidant and anti-inflammatory properties. Flavonoids have been used in the green synthesis of quantum dots and titanium dioxide nanoparticles (Alvand et al. 2019; Aslam et al. 2021). Flavonoids p-coumaric acid, catechin, and luteolin-7-glucoside are

TABLE 15.1
Different Secondary Metabolites Used in Nanoparticle Formation

Secondary Metabolite	Source	Nanoparticle Types	Applications	References
Flavonoids	Fungi, algae, mosses, citrus fruits	Quantum dots, titanium dioxide nanoparticles	Medicine, biosensors, imaging, drug delivery	Alvand et al. 2019; Aslam et al. 2021
Terpenoids	Essential oils, microorganisms, insects, marine organisms	Silicon nanoparticles, carbon nanotubes	Drug delivery, photodynamic therapy, antimicrobial coatings	Usha Rani et al. 2014; Uchida et al. 2006
Phenolic acids	Plants, microorganisms, synthetic production	Titanium dioxide nanoparticles, metallic nanoparticles	Drug delivery, cancer therapy, antimicrobial applications	Amini, and Akbari 2019; Aslam et al. 2021
Alkaloids	Plants, fungi, and some animal sources	Silicon nanoparticles, carbon nanotubes	Drug delivery, bioimaging, antimicrobial applications	Hedayati et al. 2020; Makvandi et al. 2021
Tannins	Plants, fungi, lichens, some invertebrates	Titanium dioxide nanoparticles, magnetic nanoparticles	Drug delivery systems, wound healing, antioxidant applications	Ahmad 2014; Binaeian et al. 2016
Quinones	Plants and fungi	Silicon nanoparticles, carbon nanotubes	Solar cells, catalysis, sensors	Arcudi et al. 2019
Lignans	Plants, fungi, microorganisms	Titanium dioxide nanoparticles, magnetic nanoparticles	Drug delivery, cancer therapy, imaging	Iravani and Varma 2020; Pathania et al. 2023
Saponins	Plants, marine organisms, and invertebrates	Quantum dots, silicon nanoparticles	Drug delivery, antimicrobial coatings, immune system modulation	Patel et al. 2023; Pouthika et al. 2023
Coumarins	Fungi, microorganisms, and synthetic production	Quantum dots, silicon nanoparticles, carbon nanotubes	Drug delivery, bioimaging, antimicrobial applications	Patil 2020
Essential oils	Plant extracts	Metallic nanoparticles	Aromatherapy and cosmetic applications	Obeizi et al. 2020; Fizer et al. 2022

reported for the synthesis of Ag NPs using *Withania somnifera* leaf extract (Marslin et al. 2015). In contrast, quercetin from *Ocimum sanctum* has also been used for the synthesis of them (Jain and Mehata 2017). Synthesis of Ag NPs has also been reported using naringin, hesperidin, and diosmin flavonoids. The involvement of flavonoid is also demonstrated in Au NPs synthesis using quercetin (Pal et al. 2013;

Pak et al. 2016), baicalin (Lee et al. 2016), naringin (Singh et al. 2016), and hesperetin (Krishnan et al. 2017).

15.6.2 Terpenoids

Terpenoids are a large and diverse class of secondary metabolites found in plants, microorganisms, insects, fungi, and some marine organisms. They are primarily characterised by their biosynthetic origin from isoprene units that form terpene. They have a wide range of biological activities and have been extensively studied for their medicinal properties. Terpenoids are highly effective in synthesising silicon nanoparticles and carbon nanotubes, among other materials (Usha Rani et al. 2014; Uchida et al. 2006). Sesquiterpene obtained from the leaves of *Ambrosia maritima* was used in the formation of Ag NPs and notably affected its yield (El-Kemary et al. 2016). Shankar et al. (2003) reported Au NP synthesis using terpenoids present in geranium leaves.

15.6.3 Phenolic Acids

Phenolic acids are a class of secondary metabolites widely distributed in plant sources and microorganisms and are also produced synthetically. The presence of a phenolic ring and one or more carboxylic acid groups is the characteristic feature of this class of secondary metabolite. They also possess antioxidant and anti-inflammatory properties. Phenolic acids have proven effective in creating both titanium dioxide and metallic nanoparticles (Amini and Akbari 2019; Aslam et al. 2021). The involvement of phenolics has been reported in synthesising Pd and CuO NPs using *Theobroma cocoa* seed extract (Nasrollahzadeh et al. 2015). Similarly, TiO_2 NPs were synthesised in the presence of phenolics of *Euphorbia heteradena* (Nasrollahzadeh and Sajadi 2015).

15.6.4 Alkaloids

Alkaloids are a diverse group of naturally occurring nitrogenous compounds found in plants, fungi, and some animal sources. They have a wide range of biological activities and have been primarily studied for their pharmacological properties. Alkaloids are potentially useful in synthesising silicon nanoparticles and carbon nanotubes (Hedayati et al. 2020; Makvandi et al. 2021). Synthesis of Ag_2O NPs by alkaloids from chilli has been reported by El-Shabasy et al. (2019). Alkaloids like ergoline, benzenoids, and indolizidine obtained from *Ipomoea pes-caprae* roots are reported for Ag NP synthesis (Krishanan and Maru 2006).

15.6.5 Tannins

Tannins are a class of polyphenolic compounds widely distributed in various plant sources, including fruits, bark, leaves, and seeds. They are known for their astringent taste and have been used traditionally for their medicinal properties. Tannins exhibit antioxidant, antimicrobial, and anticancer activities, among others. Tannins have been used to synthesise titanium dioxide and magnetic nanoparticles (Ahmad 2014; Binaeian et al. 2016). Their unique chemical structure and reactivity make them valuable for nanoparticle synthesis and offer opportunities for diverse applications in nanotechnology and biomedicine.

15.6.6 Quinones

Quinones are organic compounds characterised by a two-carbon carbonyl group (C=O). They are widely distributed in various natural sources, including plants and fungi. They possess many biological activities and have been studied for their antioxidant, anticancer, and antimicrobial properties. It has shown potential in synthesising silicon nanoparticles and carbon nanotubes (Arcudi et al. 2019).

15.6.7 Lignans

Lignans are phytochemicals in various plant sources, including flaxseeds, sesame seeds, and berries. They are also present in fungi and microorganisms. A prominent structural feature, the presence of a dibenzyl butane skeleton, characterises lignans. Lignans have been studied for their potential health benefits, including anti-inflammatory and anticancer properties. Lignans have shown promise in synthesising titanium dioxide and magnetic nanoparticles (Iravani and Varma 2020; Pathania et al. 2023).

15.6.8 Saponins

Saponins are a diverse group of glycosides found in various plant sources, including legumes, herbs, marine organisms, and invertebrates. They are amphiphilic, consisting of a hydrophobic aglycone portion and a hydrophilic sugar moiety. Saponins exhibit various biological activities, including antimicrobial, antifungal, and anticancer properties. Saponins were used in synthesising quantum dots and silicon nanoparticles (Patel et al. 2023; Pouthika et al. 2023). The role of saponin diosgenin in the synthesis of Ag NPs has been discussed by Hussain et al. (2014).

15.6.9 Coumarins

Coumarins are aromatic compounds found in various plant sources, including fruits, vegetables, and medicinal herbs. They can also be found in fungi and microorganisms or produced synthetically. These molecules have a benzopyrone structure and display various beneficial effects, such as antioxidant and anticancer properties. Patil (2020) has demonstrated that coumarins can potentially synthesise metallic nanoparticles in an eco-friendly way.

15.6.10 Essential Oils

Essential oils are derived from different plant sources and have complex blends of aromatic compounds. These secondary metabolites have healing and soothing effects. Thus, they are extensively used in the aromatherapy and cosmetic industry. Essential oils, including clove and tea tree, have shown involvement in synthesising metallic nanoparticles with conceivable applications in diverse fields (Obeizi et al. 2020; Fizer et al. 2022).

15.7 OTHER SOURCES FOR GREENER SYNTHESIS OF NANOPARTICLES

15.7.1 Enzymes

Many enzymes possess stabilising and reducing capabilities or properties that aid nanoparticle synthesis. Enzymes provide a sustainable and precise approach to nanoparticle production. Enzymes like alpha-amylase, lipase, cellulase, β-galactosidase, etc. have been utilised to synthesise silver nanoparticles (Sanket and Das 2021).

15.7.2 Yeast-Mediated Synthesis

Yeast, such as Baker's yeast or *Saccharomyces cerevisiae*, has been more often used for the green synthesis of nanoparticles. The yeast cells can act as reducing agents and stabilise the nanoparticles. For example, yeast-mediated synthesis has been utilised to produce nanoparticles of silver (Kowshik et al. 2002), gold (Song et al. 2009), zirconium (Tian et al. 2010), titanium oxide (Peiris et al. 2018), and selenium (Faramarzi et al. 2020).

15.8 ADVANTAGES OF GREEN SYNTHESIS

The researchers are currently working on finding new and sustainable methods for synthesising nanoparticles. They are exploring alternative approaches, such as microbial or plant-based systems, which are more environmentally friendly. It is important to note that these efforts aim to address traditional methods' limitations (Iravani 2011; Prasad and Elumalai 2011). These green synthesis approaches have a variety of advantages. Some of them are as follows:

i. Reduces harmful environmental and ecological impact: These methods utilise natural and non-toxic resources, such as plant extracts and microorganisms, as reducing and stabilising agents, eventually minimising pollution and ecological impact. Unlike traditional chemical reagents that are used in nanoparticle synthesis, these resources are biodegradable (Parveen et al. 2016).
ii. Cost-effective: The raw material used for green synthesis is inexpensive, decreasing production costs. The abundance of plant materials and microbes used in green synthesis makes them cost-effective compared to traditional chemical approaches that depend on expensive reagents and precursors (Iravani 2011).
iii. Mild reaction conditions and no toxic end products: Green synthesis methods typically occur at low room temperature and atmospheric pressure,

requiring lower energy than conventional chemical synthesis's high temperatures and pressures. These mild conditions are more sustainable and environmentally friendly, diminishing the overall carbon footprint of nanoparticle synthesis. Green synthesis methods avoid the use of hazardous and toxic chemicals that are often employed in conventional processes. As a result, the formation of toxic by-products is reduced or eliminated (Iravani 2011; Parveen et al. 2016).

iv. Controlled size and shape of synthesised nanoparticles: Green synthesis methods provide precise control over nanoparticle size, shape, and surface characteristics, imparting different nanoparticle properties. Despite the synthesis under mild conditions, the process can still be tailored to synthesise a wide range of nanoparticles with desired capabilities (Prasad and Elumalai 2011; Miu and Dinischiotu 2022).

v. Biocompatibility: The nanoparticles synthesised through green synthesis approaches exhibit enhanced biocompatibility, making them suitable for biomedical applications such as therapeutics. The use of biocompatible nanoparticles is emphasised as they have low toxicity, immunogenicity, and efficient clearance in living organisms (Iravani 2011; Parveen et al. 2016).

vi. Sustainable resources: Green synthesis methods utilise renewable resources such as microbes or plants, ensuring a continuous and eco-friendly supply of raw materials. Furthermore, the lack of toxic end products and less energy consumption align with the goal of sustainable and responsible manufacturing goals (Iravani 2011; Parveen et al. 2016).

15.9 CONCLUSION

The advent of the utilisation of secondary metabolites in the green synthesis of nanoparticles symbolises a remarkable advancement in the field of nanotechnology. By incorporating and harnessing the unique properties of these compounds derived from various sources, scientists have unlocked a sustainable and environmentally friendly approach to nanoparticle production.

The assimilation of secondary metabolites as reducing and stabilising agents enables the stable conversion of metal ions into nanoparticles while preventing aggregation or clumping. This leads to precise control over the size, shape, and surface properties of the nanoparticles generated, thus leaving room for customisation to meet specific requirements as per the use.

Moreover, using secondary metabolites for synthesis offers significant environmental advantages, such as preventing toxic ion release.

The different chemical compositions of these metabolites, as well as their abundance in different organisms like fungi, algae, and plants, open up avenues for further research and exploration to optimise the synthesis process and identify different species that can offer enhanced nanoparticle synthesis capabilities.

Lastly, integrating secondary metabolites in green synthesis not only offers a sustainable alternative for nanoparticle production but also holds great potential for advancements in diverse fields such as medicine, catalysis, sensors, and environmental remediation.

REFERENCES

Abdel-Aziz, M.S., Shaheen, M.S., El-Nekeety, A.A., Abdel-Wahhab, M.A. 2014. Antioxidant and antibacterial activity of silver nanoparticles biosynthesized using *Chenopodium murale* leaf extract. *J Saudi Chem Society* 18(4):356–363. DOI: 10.1016/j.jscs.2013.09.011

Abou El-Nour, K. M., Eftaiha, A. A., Al-Warthan, A., Ammar, R. A. 2010. Synthesis and applications of silver nanoparticles. *Arab J Chem* 3(3), 135–140. DOI: 10.1016/j.arabjc.2010.04.008

Agam, M.A., Guo, Q. 2007. Electron beam modification of polymer nanospheres. *J Nanosci Nanotechnol* 7(10):3615–3619. DOI: 10.1016/j.jscs.2013.09.011

Ahmad, T. 2014. Reviewing the tannic acid mediated synthesis of metal nanoparticles. *J Nanotechnol* 2014. DOI: 10.1155/2014/954206

Ahmed, S., Ahmad, M., Swami, B.L., Iqram, S. 2016. A review on plants extract mediated synthesis of silver nanoparticles for antimicrobial applications: A green expertise. *J Adv Res* 7(1):17–28. DOI: 10.1016/j.jare.2015.02.007

Ahmed, R.H., Mustafa, D.E. 2020. Green synthesis of silver nanoparticles mediated by traditionally used medicinal plants in Sudan. *Int Nano Lett* 10(1): 1–14. DOI: 10.1007/s40089-019-00291-9

Alvand, Z.M., Rajabi, H.R., Mirzaei, A., Masoumiasl, A., Sadatfaraji, H. 2019. Rapid and green synthesis of cadmium telluride quantum dots with low toxicity based on a plant-mediated approach after microwave and ultrasonic-assisted extraction: Synthesis, characterization, biological potentials and comparison study. *Mater Sci Eng C Mater Biol Appl* 98:535–544. DOI: 10.1016/j.msec.2019.01.010

Amini, S.M., Akbari, A. 2019. Metal nanoparticles synthesis through natural phenolic acids. *IET Nanobiotechnol* 13(8):771–777. DOI: 10.1049/iet-nbt.2018.5386

Arcudi, F., Đorđević, L., Prato, M. 2019. Design, synthesis, and functionalization strategies of tailored carbon nanodots. *Acc Chem Res* 52(8):2070–2079. DOI: 10.1021/acs.accounts.9b00249

Aslam, M., Abdullah, A.Z., Rafatullah, M. 2021. Recent development in the green synthesis of titanium dioxide nanoparticles using plant-based biomolecules for environmental and antimicrobial applications. *J Ind Eng Chem* 98:1–16. DOI: 10.1016/J.JIEC.2021.04.010

Bai, H., Maxwell, T.L., Kordesch, M.E., Balk, T.J. 2023. Physical vapor deposition and thermally induced faceting of tungsten nanoparticles. *Mater Charact* 198:112724. DOI: 10.1016/j.matchar.2023.112724

Benelli, G., Lukehart, C.M. 2017. Applications of green-synthesized nanoparticles in pharmacology, parasitology and entomology. *J Clust Sci* 28:1–2. DOI: 10.1016/j.matchar.2023.112724

Bhattacharya, D., Gupta, R.K. 2005. Nanotechnology and potential of microorganisms. *Crit Rev Biotechnol* 25(4):199–204. DOI: 10.1080/07388550500361994

Binaeian, E., Seghatoleslami, N., Chaichi, M.J., Tayebi, H.A. 2016. Preparation of titanium dioxide nanoparticles supported on hexagonal mesoporous silicate (HMS) modified by oak gall tannin and its photocatalytic performance in degradation of azo dye. *Adv Powder Technol* 27(4):1047–1055. DOI: 10.1016/j.apt.2016.03.012

Bordiwala, R.V. 2023. Green synthesis and applications of metal nanoparticles – A review article. *Results Chem* 100832. DOI: 10.1016/j.rechem.2023.100832

Choy, J.H., Jang, E.S., Won, J.H., Chung, J.H., Jang, D.J., Kim, Y.W. 2004. Hydrothermal route to ZnO nanocoral reefs and nanofibers. *Appl Phys Lett* 84(2):287–289. DOI: 10.1002/adma.200601455

Dai, Y., Liu, W., Formo, E.V., Sun, Y., Xia, Y. 2011. Ceramic nanofibers fabricated by electrospinning and their applications in catalysis, environmental science, and energy technology. *Polym Adv Technol* 22:326–338. DOI: 10.1002/pat.1839

Dan, S., Chattree, A. 2023. Nanomagnetic materials for environmental remediation. In: Giannakoudakis, D.A., Meili, L. and Anastopoulos, I. (eds), *Novel Materials for Environmental Remediation Applications*, Elsevier, 537–553. DOI: 10.1016/B978-0-323-91894-7.00001-3

Edison, T.J.I., Sethuraman, M.G. 2012. Instant green synthesis of silver nanoparticles using *Terminalia chebula* fruit extract and evaluation of their catalytic activity on reduction of methylene blue. *Process Biochem* 47(9):1351–1357. DOI: 10.1016/j.procbio.2012.04.025

Ekambaram, P., Sathali, A.A.H. 2011. Formulation and evaluation of solid lipid nanoparticles of ramipril. *J Young Pharm* 3(3):216–220. DOI: 10.4103/0975-1483.83765

Elia, P., Zach, R., Hazan, S., Kolusheva, S., Porat, Z.E., Zeiri, Y. 2014. Green synthesis of gold nanoparticles using plant extracts as reducing agents. *Int J Nanomed* 9:4007. DOI: 10.2147/IJN.S57343

El-Kemary, M., Zahran, M., Khalifa, S.A.M., El-Seedi, H.R. 2016. Spectral characterisation of the silver nanoparticles biosynthesised using *Ambrosia maritima* plant. *Micro Nano Lett* 11:311.

El-Shabasy, R., Yosri, N., El-Seedi, H., Shoueir, K., El-Kemary, M. 2019. A green synthetic approach using chili plant supported Ag/Ag_2O@P25 heterostructure with enhanced photocatalytic properties under solar irradiation. *Optik - Int J Light Elect Optics* 192: 162943.

Fakhari, S., Jamzad, M., Fard, K.H. 2019. Green synthesis of zinc oxide nanoparticles: A comparison. *Green Chem Lett Rev* 12(1):19–24. DOI: 10.1080/17518253.2018.1547925

Faramarzi, S., Anzabi, Y., Jafarizadeh-Malmiri, H. 2020. Nanobiotechnology approach in intracellular selenium nanoparticle synthesis using *Saccharomyces cerevisiae* fabrication and characterization. *Arch Microbiol* 202:1203–1209.

Fizer, M.M., Mariychuk, R.T., Fizer, O.I. 2022. Gold nanoparticles green synthesis with clove oil: Spectroscopic and theoretical study. *Appl Nanosci* 12(3):611–620. DOI: 10.1007/s13204-021-01726-6

Gavali, A., Rodrigues, J., Shimpi, N., Badani, P. 2023. An overview of the synthesis of nanomaterials. In: Cruz, J.N. (ed.), *Nanobiomaterials: Perspectives for Medical Applications in the Diagnosis and Treatment of Diseases, Materials Research Proceedings*, 145: 19–53. DOI: 10.21741/9781644902370-2

Hedayati, A., Hosseini, B., Palazon, J., Maleki, R. 2020. Improved tropane alkaloid production and changes in gene expression in hairy root cultures of two Hyoscyamus species elicited by silicon dioxide nanoparticles. *Plant Physiol Biochem* 155:416–428. DOI: 10.1016/j.plaphy.2020.07.029

Horwat, D., Zakharov, D.I., Endrino, J.L., Soldera, F. 2011. Chemistry, phase formation, and catalytic activity of thin palladium-containing oxide films synthesized by plasma-assisted physical vapor deposition. *Surf Coat Technol* 205(2): S171–S177. DOI: 10.1016/j.surfcoat.2010.12.021

Huang, G., Lu, C.H., Yang, H.H. 2019. Magnetic nanomaterials for magnetic bioanalysis. In: Wang, X. and Chen, X. (eds), *Novel Nanomaterials for Biomedical, Environmental and Energy Applications*, Elsevier, 89–109. DOI: 10.1016/B978-0-12-814497-8.00003-5

Hussain, S., Bashir, O., Khan, Z., Al-Thabaiti, S.A.J. 2014. Steroidal saponin based extracellular biosynthesis of AgNPs. *Mol Liq* 199:489.

Ijaz, I., Gilani, E., Nazir, A., Bukhari, A. 2020. Detail review on chemical, physical and green synthesis, classification, characterizations and applications of nanoparticles. *Green Chem Lett Rev* 13(3):223–245. DOI: 10.1080/17518253.2020.1802517

Iravani, S. 2011. Green synthesis of metal nanoparticles using plants. *Green Chem* 13(10): 2638–2650. DOI: 10.1039/C1GC15386B

Iravani, S., Varma, R.S. 2020. Greener synthesis of lignin nanoparticles and their applications. *Green Chem* 22(3):612–636. DOI: 10.1039/C9GC02835H

Jain, S., Mehata, M.S. 2017. Medicinal plant leaf extract and pure flavonoid mediated green synthesis of silver nanoparticles and their enhanced antibacterial property. *Sci Rep* 7: 15867.

Kargozar, S., Hoseini, S.J., Milan, P.B., Hooshmand, S., Kim, H.W., Mozafari, M. 2020. Quantum dots: A review from concept to clinic. *Biotechnol J* 15(12):2000117. DOI: 10.1002/biot.202000117

Kaur, M., Chopra, D.S. 2018. Green synthesis of iron nanoparticles for biomedical applications. *Glob J Nanomed* 4(4):68–76. DOI: 10.19080/GJN.2018.04.555643

Kaushik, S. 2021. Polymeric and ceramic nanoparticles: Possible role in biomedical applications. In: Hussain, C.M. and Thomas, S. (eds), *Handbook of Polymer and Ceramic Nanotechnology*, Springer International Publishing, 1293–1308. DOI: 10.1007/978-3-030-40513-7_39

Kawasaki, M., Nishimura, N. 2006. 1064-nm laser fragmentation of thin Au and Ag flakes in acetone for a highly productive pathway to stable metal nanoparticles. *Appl Surf Sci* 253(4):2208–2216. DOI: 10.1016/j.apsusc.2006.04.024

Khalil, M.M., Ismail, E.H., El-Baghdady, K.Z., Mohamed, D. 2014. Green synthesis of silver nanoparticles using olive leaf extract and its antibacterial activity. *Arab J Chem* 7(6):1131–1139. DOI: 10.1016/j.arabjc.2013.04.007

Khan, I., Saeed, K., Khan, I. 2019. Review nanoparticles: Properties, applications and toxicities. *Arab J Chem* 12(2):908–931. DOI: 10.1016/j.arabjc.2017.05.011

Kharissova, O.V., Dias, H.R., Kharisov, B.I., Pérez, B.O., Pérez, V.M.J. 2013. The greener synthesis of nanoparticles. *Trends Biotechnol* 31(4):240–248. DOI: 10.1016/j.tibtech.2013.01.003

Kianfar, E. 2021. Magnetic nanoparticles in targeted drug delivery: A review. *J Supercond Nov Mag.* 34(7):1709–1735. DOI: 10.1007/s10948-021-05932-9

Kowshik, M., Ashtaputre, S., Kharrazi, S., Vogel, W., Urban, J., Kulkarni, S.K., Paknikar, K.M. 2002. Extracellular synthesis of silver nanoparticles by a silver-tolerant yeast strain MKY3. *Nanotechnology* 14(1):95. DOI: 10.1088/0957-4484/14/1/321

Kralj, S., Makovec, D. 2015. Magnetic assembly of superparamagnetic iron oxide nanoparticle clusters into nanochains and nanobundles. *ACS Nano* 9(10):9700–9707. DOI: 10.1021/acsnano.5b02328

Krishanan, R., Maru, G.B. 2006. Isolation and analyses of polymeric polyphenol fractions from black tea. *Food Chem* 94:331–340.

Krishnan, G., Subramaniyan, J., Chengalvarayan Subramani, P., Muralidharan, B., Thiruvengadam, D. 2017. Hesperetin conjugated PEGylated gold nanoparticles exploring the potential role in anti-inflammation and anti-proliferation during diethylnitrosamine-induced hepatocarcinogenesis in rats. *Asian J Pharm Sci* 12:442–455. DOI: 10.1016/j.ajps.2017.04.001

Lee, D., Ko, W.-K., Hwang, D.-S., Heo, D.N., Lee, S.J., Heo, M., Lee, K.-S., Ahn, J.-Y., Jo, J., Kwon, I.K. 2016. Use of baicalin-conjugated gold nanoparticles for apoptotic induction of breast cancer cells. *Nanoscale Res Lett* 11:381. DOI: 10.1186/s11671-016-1586-3

Makvandi, P., Ashrafizadeh, M., Ghomi, M., Najafi, M., Hossein, H.H.S., Zarrabi, A., Mattoli, V., Varma, R.S. 2021. Injectable hyaluronic acid-based antibacterial hydrogel adorned with biogenically synthesized AgNPs-decorated multi-walled carbon nanotubes. *Prog Biomater* 10:77–89. DOI: 10.1007/s40204-021-00155-6

Marcano, D.C., Kosynkin, D.V., Berlin, J.M., Sinitskii, A., Sun, Z., Slesarev, A., Alemany, L.B., Lu, W., Tour, J.M. 2010. Improved synthesis of graphene oxide. *ACS Nano* 4(8):4806–4814. DOI: 10.1021/nn1006368

Marslin, G., Selvakesavan, R.K., Franklin, G., Sarmento, B., Dias, A.C. 2015. Antimicrobial activity of cream incorporated with silver nanoparticles biosynthesized from *Withania somnifera*. *Int J Nanomed.* DOI: 10:5955–5963

Mathur, P., Saini, S., Paul, E., Sharma, C., Mehtani, P. 2021. Endophytic fungi mediated synthesis of iron nanoparticles: Characterization and application in methylene blue decolorization. *CRGSC* 4:100053. DOI: 10.1016/j.crgsc.2020.100053

Mavani, K., Shah, M. 2013. Synthesis of silver nanoparticles by using sodium borohydride as a reducing agent. *Int J Eng Res Technol* 2(3):1–5.

Medici, S., Peana, M., Pelucelli, A., Zoroddu, M.A. 2021. An updated overview on metal nanoparticles toxicity. *Semin Cancer Biol* 76:17–26. DOI: 10.1016/j.semcancer.2021.06.020

Mirahadi, M., Ghanbarzadeh, S., Ghorbani, M., Gholizadeh, A., Hamishehkar, H.A. 2018. A review on the role of lipid-based nanoparticles in medical diagnosis and imaging. *Ther Deliv* 9(8):557–569. DOI: 10.4155/tde-2018-0020

Miu, B.A., Dinischiotu, A. 2022. New green approaches in nanoparticles synthesis: An overview. *Molecules* 27(19):6472. DOI: 10.3390/molecules27196472

Mukherjee, P., Ahmad, A., Mandal, D., Senapati, S., Sainkar, S.R., Khan, M.I., Parischa, R., Ajaykumar, P.V., Alam, M., Kumar, R., Sastry, M. 2001. Fungus-mediated synthesis of silver nanoparticles and their immobilization in the mycelial matrix: A novel biological approach to nanoparticle synthesis. *Nano Lett* 1(10):515–519. DOI: 10.1021/nl0155274

Murphy, C.J. 2002. Nanocubes and nanoboxes. *Science* 298(5601):2139–2141. DOI: 10.1126/science.1080007

Murthy, S.K. 2007. Nanoparticles in modern medicine: State of the art and future challenges. *Int J Nanomed* 2(2):129–141.

Nasrollahzadeh, M., Sajadi, S.M. 2015. Synthesis and characterization of titanium dioxide nanoparticles using *Euphorbia heteradena Jaub* root extract and evaluation of their stability. *Ceram Int* 41(10B):14435–14439. DOI: 10.1016/j.ceramint.2015.07.079

Nasrollahzadeh, M., Sajadi, S.M., Rostami-Vartooni, A., Bagherzadeh, M. 2015. Green synthesis of Pd/CuO nanoparticles by *Theobroma cacao* L. seeds extract and their catalytic performance for the reduction of 4-nitrophenol and phosphine-free Heck coupling reaction under aerobic conditions. *J Colloid Interface Sci* 448:106–113. DOI: 10.1016/j.jcis.2015.02.009

Nel, A., Xia, T., Madler, L., Li, N. 2006. Toxic potential of materials at the nano level. *Science* 311(5761):622–627. DOI: 10.1126/science.1114397

Obeizi, Z., Benbouzid, H., Ouchenane, S., Yılmaz, D., Culha, M., Bououdina, M. 2020. Biosynthesis of zinc oxide nanoparticles from essential oil of *Eucalyptus globulus* with antimicrobial and anti-biofilm activities. *Mater Today Comm* 25:101553. DOI: 10.1016/j.mtcomm.2020.101553

Osman, D.A.M., Mustafa, M.A. 2015. Synthesis and characterization of zinc oxide nanoparticles using zinc acetate dihydrate and sodium hydroxide. *J Nanosci Nanoeng* 1(4):248–251. DOI: 10.1088/1757-899X/599/1/012011

Pak, P.J., Go, E.B., Hwang, M.H., Lee, D.G., Cho, M.J., Joo, Y.H., Chung, N. 2016. Evaluation of the cytotoxicity of gold nanoparticle-quercetin complex and its potential as a drug delivery vesicle. *J Appl Biol Chem* 59:145–147. DOI: 10.3839/jabc.2016.026

Pal, R., Panigrahi, S., Bhattacharyya, D., Chakraborti, A.S. 2013. Characterization of citrate capped gold nanoparticle-quercetin complex: Experimental and quantum chemical approach. *J Mo. Struct* 1046:153–163. DOI: 10.1016/j.molstruc.2013.04.043

Pal, G., Rai, P., Pandey, A. 2019. Green synthesis of nanoparticles: A greener approach for a cleaner future. In: Shukla, A.K. and Iravani, S. (eds), *Green Synthesis, Characterization and Applications of Nanoparticles*, Elsevier, 1–26. DOI: 10.1016/B978-0-08-102579-6.00001-0

Parveen, K., Banse, V., Ledwani, L. 2016. Green synthesis of nanoparticles: Their advantages and disadvantages. *AIP Conf Proceed* 1724:020048. AIP Publishing LLC. DOI: 10.1063/1.4945168

Patel, A., Patel, N., Ali, A., Alim, H. 2023. Nanomaterials synthesis using saponins and their applications. In: Husen, A. (ed), *Secondary Metabolites Based Green Synthesis of Nanomaterials and Their Applications*, Springer Nature Singapore, 141–157. DOI: 10.1007/978-981-99-0927-8_7

Pathania, K., Sah, S.P., Salunke, D.B., Jain, M., Yadav, A.K., Yadav, V.G., Pawar, S.V. 2023. Green synthesis of lignin-based nanoparticles as a bio-carrier for targeted delivery in cancer therapy. *Int J Biol Macromol* 229:684–695. DOI: 10.1016/j.ijbiomac.2022.12.323

Patil, S.P., 2020. *Ficus carica* assisted green synthesis of metal nanoparticles: A mini review. *Biotechnol Rep* 28:e00569. DOI: 10.1016/j.btre.2020.e00569.

Pauluk, D., Padilha, A.K., Khalil, N.M., Mainardes, R.M. 2019. Chitosan-coated zein nanoparticles for oral delivery of resveratrol: Formation, characterization, stability, mucoadhesive properties and antioxidant activity. *Food Hydrocoll* 94:411–417. DOI: 10.1016/j.foodhyd.2019.03.042

Peiris, M., Gunasekara, T., Jayaweera, P.M., Fernando, S. 2018. TiO_2 nanoparticles from baker's yeast: A potent antimicrobial. *J Microbiol Biotechnol* 28:1664–1670.

Pouthika, K., Madhumitha, G., Roopan, S.M. 2023. Sustainable synthesis of nanoparticles using saponin-rich plants and its pharmaceutical applications. In: Bachheti, R.K. and Bachheti, A. (eds), *Secondary Metabolites from Medicinal Plants*, CRC Press, 133–150. DOI: 10.1201/9781003213727-6

Prasad, K., Elumalai, E.K. 2011. Biofabrication of Ag nanoparticles using *Moringa oleifera* leaf extract and their antimicrobial activity. *Asian Pac J Trop Biomed* 1(6):439–442. DOI: 10.1016/S2221-1691(11)60096-8

Rani, N., Singh, P., Kumar, S., Kumar, P., Bhankar, V., Kumar, K. 2023. Plant-mediated synthesis of nanoparticles and their applications: A review. *Mater Res Bull* 112233. DOI: 10.1016/j.materresbull.2023.112233

Sanket, S., Das, S.K. (2021). Role of enzymes in synthesis of nanoparticles. In: Thatoi, H., Mohapatra, S., Das, S.K. (eds), *Bioprospecting of Enzymes in Industry, Healthcare and Sustainable Environment*, Springer, 139–153. DOI: 10.1007/978-981-33-4195-1_7

Savithramma, N., Rao, M.L., Rukmini, K., Devi, P.S. 2011. Antimicrobial activity of silver nanoparticles synthesized by using medicinal plants. *Int J Chem Tech Res* 3(3):1394–1402.

Shankar, S.S., Ahmad, A., Pasricha, R., Sastry, M. 2003. Bioreduction of chloroaurate ions by geranium leaves and its endophytic fungus yields gold nanoparticles of different shapes. *J Mater Chem* 13:1822–1826.

Sharma, B., Rabinal, M.K. 2013. Ambient synthesis and optoelectronic properties of copper iodide semiconductor nanoparticles. *J Alloys Compd* 556:198–202. DOI: 10.1016/j.jallcom.2012.12.120

Singh, B., Rani, M., Singh, J., Moudgil, L., Sharma, P., Kumar, S., Saini, G.S.S., Tripathi, S.K., Singh, G., Kaura, A. 2016. Identifying the preferred interaction mode of naringin with gold nanoparticles through experimental, DFT and TDDFT techniques: Insights into their sensing and biological applications. *RSC Adv* 6:79470–79484. DOI: 10.1039/C6RA12076H

Song, J.Y., Jang, H.K., Kim, B.S. 2009. Biological synthesis of gold nanoparticles using *Magnolia kobus* and *Diopyros kaki* leaf extracts. *Process Biochem* 44(10):1133–1138. DOI: 10.1016/j.procbio.2009.06.005

Song, J.Y., Kim, B.S. 2009. Rapid biological synthesis of silver nanoparticles using plant leaf extracts. *Bioprocess Biosyst Eng* 32:79–84. DOI: 10.1007/s00449-008-0224-6

Sun, W., Sun, M., Meng, X., Zheng, Y., Li, Z., Huang, X., Humayun, M. 2023. Alkynyl carbon functionalized N-TiO_2: Ball milling synthesis and investigation of improved photocatalytic activity. *J Alloys Compd* 168826. DOI: 10.1016/j.jallcom.2023.168826

Suzuki, Y., Ishihara, H. 2021. Difference in the lipid nanoparticle technology employed in three approved siRNA (Patisiran) and mRNA (COVID-19 vaccine) drugs. *Drug Metab Pharmacokinet* 41, 100424. DOI: 10.1016/j.dmpk.2021.100424

Tian, X., He, W., Cui, J., Zhang, X., Zhou, W., Yan, S., Sun, X., Han, X., Han, S., Yue, Y. 2010. Mesoporous zirconium phosphate from yeast biotemplate. *J Colloid Interface Sci* 343:344–349. DOI: 10.1016/j.jcis.2009.11.037

Uchida, T., Ohashi, O., Kawamoto, H., Yoshimura, H., Kobayashi, K., Tanimura, M., Fujikawa, N., Nishimoto, T., Awata, K., Tachibana, M., Kojima, K. 2006. Synthesis of single-wall carbon nanotubes from diesel soot. *Jpn J Appl Phys* 45(10R):8027. DOI: 10.1143/JJAP.45.8027

Usha Rani, P., Madhusudhanamurthy, J., Sreedhar, B. 2014. Dynamic adsorption of α-pinene and linalool on silica nanoparticles for enhanced antifeedant activity against agricultural pests. *J Pest Sci* 87:191–200. DOI: 10.1007/s10340-013-0538-2

Vert, M., Doi, Y., Hellwich, K.H., Hess, M., Hodge, P., Kubisa, P., Rinaudo, M., Schue, F. 2012. Terminology for biorelated polymers and applications (IUPAC Recommendations 2012). *Pure Appl Chem* 84(2):377–410. DOI: 10.1351/PAC-REC-10-12-04

Wang, Y., Maksimuk, S., Shen, R., Yang, H. 2007. Synthesis of iron oxide nanoparticles using a freshly-made or recycled imidazolium-based ionic liquid. *Green Chem* 9(10): 1051–1056. DOI: 10.1039/B618933D

Williams, D.H., Stone, M.J., Hauck, P.R., Rahman, S.K. 1989. Why are secondary metabolites (natural products) biosynthesized? *J Nat Prod* 52(6):1189–1208. DOI: 10.1021/np50066a001

Xu, Y., Fourniols, T., Labrak, Y., Préat, V., Beloqui, A., des Rieux, A. (2022). Surface modification of lipid-based nanoparticles. *ACS Nano 16*(5), 7168–7196. DOI: 10.1021/acsnano.2c02347

Zhao, X., Liu, W., Cai, Z., Han, B., Qian, T., Zhao, D. 2016. An overview of preparation and applications of stabilized zero-valent iron nanoparticles for soil and groundwater remediation. *Water Res* 100:245–266. DOI: 10.1016/j.watres.2016.05.019

Zlatić, N., Stanković, M. 2020. Anticholinesterase, antidiabetic and anti-inflammatory activity of secondary metabolites of *Teucrium* species. In: Stankovic, M. (ed), *Teucrium Species: Biology and Applications*, Springer, 391–411. DOI: 10.1007/978-3-030-52159-2_14

16 Mushroom Cultivation and Application in Green Technology

Varsha Meshram, Khemraj, Pramod Kumar Mahish, and Nagendra Kumar Chandrawanshi

16.1 INTRODUCTION

The increasing worldwide need for foods high in protein content, combined with the constraints of conventional production techniques, has prompted the investigation of alternative methods to economically produce non-traditional protein-rich food sources (Mukherjee and Nandi 2004). Mushrooms, which fall within the category of basidiomycetes, are being recognized as a viable choice for affordable manufacturing of protein-enriched food products. The mushroom industry is a significant global sector, with production exceeding 25 million tons (Mahari et al. 2020). China is the largest mushroom producer, contributing over 20 million tons, accounting for over 80% of the world's production (Li 2012). Consequently, for every kilogram of mushrooms produced, approximately 5 kg of residual material, known as spent mushroom substrate (SMS), is generated (Lau et al. 2003). Cylindrical baglog cultivation is more advantageous than traditional wood tray cultivation, resulting in higher yields and lower contamination rates. Due to this, cylindrical baglogs provide better ventilation, which is crucial for the growth of mushrooms, and also allow for easier monitoring of the growing conditions. Differences between the methods of cultivating mushrooms in cylindrical bags and wooden trays are quite significant. In terms of the duration for spawn running, the wooden tray method usually takes less time to finish when compared to the cylindrical baglog method. This could be attributed to the larger surface area of the wooden trays, which can facilitate faster mycelial growth (Atila 2017). However, in terms of mushroom production, average mushroom weight, and biological efficiency, the cylindrical baglog method generally surpasses the performance of the wooden tray method. It is primarily due to the use of polypropylene bags in the cylindrical baglog method, which helps prevent evaporation and maintain optimal carbon dioxide levels during spawn running and fruiting. On the other hand, the increased surface area of the wooden trays can create opportunities for pathogens to enter and cause excessive loss of moisture, consequently leading to reduced mushroom yields (Oei 1996). Research has shown that the cylindrical baglog method produces more mushroom flushes and, subsequently, higher products than the wooden tray method. Nonetheless, both cultivation methods are employed globally.In Zimbabwe's mushroom sector, the wooden tray technique

DOI: 10.1201/9781003407683-16

is adopted to enhance mushroom cultivation, whereas in more advanced nations like China, the preferred approach is often the cylindrical baglog method (Mamiro et al. 2014; Utami and Susilawati 2017).

Frequently grown types of mushrooms encompass the button mushroom (*Agaricus bisporus*), shiitake mushroom (*Lentinula edodes*), and diverse oyster mushroom strains (*Pleurotus* spp.) (Arya and Rusevska 2022). Increased mushroom production leads to the accumulation of larger volumes of SMS. It is distinguished by elevated levels of external lignocellulosic enzymes, fungal mycelium, organic elements (including carbohydrates, proteins, and fats), as well as inorganic components like ammonium nitrate (Mahari et al. 2020). SMS is a by-product of mushroom production, but various valorization techniques have been developed. SMS can be used as a biocide and fungicide, biosorbent for heavy metals, total petroleum hydrocarbon, co-composting materials, alternative energy, acid mine drainage, and biofertilizer. In addition to the environmental benefits, mushroom cultivation can generate employment, making it a promising solution for long-term sustainability in agriculture. In addition, processing enterprises that produce value-added products from mushrooms can create additional income streams for farmers. Adopting new cultivation methods and valorization techniques can further enhance the growth of the mushroom industry. New technologies have been developed to make the cultivation process more efficient and cost-effective. Moreover, research can be conducted to explore new uses for SMS.

16.2 MUSHROOM CULTIVATION

Mushroom cultivation has become a prevalent practice worldwide, and there has been substantial growth in global production. According to the Food and Agriculture Organization Statistical Database (FAOSTAT), the United States of America, the Netherlands, and China are the next-largest producers of mushrooms. Worldwide, it has been reported that in 2021 a total of 8.99 million tons of mushrooms were produced. The upward trajectory in mushroom production is anticipated to persist in the coming years. Edible mushrooms are highly regarded as a nutritious food due to their significant protein, carbohydrate, fiber, vitamin, and mineral content while being low in fat (Kalac 2013; Valverde et al. 2015). Typically, mushrooms contain approximately 15–35% protein, 35–70% carbohydrates, and less than 5% fat (Rathod et al. 2021; Valverde et al. 2015). What sets certain edible mushrooms apart is their medicinal properties, as they have shown potential in combating human pathogens, cancer, diabetes, hypertension, hypercholesterolemia, and tumors (Chang et al. 2021; Cheung 2010; Kalac 2013; Rani et al. 2023; Valverde et al. 2015). Currently, there are over 50 commercially cultivated species of edible mushrooms worldwide. The most common ones found in commercial cultivation belong to the genera *Agaricus*, *Agrocybe*, *Auricularia*, *Flammulina*, *Ganoderma*, *Hericium*, *Lentinula*, *Lentinus*, *Pleurotus*, *Tremella*, and *Volvariella*. Among them, the top four globally cultivated edible mushrooms are from the genera *Lentinula* (such as shiitake and related species), *Pleurotus* (oyster mushroom), *Auricularia* (wood ear mushroom), and *Agaricus* (button mushroom and associated species) (Ma et al. 2018; Rodriguez et al. 2008).

In 2017, global mushroom production was distributed across various genera, with *Lentinula* accounting for 22%, *Pleurotus* at 19%, *Auricularia* at 18%, *Agaricus* at 15%, *Flammulina* at 11%, *Volvariella* at 5%, and the remaining 10% belonging to other genera (Ma et al. 2018). Most cultivated edible mushrooms are saprophytic fungi, which derive nutrients from decomposing organic matter. They can break down lignocellulosic materials using various enzymes, particularly lignocellulolytic enzymes. Because of this, they can use industrial and agricultural waste as a source of nutrients for growth. For this reason, mushroom cultivation has frequently been linked to the effective recycling of large amounts of agro-industrial waste (Panesar et al. 2016; Ravindran and Jaiswal 2016; Rodriguez et al. 2008; Sadh et al. 2018). Agro-industrial waste, encompassing both agricultural and industrial residues, has been widely utilized as substrates in mushroom cultivation. Typically, these waste materials contain a modest amount of nitrogen. The carbon-to-nitrogen (C/N) ratio of agro-industrial waste varies depending on the specific type, and this ratio plays a crucial role in mushroom cultivation. The various factors directly affect mycelium growth, yield, mushroom weight, and the amount of protein in the fruiting bodies of mushrooms (Grimm and Wösten 2018; Hoa et al. 2015; Royse et al. 2017). Hence, it is essential to incorporate low-nitrogen substrates for mushroom cultivation, which can be achieved by adding organic sources such as cereal bran, cereal shells, soybean meal, and manure, as well as inorganic sources like ammonium chloride and urea as nitrogen supplements (Cueva et al. 2017; Ragunathan and Swaminathan 2003). Previous studies have highlighted that the protein content in the fruiting bodies of mushrooms is influenced by both the chemical composition of substrates and the C/N ratio, in addition to the specific mushroom species being cultivated (Carrasco et al. 2018; Mirabella et al. 2014; Ragunathan and Swaminathan 2003; Wang et al. 2001).

Furthermore, the inclusion of various supplements, such as epsom salts ($MgSO_4 \cdot 7H_2O$), gypsum ($CaSO_4 \cdot 2H_2O$), and limestone (calcium carbonate ($CaCO_3$)), in the substrates has been found to promote mycelial growth and enhance the production of fruiting bodies in mushrooms (Grimm and Wösten 2018; Moonmoon et al. 2011; Royse et al. 2017). Biological efficiency (BE) is a measure used to assess the effectiveness of substrate conversion in mushroom cultivation. This has been computed by evaluating the proportion of the harvested mushroom's fresh weight to the dry weight of the cultivation substrate (Moonmoon et al. 2011). A higher BE value indicates a greater utilization of substrates for mushroom cultivation, which is desirable for profitability (Moonmoon et al. 2011; Wakchaure 2011). Using agro-industrial waste in mushroom cultivation has proven to be an effective method for producing edible proteins for human consumption (Da Silva 2016; Grimm and Wösten 2018; Sadh et al. 2018). Different cultivation methods for edible mushrooms have been practiced worldwide, resulting in variations in the chemical composition of cultivated mushrooms across various studies. These discrepancies can be ascribed to the particular mushroom species, the substrate used for cultivation, and the prevailing environmental conditions (Grimm and Wösten 2018; Mirabella et al. 2014; Sadh et al. 2018). Several research endeavors have been undertaken to investigate the capacity of mushrooms to thrive on diverse types of agro-industrial residues, encompassing materials such as wheat straw,

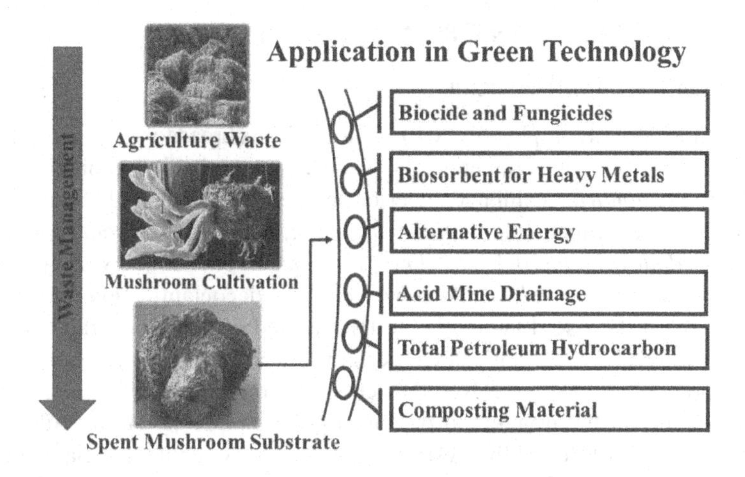

FIGURE 16.1 Several applications of SMS in green technology.

barley straw, oat straw, rice straw, corn straw, corn cob, banana leaves, sawdust, sugarcane bagasse, soy stalk, and sunflower stalk. Combinations of various agro-industrial waste materials can also be utilized in mushroom cultivation. Several applications of SMS in green technology are shown in Figure 16.1 and Table 16.1 provides the main findings regarding the growing of edible mushrooms on different agro-industrial waste materials.

TABLE 16.1
Growing of Edible Mushrooms Using Various Agro-Industrial Waste Substrates

Types of Mushrooms	Mushroom Substrate	References
P. ostreatus	Rubberwood dust, paddy straw, palm empty fruit bunches, palm-pressed fiber, Sawdust, rice bran, lime	Harith et al. 2014; Omokaro and Ogechi 2013
P. eryngii	Sweet potatoes, sawdust, wheat bran, sugarcane bagasse, rice straw	Lou et al. 2017b
Ganoderma sp.	Sawdust	Kamthan and Tiwari 2017
Lentinula edodes	Oats or cedar	Zepeda-Bastida et al. 2016
Lyophyllum decastes	Japanese cedar, sorghum, soybean pulp, rice bran, and wheat bran	Parada et al. 2012
A. subrufescens	Mixtures of sugarcane bagasse, coast cross hay, wheat bran, superphosphate fertilizer, limestone, gypsum, ammonium sulfate	Marques et al. 2014
Flammulina velutipes, *Hypsizygus marmoreus*	Sawdust, cottonseed hull, corncob, sugarcane bagasse, wheat bran, corn powder, gypsum	Zhang et al. 2018

16.3 APPLICATIONS IN GREEN TECHNOLOGY

16.3.1 Biocide and Fungicide

The pesticide pentachlorophenol (PCP) is known for its broad toxicity, which has hindered its biodegradation. Chiu et al. (1998) were among the first to observe that the SMS of *P. pulmonarius* exhibited superior performance compared to various other mushroom mycelia, including *Armillaria gallica*, *A. mellea*, *Ganoderma lucidum*, *Lentinula edodes*, *Phanerochaete chrysosporium*, *P. pulmonarius*, *Polyporus* sp., *Coprinus cinereus*, and *Volvariella volvacea*, in the decontamination of PCP shown in Table 16.3. Likewise, Law et al. (2003) reported a high biodegradation capacity, specifically 15.5 ± 1.0 mg of PCP per gram of SMS from *P. pulmonarius*. The removal process primarily involved the action of immobilized enzymes secreted by the mushroom during growth and the biosorption of PCP by chitin. Additionally, Ahlawat et al. (2010) found that SMS from *A. bisporus* and its associated microorganisms could biodegrade agricultural fungicides, namely carbendazim, and mancozeb, in laboratory settings and agricultural fields. The SMS of *A. bisporus* contained indigenous *Trichoderma* sp. fungi and *Aspergillus* sp. These microorganisms and the extracellular ligninolytic enzymes found in the SMS played a crucial role in the degradation of fungicides. However, the specific enzymes responsible for this process were not yet identified.

In a study conducted in Mexico by Cordova Juarez et al. (2011), a crude extract of SMS from *P. pulmonarius* revealed the presence of laccase, MnP, and phenol oxidase enzymes. A more recent study conducted by Gonzalez Matute et al. (2012) showed that the crude enzyme extracted from the SMS of *A. blazei* could break down metsulfuron-methyl, a sulfonylurea herbicide, and, importantly, this degradation process did not cause any significant harm to oil rape (*Brassica napus* L.) or had only minimal toxic effects on this plant species. These discoveries collectively suggest the potential for utilizing SMS obtained from mushroom industries to break down herbicides and pesticides that can harm the environment. This opens up the opportunity to use SMS to mitigate the ecotoxicological impact of these harmful chemicals. SMS in soil, particularly in areas with low organic matter content, offers several advantages. It has a beneficial impact on populations of soil microorganisms, enriches organic matter, enhances soil structure, and serves as a fertilizer. Moreover, applying SMS to soil provides essential nutrients like nitrogen (N), phosphorus (P), and potassium (K) that promote healthy plant growth, as mentioned by Ribas et al. (2009). Table 16.2 presents a compilation of different studies investigating the impact of SMS on soil's chemical and physical properties.

16.3.2 Biosorbent for Heavy Metals

While the process of heavy metal removal by SMS, as reported by Chen et al. (2005), is not primarily attributed to enzymatic reactions but rather to the biomass component, it is noteworthy to consider the utilization of SMS (specifically *L. edodes*) as a novel biosorbent for heavy metals such as cadmium, lead, and chromium. The main biomass component in *L. edodes* SMS is cellulose, accounting for

TABLE 16.2
The Impact of SMS on Soil's Chemical and Physical Properties

Soil Physicochemical Properties	Changes After Application of SMS as Soil Amendments	References
Aggregation	Induce development of a granular microstructure at 15–20 cm soil, increase soil porosity high porosity and a high fractal dimension, increase aggregate soil stability by 13–16%	Nakatsuka et al. 2016; Stewart et al. 1998a; Lopez Castro et al. 2008
Carbon content	Soil organic carbon content significantly increased with the addition of spent mushroom substrates ($P < 0.05$)	Medina et al. 2012
Organic matter	Increases soil organic matter content	Medina et al. 2012
Nutrient source	Good source of N, P, and K from *A. subrufescens* Increases soil available P content for 1.3–1.6 times	Ribas et al. 2009
Clod and surface crust formation	Reduces clod and surface crust formation by 16–31% and 18–94%, respectively	Stewart et al. 1998b
Water content	Increases soil water content by 0–7% w/w	
Temperature	Reducing diurnal temperature changes	
Infiltration rate	Increases water infiltration rate by 130–207 mm/h	
Soil respiration	Increases soil respiration rate	Medina et al. 2012
pH	No significant changes	

22.86% (g/g), followed by hemicellulose at 19.71% and lignin at 10.24%. According to the Langmuir sorption model, the estimated maximum uptake (qm) of Cd(II) by SMS was 833.33 mg/g. This value was significantly higher, approximately 56 times greater than that of chitin, *Schizomeris leibleinii*, and *Trametes versicolor*. Similarly, the maximum uptake of Pb(II) by SMS was 1,000.00 mg/g, much higher than other fungal materials. However, in the case of Cr(III) uptake, SMS demonstrated a moderate uptake capacity of 44.44 mg/g in comparison to brown seaweed biomass (34.10 mg/g) and brown seaweed *Sargassum wightii* (81.70 mg/g). In conclusion, SMS shows excellent potential as an efficient biosorbent for heavy metals, making it suitable for decontaminating polluted sites.

16.3.3 Alternative Energy

The potential application of enzymes derived from SMS extends to producing biofuels and enzymatic fuel cells, offering alternative energy sources. Nevertheless, utilizing SMS as a biofuel feedstock is appealing due to its cost-effectiveness. The lowered lignin content in SMS, resulting from the breakdown process enabled by external lignocellulosic enzymes during mushroom growth, presents a benefit for biofuel generation. It has decreased lignin content, combined with higher nitrogen and ash levels, enabling easier digestion of SMS by microbial degraders, leading to a higher yield of reducing sugars. The generated polysaccharides in this procedure function as a suitable material for hydrolysis, and the creation of SMS also serves as a kind of preliminary treatment, as highlighted by Hayes and Hayes (2009). According to Asada et al. (2011), the steam explosion and subsequent simultaneous saccharification and fermentation of SMS derived from *L. edodes* (Shiitake) resulted in bioconversion to ethanol with an impressive theoretical yield of 87.6%. In a study conducted by Oguri et al. (2011), they reported a potential application of SMS of *P. eryngii* for utilization for ethanol production. The study yielded a high percentage of total sugars (over 59.0%); upon fermentation, the ethanol yield reached an impressive 67.0%. This study indicated that SMS derived from commercially available edible mushrooms is a valuable and efficient renewable biomass resource for bioethanol production.

Additionally, there is growing interest in biogas, a renewable energy source derived from biomass. Laccase, a well-studied enzyme, has received significant attention, with many studies focusing on laccases derived from the white-rot fungus *T. versicolor* (Farneth et al. 2005; Farneth and D'Amore 2005). The extraction of laccase from SMS opens up the possibility of utilizing it in enzymatic fuel cells. However, developing a suitable cathode design and apparatus is essential for practical implementation.

16.3.4 Acid Mine Drainage

Spent substrates of mushrooms, which often contain lime, are extensively utilized as adequate organic resources for the bioremediation of acid mine drainage (AMD). Passive treatment systems are commonly employed to remediate AMD, and mushroom substrates play a prominent role in such treatment approaches. In a simplified treatment system for AMD, contaminants are directed through a chemo-bio reactive mixture, where organic substrates (such as SMS) are supplemented with sulfate-reducing bacteria for bioaugmentation, as shown in Table 16.3. Based on the findings of Cheong et al. (2010) and Newcombe and Brennan (2010), using SMS in this uncomplicated treatment system proves to be a straightforward and cost-effective technology for effectively treating AMD.

TABLE 16.3
Utilization of SMS for the Remediation of Various Pollutant Categories

Types of Pollutant	Types of SMS	References
Polycyclic aromatic hydrocarbon	*A. bisporus, P. eryngii, P. ostreatus, C. comatus*	Li et al. 2010
Biocide and fungicide	*P. Pulmonarius, A. bisporus*	Ahlawat et al. 2010; Juarez et al. 2011
Heavy metal	*L. edodes*	Chen et al. 2005
Acid mine drainage	SMS	Cheong et al. 2010; Newcombe and Brennan 2010
Petroleum	*P. pulmonarius*	Chiu et al. 2009
Co-composting material	SMS	Meng et al. 2019

16.3.5 Total Petroleum Hydrocarbon

A study by Chiu et al. (2009) demonstrated the promising potential of using SMS derived from *P. pulmonarius* to remediate soil contaminated with petroleum, oil, and grease, as well as di(2-ethylhexyl) phthalate (DEHP) shown in Table 16.3. The removal mechanism involved the hydrocarbon-degrading enzymes in SMS, namely, laccase (1.0–1.5 Umg^{-1}) and MnP (0.8–0.9 Umg^{-1}). Additionally, SMS contained a significant population of native microbes, including bacteria $(11 \pm 3) \times 10^7$ $cfug^{-1}$ and fungi $(56 \pm 9) \times 10^4$ $cfug^{-1}$, which facilitated bioaugmentation. The effectiveness of the treatment was in part linked to the macronutrients and micronutrients within SMS, which impacted the microbial population. Additionally, the favorable conditions for biodegradation were created by the elevated soil moisture, relatively lower bulk density, and increased porosity of SMS, as reported by Chiu et al. (2009), Molina-Barahona et al. (2004), and Rivera-Espinoza and Dendooven (2004).

16.3.6 Composting Material

Composting of SMS is considered the most practical and promising approach for effectively and economically recycling SMS in agricultural applications, as highlighted by Lou et al. (2017a). This process utilizes natural microorganisms to decompose and mineralize the complex organic components of SMS, such as proteins, lipids, lignocelluloses, hemicelluloses, cellulose, lignin, and other carbohydrate compounds. The result is the production of readily available soluble substances, including soluble carbon, nitrogen, potassium, and phosphorus compounds, which are beneficial for agricultural purposes, as Pergola et al. (2018) emphasized. Paredes et al. (2016) observed that compost from SMS yielded comparable lettuce growth to mineral fertilizer and displayed elevated levels of organic carbon, nitrogen, and soil constituents. Similarly, Medina et al. (2012) reported a comparable outcome, where the utilization of SMS as compost improved soil fertility, leading to a notable increase in available organic carbon, nitrogen, and phosphorus. The primary impact of SMC is not an immediate enhancement and transformation of soil quality and structure, but rather the facilitation of soluble nutrient mobilization for plant uptake, contributing to achieving soil equilibrium. While composting yields beneficial bioactive compounds and boosts the overall content of soluble nutrients for plant development, the concentration of essential nutrients remains insufficient for plant nutrition when contrasted with chemical fertilizers (Meng et al. 2017). Until now, there has been a growing focus on co-composting, which involves simultaneously composting multiple types of raw materials. Recent research has revealed that incorporating additional raw materials either at the start or during composting can expedite the overall composting timeline, boost microbial activity, augment nutrient availability for microbial proliferation, and, consequently, elevate soil quality.

In a study by Paredes et al. (2016), it was found that utilizing compost derived from SMS led to comparable lettuce yields as mineral fertilizer, along with an elevation in the overall concentrations of organic carbon, nitrogen, and phosphorus within the soil. Similarly, Medina et al. (2012) discovered that employing SMS as compost enhanced soil fertility and substantially raised the levels of available organic carbon,

TABLE 16.4
Co-composting Materials and Their Impact on the Composting Process

Co-composting Materials	Impact on the Composting Process	References
Biogas residues, pig manure, and SMS	Better tomato seedling quality compared to commercial seedling	Meng et al. 2019
Rice husks, pig manure, and SMS	Improved nutrition content and increased germination index	Meng et al. 2018a
Sewage sludge, wheat straw, and SMS	Promoted the degradation of organic matter and reduced greenhouse gas emission	Meng et al. 2017
Sucrose and SMS	Enhanced the degradation of organic matter and compost maturity	Meng et al. 2018b
Green waste, biochar, and SMS	The rate of humification and decomposition of organic waste is increased	Zhang and Sun 2014

nitrogen, and phosphorus. The primary effect of SMS-derived compost is not immediate enrichment or transformation of soil quality and structure but rather facilitating the mobilization of soluble nutrients for plant uptake and achieving soil equilibrium. Nevertheless, the quantity of crucial nutrients acquired through composting remains inadequate in comparison to chemical fertilizers, as highlighted by Meng et al. (2017). The practice of co-composting, where various raw materials are composted together, has attracted interest. Recent research has demonstrated that introducing additional raw materials during composting can expedite the composting timeline, heighten microbial activities, enrich the nutrients essential for microbial proliferation, and enhance the quality of compost derived from SMS, as indicated in Tables 16.3 and 16.4.

16.4 CONCLUSION

This chapter discusses the cultivation methods of different mushrooms and their application in green technology. The growing global demand for protein-rich food, coupled with the limitations of traditional production methods, has led to exploring alternative approaches for the cost-effective production of unconventional protein-rich food. Cylindrical baglog cultivation is more advantageous than conventional wood tray cultivation, resulting in higher yields and lower contamination rates. Increased mushroom production leads to the accumulation of larger volumes of SMS. SMS is a by-product of mushroom production, but various valorization techniques have been developed. SMS can be used as a biocide and fungicide, biosorbent for heavy metals, total petroleum hydrocarbon, co-composting materials, alternative energy, acid mine drainage, and biofertilizer. In addition to the environmental benefits, mushroom cultivation can generate employment, making it a promising solution for long-term sustainability in agriculture. In addition, processing enterprises that produce value-added products from mushrooms can create additional income streams for farmers. Adopting new cultivation methods and valorization techniques can further enhance the growth of the mushroom industry. New technologies have been developed to make the cultivation process more efficient and cost-effective. Moreover, research can be conducted to explore new uses for SMS.

ABBREVIATIONS

AMD	Acid mine drainage
BE	Biological efficiency
C/N	Carbon to nitrogen
DEHP	Di-2-ethylhexyl phthalate
FAOSTAT	Food and agriculture organization statistical database
PCP	Pentachlorophenol
SMC	Spent mushroom compost
SMS	Spent mushroom substrate

REFERENCES

Ahlawat, O. P., Gupta, P., Kumar, S., Sharma, D. K. & Ahlawat, K. (2010). Bioremediation of fungicides by spent mushroom substrate and its associated microflora. *Indian Journal of Microbiology*, *50*, 390–395.

Arya, A. & Rusevska, K. (eds.). (2022). *Biology, Cultivation and Applications of Mushrooms*. Springer Singapore.

Asada, C., Asakawa, A., Sasaki, C. & Nakamura, Y. (2011). Characterization of the steam-exploded spent Shiitake mushroom medium and its efficient conversion to ethanol. *Bioresource Technology*, *102*(21), 10052–10056.

Atila, F. (2017). Determining the effects of container types on yield and fruitbody features of *Pleurotus eryngii* strains. *International Journal of Crop Science and Technology*, *3*(1), 7–14.

Carrasco, J., Zied, D. C., Pardo, J. E., Preston, G. M. & Pardo-Gimenez, A. (2018). Supplementation in mushroom crops and its impact on yield and quality. *AMB Express*, *8*(1), 1–9.

Chang, B. V., Yang, C. P. & Yang, C. W. (2021). Application of fungus enzymes in spent mushroom composts from edible mushroom cultivation for phthalate removal. *Microorganisms*, *9*(9), 1989.

Chen, G. Q., Zeng, G. M., Tu, X., Huang, G. H. & Chen, Y. N. (2005). A novel biosorbent: Characterization of the spent mushroom compost and its application for removal of heavy metals. *Journal of Environmental Sciences*, *17*(5), 756–760.

Cheong, Y. W., Das, B. K., Roy, A. & Bhattacharya, J. (2010). Performance of a SAPS-based chemo-bioreactor treating acid mine drainage using low-DOC spent mushroom compost, and limestone as substrate. *Mine Water and the Environment*, *29*, 217–224.

Cheung, P. C. K. (2010). The nutritional and health benefits of mushrooms. *Nutrition Bulletin*, *35*(4), 292–299.

Chiu, S. W., Ching, M. L., Fong, K. L. & Moore, D. (1998). Spent oyster mushroom substrate performs better than many mushroom mycelia in removing the biocide pentachlorophenol. *Mycological Research*, *102*(12), 1553–1562.

Chiu, S. W., Gao, T., Chan, C. S. S. & Ho, C. K. M. (2009). Removal of spilled petroleum in industrial soils by spent compost of mushroom *Pleurotus pulmonarius*. *Chemosphere*, *75*(6), 837–842.

Cueva, M. B. R., Hernandez, A. & Nino-Ruiz, Z. (2017). Influence of C/N ratio on productivity and the protein contents of *Pleurotus ostreatus* grown in differents residue mixtures. *Revista de lala Facultad dede Ciencias Agrarias*, *49*(2), 331–344.

Da Silva, L. L. (2016). Adding value to agro-industrial wastes. *Industrial. Chemistry*, *2*(2).

Farneth, W. E. & D'Amore, M. B. (2005). Encapsulated laccase electrodes for fuel cell cathodes. *Journal of Electroanalytical Chemistry, 581*(2), 197–205.

Farneth, W. E., Diner, B. A., Gierke, T. D. & D'Amore, M. B. (2005). Current densities from electrocatalytic oxygen reduction in laccase/ABTS solutions. *Journal of Electroanalytical Chemistry, 581*(2), 190–196.

Grimm, D. & Wösten, H. A. (2018). Mushroom cultivation in the circular economy. *Applied Microbiology and Biotechnology, 102*, 7795–7803.

Harith, N., Abdullah, N. & Sabaratnam, V. (2014). Cultivation of *Flammulina velutipes* mushroom using various agro-residues as a fruiting substrate. *Pesquisa Agropecuaria Brasileira, 49*, 181–188.

Hayes, D.J. & Hayes, M.H. (2009). The role that lignocellulosic feedstocks and various biorefining technologies can play in meeting Ireland's biofuel targets. *Biofuels Bioproducts Biorefining, 3*, 500–520.

Hoa, H. T., Wang, C. L. & Wang, C. H. (2015). The effects of different substrates on the growth, yield, and nutritional composition of two oyster mushrooms (*Pleurotus ostreatus and Pleurotus cystidiosus*). *Mycobiology, 43*(4), 423–434.

Juarez, R. A. C., Dorry, L. L. G., Bello-Mendoza, R. & Sanchez, J. E. (2011). Use of spent substrate after *Pleurotus pulmonarius* cultivation for the treatment of chlorothalonil containing wastewater. *Journal of Environmental Management, 92*(3), 948–952.

Kalac, P. (2013). A review of chemical composition and nutritional value of wild-growing and cultivated mushrooms. *Journal of the Science of Food and Agriculture, 93*(2), 209–218.

Kamthan, R. & Tiwari, I. (2017). Agricultural wastes-potential substrates for mushroom cultivation. *European Journal of Experimental Biology, 7*(5), 31.

Lau, K. L., Tsang, Y. Y. & Chiu, S. W. (2003). Use of spent mushroom compost to bioremediate PAH-contaminated samples. *Chemosphere, 52*(9), 1539–1546.

Law, W. M., Lau, W. N., Lo, K. L., Wai, L. M. & Chiu, S. W. (2003). Removal of biocide pentachlorophenol in the water system by the spent mushroom compost of *Pleurotus pulmonarius. Chemosphere, 52*(9), 1531–1537.

Li, X., Lin, X., Zhang, J., Wu, Y., Yin, R., Feng, Y. & Wang, Y. (2010). Degradation of polycyclic aromatic hydrocarbons by crude extracts from spent mushroom substrate and possible mechanisms. *Current Microbiology, 60*, 336–342.

Li, Y. (2012). Present development situation and tendency of edible mushroom industry in China. *Mushroom Science, 18*(1), 3–9.

Lopez Castro, R. I., Delmastro, S. E. & Curvetto, N. R. (2008). Spent oyster mushroom substrate in a mix with organic soil for plant pot cultivation. *Micología Aplicada International, 20*, 17–26.

Lou, Z., Sun, Y., Bian, S., Baig, S. A., Hu, B. & Xu, X. (2017a). Nutrient conservation during spent mushroom compost application using spent mushroom substrate derived biochar. *Chemosphere, 169*, 23–31.

Lou, Z., Sun, Y., Zhou, X., Baig, S. A., Hu, B. & Xu, X. (2017b). Composition variability of spent mushroom substrates during continuous cultivation, composting process and their effects on mineral nitrogen transformation in soil. *Geoderma, 307*, 30–37.

Ma, G., Yang, W., Zhao, L., Pei, F., Fang, D. & Hu, Q. (2018). A critical review on the health promoting effects of mushrooms nutraceuticals. *Food Science and Human Wellness, 7*(2), 125–133.

Mahari, W. A. W., Peng, W., Nam, W. L., Yang, H., Lee, X. Y., Lee, Y. K., Liew, R. K., Ma, N. L., Mohammad, A., Sonne, C. & Lam, S. S. (2020). A review on valorization of oyster mushroom and waste generated in the mushroom cultivation industry. *Journal of Hazardous Materials, 400*, 123156.

Mamiro, D. P., Mamiro, P. S. & Mwatawala, M. W. (2014). Oyster mushroom (*Pleurotus* spp.) cultivation technique using re-usable substrate containers and comparison of mineral contents with common leafy vegetables. *Journal of Applied Biosciences, 80*, 7071–7080.

Marques, E. L. S., Martos, E. T., Souza, R. J., Silva, R., Zied, D. C. & Dias, E. S. (2014). Spent mushroom compost as a substrate for the production of lettuce seedlings. *Journal of Agricultural Science*, *6*(7), 138–143.

Matute, R. G., Figlas, D., Mockel, G. & Curvetto, N. (2012). Degradation of metsulfuron methyl by *Agaricus blazei* Murrill spent compost enzymes. *Bioremediation Journal*, *16*(1), 31–37.

Medina, E., Paredes, C., Bustamante, M.A., Moral, R. & Moreno-Caselles, J. (2012). Relationships between soil physico-chemical, chemical and biological properties in a soil amended with spent mushroom substrate. *Geoderma*, *173–174*, 152–161.

Meng, L., Li, W., Zhang, S., Wu, C. & Lv, L. (2017). Feasibility of co-composting of sewage sludge, spent mushroom substrate and wheat straw. *Bioresource Technology*, *226*, 39–45.

Meng, L., Zhang, S., Gong, H., Zhang, X., Wu, C. & Li, W. (2018c). Improving sewage sludge composting by addition of spent mushroom substrate and sucrose. *Bioresource Technology*, *253*, 197–203.

Meng, X., Bin, L., Chen, X., Xiaosha, L., Xufeng, Y., Xiaofen, W., Wanbin, Z., Hongliang, W. & Zongjun, C. (2018a). Effect of pig manure on the chemical composition and microbial diversity during co-composting with spent mushroom substrate and rice husks. *Bioresource Technology*, *251*, 22–30.

Meng, X., Liu, B., Xi, C., Luo, X., Yuan, X., Wang, X., Zhu, W., Wang, H. & Cui, Z. (2018b). Effect of pig manure on the chemical composition and microbial diversity during co-composting with spent mushroom substrate and rice husks. *Bioresource Technology*, *251*, 22–30.

Meng, X., Liu, B., Zhang, H., Wu, J., Yuan, X. & Cui, Z. (2019). Co-composting of the biogas residues and spent mushroom substrate: Physicochemical properties and maturity assessment. *Bioresource Technology*, *276*, 281–287.

Mirabella, N., Castellani, V. & Sala, S. (2014). Current options for the valorization of food manufacturing waste: A review. *Journal of Cleaner Production*, *65*, 28–41.

Molina-Barahona, L., Rodrıguez-Vazquez, R., Hernandez-Velasco, M., Vega-Jarquın, C., Zapata-Perez, O., Mendoza-Cantu, A. & Albores, A. (2004). Diesel removal from contaminated soils by biostimulation and supplementation with crop residues. *Applied Soil Ecology*, *27*(2), 165–175.

Moonmoon, M., Shelly, N. J., Khan, M. A., Uddin, M. N., Hossain, K., Tania, M. & Ahmed, S. (2011). Effects of different levels of wheat bran, rice bran and maize powder supplementation with saw dust on the production of shiitake mushroom (*Lentinus edodes* (Berk.) Singer). *Saudi Journal of Biological Sciences*, *18*(4), 323–328.

Mukherjee, R. & Nandi, B. (2004). Improvement of in vitro digestibility through biological treatment of water hyacinth biomass by two *Pleurotus* species. *International Biodeterioration & Biodegradation*, *53*(1), 7–12.

Nakatsuka, H., Oda, M., Hayashi, Y. & Tamura, K. (2016). Effects of fresh spent mushroom substrate of *Pleurotus ostreatus* on soil micromorphology in Brazil. *Geoderma*, *269*, 54–60.

Newcombe, C. E. & Brennan, R. A. (2010). Improved passive treatment of acid mine drainage in mushroom compost amended with crab-shell chitin. *Journal of Environmental Engineering*, *136*(6), 616–626.

Oei, P. (1996). Mushroom cultivation with special emphasis on appropriate techniques for developing countries. Tool Publication, Leiden, The Netherlands. ISBN 13: 9789070857363.

Oguri, E., Takimura, O., Matsushika, A., Inoue, H. & Sawayama, S. (2011). Bioethanol production by *Pichia stipitis* from enzymatic hydrolysates of corncob-based spent mushroom substrate. *Food Science and Technology Research*, *17*(4), 267–272.

Omokaro, O. & Ogechi, A. A. (2013). Cultivation of mushroom (*Pleurotus ostreatus)* and the microorganisms associated with the substrate used. *E-Journal of Science & Technology*, *8*(4).

Panesar, P. S., Kaur, R., Singla, G. & Sangwan, R. S. (2016). Bio-processing of agro-industrial wastes for production of food-grade enzymes: Progress and prospects. *Applied Food Biotechnology, 3*(4), 208–227.

Parada, R. Y., Murakami, S., Shimomura, N. & Otani, H. (2012). Suppression of fungal and bacterial diseases of cucumber plants by using the spent mushroom substrate of *Lyophyllum decastes* and *Pleurotus eryngii*. *Journal of Phytopathology, 160*(7–8), 390–396.

Paredes, C., Medina, E., Bustamante, M. A. & Moral, R. (2016). Effects of spent mushroom substrates and inorganic fertilizer on the characteristics of a calcareous clayey-loam soil and lettuce production. *Soil Use and Management, 32*(4), 487–494.

Pergola, M., Persiani, A., Maria, A. P., Di Meo, Di. V., Pastore, V., D'Adamo, C. & Celano, G. (2018). Composting: The way for a sustainable agriculture. *Applied Soil Ecology,123*, 744–750.

Ragunathan, R. & Swaminathan, K. (2003). Nutritional status of *Pleurotus* spp. grown on various agro-wastes. *Food Chemistry, 80*(3), 371–375.

Rani, M., Mondal, S.M., Kundu, P., Thakur, A., Chaudhary, A., Vashistt, J. & Shankar, J. (2023). Edible mushroom: Occurrence, management and health benefits. *Food Materials Research, 3*, 21.

Rathod, M. G., Gadade, R. B., Thakur, G. M. & Pathak, A. P. (2021). Oyster mushroom: Cultivation, bioactive significance and commercial status. *Front. Life Science, 2*(21), 21–30.

Ravindran, R. & Jaiswal, A. K. (2016). Exploitation of food industry waste for high-value products. *Trends in Biotechnology, 34*(1), 58–69.

Ribas, L. C. C., De Mendonça, M. M., Camelini, C. M. & Soares, C. H. L. (2009). Use of spent mushroom substrates from *Agaricus subrufescens* (syn. *A. blazei, A. brasiliensis*) and *Lentinula edodes* productions in the enrichment of a soil-based potting media for lettuce (*Lactuca sativa*) cultivation: Growth promotion and soil bioremediation. *Bioresource Technology, 100*(20), 4750–4757.

Rivera-Espinoza, Y. & Dendooven, L. (2004). Dynamics of carbon, nitrogen and hydrocarbons in diesel-contaminated soil amended with biosolids and maize. *Chemosphere, 54*(3), 379–386.

Rodriguez, G., Lama, A., Rodriguez, R., Jimenez, A., Guillen, R. & Fernandez-Bolanos, J. (2008). Olive stone an attractive source of bioactive and valuable compounds. *Bioresource Technology, 99*(13), 5261–5269.

Royse, D. J., Baars, J. & Tan, Q. 2017. Current overview of mushroom production in the world. *Edible and MedicinalMushrooms: Technology and Applications*. John Wiley & Sons Ltd. D, Hoboken, 5–13.

Sadh, P. K., Duhan, S. & Duhan, J. S. (2018). Agro-industrial wastes and their utilization using solid state fermentation: A review. *Bioresources and Bioprocessing, 5*(1), 1–15.

Stewart, D. P. C., Cameron, K. C. & Cornforth, I. S. (1998a). Effects of spent mushroom substrate on soil chemical conditions and plant growth in an intensive horticultural system: A comparison with inorganic fertiliser. *Soil Research, 36*(2), 185–198.

Stewart, D. P. C., Cameron, K. C., Cornforth, I. S. & Sedcole, J. R. (1998b). Effects of spent mushroom substrate on soil physical conditions and plant growth in an intensive horticultural system. *Soil Research, 36*(6), 899–912.

Utami, C. P. & Susilawati, P. R. (2017, August). Rice straw addition as sawdust substitution in oyster mushroom (*Pleurotus ostreatus*) planted media. In *AIP Conference Proceedings, 1868*(1), 090002. AIP Publishing LLC.

Valverde, M. E., Hernandez-Perez, T. & Paredes-Lopez, O. (2015). Edible mushrooms: Improving human health and promoting quality life. *International Journal of Microbiology, 2015*, 1–14.

Wakchaure, G. C. (2011). Production and marketing of mushrooms: Global and national scenario. In *Mushrooms-Cultivation, Marketing and Consumption*, by M. Singh, B. Vijay, S. Kamal & G. C. Wakchaure (eds.). Directorate of Mushroom Research, Solan, 15–22.

Wang, D., Sakoda, A. & Suzuki, M. (2001). Biological efficiency and nutritional value of *Pleurotus ostreatus* cultivated on spent beer grain. *Bioresource Technology*, *78*(3), 293–300.

Zepeda-Bastida, A., Ojeda-Ramirez, D., Soto-Simental, S., Rivero-Perez, N. & Ayala-Martínez, M. (2016). Comparison of antibacterial activity of the spent substrate of *Pleurotus ostreatus* and *Lentinula edodes*. *Journal of Agricultural Science*, *8*(4), 43–49.

Zhang, B., Yan, L., Li, Q., Zou, J., Tan, H., Tan, W., Peng, X. L. & Zhang, X. (2018). Dynamic succession of substrate-associated bacterial composition and function during *Ganoderma lucidum* growth. *PeerJ*, *6*, e4975.

Zhang, L. & Sun, X. (2014). Changes in physical, chemical, and microbiological properties during the two-stage co-composting of green waste with spent mushroom compost and biochar. *Bioresource Technology*, *171*, 274–284.

17 Yeast System as Biofuel

Garima, Sumanpreet Kaur, and Raman Thakur

17.1 INTRODUCTION

Bioethanol is a form of renewable energy derived from the fermentation of organic matter, specifically plant-based materials such as corn, sugarcane, or switchgrass, through the use of microorganisms such as yeast (Bušić et al. 2018). Commonly employed as a petrol supplement or as a vehicular fuel, it is frequently amalgamated with petrol in diverse ratios. Bioethanol is considered a sustainable source of energy due to its production from renewable plant-based resources that can be cultivated and harvested on an annual basis. Compared to petrol, it is widely acknowledged that this fuel has a superior environmental profile owing to its lower emission of greenhouse gases and pollutants upon combustion (Jeswani et al. 2020). Furthermore, the utilization of bioethanol has the potential to mitigate the consumption of non-renewable energy sources and decrease reliance on imported petroleum. Notwithstanding its potential benefits, the production of bioethanol may entail certain drawbacks. The production process of bioethanol may necessitate a considerable quantity of water and energy (Mahmud et al. 2022). Furthermore, the utilization of specific crops for bioethanol production may create competition with food production and could result in environmental issues such as deforestation if not appropriately regulated. In general, the utilization of bioethanol as an energy source presents a range of benefits and drawbacks, and its implementation is contingent upon several variables such as the accessibility of feedstocks, economic feasibility, and ecological implications (Barr et al. 2021). Bioethanol is classified as first- and second-generation bioethanol and biofuels. Among these, the bioethanol of the first generation is usually produced by the use of food crops, such as corn, sugarcane, and wheat. The aforementioned crops undergo processing procedures to extract sugars or starches, subsequently subjected to fermentation processes for the production of ethanol (Krishnan et al. 2020). The utilization of first-generation bioethanol has garnered significant criticism due to its potential to compete with food production, leading to potential food shortages and increased food prices. Bioethanol of the second generation is produced utilizing non-food plant matter such as *Jatropha curcas* (Gangwar & Shankar 2020) including agricultural waste, wood chips, and grasses (Bušić et al. 2018) and with the application in-silico techniques (Thakur & Shankar 2022). Advanced technologies, such as cellulosic ethanol, are utilized to produce this particular type of bioethanol. The process involves breaking down the resilient fibers of the plant material to extract the sugars. The utilization of non-food crops and waste materials in second-generation bioethanol production presents a notable advantage as it mitigates competition with food production and has the potential to offer superior environmental benefits (Robak and Balcerek, 2018). A third-generation bioethanol variant derived from algae is also available. Algae cultivation takes place in either ponds or bioreactors, and subsequent

DOI: 10.1201/9781003407683-17

harvesting is conducted to extract oils that are subsequently transformed into biofuels, such as bioethanol. The production of third-generation bioethanol is currently in its nascent phase and has not yet been implemented on a commercial scale (Powar et al. 2022).

17.2 YEAST STRAINS USED FOR BIOFUEL PRODUCTION

17.2.1 Saccharomyces Cerevisiae

Saccharomyces cerevisiae is a commonly utilized yeast strain in the generation of bioethanol. This particular strain is well-known for its exceptional capacity to withstand high levels of ethanol and effectively convert a diverse range of sugars and starches through the process of fermentation (Ruchala et al. 2020).

17.2.2 Schizosaccharomyces Pombe

Schizosaccharomyces pombe is a yeast strain that exhibits proficiency in ethanol production at low temperatures, providing it a suitable candidate for bioethanol production in regions with cold climates (Choi et al. 2010).

17.2.3 Kluyveromyces Marxianus

Kluyveromyces marxianus is another yeast strain that exhibits the ability to ferment a diverse array of substrates, including lactose and xylose, which are typically considered resistant to fermentation by other yeast strains (Bilal et al. 2022).

17.2.4 Candida Tropicalis

Candida tropicalis is a yeast strain that exhibits proficiency in generating ethanol from lignocellulosic materials, including agricultural waste and wood chips. These materials are frequently employed in the production of second-generation bioethanol (Shariq et al. 2019).

17.2.5 Pichia Stipitis

Pichia stipitis is a yeast strain that exhibits the ability to metabolize lignocellulosic compounds, with an emphasis on xylose. This characteristic renders it a highly desirable candidate for second-generation bioethanol production (Ergün et al. 2022).

The selection of a particular strain of yeast is dependent upon various factors including the substrate type, ethanol yield objectives, and the environmental conditions of the biofuel production system, as each strain possesses distinct characteristics and benefits.

17.3 RECOMBINANT YEAST STRAINS FOR BIOFUEL

The utilization of recombinant yeast strains has been widely employed in the manufacture of various biofuels such as ethanol, butanol, and isobutanol. The strains have undergone genetic modification to augment their metabolic potential and enhance their proficiency in generating biofuels from diverse feedstocks (Steensels et al. 2014). Several instances of recombinant yeast strains employed in the production of biofuels are presented below: *Saccharomyces cerevisiae* is a frequently utilized yeast strain in the production of ethanol. Novel strains of *S. cerevisiae* have been genetically engineered to exhibit improved xylose metabolism, thereby facilitating the conversion of lignocellulosic biomass into ethanol (Ruchala et al. 2020). The yeast strain *Kluyveromyces marxianus* has undergone genetic modifications to enhance its capacity to produce ethanol from glucose and as well as xylose that led to increasing yields. The recombinant strain has demonstrated increased tolerance to elevated temperatures, rendering it a viable contender for the production of biofuels in tropical areas (Bilal et al. 2022). The yeast strain known as *Pichia stipitis* has undergone genetic modification to enable the production of isobutanol, a biofuel that exhibits high energy density and is compatible with pre-existing infrastructure. The findings indicate that the recombinant *P. stipitis* strain exhibits a notable capacity to generate substantial quantities of isobutanol from xylose (Ergün et al. 2022). The *Candida tropicalis* yeast strain has undergone genetic modification to facilitate the production of butanol, which is a highly promising biofuel. The research findings indicate that the genetically modified variant of C. *tropicalis* has the ability to generate substantial quantities of butanol through the utilization of glucose and xylose (Maria et al. 2019). In general, the utilization of recombinant yeast strains holds promise for transforming the biofuel production landscape through enhancements in productivity, cost-effectiveness, and the expansion of the spectrum of viable feedstocks for utilization.

17.4 TYPES OF BIOFUELS

Biofuel is a type of fuel derived from biomass within a relatively brief period, as opposed to the protracted natural processes that underlie the creation of fossil fuels. This fuel can be produced using various sources such as plants, agricultural, domestic, or industrial biowaste. The efficacy of biofuel in addressing climate change exhibits significant variability, encompassing emissions that are on par with those of fossil fuels. The predominant application of biofuels is in the realm of transportation; however, they have the potential to be utilized for the purposes of heating and electricity generation as well (Hu et al. 2020; Pontrelli et al. 2018). According to Letcher (2021), biofuels and bioenergy are commonly recognized as sustainable energy sources. First-generation biofuels having bioethanol, also known as conventional biofuels, usually are produced from food crops (Hu et al. 2020). The conversion of complex sugars present in food crops or the oil content of crops into bioethanol is achieved through the utilization of transesterification or yeast fermentation techniques. Second-generation bioethanol-based biofuels, also referred to as sustainable biofuels, are produced using waste materials as a means of addressing the challenge of balancing food and fuel production. The aforementioned materials, namely rice crop-based materials such as straw, husk, wood chips, and sawdust, are obtained from agricultural and forestry practices (Hu et al. 2020).

17.5 PROCESS OF BIOFUEL PRODUCTION BY YEAST

The generation of biofuel by yeast entails a series of sequential steps. The initial step in the process involves the preparation of the feedstock, which entails subjecting it to milling, grinding, or pressing techniques to extract the sugars or starches that will serve as the substrate for the yeast (Malik et al. 2020). Common examples of feedstock include maize and sugarcane. The next step in the process involves the introduction of the prepared substrate into a fermenter vessel, where it is combined with yeast and additional nutrients, including nitrogen and phosphorus. This process is commonly referred to as fermentation. The conversion of complex sugars by yeast fermentation yields into ethanol and carbon dioxide. The fermenter's temperature and pH are meticulously regulated to guarantee the most favorable circumstances for yeast proliferation and bioethanol synthesis (Maicas et al. 2020). After the process of fermentation has concluded, the amalgamation of water and ethanol is isolated from the residual impurities and solids through the method of distillation. The mixture of ethanol and water is subjected to heating until the temperature reaches the point of vaporization for ethanol, following which the resulting vapor is subsequently condensed into a liquid state. The aforementioned procedure facilitates the segregation of ethanol from water and sundry contaminants, thereby augmenting the ethanol concentration in the ultimate output. The ultimate stage in the process of biofuel production involves dehydration, which entails the elimination of any residual water from the ethanol through the utilization of a molecular sieve or other desiccant. This process results in an elevated ethanol concentration of 99% or greater, rendering it appropriate for utilization as a fuel source. The biofuel production process by yeast is subject to variations in its specific conditions and steps, contingent upon the feedstock type and the intended product. Nonetheless, the following steps offer a comprehensive outline of said process (Tse et al. 2021).

17.6 SUBSTRATE USED BY YEAST FOR PRODUCTION OF BIOETHANOL

Yeast has the capacity to utilize diverse substrates for the production of bioethanol, encompassing: Glucose is the predominant substrate utilized for the production of bioethanol via yeast fermentation. Glucose can be procured from diverse sources, including corn starch, sugarcane juice, or molasses (Pavlečić et al. 2010). Fructose, a saccharide present in fruits and honey, has the potential to serve as a substrate for yeast-mediated bioethanol synthesis. The disaccharide sugar such as sucrose serves as a prevalent substrate for bioethanol production in Brazil. This is due to the fact that sugarcane is the primary feedstock in the region. Lactose, a saccharide present in milk, has the potential to serve as a substrate for yeast-mediated bioethanol synthesis. Cellulosic biomass, comprising materials such as wood chips, corn stover, and switchgrass, can serve as a viable substrate for the production of bioethanol through yeast fermentation. Cellulosic biomass comprises cellulose, hemicellulose, and lignin, which can undergo saccharification to yield fermentable sugars that can be utilized for ethanol production. Bioethanol production by yeast can utilize waste materials, including agricultural waste and food waste, as substrates. The selection of substrate is contingent upon various factors, including but not limited to the accessibility, expenditure, and regulatory measures imposed by the governing authorities. Various substrates and sugars can be utilized by yeast, contingent on the particular strain and genetic alterations (Bušić et al. 2018; Pavlečić et al. 2010).

17.7 FACTORS AFFECTING YEAST BIOFUEL PRODUCTION

Cellulosic biomass, particularly sawdust, is a significant origin of glucose, which is a type of dextrose carbohydrate. This resource is utilized for biofuel production in both developed and developing nations. Amaefule et al. looked at how the highlighted factors affected two distinct *Saccharomyces* species (*S. cerevisiae* and *S. chevaelri*) obtained from different sources. The bakery mixture and palm wine were frequently utilized substances for this reason. Sawdust is a resource that is generated in significant quantities through agricultural practices. The environmental impact of agricultural disposal practices is significant on a global scale (Amaefule et al. 2023). The production of biofuels from these sources is considered a sustainable practice due to its simplicity, cost-effectiveness, and high potential for generation. According to Pérez-Carrillo et al., different factors that can influence bioethanol production such as temperature, sugar concentration, pH, fermentation period, agitation rate, and inoculum size should be taken into account. Elevated temperatures possess the capacity to cause the denaturation of enzymes, thereby restricting their activity (Pérez-Carrillo et al. 2011). According to Zhang et al., the temperature range of 20–35°C is considered optimal for biomass fermentation (Zhang et al. 2011). According to Pérez-Carrillo et al., the optimal bioethanol yield was achieved at a concentration of 150 g/L. The bioethanol production process is affected by the pH of the broth due to its impact on various factors such as contamination of bacteria, yeast overgrowth, fermentation rate, and the by-product of fermentation (Pérez-Carrillo et al. 2011). According to Tesfaw and Assefa, a pH range between 4.0 and 5.0 is ideal for biomass fermentation using *Saccharomyces cerevisiae.* A long incubation period is required if the pH is below 4.0; nevertheless, when the pH is over 5.0, ethanol concentration is significantly reduced. An additional factor to take into account when optimizing the production of bioethanol is the level of agitation. Increasing agitation rate yields more ethanol (Tesfaw & Assefa 2014). The agitation rate commonly utilized for yeast cell fermentation ranges between 150 and 200 rpm. According to Pérez-Carrillo et al., an elevated agitation rate could potentially have a negative impact on the metabolic activity of cells (Pérez-Carrillo et al. 2011).

17.8 SYNTHETIC BIOLOGY IN BIOFUEL PRODUCTION AND YEAST STRAIN IMPROVEMENT

The price of petroleum and coal and the concentration of carbon dioxide in the atmosphere exhibit a persistent upward trend. Due to these, researchers have started the development of bio-based fuels as a promising alternative energy option. It is imperative that combustion engines are designed to be compatible with biofuels. Lignocellulosic biomass is considered the predominant source of biofuel due to its widespread availability and cost-effectiveness. According to Tsai et al., *S. cerevisiae* is one of the yeast that is prominently used for the production of ethanol. One of the main obstacles encountered in the process of commercializing biological fuels and chemicals is the inadequate bridging of the gap between laboratory research and the market (Tsai et al. 2015). As per the works of Peralta-Yahya et al. and Nielsen, synthetic biology involves the amalgamation of biological components and designs, which have been derived from the extensive data produced by transcriptomics, proteomics, metabolomics, and fluxomics. This has facilitated the development of novel synthetic circuits (Peralta-Yahya et al. 2012; Nielsen 2015). The challenge of synthetic biology involves the creation of host systems capable of supporting multiple complex metabolic regulatory circuits through the design of synthetic circuits using genetic components. The expeditious advancement of methods for sequencing DNA has facilitated the ability to construct and engineer artificial biological apparatus (Unkles et al. 2014). The design of synthetic microbial host cells represents a highly demanding facet of genetic engineering in the microbial domain. The application of synthetic biology is rapidly growing in the generation of fuels and chemicals. In contemporary times, several technological platforms have been developed to compile a repository of DNA fragments into artificial pathways or circuits. These pathways or circuits are subsequently incorporated into the genetic makeup of the host organism, thereby altering the pre-existing genes. The utilization of synthetic biology and metabolic engineering techniques has resulted in enhanced ethanol production in *Escherichia coli*, *S. cerevisiae*, and *Zymomonas mobilis*, as reported by (Chubukov et al. 2016). Hence, the significance of studies and findings in synthetic biology cannot be overstated, as they are imperative in expediting the development and optimization of novel pathways. The organism *S. cerevisiae* has been extensively researched and serves as a model system for investigations in the fields of cell and molecular biology. This has been demonstrated in several studies (Blazeck et al. 2012; Bao et al. 2015; Blount et al. 2012a,b). Several synthetic biology and metabolic processes engineering projects have been initiated in *S. cerevisiae* to improve the production of biofuels, as reported by Tsai et al. (2015). Therefore, the utilization of yeast as a fundamental basis for the creation of synthetic biology techniques can be advantageous, and this approach can be implemented to enhance the current status of biofuel generation. The primary limitation of *S. cerevisiae* pertains to its inability to metabolize diverse substrates, such as glycerol and xylose (Garcia Sanchez et al. 2010; Swinnen et al. 2013). Caseta et al. suggested that the utilization of *S. cerevisiae* in biofuel applications requiring elevated temperatures (>34°C) is not recommended. Non-conventional yeast species confer several advantages over *S. cerevisiae* with respect to their biology, metabolic pathways, and regulation. Yeast species that have been subjected to thorough research and are considered favorable for production purposes are *Yarrowia lipolytica*, *Hansenula polymorpha*, *Pichia pastoris*, and *Kluyveromyces lactis*. According to Wagner and Alper, yeast systems have developed highly efficient mechanisms to endure unfavorable environmental conditions (Wagner & Alper 2016). Several yeast species have undergone independent evolution from *S. cerevisiae*, exhibiting diverse genetic characteristics and growth behaviors that enable them to withstand different types of stressors including biotic or abiotic (Souciet et al. 2009; Tiwari et al. 2015). However, the characterization of several non-conventional yeast species remains to be accomplished. The development of microorganisms for the purpose of producing desired fuels necessitates consideration of several key factors, including but not limited to enhancing output, utilizing diverse substrates, and ensuring cost-effectiveness, optimal efficiency, and streamlined downstream processes. In order to enhance fuel production, it is imperative to consider additional factors such as fuel tolerance, inhibitor tolerance, and thermotolerance, among others. The regulation of redox balance poses a significant obstacle in the development of strains capable of generating substantial fuel outputs. In contrast to *S. cerevisiae*, non-traditional yeasts have been employed as microorganisms in the industrial sector for the production of biofuel (Ruyters et al. 2015).

Saccharomyces cerevisiae has been widely employed in the food and biotechnology sectors due to its Generally Recognized as Safe (GRAS) status for broad-scale usage. The molecular and cell biology of *S. cerevisiae* has been widely studied for numerous years, making it a prominent model eukaryotic system. Consequently, yeast serves as a valuable microorganism platform for conducting synthetic biology experiments. The utilization of *S. cerevisiae* as a synthetic biological framework offers several advantages. This is due to the exceptional homologous recombination capacity of *S. cerevisiae*, which facilitates the in vivo synthesis of lengthy DNA pieces (Thakur & Shankar 2017). The entity in question is widely recognized for its proficiency in genetic manipulation, having generated a multitude of genetic instruments that facilitate the overexpression and knockdown of target genes (de Jong et al. 2012). In comparison to *E. coli*, *S. cerevisiae* possesses a diverse array of organelles that provide distinct environments for biosynthesis that can be compartmentalized. Finally, it should be noted that *S. cerevisiae* exhibits a notable capacity to withstand the presence of various substances and deleterious inhibitors commonly encountered in cellulosic hydrolysates. Consequently, *S. cerevisiae* has become the predominant engineered host utilized for the production of biofuels and chemicals. According to Nielsen, a range of valuable chemical compounds are generated from acetyl coenzyme A (acetyl-CoA) during yeast fermentations (Nielsen 2014).

The metabolism of acetyl-CoA in yeast is quite complex since it is synthesized in four different compartments (Pronk et al. 1996). In addition, the cytosol synthesizes acetaldehyde, an intermediate in the conversion of pyruvate to ethanol. The yield of product from glucose may be impacted by the additional need for two ATP equivalents for the cytosolic route to acetyl-CoA. Adding another level of intricacy, glucose inhibits both the respiratory system and the tricarboxylic acid (TCA) cycle (Pronk et al. 1996). ADH activity may be eliminated to restrict the conversion of acetaldehyde to ethanol. Another obstacle to the rerouting of flow toward cytosolic acetyl-CoA is the tightly regulated acetyl-CoA synthetase (ACS) activity (Figure 17.1). To prevent the conversion of acetaldehyde to ethanol, it may seem straightforward to simply delete ADH activity. However, yeast contains a very large number of ADH enzymes, and many of the specific product pathways, such as butanol biosynthesis, may also depend on ADH activity (de Jong et al. 2012). The fact that the activity of ACS, which directs flow toward cytosolic acetyl-CoA, is tightly controlled complicates matters further (Figure 17.1). During glycolysis, glucose undergoes a series of enzymatic reactions resulting in the production of pyruvate, which can subsequently be utilized as a substrate for mitochondrial respiration. The conversion of pyruvate

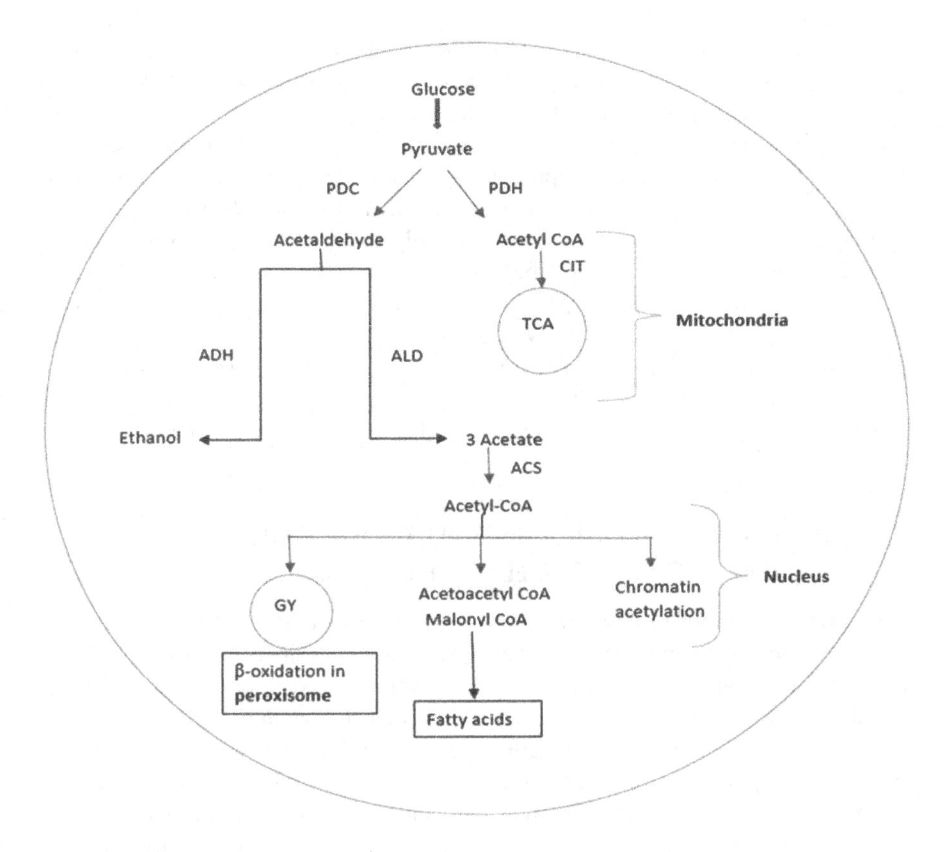

FIGURE 17.1 An overview of yeast acetyl-CoA metabolism.

to acetyl-CoA is facilitated by the pyruvate dehydrogenase complex (PDH). The process of acetyl-CoA oxidation is continued through the tricarboxylic acid (TCA) cycle, where the initial step is catalyzed by citrate synthase (CIT). The conversion of pyruvate to acetaldehyde is facilitated by pyruvate decarboxylase (PDC), an enzyme present in the cytosol. Subsequently, acetaldehyde can be further metabolized to either ethanol via ADH or to acetate via aldehyde dehydrogenase (ALD). Acetate has the ability to penetrate the nucleus and peroxisome, undergoing a conversion to acetyl-CoA through the catalytic activity of ACS. This metabolic process also occurs in the cytoplasm. Acetyl-CoA is utilized for the process of histone acetylation within the nucleus; however, it is not employed for this purpose within the peroxisome.

The presence of two ACS-encoding genes, namely Acs1 and Acs2, has been identified in yeast. The localization of their encoded proteins is contingent upon the carbon source, although both have the potential to be active within the cytosol, as per the findings of Chen et al. (2012). The expression of Acs1 is inhibited by glucose, whereas Acs2 is upregulated during glucose-dependent growth and plays a crucial function in facilitating the supply of cytosolic acetyl-CoA for the synthesis of fatty acids and sterols. While acetylation has been demonstrated to regulate posttranscriptional processes of ACS in bacteria, no comparable regulation has been observed in yeast, as per findings by Chen et al. (2012). As per the comprehension, *Salmonella enterica* in a deregulated form was employed efficaciously to enhance the yield toward sesquiterpenes that are generated from acetyl-CoA through the overexpression of ALD, specifically ALD6 (de Jong et al. 2012). To enhance the production of sesquiterpenes, the approach was reinforced by upregulating ERG10, an enzyme that facilitates the transformation of acetyl-CoA to acetoacetyl-CoA, eliminating either CIT2 or MLS1 to impede the entry of acetyl-CoA into the glyoxylate cycle, and overexpressing ADH2, an enzyme that catalyzes the conversion of ethanol to acetaldehyde (Chen et al. 2013). This finding highlights the significance of developing a platform strain to augment the production of acetyl-CoA. The aforementioned methodology was demonstrated to be efficacious in augmenting the synthesis of various commodities, including 1-butanol (Krivoruchko et al. 2013), polyhydroxybutyrates (Kocharin et al. 2012), 3-hydroxypropionic acid (Chen et al. 2014), and fatty acyl ethyl esters.

17.9 RECOMBINATION TOOLS AND TECHNIQUES ENSURE BETTER BIOFUEL PRODUCTION

The careful choice of promoter components plays a pivotal role in the facilitation of heterologous protein expression, hence serving as a critical determinant in the development of microbe-based cell factories. The complete comprehension of the metabolic regulation of said promoters remains elusive, with a limited number of such promoters identified in non-traditional yeast strains. The eukaryotic essential promoter components are considered to be a vital area for the attachment of transcription factors along with additional regulatory variables that participate in the regulation of transcription. Hence, it is imperative to design promoters that are suitably customized for organisms lacking well-defined genetic tools, with the aim of converting these organisms into hosts for the manufacture of fuels and chemicals on an industrial scale. In contrast to other non-conventional yeast species, *S. cerevisiae* has expeditiously implemented the synthetic promoter approach. The existence of multiple genome sequences of non-traditional yeasts, including *Y. lipolytica*, has stimulated fundamental and practical investigations aimed at comprehending the genotypic and phenotypic traits, in addition to facilitating the advancement of metabolic engineering methodologies for the purpose of biofuel production. The primary objective of activator engineering research is to attain precise and adjustable control over metabolic regulatory pathways. In pursuit of this goal, a considerable amount of research has focused on the manipulation of regulatory sequences located upstream (Hartner et al. 2008; Xuan et al. 2009), manipulating core promoter elements (Blazeck et al. 2012), or introduction of random mutations into core regulatory elements (Berg et al. 2013). Modifications to the core promoter regions, as well as upstream elements, such as binding regions for transcription factors, play a crucial role in gene regulation and have the potential to enhance the strength of promoters. The TetR protein, which has been the subject of extensive research and application in molecular biology, holds its capacity to bind to DNA regulatory regions in yeast, while also exhibiting sensitivity to chemical persuaders such as anhydrotetracycline or doxycycline. The regulation of distinct genes in diverse yeasts has been facilitated through the utilization of promoters that possess TetR-binding sites, as reported by Blount et al. (2012b). Nonetheless, the definition of non-conventional yeast promoters remains inadequate. Nevertheless, given that most non-conventional yeast promoters exhibit cross-recognition capabilities, insights from *S. cerevisiae* could be extrapolated for the vast majority of highly competent unconventional yeast systems. The commercial production of recombinant proteins in *Kluyveromyces lactis* employs the lactose-inducible promoter PLAC4. The production of proteins occurs within the culture fluid, rendering the purification process straightforward. *K. lactis* is considered a favorable host for the production of heterologous proteins owing to its highly efficient protein secretion system and its ability to achieve significant biomass in submerged culture settings (Gellissen & Hollenberg 1997; Van den Berg et al. 1990). Madhavan and Sukumaran developed a hybrid promoter technique in *K. lactis* to enhance the production of recombinant proteins in yeast (Madhavan & Sukumaran 2014). The methodology employed in this study entailed the fusion of the core promoter components of *Trichoderma reesei* cellobiohydrolasel (cbh1) with the lac4 promoter of *K. lactis*. The incorporation of substantial core promoter sequences for the construction of synthetic promoters with high activity levels may enhance the probability of modifying core promoter constituents across various species classifications. As per the findings of Vogl et al., *Pichia pastoris* is an extensively employed platform for protein synthesis, particularly for the production of enzymes and pharmaceuticals. The alcohol oxidase 1 gene (pAOX1) is a widely reported and tightly regulated promoter in *P. pastoris*. It exhibits a high level of activity and is induced by methanol (Vogl et al. 2013). Extensive research has been undertaken on the examination of promoter regulatory elements and transcription factors. Berg et al. detected that a randomly modified library of pAOX1 exhibited heightened activity and diminished glucose regulation in comparison to its wild counterpart (Berg et al. 2013). Qin et al have documented the creation of a series of promoters derived from the inherent pGAP (glyceraldehyde-3-phosphate dehydrogenase promoter). These promoters demonstrate a spectrum of activity ranging from 0.01-fold to 19.6-fold greater than that of the original pGAP (Qin et al. 2011). Blazeck and colleagues reported that the inclusion of synthetic upstream activator sequences within core promoter regions can enhance the levels of gene expression (Blazeck and colleagues 2012). The use of the clustered regularly interspaced short palindromic repeat (CRISPR-Cas) mechanism in diverse host species has exerted a substantial impact on the domain of synthetic biology (Cong et al. 2013). The utilization of nuclease-deficient CRISPR for RNA-guided gene regulation has been observed. The utilization of synthetic single-stranded guide RNAs (sgRNAs) in conjunction with Cas9 (dCas9) has been documented as an effective method for regulating gene expression within yeast by targeting specific genes for silencing on the genome. The technique employed by Bao et al. (2015) facilitated the expeditious and efficient generation of multiple concurrent directed insertion and double-stranded breaks in the model organism in *S. cerevisiae*. The protein Mxi1 was associated with the dCas9 protein in order to enhance the interaction with the histone deacetylase Sin3p homologue. This particular histone deacetylase is a component of the silencer gene complex seen in yeast. In research conducted by Gilbert et al. (2013), it has been seen that the Mxi1 fusion protein exhibited a significant downregulation of the TEF promoter of *S. cerevisiae* by 53-fold. On the other hand, dCas9 and sgRNAs were found to repress it by approximately 10-fold. Horwitz et al. employed this approach to enhance the yield of muconic acid in a commercially relevant non-traditional *K. lactis* strain. The integration of six genes at three distinct loci was successfully accomplished, hence enabling the establishment of a pathway for the synthesis of precursors necessary for muconic acid biosynthesis (Horwitz et al. 2015).

The utilization of CRISPR-Cas9 has facilitated gene deletion and gene cassette insertions in *S. cerevisiae* and *K. marxianus*, resulting in enhanced bioethanol and 2-phenyl ethanol production, respectively. Lakhawat et al. reported that the genomes of multiple bacterial strains were modified to enhance their ability to synthesize butanol and ethanol (Lakhawat et al. 2022). CRISPR-Cas9 modified microalgae have demonstrated an increase in overall lipid content, a crucial component in the production of biofuels. The utilization of CRISPR-Cas9 has emerged as the favored approach for the manipulation of metabolic pathways and genomes in organisms, with the aim of generating industrial biofuels on a global scale (Lakhawat et al. 2022). The biosynthesis pathways of lignin in the context of plant-derived biofuel production pose a hindrance to the efficient liberation of fermentable sugars, thereby impeding the optimal generation of biofuels. The promising role of CRISPR-Cas9 has been demonstrated in reducing the quantity of lignin in various plant species, including barley, switchgrass, and rice straw.

17.10 GLOBAL USE OF BIOFUEL

The global production of biofuels has recently experienced a significant increase, reaching billions of metric tonnes. The projected increase in demand for biofuels is estimated to be 28% or 41 billion liters from 2021 to 2026. Approximately 20% of the observed demand trend can be attributed to the restoration of demand levels that existed prior to the commencement of the COVID-19 epidemic. Government policies are the main catalyst for the continued growth of the biofuel industry. However the use of ethanol is depends upon the various factors such as government policies on its production, use of conventional fuel and associated cost factors, including the total demand for gasoline, associated expenses, and the specific policy design (https://www.iea.org). Throughout the projected time frame, the amalgamation of these factors propels the production of biofuels in Asia beyond that of Europe. The increasing demand for renewable diesel, also known as hydrogenated vegetable oil (HVO), on the European market has been attributed to the implementation of policies in both the United States and Europe, resulting in a threefold increase in demand. The uncertainty surrounding the factors that influence the demand for biofuels is noteworthy. In light of the present exorbitant cost of feedstock, certain governments have opted to ease or postpone biofuel blending regulations, thereby reducing the overall demand. Other key global economies revolve around various pressing issues and have the potential to increase biofuel demand growth by more than twofold in the accelerated case, as stated by the source (https://www.iea.org). The escalation of prices has resulted in a deceleration of biofuel production. However, our projections indicate that the demand for biofuels, which experienced a downturn in 2020 during the COVID-19 pandemic, will recover in 2021. The prices of feedstock and biofuels in Brazil, Argentina, Colombia, and Indonesia are experiencing an upward trend. One of the cost-cutting strategies involves the implementation of temporary reductions or deferrals of blending mandates. According to statistical analysis, the projected outcome of the aforementioned actions is a reduction in demand by 3%, equivalent to 5 billion liters, in the year 2021, in contrast to a hypothetical situation in which mandates remained unaltered or were augmented as first projected. As of August 2021, biofuel prices in the United States, Europe, Brazil, and Indonesia have experienced an increase ranging from 70% to 150% created on the precise market and kind of fuel, when compared to average pricing in 2019. During the same time frame, there was a 40% increase in crude oil prices indicating a slight decline in the growth rate of biofuel consumption. Despite a decrease in growth this year, the recovery is characterized by disparities. According to a source from www.biofuelorg.uk, the demand for ethanol in the year 2021 remains 4% lower than that of 2019 and is projected to not achieve complete recuperation until the year 2023. The reduced production of ethanol in 2021 can be attributed to the elevated costs of ethanol in Brazil and the decreased consumption of petrol in the United States as compared to the levels observed in 2019. In 2021, the International Energy Agency (IEA) reported that there will be a notable demand for fuel in both the United States and Europe by 2023 that was disturbed by COVID-19 pandemic. However, it remained notably lower than the levels observed in 2019. The decrease in demand can be attributed to the rise in energy efficiency, a notable increase in electric vehicle purchases, and shifts in consumer behavior. The present policies dictate that a decrease in petrol consumption results in a corresponding reduction in the production of ethanol. By 2023, it is anticipated that the recuperation of ethanol demand in Brazil, coupled with an increase in demand in Asia, will effectively counterbalance the decline in the USA and Europe. In the year 2021, the demand for ethanol has exhibited a 4% decrease as compared to the demand observed in 2019. However, there has been a significant expansion in the demand for diesel and biojet, surpassing the levels observed in 2019. It is noteworthy that the demand for biojet has increased from a relatively lower base. According to a source from www.biofuelorg.uk, there has been a 15% increase in the aggregate demand for these fuels in the year 2021, which amounts to 7 billion liters more than the demand observed in 2019. The growth witnessed in this context can be mostly attributed to the demand for renewable fuels in the United States and biodiesel in Asia. Throughout the projected time frame, it is anticipated that Asia will exceed the overall biofuel production output of Europe, owing to robust domestic policies, an increase in demand for liquid fuel, and export-oriented manufacturing. According to a source from www.biofuelorg.uk, it is projected that Asian countries will contribute to approximately 33% of newly produced goods during the specified period. The principal cause of this expansion can be ascribed to the biodiesel blend objectives in Indonesia and Malaysia, along with India's ethanol strategy. The demand for biofuels in North America is projected to experience a significant increase by 2026, with a substantial portion of this growth attributable to the resurgence in demand subsequent to the COVID-19 pandemic-induced decline, which accounts for approximately 40% of the overall increase. According to a source from www.biofuelorg.uk, the United States and Brazil remain the principal hubs of biofuel production and

TABLE 17.1
Represented the Bioethanol Production Prediction and Consumption of Biofuel by 2024

Countries	Bioethanol Production Prediction	Consumption of Biofuel by 2024
Brazil	31%	29%
USA	42%	41%
EU	7%	8%
China	7%	7%
India	2%	2%
Thailand	2%	2%
Others	9%	11%

consumption. According to the data from 2016, the worldwide production of bioethanol amounted to 100.2 billion liters (WBA, 2017). The production of bioethanol on an annual basis is exhibiting a steady upward trend. According to predictions, it is anticipated that there will be an increase in the global production and consumption of bioethanol to reach approximately 134.5 billion liters by the year 2024 (www.biofuelorg.uk), as depicted in Table 17.1.

17.11 CONCLUSION

The microorganisms known as yeasts are of the utmost significance in the process of bioethanol production other than used in the fermentation industry or promoting the healthy gut mycobiota composition (Shankar 2021). They are responsible for the fermentation of sugars into ethanol. Numerous yeast strains have been identified globally, possessing the capacity to produce ethanol from various feedstocks. The process of fermentation demonstrated a noteworthy impact on the production of ethanol. The continuous simultaneous saccharification and fermentation (SSF) technique has demonstrated its capacity to generate elevated ethanol concentrations while maintaining high productivity. The escalation of fuel expenses has led to a surge in the quest for a substitute for conventional fossil fuels. The aforementioned notion gives rise to the notion of biofuels, which involves the manipulation of microbial cellular metabolism to generate fuels and chemicals. The application of well-established synthetic biology methodologies in the model organism *S. cerevisiae* holds promise for improving the metabolic pathways involved in biofuel synthesis in non-conventional yeast species by means of rewiring. To encapsulate, the production of biofuels by means of genetically modified yeast via synthetic biology necessitates the implementation of various synthetic biology methodologies and frameworks. The utilization of this technology possesses the capability to enhance the prognostication of metabolically advantageous routes, genetic constructs, and metabolite screening. The progressions in several "omics" technologies have the potential to function as a valuable instrument for the prospective synthetic biology of atypical yeast. The evidence presented indicates that bioethanol exhibits promise as a feasible solution to mitigate the prevailing fuel issue. Significant progress has been achieved in the field of sustainable biomass pre-treatment, cellulase generation, co-fermentation of pentose and hexose sugars, and bioethanol separation and purification in the past few decades. Nevertheless, despite the aforementioned advancements, bioethanol continues to face economic challenges in its competitiveness with fossil fuels, save for instances where bioethanol is derived from sugar cane in Brazil. The primary obstacle that persists is devising strategies to decrease the manufacturing expenses associated with bioethanol. Hence, there is a requirement for the implementation of the biorefinery concept to enable the comprehensive utilization of renewable feedstocks and the production of coproducts with greater value (such as bio-based materials derived from lignin); this would lead to a decrease in the expenses associated with the manufacturing of bioethanol. This phenomenon is expected to enhance the economic competitiveness of bioethanol relative to fossil fuels.

REFERENCES

Amaefule, D., Nwakaire, J., Ogbuagu, N., Anyadike, C., Ogenyi, C., Ohagwu, C. and Egbuhuzor, O. 2023. Effect of production factors on the bioethanol yield of tropical sawdust. *International Journal of Energy Research* 2023. doi: 10.1155/2023/9983840

Bao, Z., Xiao, H., Liang, J., et al. 2015. Homology integrated CRISPR-Cas (HI-CRISPR) system for one-step multigene disruption in saccharomyces cerevisiae. *ACS Synthetic Biology* 4:585–594.

Barr, M.R., Volpe, R., and Kandiyoti, R. 2021. Liquid biofuels from food crops in transportation – A balance sheet of outcomes. *Chemical Engineering Science: X* 10:100090.

Berg, L., Strand, T. A., Valla, S., and Brautaset, T. 2013. Combinatorial mutagenesis and selection to understand and improve yeast promoters. *BioMed Research International* 2013:926985.

Bilal, M., Ji, L., Xu, Y., Xu, S., Lin, Y., Iqbal, H., and Cheng, H. 2022. Bioprospecting *Kluyveromyces marxianus* as a robust host for industrial biotechnology. *Frontiers in Bioengineering and Biotechnology* 10:851768.

Blazeck, J., Garg, R., Reed, B., and Alper, H. S. 2012. Controlling promoter strength and regulation in Saccharomyces cerevisiae using synthetic hybrid promoters. *Biotechnology and Bioengineering* 109:2884–2895.

Blount, B. A., Weenink, T., and Ellis, T. 2012a. Construction of synthetic regulatory networks in yeast. *FEBS Letter* 586:2112–2121.

Blount, B. A., Weenink, T., Vasylechko, S., and Ellis, T. 2012b. Rational diversification of a promoter providing fine-tuned expression and orthogonal regulation for synthetic biology. *PLoS One* 7:e33279.

Bušić, A., Marđetko, N., Kundas, S., Morzak, G., Belskaya, H., Ivančić Šantek, M., Komes, D., Novak, S., and Šantek, B. 2018. Bioethanol production from renewable raw materials and its separation and purification: a review. *Food Technology and Biotechnology* 56(3):289–311.

Chen, Y., Bao, J., Kim, I. K., Siewers, V., and Nielsen, J. 2014. Coupled incremental precursor and co-factor supply improves 3-hydroxypropionic acid production in *Saccharomyces cerevisiae. Metabolic Engineering* 22:104–109.

Chen, Y., Daviet, M., Schalk, M., Siewers, V., and Nielsen, J. 2013. Establishing a platform cell factory through engineering of yeast acetyl-CoA metabolism. *Metabolic Engineering* 15:48–54.

Chen, Y., Siewers, V., and Nielsen, J. 2012. Profiling of cytosolic and peroxisomal acetyl-CoA metabolism in *Saccharomyces cerevisiae. PLoS One* 7:e42475.

Choi, G. W., Um, H. J., Kim, M. N., Kim, Y., Kang, H. W., Chung, B. W., and Kim, Y. H. 2010. Isolation and characterization of ethanol-producing *Schizosaccharomyces pombe* CHFY0201. *Journal of Microbiology and Biotechnology* 20(4):828–834.

Chubukov, V., Mukhopadhyay, A., Petzold, C. J., Keasling, J. D., and Martín, H. G. 2016. Synthetic and systems biology for microbial production of commodity chemicals. *NPJ Systems Biology and Applications* 2:16009.

Cong, L., Ran, F. A., Cox, D., et al. 2013. Multiplex genome engineering using CRISPR/Cas systems. *Science* 339:819–823.

de Jong, B., Siewers, V., and Nielsen, J. 2012. Systems biology of yeast: Enabling technology for development of cell factories for production of advanced biofuels. *Current Opinion in Biotechnology* 23(4):624–630.

Ergün, B. G., Laçın, K., Çaloğlu, B., and Binay, B. 2022. Second generation Pichia pastoris strain and bioprocess designs. *Biotechnology for Biofuels and Bioproducts* 15(1):1–19.

Gangwar, M., and Shankar, J. 2020. Molecular mechanisms of floral biology of *Jatropha curcas*: Opportunities and challenges as an energy crop. *Frontiers in Plant Science* 11:609.

Garcia Sanchez, R., Karhumaa, K., Fonseca, C., et al. 2010. Improved xylose and arabinose utilization by an industrial recombinant *Saccharomyces cerevisiae* strain using evolutionary engineering. *Biotechnology for Biofuels and Bioproducts* 3:13.

Gellissen, G., and Hollenberg, C. P. 1997. Application of yeasts in gene expression studies: A comparison of *Saccharomyces cerevisiae*, *Hansenula polymorpha* and *Kluyveromyces lactis* – A review. *Gene* 190:87–97.

Gilbert, L. A., Larson, M. H., Morsut, L., et al. 2013. CRISPR-mediated modular RNA-guided regulation of transcription in eukaryotes. *Cell* 154:442–451.

Hartner, F. S., Ruth, C., Langenegger, D., et al. 2008. Promoter library designed for fine-tuned gene expression in Pichia pastoris. *Nucleic Acids Research* 36:e76.

Horwitz, A. A., Walter, J. M., Schubert, M. G., et al. 2015. Efficient multiplexed integration of synergistic alleles and metabolic pathways in yeasts via CRISPR-Cas. *Cell Systems* 1:1–9.

Hu, Y., Bassi, A., Xu, C. 2020. Energy from Biomass. In Letcher, T. M. (ed) *Future Energy* (3rd ed.). Elsevier, pp. 447–471.

Jeswani, H.K., Chilvers, A. and Azapagic, A. 2020. Environmental sustainability of biofuels: A review. *Proceedings of the Royal Society A*, 476(2243), 20200351.

Kocharin, K., Chen, Y., Siewers, V., and Nielsen, J. 2012. Engineering of acetyl-CoA metabolism for the improved production of polyhydroxybutyrate in *Saccharomyces cerevisiae. AMB Express* 2:52.

Krishnan, S., Ahmad, M. F., Zainuddin, N. A., Din, M. F. M., Rezania, S., Chelliapan, S., Taib, S. M., Nasrullah, M., and Wahid, Z. A., 2020. Bioethanol production from lignocellulosic biomass (water hyacinth): A biofuel alternative. In Singh, L., Yousuf, A., Mahapatra, D. M. (eds) *Bioreactors*. Elsevier, pp. 123–143.

Krivoruchko, A., Serrano-Amatriain, C., Chen, Y., Siewers, V., and Nielsen, J. 2013. Improving biobutanol production in engineered *Saccharomyces cerevisiae* by manipulation of acetyl-CoA metabolism. *Journal of Industrial Microbiology and Biotechnology* 40:1051–1056.

Lakhawat, S. S., Malik, N., Kumar, V., Kumar, S., and Sharma, P. K. 2022. Implications of CRISPR-Cas9 in developing next generation biofuel: A mini-review. *Current Protein and Peptide Science* 23:574–584.

Lin, Y., Zhang, W., Li, C., Sakakibara, K., Tanaka, S., and Kong, H. 2012. Factors affecting ethanol fermentation using Saccharomyces cerevisiae BY4742. *Biomass and Bioenergy* 47:395–401.

Madhavan, A., and Sukumaran, R. K. 2014. Promoter and signal sequence from filamentous fungus can drive recombinant protein production in the yeast *Kluyveromyces lactis*. *Bioresource Technology* 165:302–308.

Mahmud, S., Haider, A. R., Shahriar, S. T., Salehin, S., Hasan, A. M. and Johansson, M. T. 2022. Bioethanol and biodiesel blended fuels – Feasibility analysis of biofuel feedstocks in Bangladesh. *Energy Reports* 8:1741–1756.

Maicas, S. 2020. The role of yeasts in fermentation processes. *Microorganisms* 8(8):1142.

Malik, K., Salama, E. S., Kim, T. H. and Li, X. 2020. Enhanced ethanol production by Saccharomyces cerevisiae fermentation post acidic and alkali chemical pretreatments of cotton stalk lignocellulose. *International Biodeterioration & Biodegradation* 147:104869.

Nielsen, J. 2014. Synthetic biology for engineering acetyl coenzyme A metabolism in yeast. *mBio* 5(6):e02153–14.

Nielsen, J. 2015. Yeast cell factories on the horizon. *Science* 349:1050–1051.

Pavlečić, M., Vrana, I., Vibovec, K., Ivančić Šantek, M., Horvat, P. and Šantek, B. 2010. Ethanol production from different intermediates of sugar beet processing. *Food Technology and Biotechnology* 48(3):362–367.

Peralta-Yahya, P. P., Zhang, F., del Cardayre, S. B., and Keasling, J. D. 2012. Microbial engineering for the production of advanced biofuels. *Nature* 488:320–328.

Pérez-Carrillo, E., Luisa Cortés-Callejas, M., Sabillón-Galeas, L. E., Montalvo-Villarreal, J. L., Canizo, J. R., Georgina Moreno-Zepeda, M., and Serna-Saldivar, S. O. 2011. Detrimental effect of increasing sugar concentrations on ethanol production from maize or decorticated sorghum mashes fermented with Saccharomyces cerevisiae or *Zymomonas mobilis*: Biofuels and environmental biotechnology. *Biotechnology Letters* 33:301–307.

Pontrelli, S., Fricke, R. C., Sakurai, S. S., et al. 2018. Directed strain evolution restructures metabolism for 1-butanol production in minimal media. *Metabolic Engineering* 49:153–163.

Powar, R. S., Yadav, A. S., Ramakrishna, C. S., Patel, S., Mohan, M., Sakharwade, S. G., Choubey, M., Ansu, A. K., and Sharma, A. 2022. Algae: A potential feedstock for third generation biofuel. *Materials Today: Proceedings* 63:A27–A33.

Pronk, J. T., Steensma, H. Y., and van Dijken, J. P. 1996. Pyruvate metabolism in *Saccharomyces cerevisiae*. *Yeast* 12:1607–1633.

Qin, X., Qian, J., Yao, G., Zhuang, Y., Zhang, S., and Chu, J. 2011. GAP promoter library for fine-tuning of gene expression in *Pichia pastoris*. *Applied and Environmental Microbiology* 77:3600–3608.

Robak, K. and Balcerek, M. 2018. Review of second generation bioethanol production from residual biomass. *Food Technology and Biotechnology* 56(2):174.

Ruchala, J., Kurylenko, O. O., Dmytruk, K. V. and Sibirny, A. A. 2020. Construction of advanced producers of first-and second-generation ethanol in *Saccharomyces cerevisiae* and selected species of non-conventional yeasts (Scheffersomyces stipitis, Ogataea polymorpha). *Journal of Industrial Microbiology and Biotechnology* 47(1):109–132.

Ruyters, S., Mukherjee, V., Verstrepen, K. J., Thevelein, J. M., Willems, K. A., and Lievens, B. 2015. Assessing the potential of wild yeasts for bioethanol production. *Journal of Industrial Microbiology and Biotechnology* 42:39–48.

Shankar, J. 2021. Food habit associated mycobiota composition and their impact on human health. *Frontiers in Nutrition* 8:773577.

Shariq, M. and Sohail, M. 2019. Application of *Candida tropicalis* MK-160 for the production of xylanase and ethanol. *Journal of King Saud University-Science* 31(4):1189–1194.

Souciet, J. L., Dujon, B., Gaillardin, C., et al. 2009. Comparative genomics of protoploid Saccharomycetaceae. *Genome Research* 19:1696–1709.

Steensels, J., Snoek, T., Meersman, E., Nicolino, M.P., Voordeckers, K., and Verstrepen, K.J. 2014. Improving industrial yeast strains: Exploiting natural and artificial diversity. *FEMS Microbiology Reviews* 38(5):947–995.

Swinnen, S., Klein, M., Carrillo, M., McInnes, J., Nguyen, H. T. T., and Nevoigt, E. 2013. Re-evaluation of glycerol utilization in Saccharomyces cerevisiae: Characterization of an isolate that grows on glycerol without supporting supplements. *Biotechnology for Biofuels* 6:1–12.

Tesfaw, A., and Assefa, F. 2014. Current trends in bioethanol production by Saccharomyces cerevisiae: Substrate, inhibitor reduction, growth variables, coculture, and immobilization. *International Scholarly Research* Notices. doi: 10.1155/2014/532852

Thakur, C., and Shankar, J. 2022. Bioinformatics Integration to Biomass Waste Biodegradation and Valorization. In: Verma, P. (ed) *Enzymes in the Valorization of Waste*. CRC Press, Boca Raton.

Thakur, R., and Shankar, J. 2017. Strategies for Gene Expression in Prokaryotic and Eukaryotic System. In: Kalia, V., Saini, A. (eds) *Metabolic Engineering for Bioactive Compounds*. Springer, Singapore.

Tiwari, S., Thakur, R., and Shankar, J. (2015). Role of heat-shock proteins in cellular function and in the biology of fungi. *Biotechnology Research International* 2015. doi: 10.1155/2015/132635

Tsai, C. S., Kwak, S., Turner, T. L., and Jin, Y. S. 2015. Yeast synthetic biology toolbox and applications for biofuel production. *FEMS Yeast Research* 15:1–15.

Tse, T. J., Wiens, D. J., and Reaney, M. J. 2021. Production of bioethanol – A review of factors affecting ethanol yield. *Fermentation* 7(4):268.

Unkles, S. E., Valiante, V., Mattern, D. J., and Brakhage, A. A. 2014. Synthetic biology tools for bioprospecting of natural products in eukaryotes. *Chemistry and Biology* 21:502–508.

Van den Berg, J. A., Van der Laken, K. J., Van Ooyen, A. J., et al. 1990. *Kluyveromyces* a host for heterologous gene expression: Expression and secretion of prochymosin. *Biotechnology* 8:135–139.

Vogl, T., Hartner, F. S., and Glieder, A. 2013. New opportunities by synthetic biology for biopharmaceutical production in *Pichia pastoris. Current Opinion in Biotechnology* 24:1094–1101.

Wagner, J. M., and Alper, H. S. 2016. Synthetic biology and molecular genetics in non-conventional yeasts: Current tools and future advances. *Fungal Genetics and Biology* 89:126–136.

WBA (World Bioenergy Association). 2017. WBA Global Bioenergy Statistics. https://www.worldbioenergy.org/uploads/WBA%20GBS%202017_hq.pdf

Xuan, Y., Zhou, X., Zhang, W., Zhang, X., Song, Z., and Zhang, Y. 2009. An upstream activation sequence controls the expression of AOX1 gene in *Pichia pastoris. FEMS Yeast Research* 9:1271–1282.

Zhang, J., Fang, X., Zhu, X. L., et al. 2011. Microbial lipid production by the oleaginous yeast *Cryptococcus curvatus* O3 grown in fed-batch culture. *Biomass and Bioenergy* 35(5):1906–1911.

Zhang, C. M., Jiang, L., Mao, Z. G., Zhang, J. H., and Tang, L. 2011. Effects of propionic acid and pH on ethanol fermentation by Saccharomyces cerevisiae in cassava mash. *Applied Biochemistry and Biotechnology* 165(3–4):883–891.

Index

Printed in the United States
by Baker & Taylor Publisher Services